D0166151

Understanding Social Problems

Second Edition

Linda A. Mooney
David Knox
Caroline Schacht

East Carolina University

Belmont, CA • Albany, NY • Boston • Cincinnati • Johannesburg • London • Madrid • Melbourne
Mexico City • New York • Pacific Grove, CA • Scottsdale, AZ • Singapore • Tokyo • Toronto

Publisher: Eve Howard
Assistant Editor: Ari Levenfeld
Editorial Assistant: Bridget Schulte
Marketing Assistant: Kelli Gaslin
Project Editor: Jerilyn Emori
Print Buyer: Karen Hunt
Permissions Editors: Susan Walter; Sherry Hoesly/
The Permissions Group
Production: Marcia Craig, Graphic World Publishing Services
Interior and Cover Designer: Hespenheide Design
Photo Researcher: Sue Howard
Copy Editor: Lois Lasater
Illustrations: Graphic World Illustration Services
Compositor: Graphic World, Inc.
Cover Photo Collage: Sandy Young
Printer/Binder: World Color Book Services/Taunton

Printed in the United States of America
1 2 3 4 5 6 7 03 02 01 00 99

For permission to use material from this text, contact us by:

web www.thomsonrights.com
fax 1-800-730-2215
phone 1-800-730-2214

Library of Congress Cataloging-in-Publication Data
Mooney, Linda A.
Understanding social problems / Linda A. Mooney, David Knox, Caroline Schacht. — 2nd ed.
p. cm.
Includes bibliographical references and index.
ISBN 0-534-56511-5
1. Social problems—United States.
2. United States—Social conditions—1980– I. Knox, David.
II. Schacht, Caroline. III. Title.
HN59.2.M66 1999
361.1'0973—dc21

Wadsworth/Thomson Learning
10 Davis Drive
Belmont, CA 94002
USA
www.wadsworth.com

International Headquarters
Thomson Learning
290 Harbor Drive, 2nd Floor
Stamford, CT 06902-7477
USA

UK/Europe/Middle East
Thomson Learning
Berkshire House
168-173 High Holborn
London WC1V 7AA
United Kingdom

Asia
Thomson Learning
60 Albert Street #15-01
Albert Complex
Singapore 189969

Canada
Nelson/Thomson Learning
1120 Birchmount Road
Scarborough, Ontario M1K 5G4
Canada

Brief Contents

Contents

CHAPTER 1 THINKING ABOUT SOCIAL PROBLEMS 1

SECTION 1 PROBLEMS OF WELL-BEING 28

CHAPTER 2 ILLNESS AND THE HEALTH CARE CRISIS 30

CHAPTER 4 CRIME AND VIOLENCE 84

CHAPTER 6 THE YOUNG AND THE OLD 140

CHAPTER 7 GENDER INEQUALITY 168

CHAPTER 8 RACE AND ETHNIC RELATIONS 195

CHAPTER 9 SEXUAL ORIENTATION 223

CHAPTER 15 POPULATION AND ENVIRONMENTAL PROBLEMS 393

15

Preface

Violence in the home, school, and street; impoverished living conditions among millions of people throughout the world; increasing levels of environmental pollution and depletion of the earth's natural resources; persistent conflict between and within nations; ongoing oppression of minorities; and the widening gap between the "haves" and the "have-nots" paint a disturbing picture of our modern world. In *A Guide for the Perplexed,* E.F. Schumacher questions whether a "turning around will be accomplished by enough people quickly enough to save the modern world"? (quoted in Safransky 1990, p.115). Schumacher notes that "this question is often asked, but whatever the answer given to it will mislead. The answer 'yes' would lead to complacency; the answer 'no' to despair. It is desirable to leave these perplexities behind us and get down to work."

In *Understanding Social Problems,* we "get down to work" by examining how the social structure and culture of society contribute to social problems and their consequences. Understanding the social forces that contribute to social problems is necessary for designing strategies for action—programs, policies and other interventions intended to ameliorate the social problem.

ACADEMIC FEATURES OF THE NEW EDITION

In response to feedback from teachers, reviewers, and students who have read the first edition of *Understanding Social Problems* (1997), we have retained several features and added several others:

Strong Integrative Theoretical Foundation. The three major sociological approaches, structural functionalism, symbolic interactionism, and conflict theory, are introduced in the first chapter and discussed and applied, where appropriate, to various social problems throughout the text. Other theories of social problems, as well as feminist approaches, are also presented where appropriate.

Emphasis on the Structure and Culture of Society. As noted above, the text emphasizes how the social structure and culture of society contribute to and maintain social problems, as well as provide the basis for alternative solutions.

Review of Basic Sociological Terms. An overview of basic sociological terms and concepts is presented in the first chapter. This overview is essential for students who have not taken an introductory course and is helpful, as a review, for those who have. Additionally, Appendix A details "Methods of Data Analysis."

Unique Organization. The order of the 16 chapters reflects a progression from a micro to a macro level of analysis, focusing first on problems of health care, drug use, and crime and then broadening to the wider concerns of science and technology, population growth and environmental problems, and conflict around the world.

Two chapters merit special mention: "Sexual Orientation" (Chapter 9) and "Science and Technology" (Chapter 14). Whereas traditional texts discuss sexual orientation under the rubric of "deviance," this topic is examined in the section on problems of human diversity along with the related issues of age, gender, and racial and ethnic inequality. The chapter on science and technology includes such topics as biotechnology, the computer revolution, and the information highway. This chapter emphasizes the transformation of society through scientific and technological innovations, the societal costs of such innovations, and issues of social responsibility. This chapter is particularly relevant to college students, many of whom have never known a world without computers.

Expanded Coverage of Global Issues. In this second edition, we place a greater emphasis on examining social problems from a global perspective. Each chapter contains a heading entitled "The Global Context," and the number and scope of references to international issues has been expanded.

Unique Chapter Format. Each chapter follows a similar format: the social problem is defined, the theoretical explanations are discussed, the consequences of the social problem are explored, and the alternative solutions and policies are examined. A concluding section assesses the current state of knowledge for each social problem.

Standard and Cutting Edge Topics. In addition to problems that are typically addressed in social problems courses and texts, new and emerging topics are examined. Topics new to the second edition include the threat of nuclear terrorism, DNA testing in criminal investigations, telemedicine, divorce law reform (such as the Covenant Marriage law in Louisiana), cloning, the Census Bureau's new multiracial classification, bisexuality and biphobia, effects of welfare reform, and environmental justice.

PEDAGOGICAL FEATURES OF THE NEW EDITION

Student-friendly Presentation. To enhance the book's appeal to students, the second edition includes expanded information relevant to the college population. In Chapter 1, for example, we present data on the beliefs of college students about various social problems. Chapter 2 has a new section on "Health Concerns of College Students" and in Chapter 3 we have added a section on binge drinking and other alcohol-related problems on campus. Further, Chapter 5 contains poems written by college students about the divorce of their parents and in Chapter 12, "Problems in Education," students may complete a "Student Alienation Scale."

Self & Society. Each chapter includes a social survey designed to help students assess their own attitudes, beliefs, knowledge, or behavior regarding some aspect of a social problem. Students may also compare their responses with those from a larger sample. Examples include a Criminal Activities Survey, a Family Functioning Scale, and an AIDS Knowledge Scale.

National and International Data. National statistics and research data are presented throughout the text and offset as "National Data." Similar inserts called "International Data" present data from nations around the world.

Consideration Sections. Sections labeled "Consideration" provide unique examples, insights, implications, explanations, and applications of material presented in the text. These sections are designed to illuminate a previous point in a thought-provoking way.

The Human Side. To personalize the information being discussed, each chapter includes a feature entitled "The Human Side." These features describe personal experiences of individuals who have been affected by the social problem under discussion. Examples include an interview with a former racist Skinhead, a witness' description of an execution, personal stories about "managed care," and a Gulf War veteran's parents' description of their son's death.

Focus on Social Problems Research. Also new to the second edition, offset boxes called "Focus on Social Problems Research" present examples of significant research. These boxes demonstrate for students the sociological enterprise from theory and data collection, to results and conclusions. Examples of research topics covered include variables which affect college students' openness to diversity (Chapter 8), survival strategies of low income single mothers (Chapter 10), and the socially constructed world of computer hackers (Chapter 14).

Focus on Technology. Also new to the second edition are offset boxes called "Focus on Technology." These boxes present information on how technology may contribute to social problems and their solutions. For example, in Chapter 4, Crime and Violence, the Focus on Technology feature highlights the use of DNA testing in criminal investigations. In Chapter 12, Problems in Education, distance learning and the "new education" are examined.

Is It True? Each chapter begins with five true-false items to stimulate student interest and thinking.

Critical Thinking. Each chapter ends with a brief section called "Critical Thinking" that raises several questions related to the chapter topic. These questions invite the student to use critical thinking skills in applying the information discussed in the chapters.

InfoTrac College Edition. New to the second edition is the inclusion of InfoTrac College Edition—an online reference service that allows students to search for articles by subject, title, and author in over 700 journals. Suggested articles are listed at the end of each chapter along with questions to be answered by students from the assigned readings.

Worldwide Web Home Page. As an additional pedagogical tool, *Understanding Social Problems* has its own home page on the Worldwide Web. Students and faculty can access relevant research studies, statistics, and theoretical links as suggested at the end of each chapter in an "Internet" section. For example, in Chapter 13, Cities in Crisis, faculty and students are referred to

Internet links on smart growth, the new urbanism, sustainable communities, and road rage.

In addition to chapter by chapter links, several additional features have been incorporated in the text's home page. For faculty, the *Understanding Social Problems* home page now contains an online instructor's manual and test bank, both password protected. For students, the home page now includes chapter outlines as well as the InfoTrac College Edition articles and questions listed at the end of each text chapter. Links to information about employment opportunities and sociology as a career are also listed. Students and faculty can send questions, comments, or suggestions directly to the authors and contact Wadsworth Publishing Company concerning book adoption.

NEW TO THIS EDITION—A CHAPTER BY CHAPTER LOOK

In addition to the academic and pedagogical features noted above, *Understanding Social Problems'* content areas have been significantly revised. Over thirty new citations have been added to every chapter. Further, in addition to expanded coverage of important topics from the first edition, we have added new areas of research and theorizing and have expanded issues concerning race, gender, and social class. A partial list of new or expanded topics follows:

Chapter 1: Thinking About Social Problems. Collaborative research as an emerging methodology, student activism on college campuses, and "sociological mindfulness."

Chapter 2: Illness and the Health Care System. A new approach to measuring the health status of a population—disability-adjusted life year (DALY), health concerns of college students, managed care, and telemedicine.

Chapter 3: Alcohol and Other Drugs. Survey data on college student drinking and "binging" behavior, teenagers' use of inhalants, the medical use of marijuana debate, and tobacco suits and settlements.

Chapter 4: Crime and Violence. The 1998 International Crime Control Act, pornography on the Internet, children who kill, and innovative criminal justice initiatives (e.g., community policing).

Chapter 5: Family Problems. Results of the First National Survey of Parents' Political Priorities, college students' experiences with parental divorce; premarital education proposals and policies, and divorce law reform.

Chapter 6: The Young and the Old. Children's Health Insurance Program (CHIP), children and poverty, child prostitution, physician-assisted suicide, and social security reform.

Chapter 7: Gender Inequality. Fathers' pre-natal rights; the "new feminist;" men, women and computer technologies; and the men's movement.

Chapter 8: Race and Ethnic Relations. The social construction of race, census guidelines for establishing mixed-race identity, and the religious racial reconciliation movement.

Chapter 9: Sexual Orientation. Biphobia, gay and lesbian youth, socio-legal issues relating to sexual orientation including same-sex marriage, ENDA, and sodomy laws.

Chapter 10: The Haves and the Have-Nots. The Human Poverty Index (HPI), the extent and consequences of economic inequality, welfare reform, the working poor, living wage laws, the role of charity and non-profit organizations in poverty alleviation.

Chapter 11: Work and Unemployment. Child labor and child labor laws, societal attitudes toward corporate America, labor unions and the development of the Labor Party, job burnout, corporate multinationalism, contingent workers, cumulative trauma disorders and ergonomics in the workplace.

Chapter 12: Problems in Education. Distance learning, the charter school movement, National Education Standards, the quality of teaching and teachers in the U.S., and the "dumbing down" of American education.

Chapter 13: Cities in Crisis. Urban sprawl, the urban "three-headed monster" (HIV/AIDS, drug addiction, and crime), road rage, the new urbanism, smart growth, regionalism, neighborhood transit, and mixed-use neighborhoods.

Chapter 14: Science and Technology. Postmodernism; the demographics of Worldwide Web users; the debate over partial birth abortions and cloning; deskilling versus upskilling; e-commerce; and science, ethics and the law.

Chapter 15: Population and Environmental Problems. Falling population levels in Europe, threatened biodiversity, environmental justice issues, environmentally induced illnesses, and "greenwashing."

Chapter 16: Conflict around the World. Nuclear terrorism, Gulf War Syndrome, the Comprehensive Test Ban Treaty, Operation Desert Fox, bombings of American embassies, and the increased threat of nuclear war.

SUPPLEMENTS

Understanding Social Problems comes with a full complement of supplements designed with both faculty and students in mind. Supplements include:

Web Site

Virtual Society web site: The Wadsworth Sociology Resource Center **http://sociology.wadsworth.com** at the Wadsworth Sociology Resource Center, you can find surfing lessons (tips to find information on the web), a career center, links to great sociology web sites, and many other selections.

Supplements for the Instructor

Instructor's Manual with Test Bank
0-534-56513-1
Offers the instructor chapter specific lecture outlines, learning objectives, key terms, classroom activities, student projects, and Internet assignments. There are multiple-choice and true-false questions for each chapter, all with page references. Also includes short answer questions and essay questions for each chapter. Available upon adoption.

Transparency Acetates for Understanding Social Problems
0-534-56516-6
More than 50 book specific acetates to help facilitate the instructor's presentations, available free upon adoption.

CNN Social Problems Video
0-534-54119-4
Consists of footage from stories originally broadcast on CNN within the last several years. Each segment is introduced to place the segment into a proper sociological framework.

Thomson World Class Learning Testing Tools
Windows (0-534-56514-X) and Macintosh 0-534-56515-8
This fully integrated suite of test creation, delivery, and classroom management tools includes Thomson World Class Test, Test Online, and World Class Management software. Thomson World Class Testing Tools allows professors to deliver tests via print, floppy, hard drive, LAN, or Internet. With these tools, professors can create cross-platform exam files from publisher files or existing WESTest 3.2 test banks, edit questions, create questions, and provide their own feedback to objective test questions—enabling the system to work as a tutorial or an examination. In addition, professors can create tests that include multiple-choice, true/false, or matching questions. Professors can also track the progress of an entire class or an individual student. Testing and tutorial results can be integrated into the class management tool which offers scoring, gradebook, and reporting capabilities. Call-in testing is also available.

Supplements for the Student

Study Guide
0-534-56512-3
Includes learning objectives, detailed chapter outlines, fill-in-the-blank test items using key terms from the chapter, and 20 multiple choice and true-false practice test questions with an answer key.

ACKNOWLEDGMENTS

This text reflects the work of many people. We would like to thank the following people for their contributions to the development of this text: Eve Howard, Barbara Yien, Ari Levenfeld, Jerilyn Emori, Marcia Craig, Susan Walters, Karen Hunt, Sue Howard, and Lois Lasater.

We would also like to acknowledge the support and assistance of Blair Carr, Margaret and Thomas Mooney, Marieke Van Willigan, Bob Edwards, Richard Caston, Lakisha Sturdivant, Jon Beckert, Chris Cooper, and Jasper Register.

To each we are grateful.

Additionally, we are indebted to those who read the manuscript in its various drafts and provided valuable insights and suggestions, many of which have been incorporated into the final manuscript:

David Allen
University of New OrleansLinda Nyce

Walter Carroll
Bridgewater State College

Verghese Chirayath
John Carroll University

Kimberly Clark
DeKalb College–Central Campus

William Cross
Illinois College

Doug Degher
Northern Arizona University

Jane Ely
State University of New York
Stony Brook

Joan Ferrante
Northern Kentucky University

Julia Hall
Drexel University

Nancy Kleniewski
University of Massachusetts, Lowell

Daniel Klenow
North Dakota State University

Mary Ann Lamanna
University of Nebraska

Phyllis Langton
George Washington University

Lionel Maldanando
California State University
San Marcos

Peter Meiksins
Cleveland State University

Linda Nyce
Bluffton College

James Peacock
University of North Carolina

Cynthia Reynaud
Louisiana State University

Mareleyn Schneider
Yeshiva University

Paula Snyder
Columbus State Community
College

Rose Weitz
Arizona State University

Bob Weyer
County College of Morris

Mark Winton
University of Central Florida

Diane Zablotsky
University of North Carolina

We are also grateful to the reviewers of the first edition:

David Allen
University of New Orleans

Patricia Atchison
Colorado State University

Roland Chilton
University of Massachusetts

Barbara Costello
Mississippi State University

Robert Gliner
San Jose State University

Millie Harmon
Chemeketa Community College

Sylvia Jones
Jefferson Community College

Judith Mayo
Arizona State University

Madonna Harrington-Meyer
University of Illinois

Clifford Mottaz
University of Wisconsin—
River Falls

Ed Ponczek
William Rainey Harper College

Rita Sakitt
Suffolk County Community College

Lawrence Stern
Collin County Community College

John Stratton
University of Iowa

Joseph Trumino
St. Vincent's College of
St. John's University

Joseph Vielbig
Arizona Western University

Oscar Williams
Diablo Valley College

Finally, we are interested in ways to improve the text and invite your feedback and suggestions for new ideas and material to be included in subsequent editions.

Linda A. Mooney, David Knox, Caroline Schacht
Department of Sociology
East Carolina University
Greenville, NC 27858

E-mail addresses:

MOONEYL@MAIL.ECU.EDU
KNOXD@MAIL.ECU.EDU
SCHACHTC@MAIL.ECU.EDU

CHAPTER ONE

Thinking about Social Problems

- What Is a Social Problem?
- Elements of Social Structure and Culture
- *Self & Society: Personal Beliefs about Various Social Problems*
- The Sociological Imagination
- Theoretical Perspectives
- Social Problems Research
- *Focus on Social Problems Research: Collaborative Research as an Emerging Research Model*
- Goals of the Text
- *The Human Side: College Student Activism*
- Understanding Social Problems

Is It True?

Is It True?

1. An annual study of social well-being in America revealed that since 1970 social conditions have steadily improved in our society.
2. Prior to the nineteenth century, it was considered a husband's legal right and marital obligation to discipline and control his wife through the use of physical force.
3. In seventeenth- and eighteenth-century England, tea drinking was considered a social problem.
4. Questions involving values, religion, and morality can be answered only through scientific research.
5. In a national survey of first-year college students in the United States, most students agreed with the statement: "Realistically, an individual person can do little to bring about changes in our society."

ANSWERS: 1 = F; 2 = T; 3 = T; 4 = F; 5 =

It was the best of times, it was the worst of times; it was the age of wisdom, it was the age of foolishness; it was the epoch of belief, it was the epoch of incredulity; it was the season of light, it was the season of darkness; it was the spring of hope, it was the winter of despair.

Charles Dickens
Novelist

Researchers at Fordham University conduct an annual study called "The Index of Social Health." This study evaluates the cumulative effect on Americans of 16 major social problems, including crime, unemployment, drug abuse, suicide rates, homicide rates, and child abuse. According to analyses of these 16 social indicators, the nation's social health has improved in 2 of the past 7 years (Kannapell 1997). However, of the 8 years with the lowest levels of social health since 1970, 6 have been in the 1990s (Kannapell 1997).

A global perspective on social problems is even more troubling. In 1990 the United Nations Development Programme published its first annual *Human Development Report,* which measured the well-being of populations around the world according to a "human development index" (HDI). This index measures three basic dimensions of human development—longevity, knowledge (i.e., educational attainment), and a decent standard of living. Recent data show that "the human development index declined in the past year in 30 countries, more than in any year since the *Human Development Report* was first issued in 1990" (*Human Development Report* 1997, 3).

International Data

A quarter of the world's population live in severe poverty.

(*Human Development Report* 1997)

Problems related to poverty and malnutrition, inadequate education, acquired immunodeficiency syndrome (AIDS) and other sexually transmitted diseases (STDs), inadequate health care, crime, conflict, oppression of minorities, environmental destruction, and other social issues are both national and international concerns. Such problems present both a threat and a challenge to our national and global society.

The primary goal of this text is to facilitate increased awareness and understanding of problematic social conditions in U.S. society and throughout the world. Although the topics covered in this text vary widely, all chapters share common objectives: to explain how social problems are created and maintained; to indicate how they affect individuals, social groups, and societies as a whole; and to examine programs and policies for change. We begin by looking at the nature of social problems.

WHAT IS A SOCIAL PROBLEM?

There is no universal, constant, or absolute definition of what constitutes a social problem. Rather, social problems are defined by a combination of objective and subjective criteria that vary across societies, among individuals and groups within a society, and across historical time periods.

Social problems are fundamentally products of collective definition. . . . A social problem does not exist for society unless it is recognized by that society to exist.

—Herbert Blumer
Sociologist

Objective and Subjective Elements of Social Problems

Although social problems take many forms, they all share two important elements: an objective social condition and a subjective interpretation of that social condition. The **objective element** of a social problem refers to the existence of a social condition. We become aware of social conditions through our own life experience, through the media, and through education. We see the homeless, hear gunfire in the streets, and see battered

Oil-rig worker Joseph Oncal filed a federal sexual harassment suit against his employer, claiming that three male co-workers held him down in a shower and shoved a bar of soap between his buttocks and one co-worker threatened rape. In March 1998, the Supreme Court ruled in the Joseph Oncal case that men sexually harassing men (and women who harass women) constitutes illegal discrimination and a violation of the 1964 Civil Rights Act. Prior to this ruling, same-sex harassment was not considered a social problem.

women in hospital emergency rooms. We read about employees losing their jobs as businesses downsize and factories close. In television news reports we see the anguished faces of parents whose children have been killed by violent youth.

The **subjective element** of a social problem refers to the belief that a particular social condition is harmful to society, or to a segment of society, and that it should and can be changed. We know that crime, drug addiction, poverty, racism, violence, and pollution exist. These social conditions are not considered social problems, however, unless at least a segment of society believes that these conditions diminish the quality of human life.

By combining these objective and subjective elements, we arrive at the following definition: A **social problem** is a social condition that a segment of society views as harmful to members of society and in need of remedy.

CONSIDERATION

Between 1933 and 1945, more than 12 million Jews, homosexuals, people with mental and physical disabilities, and gypsies were exterminated in death camps by Hitler's Nazi regime. Hitler's goal was to rid German society of "inferior" people so he could build a "master Aryan race." Why did the German population accept what Hitler was doing? Why didn't they view the extermination of millions of people as a social problem? One reason is that, to some degree, Hitler kept the existence of the extermination camps hidden from the German population. Thus, to some extent, the objective condition was unknown. In addition, Hitler worked hard to influence how those Germans who were aware of the camps subjectively interpreted this social condition. Hitler's propaganda suggested that these were labor camps—not death camps. Germans who were aware of what was actually happening in the camps were encouraged to view the mass extermination as important and necessary for the future of Aryan society, rather than as a harmful social condition that should be stopped.

Albert Einstein once said to me: "Two things are infinite: the universe and human stupidity." But what is much more widespread than the actual stupidity is the *playing* stupid, turning off your ear, not listening, not seeing . . . playing helpless.

—Fritz Perls
Gestalt therapist

Variability in Definitions of Social Problems

Individuals and groups frequently disagree about what constitutes a social problem. For example, some Americans view the availability of abortion as a

While some individuals view the availability of abortion as a social problem, others view restrictions on abortion as a social problem. The disagreement can lead to violence, destruction, and murder. In 1998, Dr. Barnett Slepian, who had performed abortions at a Buffalo, New York, clinic, was shot and killed by an antiabortion activist.

Suppose someone were to say, "Imagine this butterfly exactly as it is, but ugly instead of beautiful."

–Ludwig Wittgenstein
Philosopher

It is easy to find things in the real world that are not perfect and to say that something should be done to correct the problems. However, for every ten people who see the same problem, there will be ten different ideal solutions.

–Randall G. Holcombe
Chairman of the Research Advisory Council of the James Madison Institute for Public Policy Studies

social problem, while others view restrictions on abortion as a social problem. Similarly, some Americans view homosexuality as a social problem, while others view prejudice and discrimination against homosexuals as a social problem. Such variations in what is considered a social problem are due to differences in values, beliefs, and life experiences.

Definitions of social problems vary not only within societies, but across societies and historical time periods as well. For example, prior to the nineteenth century, it was a husband's legal right and marital obligation to discipline and control his wife through the use of physical force. Today, the use of physical force is regarded as a social problem rather than a marital right.

Tea drinking is another example of how what is considered a social problem can change over time. In seventeenth- and eighteenth-century England, tea drinking was regarded as a "base Indian practice" that was "pernicious to health, obscuring industry, and impoverishing the nation" (Ukers 1935, cited in Troyer and Markle 1984). Today, the English are known for their tradition of drinking tea in the afternoon.

Because social problems can be highly complex, it is helpful to have a framework within which to view them. Sociology provides such a framework. Using a sociological perspective to examine social problems requires a knowledge of the basic concepts and tools of sociology. In the remainder of this chapter, we discuss some of these concepts and tools: social structure, culture, the "sociological imagination," major theoretical perspectives, and types of research methods.

ELEMENTS OF SOCIAL STRUCTURE AND CULTURE

> Sometimes a cigar is just a cigar.
>
> —Sigmund Freud
> Founder of psychoanalysis

Although society surrounds us and permeates our lives, it is difficult to "see" society. By thinking of society in terms of a picture or image, however, we can visualize society and therefore better understand it. Imagine that society is a coin with two sides: on one side is the structure of society, and on the other is the culture of society. Although each "side" is distinct, both are inseparable from the whole. By looking at the various elements of social structure and culture, we can better understand the root causes of social problems.

Elements of Social Structure

The *structure* of a society refers to the way society is organized. Society is organized into different parts: institutions, social groups, statuses, and roles.

Institutions An **institution** is an established and enduring pattern of social relationships. The five traditional institutions are family, religion, politics, economics, and education, but some sociologists argue that other social institutions, such as science and technology, mass media, medicine, sport, and the military, also play important roles in modern society.

Many social problems are generated by inadequacies in various institutions. For example, unemployment may be influenced by the educational institution's failure to prepare individuals for the job market and by alterations in the structure of the economic institution.

Social Groups Institutions are made up of social groups. A **social group** is defined as two or more people who have a common identity, interact, and form a social relationship. For example, the family in which you were reared is a social group that is part of the family institution. The religious association to which you may belong is a social group that is part of the religious institution.

Social groups may be categorized as primary or secondary. **Primary groups**, which tend to involve small numbers of individuals, are characterized by intimate and informal interaction. Families and friends are examples of primary groups. **Secondary groups**, which may involve small or large numbers of individuals, are task-oriented and characterized by impersonal and formal interaction. Examples of secondary groups include employers and their employees, and clerks and their customers.

Statuses Just as institutions consist of social groups, social groups consist of statuses. A **status** is a position a person occupies within a social group. The statuses we occupy largely define our social identity. The statuses in a family may consist of mother, father, stepmother, stepfather, wife, husband, child, and so on. Statuses may be either ascribed or achieved. An **ascribed status** is one that society assigns to an individual on the basis of factors over which the individual has no control. For example, we have no control over the sex, race, ethnic background, and socioeconomic status into which we are born. Similarly, we are assigned the status of "child," "teenager," "adult," or "senior citizen" on the basis of our age—something we do not choose or control.

An **achieved status** is assigned on the basis of some characteristic or behavior over which the individual has some control. Whether or not you achieve the status of college graduate, spouse, parent, bank president, or prison

When I fulfill my obligations as a brother, husband, or citizen, when I execute contracts, I perform duties that are defined externally to myself. . . . Even if I conform in my own sentiments and feel their reality subjectively, such reality is still objective, for I did not create them; I merely inherited them.

—Emile Durkheim
Sociologist

inmate depends largely on your own efforts, behavior, and choices. One's ascribed statuses may affect the likelihood of achieving other statuses, however. For example, if you are born into a poor socioeconomic status, you may find it more difficult to achieve the status of "college graduate" because of the high cost of a college education.

Every individual has numerous statuses simultaneously. You may be a student, parent, tutor, volunteer fundraiser, female, and Hispanic. A person's **master status** is the status that is considered the most significant in a person's social identity. Typically, a person's occupational status is regarded as his or her master status. If you are a full-time student, your master status is likely to be "student."

Roles Every status is associated with many **roles**, or the set of rights, obligations, and expectations associated with a status. Roles guide our behavior and allow us to predict the behavior of others. As a student, you are expected to attend class, listen and take notes, study for tests, and complete assignments. Because you know what the role of teacher involves, you can predict that your teacher will lecture, give exams, and assign grades based on your performance on tests.

A single status involves more than one role. For example, the status of prison inmate includes one role for interacting with prison guards and another role for interacting with other prison inmates. Similarly, the status of nurse involves different roles for interacting with physicians and with patients.

Elements of Culture

Man is made by his belief. As he believes, so he is.

—Bhagavad Gita

Whereas social structure refers to the organization of society, culture refers to the meanings and ways of life that characterize a society. The elements of culture include beliefs, values, norms, sanctions, and symbols.

Beliefs **Beliefs** refer to definitions and explanations about what is assumed to be true. The beliefs of an individual or group influence whether that individual or group views a particular social condition as a social problem. Does secondhand smoke harm nonsmokers? Are nuclear power plants safe? Does violence in movies and on television lead to increased aggression in children? Our beliefs regarding these issues influence whether we view the issues as social problems. Beliefs not only influence how a social condition is interpreted, they also influence the existence of the condition itself. For example, men who believe that when a woman says "no," she really means "yes" or "maybe" are more likely to commit rape and sexual assault than men who do not have these beliefs (Frank 1991). The *Self & Society* feature in this chapter allows you to assess your own beliefs about various social issues and compare your beliefs with a national sample of first-year college students.

When people cherish some set of values and do not feel any threat to them, they experience well-being. When they cherish values but do feel them to be threatened, they experience a crisis—either as a personal trouble or as a public issue.

—C. Wright Mills
Sociologist

Values **Values** are social agreements about what is considered good and bad, right and wrong, desirable and undesirable. Frequently, social conditions are viewed as social problems when the conditions are incompatible with or contradict closely held values. For example, poverty and homelessness violate the value of human welfare; crime contradicts the values of honesty, private property, and nonviolence; racism, sexism, and heterosexism violate the values of equality and fairness.

Values play an important role not only in the interpretation of a condition as a social problem, but also in the development of the social condition itself.

Personal Beliefs about Various Social Problems

Indicate whether you agree or disagree with each of the following statements:

Statement	Agree	Disagree
1. The federal government is not doing enough to control environmental pollution.	___	___
2. The federal government should raise taxes to reduce the deficit.	___	___
3. There is too much concern in the courts for the rights of criminals.	___	___
4. Abortion should be legal.	___	___
5. The death penalty should be abolished.	___	___
6. The activities of married women are best confined to the home and family.	___	___
7. Marijuana should be legalized.	___	___
8. It is important to have laws prohibiting homosexual relationships.	___	___
9. Employers should be allowed to require drug testing of employees or job applicants.	___	___
10. The federal government should do more to control the sale of handguns.	___	___
11. A national health care plan is needed to cover everybody's medical costs.	___	___
12. Racial discrimination is no longer a major problem in America.	___	___
13. Realistically, an individual can do little to bring about changes in our society.	___	___
14. Wealthy people should pay a larger share of taxes than they do now.	___	___
15. Affirmative action in college admissions should be abolished.	___	___
16. Same-sex couples should have the right to legal marital status.	___	___

Percentage* of First-Year College Students Agreeing with Belief Statements

	Percentage Agreeing in 1997		
Statement Number	***Total***	***Women***	***Men***
1. Pollution control	81	84	77
2. Tax increase to reduce deficit	22	21	24
3. Too much concern for criminals' rights	70	70	71
4. Abortion rights	54	53	54
5. Abolishment of death penalty	24	26	21
6. Women's activities confined to home	25	20	32
7. Legalization of marijuana	35	32	39
8. Laws prohibiting gay relationships	34	24	46
9. Employers' right to drug test	78	81	75
10. Federal control of handgun sales	81	90	71
11. Need for national health care plan	72	77	67
12. Racial discrimination not a problem	20	16	24
13. Individuals can't influence social change	33	29	37
14. Wealthy should pay higher taxes	63	63	64
15. Affirmative action abolished in college	50	44	57
16. Legal right of same-sex couples to marry	50	58	40

*Percentages are rounded.

SOURCE: L.J. Sax, A.W. Astin, W.S. Korn, K.M. Mahoney. 1997. *The American Freshman: National Norms for Fall 1997*. Los Angeles: Higher Education Research Institute, UCLA. Copyright © 1997 by the Regents of the University of California. Used by permission.

Sylvia Ann Hewlett (1992) explains how the American values of freedom and individualism are at the root of many of our social problems:

> **There are two sides to the coin of freedom. On the one hand, there is enormous potential for prosperity and personal fulfillment; on the other are all the hazards of untrammeled opportunity and unfettered choice. Free markets can produce grinding poverty as well as spectacular wealth; unregulated industry can create dangerous levels of pollution as well as rapid rates of growth; and an unfettered drive for personal fulfillment can have disastrous effects on families and children. Rampant individualism does not bring with it sweet freedom; rather, it explodes in our faces and limits life's potential. (pp. 350–51)**

Absent or weak values may contribute to some social problems. For example, many industries do not value protection of the environment and thus contribute to environmental pollution.

National Data

A national survey of first-year college and university students revealed that 36 percent consider "influencing social values" to be one of their essential or very important objectives in life. Women were more likely than men to identify this objective as essential or very important (40 percent of women versus 32 percent of men).

("This Year's Freshman: A Statistical Profile" 1999)

Norms and Sanctions

Norms are socially defined rules of behavior. Norms serve as guidelines for our behavior and for our expectations of the behavior of others.

There are three types of norms: folkways, laws, and mores. **Folkways** refer to the customs and manners of society. In many segments of our society, it is customary to shake hands when being introduced to a new acquaintance, to say "excuse me" after sneezing, and to give presents to family and friends on their birthdays. Although no laws require us to do these things, we are expected to do them because they are part of the cultural traditions, or folkways, of the society in which we live.

Laws are norms that are formalized and backed by political authority. A person who eats food out of a public garbage container is violating a folkway; no law prohibits this behavior. However, throwing trash onto a public street is considered littering and is against the law.

Some norms, called **mores**, have a moral basis. Violations of mores may produce shock, horror, and moral indignation. Both littering and child sexual abuse are violations of law, but child sexual abuse is also a violation of our mores because we view such behavior as immoral.

All norms are associated with **sanctions**, or social consequences for conforming to or violating norms. When we conform to a social norm, we may be rewarded by a positive sanction. These may range from an approving smile to a public ceremony in our honor. When we violate a social norm, we may be punished by a negative sanction, which may range from a disapproving look to the death penalty or life in prison. Most sanctions are spontaneous expressions of approval or disapproval by groups or individuals—these are referred to as informal sanctions. Sanctions that are carried out according to some recognized or formal procedure are referred to as formal sanctions. Types of sanctions, then, include positive informal sanctions, positive formal sanctions, negative informal sanctions, and negative formal sanctions (see Table 1.1).

Symbols

A **symbol** is something that represents something else. Without symbols, we could not communicate with each other or live as social beings.

The symbols of a culture include language, gestures, and objects whose meaning is commonly understood by the members of a society. In our society, a red ribbon tied around a car antenna symbolizes Mothers Against Drunk

Table 1.1 Types and Examples of Sanctions

	Positive	Negative
Informal	Being praised by one's neighbors for organizing a neighborhood recycling program.	Being criticized by one's neighbors for refusing to participate in the neighborhood recycling program.
Formal	Being granted a citizen's award for organizing a neighborhood recycling program.	Being fined by the city for failing to dispose of trash properly.

Driving, a peace sign symbolizes the value of nonviolence, and a white hooded robe symbolizes the Ku Klux Klan. Sometimes people attach different meanings to the same symbol. The Confederate flag is a symbol of Southern pride to some, a symbol of racial bigotry to others.

The elements of the social structure and culture just discussed play a central role in the creation, maintenance, and social response to various social problems. One of the goals of taking a course in social problems is to develop an awareness of how the elements of social structure and culture contribute to social problems. Sociologists refer to this awareness as the "sociological imagination" or "sociological mindfulness."

> Freedom is what you do with what's been done to you.
>
> —Jean-Paul Sartre
> Philosopher

THE SOCIOLOGICAL IMAGINATION

The **sociological imagination**, a term developed by C. Wright Mills (1959), refers to the ability to see the connections between our personal lives and the social world in which we live. When we use our sociological imagination, we are able to distinguish between "private troubles" and "public issues" and to see connections between the events and conditions of our lives and the social and historical context in which we live.

For example, that one man is unemployed constitutes a private trouble. That millions of people are unemployed in the United States constitutes a public issue. Once we understand that personal troubles such as HIV infection, criminal victimization, and poverty are shared by other segments of society, we can look for the elements of social structure and culture that contribute to these public issues and private troubles. If the various elements of social structure and culture contribute to private troubles and public issues, then society's social structure and culture must be changed if these concerns are to be resolved.

Rather than viewing the private trouble of being unemployed as being due to an individual's faulty character or lack of job skills, we may understand unemployment as a public issue that results from the failure of the economic and political institutions of society to provide job opportunities to all citizens. Technological innovations emerging from the Industrial Revolution led to individual workers being replaced by machines. During the economic recession of the 1980s, employers fired employees so the firm could stay in business. Thus, in both these cases, social forces rather than individual skills largely determined whether a person was employed or not.

Another concept similar to "sociological imagination" is "**sociological mindfulness.**" According to sociologist Michael Schwalbe (1998), sociological mindfulness is a way of paying attention to the social world. What do we see if we practice sociological mindfulness? Schwalbe (1998) answers:

> **We see, for example, how the social world is created by people; . . . how people's behavior is a response to the conditions under which they live; how social life consists of patterns . . . how power is exercised; how inequalities are created and maintained. (p. 4)**
>
> **Being sociologically mindful also means paying attention to the hardships and options other people face. If we understand how others' circumstances differ from ours, we are more likely to show compassion for them and to grant them the respect they deserve as human beings. We are also less likely to condemn them unfairly . . . (p. 5)**
>
> **Part of being sociologically mindful of the constructedness of the social world is seeing the possibility of changing it. This means recognizing the possibility of acting differently, of choosing not to support arrangements that are harmful or unjust (p. 23) . . . Being sociologically mindful, we thus see that human beings are both social products and social forces. (p. 25)**

THEORETICAL PERSPECTIVES

> The most incomprehensible thing about the world is the fact that it is comprehensible.
>
> –Albert Einstein
> Scientist

Theories in sociology provide us with different perspectives with which to view our social world. A perspective is simply a way of looking at the world. A theory is a set of interrelated propositions or principles designed to answer a question or explain a particular phenomenon; it provides us with a perspective. Sociological theories help us to explain and predict the social world in which we live.

Sociology includes three major theoretical perspectives: the structural-functionalist perspective, the conflict perspective, and the symbolic interactionist perspective. Each perspective offers a variety of explanations about the causes of and possible solutions for social problems (Rubington and Weinberg 1995).

Structural-Functionalist Perspective

> Some see the glass half-empty, some see the glass half-full. I see the glass as too big.
>
> –George Carlin
> Comedian

The structural-functionalist perspective is largely based on the works of Herbert Spencer, Emile Durkheim, Talcott Parsons, and Robert Merton. According to **structural-functionalism,** society is a system of interconnected parts that work together in harmony to maintain a state of balance and social equilibrium for the whole. For example, each of the social institutions contributes important functions for society: family provides a context for reproducing, nurturing, and socializing children; education offers a way to transmit a society's skills, knowledge, and culture to its youth; politics provides a means of governing members of society; economics provides for the production, distribution, and consumption of goods and services; and religion provides moral guidance and an outlet for worship of a higher power.

The structural-functionalist perspective emphasizes the interconnectedness of society by focusing on how each part influences and is influenced by other parts. For example, the increase in single-parent and dual-earner families has contributed to the number of children who are failing in school because parents have become less available to supervise their children's homework. Due to changes in technology, colleges are offering more technical programs, and many adults are returning to school to learn new skills that are required in the workplace. The increasing number of women in the workforce has contributed to the formulation of policies against sexual harassment and job discrimination.

CONSIDERATION

In viewing society as a set of interrelated parts, structural-functionalists also note that proposed solutions to a social problem may cause additional social problems. For example, racial imbalance in public schools led to forced integration, which in turn generated violence and increased hostility between the races. The use of plea bargaining was adopted as a means of dealing with overcrowded court dockets but resulted in "the revolving door of justice." Urban renewal projects often displaced residents and broke up community cohesion.

Structural-functionalists use the terms "functional" and "dysfunctional" to describe the effects of social elements on society. Elements of society are functional if they contribute to social stability and dysfunctional if they disrupt social stability. Some aspects of society may be both functional and dysfunctional for society. For example, crime is dysfunctional in that it is associated with physical violence, loss of property, and fear. But, according to Durkheim and other functionalists, crime is also functional for society because it leads to heightened awareness of shared moral bonds and increased social cohesion.

Sociologists have identified two types of functions: manifest and latent (Merton 1968). **Manifest functions** are consequences that are intended and commonly recognized. **Latent functions** are consequences that are unintended and often hidden. For example, the manifest function of education is to transmit knowledge and skills to society's youth. But public elementary schools also serve as baby-sitters for employed parents, and colleges offer a place for young adults to meet potential mates. The baby-sitting and mate selection functions are not the intended or commonly recognized functions of education—hence, they are latent functions.

Structural-Functionalist Theories of Social Problems

Two dominant theories of social problems grew out of the structural-functionalist perspective: social pathology and social disorganization.

> Everybody should live a good and productive life. When there are impediments to that, we as a society have a responsibility to help.
>
> —Tipper Gore
> Social activist

Social Pathology

According to the social pathology model, social problems result from some "sickness" in society. Just as the human body becomes ill when our systems, organs, and cells do not function normally, society becomes "ill" when its parts (i.e., elements of the structure and culture) no longer perform properly. For example, problems such as crime, violence, poverty, and juvenile delinquency are often attributed to the breakdown of the family institution, the decline of the religious institution, and inadequacies in our economic, educational, and political institutions.

Social "illness" also results when members of a society are not adequately socialized to adopt its norms and values. Persons who do not value honesty, for example, are prone to dishonesties of all sorts. Early theorists attributed the failure in socialization to "sick" people who could not be socialized. Later theorists recognized that failure in the socialization process stemmed from "sick" social conditions, not "sick" people. To prevent or solve social problems, members of society must receive proper socialization and moral education, which may be accomplished in the family, schools, churches, workplace, and/or through the media.

Social Disorganization According to the social disorganization view of social problems, rapid social change disrupts the norms in a society. When norms become weak or are in conflict with each other, society is in a state of **anomie** or normlessness. Hence, people may steal, physically abuse their spouse or children, abuse drugs, rape, or engage in other deviant behavior because the norms regarding these behaviors are weak or conflicting. According to this view, the solution to social problems lies in slowing the pace of social change and strengthening social norms. For example, although the use of alcohol by teenagers is considered a violation of a social norm in our society, this norm is weak. The media portray young people drinking alcohol, teenagers teach each other to drink alcohol and buy fake identification cards (IDs) to purchase alcohol, and parents model drinking behavior by having a few drinks after work or at a social event. Solutions to teenage drinking may involve strengthening norms against it through public education, restricting media depictions of youth and alcohol, imposing stronger sanctions against the use of fake IDs to purchase alcohol, and educating parents to model moderate and responsible drinking behavior.

Conflict Perspective

Whereas the structural-functionalist perspective views society as comprising different parts working together, the **conflict perspective** views society as comprising different groups and interests competing for power and resources. The conflict perspective explains various aspects of our social world by looking at which groups have power and benefit from a particular social arrangement.

The origins of the conflict perspective can be traced to the classic works of Karl Marx. Marx suggested that all societies go through stages of economic development. As societies evolve from agricultural to industrial, concern over meeting survival needs is replaced by concern over making a profit, the hallmark of a capitalist system. Industrialization leads to the development of two classes of people: the bourgeoisie, or the owners of the means of production (e.g., factories, farms, businesses), and the proletariat, or the workers who earn wages.

The division of society into two broad classes of people—the "haves" and the "have-nots"—is beneficial to the owners of the means of production. The workers, who may earn only subsistence wages, are denied access to the many resources available to the wealthy owners. According to Marx, the bourgeoisie use their power to control the institutions of society to their advantage. For example, Marx suggested that religion serves as an "opiate of the masses" in that it soothes the distress and suffering associated with the working-class lifestyle and focuses the workers' attention on spirituality, God, and the afterlife rather than on such worldly concerns as living conditions. In essence, religion diverts the workers so that they concentrate on being rewarded in heaven for living a moral life rather than on questioning their exploitation.

Conflict Theories of Social Problems

There are two general types of conflict theories of social problems: Marxist and non-Marxist. Marxist theories focus on social conflict that results from economic inequalities; non-Marxist theories focus on social conflict that results from competing values and interests among social groups.

> Underlying virtually all social problems are conditions caused in whole or in part by social injustice.
>
> —Pamela Ann Roby
> Sociologist, University of California, Santa Cruz

Marxist Conflict Theories According to contemporary Marxist theorists, social problems result from class inequality inherent in a capitalistic system. A system of "haves" and "have-nots" may be beneficial to the "haves" but often translates into poverty for the "have-nots." As we shall explore later in this text, many social problems, including physical and mental illness, low educational achievement, and crime, are linked to poverty.

In addition to creating an impoverished class of people, capitalism also encourages "corporate violence." Corporate violence may be defined as actual harm and/or risk of harm inflicted on consumers, workers, and the general public as a result of decisions by corporate executives or managers. Corporate violence may also result from corporate negligence, the quest for profits at any cost, and willful violations of health, safety, and environmental laws (Hills 1987). Our profit-motivated economy encourages individuals who are otherwise good, kind, and law-abiding to knowingly participate in the manufacturing and marketing of defective brakes on American jets, fuel tanks on automobiles, and contraceptive devices (intrauterine devices [IUDs]). The profit motive has also caused individuals to sell defective medical devices, toxic pesticides, and contaminated foods to developing countries. Blumberg (1989) suggests that "in an economic system based exclusively on motives of self-interest and profit, such behavior is inevitable" (p. 106).

Marxist conflict theories also focus on the problem of **alienation**, or powerlessness and meaninglessness in people's lives. In industrialized societies, workers often have little power or control over their jobs, which fosters a sense of powerlessness in their lives. The specialized nature of work requires workers to perform limited and repetitive tasks; as a result, the workers may come to feel that their lives are meaningless.

Alienation is bred not only in the workplace, but also in the classroom. Students have little power over their education and often find the curriculum is not meaningful to their lives. Like poverty, alienation is linked to other social problems, such as low educational achievement, violence, and suicide.

Marxist explanations of social problems imply that the solution lies in eliminating inequality among classes of people by creating a classless society. The nature of work must also change to avoid alienation. Finally, stronger controls must be applied to corporations to ensure that corporate decisions and practices are based on safety rather than profit considerations.

Non-Marxist Conflict Theories Non-Marxist conflict theorists such as Ralf Dahrendorf are concerned with conflict that arises when groups have opposing values and interests. For example, antiabortion activists value the life of unborn embryos and fetuses; prochoice activists value the right of women to control their own body and reproductive decisions. These different value positions reflect different subjective interpretations of what constitutes a social problem. For antiabortionists, the availability of abortion is the social problem; for prochoice advocates, restrictions on abortion are the social problem. Sometimes the social problem is not the conflict itself, but rather the way that conflict is expressed. Even most prolife advocates agree that shooting doctors who perform abortions and blowing up abortion clinics constitute unnecessary violence and lack of respect for life. Value conflicts may occur between diverse categories of people, including nonwhites versus whites, heterosexuals versus homosexuals, young versus old, Democrats versus Republicans, and environmentalists versus industrialists.

Solutions to the problems that are generated by competing values may involve ensuring that conflicting groups understand each other's views, resolving differences through negotiation or mediation, or agreeing to disagree. Ideally, solutions should be win-win; both conflicting groups are satisfied with the solution. However, outcomes of value conflicts are often influenced by power; the group with the most power may use its position to influence the outcome of value conflicts. For example, when Congress could not get all states to voluntarily increase the legal drinking age to 21, it threatened to withdraw federal highway funds from those that would not comply.

Symbolic Interactionist Perspective

Both the structural-functionalist and the conflict perspectives are concerned with how broad aspects of society, such as institutions and large social groups, influence the social world. This level of sociological analysis is called **macro sociology:** it looks at the "big picture" of society and suggests how social problems are affected at the institutional level.

> Each to each a looking glass, Reflects the other that doth pass.
>
> –Charles Horton Cooley
> Sociologist

Micro sociology, another level of sociological analysis, is concerned with the social psychological dynamics of individuals interacting in small groups. **Symbolic interactionism** reflects the micro sociological perspective and was largely influenced by the work of early sociologists and philosophers such as Max Weber, George Simmel, Charles Horton Cooley, G. H. Mead, W. I. Thomas, Erving Goffman, and Howard Becker. Symbolic interactionism emphasizes that human behavior is influenced by definitions and meanings that are created and maintained through symbolic interaction with others.

Sociologist W. I. Thomas ([1931] 1966) emphasized the importance of definitions and meanings in social behavior and its consequences. He suggested that humans respond to their definition of a situation rather than to the objective situation itself. Hence, Thomas noted that situations we define as real become real in their consequences.

Symbolic interactionism also suggests that our identity or sense of self is shaped by social interaction. We develop our self-concept by observing how others interact with us and label us. By observing how others view us, we see a reflection of ourselves that Cooley calls the "looking glass self."

Lastly, the symbolic interaction perspective has important implications for how social scientists conduct research. The German sociologist Max Weber (1864–1920) argued that in order to understand individual and group behavior, social scientists must see the world from the eyes of that individual or group. Weber called this approach *Verstehen,* which in German means "empathy." *Verstehen* implies that in conducting research, social scientists must try to understand others' view of reality and the subjective aspects of their experiences, including their symbols, values, attitudes, and beliefs.

Symbolic Interactionist Theories of Social Problems

A basic premise of symbolic interactionist theories of social problems is that a condition must be defined or recognized as a social problem in order for it to be a social problem. Based on this premise, Herbert Blumer (1971) suggested that social problems develop in stages. First, social problems pass through the stage of "societal recognition"—the process by which a social problem, for example, drunk driving, is "born." Second, "social legitimation" takes place when the social problem achieves recognition by the larger community, including the media, schools, and churches. As the visibility of traffic fatalities associated with alcohol increased, so did the legitimation of drunk driving as a social problem. The next stage in the development of a social problem involves "mobilization for action," which occurs when individuals and groups, such as Mothers Against Drunk Driving, become concerned about how to respond to the social condition. This mobilization leads to the "development and implementation of an official plan" for dealing with the problem, involving, for example, highway checkpoints, lower legal blood-alcohol levels, and tougher drunk driving regulations.

Blumer's stage development view of social problems is helpful in tracing the development of social problems. For example, although sexual harassment and date rape have occurred throughout this century, these issues did not begin to receive recognition as social problems until the 1970s. Social legitimation of these problems was achieved when high schools, colleges, churches, employers, and the media recognized their existence. Organized social groups mobilized to develop and implement plans to deal with these problems. For example, groups successfully lobbied for the enactment of laws against sexual harassment and the enforcement of sanctions against violators of these laws. Groups also mobilized to provide educational seminars on date rape for high school and college students and to offer support services to victims of date rape.

Some disagree with the symbolic interactionist view that social problems exist only if they are recognized. According to this view, individuals who were victims of date rape in the 1960s may be considered victims of a problem, even though date rape was not recognized at that time as a social problem.

Labeling theory, a major symbolic interactionist theory of social problems, suggests that a social condition or group is viewed as problematic if it is labeled as such. According to labeling theory, resolving social problems sometimes involves changing the meanings and definitions that are attributed to people and situations. For example, as long as teenagers define drinking alcohol as "cool" and "fun," they will continue to abuse alcohol. As long as our society defines providing sex education and contraceptives to teenagers as inappropriate or immoral, the teenage pregnancy rate in our country will continue to be higher than in other industrialized nations.

Table 1.2 summarizes and compares the major theoretical perspectives, their criticisms, and social policy recommendations as they relate to social problems. The study of social problems is based on research as well as theory, however. Indeed, research and theory are intricately related. As Wilson (1983) states,

> **Most of us think of theorizing as quite divorced from the business of gathering facts. It seems to require an abstractness of thought remote from the practical activity of empirical research. But theory building is not a separate activity within sociology. Without theory, the empirical**

Table 1.2 **Comparison of Theoretical Perspectives**

	Structural-Functionalism	Conflict Theory	Symbolic Interactionism
Representative Theorists	Emile Durkheim Talcott Parsons Robert Merton	Karl Marx Ralf Dahrendorf	George H. Mead Charles Cooley Erving Goffman
Society	Society is a set of interrelated parts; cultural consensus exists and leads to social order; natural state of society—balance and harmony.	Society is marked by power struggles over scarce resources; inequities result in conflict; social change is inevitable; natural state of society—imbalance.	Society is a network of interlocking roles; social order is constructed through interaction as individuals, through shared meaning, make sense out of their social world.
Individuals	Individuals are socialized by society's institutions; socialization is the process by which social control is exerted; people need society and its institutions.	People are inherently good but are corrupted by society and its economic structure; institutions are controlled by groups with power; "order" is part of the illusion.	Humans are interpretative and interactive; they are constantly changing as their "social beings" emerge and are molded by changing circumstances.
Cause of Social Problems?	Rapid social change: social disorganization that disrupts the harmony and balance; inadequate socialization and/or weak institutions.	Inequality; the dominance of groups of people over other groups of people; oppression and exploitation; competition between groups.	Different interpretations of roles; labeling of individuals, groups, or behaviors as deviant; definition of an objective condition as a social problem.
Social Policy/ Solutions	Repair weak institutions; assure proper socialization; cultivate a strong collective sense of right and wrong.	Minimize competition; create an equitable system for the distribution of resources.	Reduce impact of labeling and associated stigmatization; alter definitions of what is defined as a social problem.
Criticisms	Called "sunshine sociology"; supports the maintenance of the status quo; needs to ask "functional for whom?" Does not deal with issues of power and conflict; incorrectly assumes a consensus.	Utopian model; Marxist states have failed; denies existence of cooperation and equitable exchange. Can't explain cohesion and harmony.	Concentrates on micro issues only; fails to link micro issues to macro-level concerns; too psychological in its approach; assumes label amplifies problem.

researcher would find it impossible to decide what to observe, how to observe it, or what to make of the observations . . . . (p. 1)

In science (as in everyday life) things must be believed in order to be seen as well as seen in order to be believed.

—Walter L. Wallace
Social scientist

SOCIAL PROBLEMS RESEARCH

Most students taking a course in social problems will not become researchers or conduct research on social problems. Nevertheless, we are all consumers of research that is reported in the media. Politicians, social activist groups, and organizations attempt to justify their decisions, actions, and positions by citing research results. As consumers of research, it is important to understand that our personal experiences and casual observations are less reliable than generalizations based on systematic research. One strength of scientific research is that it is subjected to critical examination by other researchers. The more you understand how research is done, the better able you will be to critically examine and question research, rather than to passively consume research findings. The remainder of this section discusses the stages of conducting a research study and the various methods of research used by sociologists.

FOCUS ON SOCIAL PROBLEMS RESEARCH

Collaborative Research as an Emerging Research Model

Collaborative research (also known as *participatory research*), refers to a research approach in which community activists and academicians work together in all phases of research, including identifying the research question, developing the research design, collecting and analyzing the data, presenting the results, and working with policy makers and practitioners in designing programs and policies.

Collaborative research is a win-win endeavor, benefiting professional researchers as well as community groups and organizations. Community groups often lack resources to conduct quality research. Through collaboration with university researchers, community activists and organizations gain access to financial and technical resources and discover the usefulness of using systematic research methods to analyze social problems and identify, implement, and evaluate solutions. Collaborative research provides "a means of putting research capabilities in the hands of the deprived and disenfranchised people so that they can transform their lives for themselves" (Park 1993, 1).

Carlos DeJesus, Executive Director of Latinos United, provides an example of the benefits of collaborative research for community organizations:

> We knew that our community had minimal access to publicly assisted housing resources, but we did not know the extent of our underrepresentation, and did not have the resources with which to do a thorough analysis. We established a relationship with the Policy Research Action Group (PRAG), a network of universities and community organizations, which provided us with graduate students as research associates. In short order, we were able to document that Chicago has 110,000 units of publicly assisted housing units and that less than 3 percent of these units housed Latino families, even though Latinos compromise 25 percent of the eligible population. . . . This data augmented our findings that the Chicago Housing Authority (CHA) had engaged in policies and practices that were discriminatory against Latinos.
>
> Latinos United, along with four other community organizations and five community residents, filed and won a class action lawsuit against HUD [U.S. Department of Housing and Urban Development] and CHA. The net result is the elimination of these discriminatory policies and practices, and an estimated $210 million in housing resources for eligible Latino families over the next 10 years. (DeJesus 1997, xxiv).

Through collaboration with community organizations and activists, academic researchers gain new insights into social problems and find new ways to use their skills and contribute to public well-being. Because collaborative research projects investigate the concerns of community organizations and practitioners, "the results are more likely to have relevance in solving pressing problems in today's society. . . . " (Nyden et al. 1997, 4).

The collaborative research model not only helps achieve solutions to specific social problems, it also contributes to the redistribution of power in the creation of legitimate, scientific knowledge. The production of scientific knowledge has been largely controlled by researchers who are academic "experts" in their field, well-versed in theories, and who have the technical knowledge and financial resources to conduct research. Research agendas and the production of legitimate, scientifically produced knowledge have been shaped and controlled by universities and by corporations and government agencies that have the economic resources to hire researchers and fund research projects. By involving practitioners, policy makers, and social activists in the research process, collaborative research offers a means to break the university's monopoly on the production of knowledge and empowers community members to shape research based on the concerns of individuals and organizations facing the issues on a daily basis.

Combining social science with social activism has received institutional support, as evidenced by the formation of the Policy Research Action Group (PRAG)—a collaborative partnership between four universities in the Chicago area and more than 20 community organizations. Could this emerging collaborative research model become the dominant model of social science research in the coming decades? Given the movement toward downsizing and the fierce competition over scarce research funding dollars, there are certainly economic pressures encouraging community organizations and academic institutions to pool their resources. Publicly funded academic institutions are also under pressure to make direct contributions to social well-being and thus justify their public funding. Beyond economic advantages of collaborative research, Brown University sociologist Phil Brown (1997) notes, "our task of integrating social science and social activism can help retain and rekindle the spirit of human betterment that has long been a part of sociology" (p. 101).

SOURCES

Brown, Phil. 1997. "Social Science and Environmental Activism: A Personal Account." In *Building Community: Social Science in Action,* ed. P. Nyden, A. Figert, M. Shibley, and D. Burrows, pp. 98–102. Thousand Oaks, Calif: Pine Forge Press.

DeJesus, Carlos R. 1997. From *Building Community: Social Science in Action,* by DeJesus, Copyright 1997 by Pine Forge Press. Reprinted by permission of Pine Forge Press.

Nyden, Philip, Anne Figert, Mark Shibley, and Darryl Burrows. 1997. "University-Community Collaborative Research: Adding Chairs at the Research Table." In *Building Community: Social Science in Action,* ed. P. Nyden, A. Figert, M. Shibley, and D. Burrows, pp. 3-13. Thousand Oaks, Calif: Pine Forge Press.

Park, Peter. 1993. "What Is Participatory Research? A Theoretical and Methodological Perspective." In *Voices of Change: Participatory Research in the United States and Canada,* ed. P. Park, M. Brydon-Miller, B. Hall, and T. Jackson, pp. 1-19. Toronto, Ontario: The Ontario Institute for Studies in Education Press.

Stages of Conducting a Research Study

Sociologists progress through various stages in conducting research on a social problem. This section describes the first four stages: formulating a research question, reviewing the literature, defining variables, and formulating a hypothesis.

Formulating a Research Question

A research study usually begins with a research question. Where do research questions originate? How does a particular researcher come to ask a particular research question? In some cases, researchers have a personal interest in a specific topic because of their own life experience. For example, a researcher who has experienced spouse abuse may wish to do research on such questions as "What factors are associated with domestic violence?" and "How helpful are battered women's shelters in helping abused women break the cycle of abuse in their lives?" Other researchers may ask a particular research question because of their personal values—their concern for humanity and the desire to improve human life. Researchers who are concerned about the spread of human immunodeficiency virus (HIV) infection and AIDS may conduct research on such questions as "How does the use of alcohol influence condom use?" and "What educational strategies are effective for increasing safer sex behavior?" Researchers may also want to test a particular sociological theory, or some aspect of it, in order to establish its validity or conduct studies to evaluate the effect of a social policy or program. Research questions may also be formulated by the concerns of community groups and social activist organizations in collaboration with academic researchers. Government and industry also hire researchers to answer questions such as "How many children are victimized by episodes of violence at school?" and "What types of computer technologies can protect children against being exposed to pornography on the Internet?"

> Science is meaningless because it gives no answer to the question, the only question of importance for us: "What shall we do and how shall we live?"
>
> —Leo Tolstoy
> Novelist

CONSIDERATION

Many questions involving morals, values, and religion cannot be answered through scientific research. For example, scientific research cannot determine whether the death penalty or anything else is moral or immoral. Scientific research can, however, reveal information that may support or cause us to question our own moral judgments. Research can tell us how various segments of the population view capital punishment and what social and personal factors are associated with the different views. Research may also identify some of the social and economic consequences of allowing versus prohibiting capital punishment.

Reviewing the Literature

After a research question is formulated, the researcher reviews the published material on the topic to find out what is already known about it. Reviewing the literature also provides researchers with ideas about how to conduct their research and helps them formulate new research questions. A literature review also serves as an evaluation tool, allowing a comparison of research findings and other sources of information, such as expert opinions, political claims, and journalistic reports.

Defining Variables

A **variable** is any measurable event, characteristic, or property that varies or is subject to change. Researchers must operationally define the variables they study. An **operational definition** specifies how a variable is to be measured. For example, an operational definition of the variable "religiosity" might be the number of times the respondent reports going to church or synagogue. Another operational definition of "religiosity" might be the respondent's answer to the question, "How important is religion in your life?" (1 = not important, 2 = somewhat important, 3 = very important).

Operational definitions are particularly important for defining variables that cannot be directly observed. For example, researchers cannot directly observe concepts such as "mental illness," "sexual harassment," "child neglect," "job satisfaction," and "drug abuse." Nor can researchers directly observe perceptions, values, and attitudes.

Formulating a Hypothesis After defining the research variables, researchers may formulate a **hypothesis,** which is a prediction or educated guess about how one variable is related to another variable. The **dependent variable** is the variable that the researcher wants to explain; that is, it is the variable of interest. The **independent variable** is the variable that is expected to explain change in the dependent variable. In formulating a hypothesis, the researcher predicts how the independent variable affects the dependent variable. For example, Peterson and Krivo (1993) hypothesized that residential segregation, and the associated social isolation, affects the probability of homicide victimization of African Americans. Their research found that as racial segregation increases, the probability of homicide victimization of blacks also increases. In this example, the independent variable is residential segregation; the dependent variable is homicide victimization.

CONSIDERATION

Some social problems act as independent variables in the production of other social problems. Social problems that produce many other social problems are called "primary social problems" (Manis 1974). For example, poverty is a social problem that leads to the secondary social problem of slum neighborhoods, which, in turn, lead to the social problems of juvenile delinquency and addiction. In this example, "slum neighborhoods" acts as both an independent and a dependent variable.

In studying social problems, researchers often assess the effects of several independent variables on one or more dependent variables. For example, Jekielek (1998) examined the impact of parental conflict and marital disruption (two independent variables) on the emotional well-being of children (the dependent variable). Her research found that both parental conflict and marital disruption (separation or divorce) negatively affect children's emotional well-being. However, children in high-conflict intact families exhibit lower levels of well-being than children who have experienced high levels of parental conflict but whose parents divorce or separate.

Methods of Data Collection

After identifying a research topic, reviewing the literature, and developing hypotheses, researchers decide which method of data collection to use. Alternatives include experiments, surveys, field research, and secondary data.

Experiments **Experiments** involve manipulating the independent variable in order to determine how it affects the dependent variable. Experiments require one or more experimental groups that are exposed to the experimental treatment(s) and a control group that is not exposed. After the researcher randomly assigns participants to either an experimental or a control group, she or he measures the dependent variable. After the experimental groups are exposed to the treatment, the research measures the dependent variable again. If participants have been randomly assigned to the different groups, the researcher may conclude that any difference in the dependent variable among the groups is due to the effect of the independent variable.

An example of a "social problems" experiment on poverty would be to provide welfare payments to one group of unemployed single mothers (experimental group) and no such payments to another group of unemployed single mothers (control group). The independent variable would be welfare payments; the dependent variable would be employment. The researcher's hypothesis would be that mothers in the experimental group would be less likely to have a job after 12 months than mothers in the control group.

The major strength of the experimental method is that it provides evidence for causal relationships; that is, how one variable affects another. A primary weakness is that experiments are often conducted on small samples, usually in artificial laboratory settings; thus, the findings may not be generalized to other people in natural settings.

My latest survey shows that people don't believe in surveys.

—Laurence Peter
Humorist

Surveys **Survey research** involves eliciting information from respondents through questions. An important part of survey research is selecting a sample of those to be questioned. A **sample** is a portion of the population, selected to be representative so that the information from the sample can be generalized to a larger population. For example, instead of asking all abused spouses about their experience, you could ask a representative sample of them and assume that those you did not question would give similar responses. After selecting a representative sample, survey researchers either interview people, ask them to complete written questionnaires, or elicit responses to research questions through computers.

1. *Interviews*. In interview survey research, trained interviewers ask respondents a series of questions and make written notes about or tape-record the respondents' answers. Interviews may be conducted over the telephone or face-to-face. The 1997 Gallup Poll Social Audit of Black/White Relations involved telephone interviews with over 3,000 U.S. adults. One of the questions interviewers asked is whether respondents thought that relations between blacks and whites will always be a problem in the United States, or whether a solution will eventually be worked out. Over half of both blacks and whites (58 percent of blacks and 54 percent of whites) said that U.S. black/white relations will "always be a problem" (Gallup Organization 1997).

One advantage of interview research is that researchers are able to clarify questions for the respondent and follow up on answers to particular questions. Researchers often conduct face-to-face interviews with groups of individuals who might otherwise be inaccessible. For example, some AIDS-related research attempts to assess the degree to which individuals engage in behavior that places them at high risk for transmitting or contracting HIV. Street youth and intravenous drug users, both high-risk groups for HIV infection, may not have a telephone or address due to their transient lifestyle (Catania et al. 1990).

These groups may be accessible, however, if the researcher locates their hangouts and conducts face-to-face interviews. Research on homeless individuals may also require a face-to-face interview survey design.

The most serious disadvantages of interview research are cost and the lack of privacy and anonymity. Respondents may feel embarrassed or threatened when asked questions that relate to personal issues such as drug use, domestic violence, and sexual behavior. As a result, some respondents may choose not to participate in interview research on sensitive topics. Those who do participate may conceal or alter information or give socially desirable answers to the interviewer's questions (e.g., "No, I do not use drugs").

2. *Questionnaires.* Instead of conducting personal or phone interviews, researchers may develop questionnaires that they either mail or give to a sample of respondents. Questionnaire research offers the advantages of being less expensive and time-consuming than face-to-face or telephone surveys. In addition, questionnaire research provides privacy and anonymity to the research participants. This reduces the likelihood that they will feel threatened or embarrassed when asked personal questions and increases the likelihood that they will provide answers that are not intentionally inaccurate or distorted.

> When I was younger I could remember anything—whether it happened or not.
>
> –Mark Twain
> American humorist and writer

The major disadvantage of mail questionnaires is that it is difficult to obtain an adequate response rate. Many people do not want to take the time or make the effort to complete and mail a questionnaire. Others may be unable to read and understand the questionnaire.

3. *"Talking" Computers.* A new method of conducting survey research is asking respondents to provide answers to a computer that "talks." Romer et al. (1997) found that respondents rated computer interviews about sexual issues more favorably than face-to-face interviews and that the former were more reliable. Such increased reliability may be particularly valuable when conducting research on drug use, deviant sexual behavior, and sexual orientation as respondents reported the privacy of computers as a major advantage (Romer et al. 1997).

Field Research **Field research** involves observing and studying social behavior in settings in which it occurs naturally. Two types of field research are participant observation and nonparticipant observation.

> Feminists in all disciplines have demonstrated that objectivity has about as much substance as the emperor's new clothes.
>
> –Connie Miller
> Feminist scholar

In participant observation research, the researcher participates in the phenomenon being studied in order to obtain an insider's perspective of the people and/or behavior being observed. Coleman (1990), a middle-class white male, changed clothes to live on the streets of New York as a homeless person for 10 days. In nonparticipant observation research, the researcher observes the phenomenon being studied without actively participating in the group or the activity. For example, Dordick (1997) studied homelessness by observing and talking with homeless individuals in a variety of settings, but she did not live as a homeless person as part of her research.

Sometimes sociologists conduct in-depth detailed analyses or case studies of an individual, group, or event. For example, Skeen (1991) conducted case studies of a prostitute and her adjustment to leaving the profession, an incest survivor, and a person with AIDS.

The main advantage of field research on social problems is that it provides detailed information about the values, rituals, norms, behaviors, symbols, beliefs, and emotions of those being studied. A potential problem with field research is that the researcher's observations may be biased (e.g., the researcher becomes too involved in the group to be objective). In addition, because field research is usually based on small samples, the findings may not be generalizable.

Secondary Data Research Sometimes researchers analyze secondary data, which are data that have already been collected by other researchers or government agencies or that exist in forms such as historical documents, police reports, hospital records, and official records of marriages, births, and deaths. For example, Roberts (1993) used nationally representative data from the 1989 National Opinion Research Center (NORC) General Social Survey to examine the effect of "unhealthy" and "dangerous" workplaces on such psychosocial variables as depression and trust. Results indicated that working in unhealthy conditions was related to higher levels of unhappiness and depression and lower levels of trust and confidence in big business, corporations, government, and science.

A major advantage of using secondary data in studying social problems is that the data are readily accessible, so researchers avoid the time and expense of collecting their own data. Secondary data are also often based on large representative samples. The disadvantage of secondary data is that the researcher is limited to the data already collected.

GOALS OF THE TEXT

This text approaches the study of social problems with several goals in mind.

1. *Provide an integrated theoretical background.* This text reflects an integrative theoretical approach to the study of social problems. More than one theoretical perspective can be used to explain a social problem because social problems usually have multiple causes. For example, youth crime is linked to (1) an increased number of youth living in inner-city neighborhoods with little or no parental supervision (social disorganization), (2) young people having no legitimate means of acquiring material wealth (anomie theory), (3) youth being angry and frustrated at the inequality and racism in our society (conflict theory), and (4) teachers regarding youth as "no good" and treating them accordingly (labeling theory).

> The gulf between knowledge and truth is infinite.
>
> —Henry Miller
> Novelist

2. *Encourage the development of a sociological imagination.* A survey study of over 2,000 Americans found that the majority believe that they are in control of their own lives (Mirowsky, Ross, and Van Willigen 1996). Ninety-three percent agreed with "I am responsible for my own success." More than two-thirds of the sample agreed with statements claiming responsibility for misfortunes and failures. However, a major insight of the sociological perspective is that various structural and cultural elements of society have far-reaching effects on individual lives and social well-being. This insight, known as the sociological imagination or sociological mindfulness, enables us to understand how social forces underlie personal misfortunes and failures as well as contribute to personal successes and achievements. Each chapter in this text emphasizes how structural and cultural factors contribute to social problems. This emphasis encourages you to develop your sociological imagination by recognizing how structural and cultural factors influence private troubles and public issues.

> What are traditionally targeted as social problems (e.g., drug addicts, delinquents) are the measure of the social problem, not the substance or cause of it.
>
> —Buford Rhea
> Sociology Professor Emeritis
> East Carolina University

3. *Provide global coverage of social problems.* The modern world is often referred to as a "global village." The Internet and fax machines connect individuals around the world, economies are interconnected, environmental destruction in one region of the world affects other regions of the world, and diseases

cross national boundaries. Understanding social problems requires an awareness of how global trends and policies affect social problems. Many social problems call for collective action involving countries around the world; efforts to end poverty, protect the environment, control population growth, and reduce the spread of HIV are some of the social problems that have been addressed at the global level. Each chapter in this text includes coverage of global aspects of social problems. We hope that attention to the global aspects of social problems broadens students' awareness of pressing world issues.

> In many respects the world is sailing through the current era of globalization with neither compass nor map.
>
> –*Human Development Report 1997*
> United Nations Development Programme

4. *Provide an opportunity to assess personal beliefs and attitudes.* Each chapter in this text contains a section called *Self and Society,* which offers you an opportunity to assess your attitudes and beliefs regarding some aspect of the social problem discussed. Earlier in this chapter, the *Self and Society* feature allowed you to assess your beliefs about a number of social problems and compare your beliefs with a national sample of first-year college students.

5. *Emphasize the human side of social problems.* Each chapter in this text contains a feature called *The Human Side,* which presents personal stories of how social problems have affected individual lives. By conveying the private pain and personal triumphs associated with social problems, we hope to elicit a level of understanding and compassion that may not be attained through the academic study of social problems alone. This chapter's *The Human Side* presents stories about how college students, disturbed by various social conditions, have participated in social activism.

> But this dark is deep: now I warm you with my blood, listen to this flesh.
> It is far truer than poems.
>
> –Marina Tsvetayeva

6. *Encourage students to take prosocial action.* Individuals who understand the factors that contribute to social problems may be better able to formulate interventions to remediate those problems. Recognizing the personal pain and public costs associated with social problems encourages some to initiate social intervention.

Individuals can make a difference in society by the choices they make. Individuals may choose to vote for one candidate over another, demand the right

One way to affect social change is through demonstrations. A U. S. survey of first-year college and university students revealed that 43% reported having participated in organized demonstrations in the last year (American Council on Education and University of California, 1997). This photo depicts student demonstrators at The University of California, Berkeley, protesting the anti-affirmative action measure, Proposition 209.

In a certain sense, every single human soul has more meaning and value than the whole of history.

—Nicholas Berdyaev
Philosopher

Activism pays the rent on being alive and being here on the planet. . . . If I weren't active politically, I would feel as if I were sitting back eating at the banquet without washing the dishes or preparing the food. It wouldn't feel right.

—Alice Walker
Novelist

THE HUMAN SIDE

COLLEGE STUDENT ACTIVISM

Some people believe that in order to promote social change one must be in a position of political power and/or have large financial resources. However, the most important prerequisite for becoming actively involved in improving levels of social well-being may be genuine concern and dedication to a social "cause." The following vignettes provide a sampler of college student activism—college students making a difference in the world.

- In May 1989, hundreds of Chinese college students protested in Tiananmen Square in Beijing, China, because Chinese government officials would not meet with them to hear their pleas for a democratic government. These students boycotted classes and started a hunger strike. On June 4, 1989, thousands of students and other protesters were massacred or arrested in Tiananmen Square.
- In 1971, a student organization at Michigan State University, the Gay Liberation Movement, started a campaign for gay civil rights protections in East Lansing, Michigan. The group supported two city council candidates who pledged support for gay rights. In collaboration with another human rights organization, the Gay Liberation Movement presented proposals to the city's Human Relations Commission to add both gender and sexual orientation as protected classifications to the city's civil rights laws. In 1972, East Lansing, Michigan, became the first U.S. community to adopt a law or policy that legally banned discrimination on the basis of sexual orientation (Button, Rienzo, and Wald 1997).
- While a student at George Washington University, Ross Misher started an organization called Students Against Handgun Violence. When Ross was 13, his father was shot and killed by a coworker who had purchased a handgun during his lunch hour and returned to shoot Ross's father before killing himself (Lewis 1991).
- While a zoology major at the University of Colorado, Jeff Galus began the Animal Rights Student Group. This organization focuses on informing the public abut how animals are treated in research and what corporations use animals in testing their products.
- Students from the University of North Carolina at Chapel Hill and other college campuses have joined forces with the organization UNITE to protest contracts with Nike—a company that is criticized for its labor practices and working conditions. Such protests have resulted in Nike's reduction in child labor and commitment to use less toxic glues.

Students who are interested in becoming involved in student activism, or who are already involved, might explore the website for the Center for Campus Organizing (1998)—a national organization that supports social justice activism and investigative journalism on campuses nationwide. The organization was founded on the premise that students and faculty have played

THE HUMAN SIDE

COLLEGE STUDENT ACTIVISM *(continued)*

critical roles in larger social movements for social justice in our society, including the Civil Rights movement, the anti-Vietnam War movement, the Anti-Apartheid movement, the women's rights movement, and the environmental movement.

SOURCES:

Button, James W., Barbara A. Rienzo, and Denneth D. Wald. 1997. *Private Lives, Public Conflicts: Battles over Gay Rights in American Communities*. Washington, DC: Congressional Quarterly Inc.

Center for Campus Organizing. 1998. **http://www.cco.org/about.html** (12/8/98)

Lewis, Barbara A. 1991. *The Kid's Guide to Social Action,* Minneapolis, Minn.: Free Spirit Publishing, Inc. pp. 110-11. **www. 1998. http://www.colorado.edu/StudentGroups/animal rights/Galus@UCSU. Colorado.edu.**

www. 1998. http/www.compugraph.com/clr/alerts/alerts/campusactivism-q-a.html

Most politicians will not stick their necks out unless they sense grass-roots support. . . . Neither you nor I should expect someone else to take our responsibility.

—Katharine Hepburn
Actress

to reproductive choice or protest government policies that permit it, drive drunk or stop a friend from driving drunk, repeat a racist or sexist joke or chastise the person who tells it, and practice safe sex or risk the transmission of sexually transmitted diseases. Individuals can also "make a difference" by addressing social concerns in their occupational role, as well as through volunteer work.

Although individual choices make an important impact, collective social action often has a more pervasive effect. For example, while individual parents discourage their teenage children from driving under the influence of alcohol, Mothers Against Drunk Driving contributed to the enactment of national legislation that potentially will influence every U.S. citizen's decision about whether to use alcohol while driving.

Schwalbe (1998) reminds us that we don't have to join a group or organize a protest to make changes in the world.

> **We *can* change a small part of the social world single-handedly. If we treat others with more respect and compassion, if we refuse to participate in re-creating inequalities even in little ways, if we raise questions about official representation of reality, if we refuse to work in destructive industries, then we are making change. (p. 206)**

National Data

In a national survey of first-year college and university students, one-third agreed with the statement, "Realistically, an individual can do little to bring about changes in our society." Thus, two-thirds of first-year U.S. college students believe that individuals *can* bring about social change.

("This Year's Freshman: A Statistical Profile" 1999).

UNDERSTANDING SOCIAL PROBLEMS

At the end of each chapter to follow, we offer a section entitled *Understanding* in which we reemphasize the social origin of the problem being discussed, the consequences, and the alternative social solutions. It is our hope that the reader will end each chapter with a "sociological imagination" view of the problem and how, as a society, we might approach a solution.

Sociologists have been studying social problems since the Industrial Revolution at the turn of the twentieth century. Industrialization brought about massive social changes: the influence of religion declined; families became smaller and moved from traditional, rural communities to urban settings. These and other changes have been associated with increases in crime, pollution, divorce, and juvenile delinquency. As these social problems became more

National Data

In a survey of first-year university students, 74 percent reported that they had performed volunteer work in the last year. Women were more likely than men to have participated in volunteer work—78 percent versus 70 percent.

("This Year's Freshman: A Statistical Profile" 1999).

Resolve to create a good future. It's where you'll spend the rest of your life.

—Charles Franklin Kettering
American industrialist

Although the world is very full of suffering, it is also full of the overcoming of it.

—Helen Keller
Social activist

widespread, the need to understand their origins and possible solutions became more urgent. The field of sociology developed in response to this urgency. Social problems provided the initial impetus for the development of the field of sociology and continue to be a major focus of sociology.

There is no single agreed-upon definition of what constitutes a social problem. Most sociologists agree, however, that all social problems share two important elements: an objective social condition and a subjective interpretation of that condition. Each of the three major theoretical perspectives in sociology—structural-functionalist, conflict, and symbolic interactionist—has its own notion of the causes, consequences, and solutions of social problems.

CRITICAL THINKING

1. People increasingly are using information technologies as a means of getting their daily news. As a matter of fact, some research indicates that news on the Internet is beginning to replace television news as the primary source of information among computer users (see Chapter 14). What role does the media play in our awareness of social problems, and will definitions of social problems change as sources of information change?
2. Each of you occupy several social statuses, each one carrying an expectation of role performance, that is, what you should and shouldn't do given your position. List five statuses you occupy, the expectations of their accompanying roles, and any role conflict that may result. What types of social problems are affected by role conflict?
3. Definitions of social problems change over time. Identify a social condition that is now widely accepted which might be viewed as a social problem in the future.
4. How would each of the three major sociological perspectives analyze the impeachment procedures initiated against President Clinton?

KEY TERMS

achieved status
alienation
anomie
ascribed status
beliefs
conflict perspective
dependent variable
experiment
field research
folkway
hypothesis
independent variable
institution
labeling theory
latent function
law
macro sociology
manifest function
master status
micro sociology
mores
norm
objective element
operational definition
primary group
role
sanction
sample
secondary group
social group
social problem
sociological imagination
sociological mindfulness
status
structural-functionalism
subjective element
survey research
symbol
symbolic interactionism
values
variable

INTERNET

You can find more information on sociological theories, the American Sociological Association, research methods, and student activism at the *Understanding Social Problems* web site at **http://sociology.wadsworth.com**.

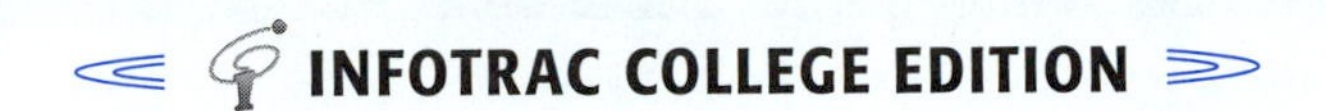

INFOTRAC COLLEGE EDITION

Either from the Wadsworth sociology resource center at **http://sociology.wadsworth.com** or directly from your web browser, you may access InfoTrac College Edition, an online university library that includes over 700 popular and scholarly journals in which you can find articles related to the topics in this chapter. Suggested articles and questions relating to these articles are listed below.

Affinnih, Yahya H. 1997. "A Critical Reexamination of Alienation." *The Social Science Journal* 34:383–88.

1. Does the author believe that alienation is a useful concept in sociology and, if so, why?
2. According to the author, what two categories can alienation articles be divided into?
3. Describe what kinds of alienation exist in today's society.

Sapon-Shevin, Mara, Anne Dobbelaere, Cathleen Corrigan, Kathleen Goodman, and Mary Mastin. 1998. "Everyone Here Can Play." *Educational Leadership* 56:42–46.

1. What book was the basis of the research project detailed in this article?
2. What changes were seen on the 1st grade level with this new program?
3. What was the main problem or challenge in implementing this program?

Wagner, Cynthia G. 1998. "This Just In—Old Statistics." *The Futurist* 32:6–8.

1. What is the general thesis of the article, that is, what is the author arguing?
2. Which of the social indicators listed is the most outdated?
3. What recommendations does the author make in dealing with outdated statistics?

Section 1
Problems of Well-Being

Section 1 deals with problems that are often regarded as private rather than public issues; that is, they are viewed as internally caused or as a function of individual free will. People often respond to these problems by assuming that the problem is the victims' fault—that in some way they have freely chosen their plight. In this set of problems, blame is most often attached to the individuals themselves. Thus, the physically and mentally ill (Chapter 2), the drunk and the drug addict (Chapter 3), the criminal and the delinquent (Chapter 4), and the divorced person and the child abuser (Chapter 5) are thought to be bad, or weak, or immoral, or somehow different from the average person. Consider the following scenarios:

1. A man on a fixed income without health insurance decides not to go to a doctor in order to save money to pay the rent and buy food for his children. When he becomes sick, his illness is blamed on his decision not to go to the doctor. As sociologists, we would say that the man did not want to be sick, but rather chose what he perceived as the least of several unfortunate alternatives. In this case, factors that underlie his illness include poverty, the structure of medical care in the United States, the rising cost of health insurance, and the value system that stresses parental responsibility and sacrifice.

2. A teenager from an urban lower-class neighborhood decides to sell drugs rather than stay in school or get a regular job. Such a teenager is generally viewed as being "weak" or having "low" morals. Sociologists view such a person as a lower-class, poorly educated individual with few alternatives in a society that values success. Raised in an environment where the most successful role models are often criminals, legitimate opportunities are few, traditional norms and values are weak, and peer pressure to use and sell drugs is strong, what are his choices? He can pump gas or serve fast food for minimum wage, or he can sell drugs for as much as $5,000 a week.

3. A mother comes home from work and finds her children playing and the house in disorder. She had told the children to clean the house while she was gone. She decides they need to be whipped with a belt because of their disobedience. The physical abuse she engages in is viewed as a reflection of her mental instability and her inability to control her temper. Research indicates, however,

that fewer than 10 percent of identified child abusers are severely psychologically impaired.

If being mentally unstable does not explain the majority of child abuse cases, what does explain them? A history of being abused as a child is the strongest independent predictor of who will be a child abuser as an adult. Additionally, the culture of society includes a myriad of beliefs that contribute to child abuse: acceptance of corporal punishment of children and the ambiguity surrounding what constitutes appropriate discipline, the belief that parental control is an inalienable right, and the historical and lingering belief that children (as well as women) are property.

4. A college student drinks alcohol daily and often cuts classes. Although the general public views such behavior as a personal weakness, sociologists emphasize the role of the individual's socialization and society. For example, a disproportionate number of individuals with drinking problems were reared in homes where one or both parents drank heavily. In the general culture, media portrayals of drinking as desirable, fun, glamorous, and a source of status further promote drinking. College culture itself often emphasizes bars and drinking parties as primary sources of recreation and affiliation.

These examples illustrate that many behaviors result more from social factors than from individual choice. To the degree that individuals do make choices, these choices are socially determined in that the structure and culture of society limit and influence individual choices. For example, customers in a restaurant cannot choose anything they want to eat; they are limited to what is on the menu. Sociologically, one's social status—black, white, male, female, young, old, rich, poor—determines one's menu of life choices.

In each of the above examples, the alternatives were limited by the individual's position in the social structure of society and by the cultural and subcultural definitions of appropriate behavior. While conflict theorists, structural-functionalists, and symbolic interactionists may disagree as to the relative importance and mechanisms of the shared structure and culture of society in determining the problems identified, all would agree that society, not the individual, is the primary source of the solutions. In this and the following sections, we emphasize the importance of the social structure and culture of society as both the sources of and solutions to social problems.

CHAPTER TWO

Illness and the Health Care Crisis

- The Global Context: Patterns of Health and Disease
- Sociological Theories of Illness and Health Care
- Health Concerns of U.S. College Students
- *Self & Society: AIDS Knowledge Scale*
- Social Factors Associated with Health and Illness
- Problems in U.S. Health Care
- Strategies for Action: Improving Health and Health Care
- *The Human Side: Managed Care Horror Stories*
- Understanding Illness and Health Care
- *Focus on Technology: The Emerging Field of Telemedicine*

Is It True?

Is It True?

1. Individuals in the United States can expect to live longer than citizens of any other country.
2. Compared to other industrialized countries, the United States has the lowest infant death rate.
3. Accidents are the number one cause of death in the United States among individuals aged 15 to 25.
4. In the United States, at least one person in four will have a sexually transmitted infection in his or her lifetime.
5. In 1999, 16 percent of the U.S. population had no medical insurance coverage.

Answers: 1 = F; 2 = F; 3 = T; 4 = T; 5 = T

It is ironic that in some parts of the world hundreds of millions of people suffer daily from a lack of basic health care while in other parts millions of people spend money on things that are not healthy. Think what a billion dollars could do to help immunize people against deadly diseases in developing countries. A billion dollars is not much money—it is what Americans spend on beer every 12 days and what Europeans spend on cigarettes every five days.

David Wright
Telemedicine and Developing Countries

In August 1997, Jeanne Calment, then the oldest woman in the world, died in France at the age of 122. When she was born in 1875, Thomas Edison had not yet discovered electricity; before she died, photographs from the planet Mars had been transmitted to Earth. During Jeanne Calment's lifetime, the world changed in unimaginable ways. One of the most profound changes over the last century has been the increase in the average length of life. Since the end of World War II, longevity of life in most developed and developing countries has increased by almost 25 years—the greatest increase seen in the history of humankind (LaPorte 1997).

Despite overall improvements in living conditions and medical care, health problems and health care delivery are major concerns of individuals, families, communities, and nations. Although technological advances in health care are increasing, disparities in health and access to health care between and within countries are also increasing (Creese, Martin, and Visschedijk 1998).

In this chapter, we review health concerns in the United States and throughout the world. The World Health Organization (1946) defines **health** as "a state of complete physical, mental, and social well-being" (p. 3). Sociologists are concerned with how social forces affect and are affected by health and illness, why some social groups suffer more illness than others, and how health and health care can be improved.

Health is a crossroads. It is where biological and social factors, the individual and the community, the social and the economic policy all converge. . . . Health is a means to personal and collective advancement. It is, therefore, an indicator of success achieved by a society and its institutions of government in promoting well-being, which is the ultimate meaning of development.

—Julio Frenk

THE GLOBAL CONTEXT: PATTERNS OF HEALTH AND DISEASE

The study of the distribution of disease within a population is called **epidemiology.** The field of epidemiology incorporates several disciplines, including public health, medicine, biology, and sociology. Sociologists who are **epidemiologists** are concerned with the social origins and distribution of health problems in a population and how patterns of illness and disease vary between and within societies. Next, we look at global patterns of morbidity, longevity, mortality, and disease burden.

Patterns of Morbidity

Morbidity refers to acute and chronic illnesses and diseases and the symptoms and impairments they produce. **Acute conditions** are short-term; by definition they can last no more than three months. **Chronic conditions** are long-term health problems. The rate of serious morbidity in a population provides one measure of the health of that population. Morbidity may be measured according to the incidence and prevalence of specific illnesses and diseases. **Incidence** refers to the number of *new cases* of a specific health problem with a given population during a specified time period. **Prevalence** refers to the *total number of cases* of a specific health problem within a population that exist at a given time. For example, the incidence of HIV infection worldwide was 5.8 million in 1997, meaning that there were 5.8 million people newly infected with HIV in 1997. In the same year,

Female infants and young children in India, like the child on the left in the photo, are more likely than males to be denied food and medical care.

National Data

Nearly half of noninstitutionalized persons aged 15 to 54 report having had at least one mental disorder and nearly 30 percent report having a mental disorder in the previous 12 months. The most common mental disorders in the United States are depression, alcohol dependence, and phobias (irrational fears).

SOURCE: Kessler et al. 1994

International Data

In the next 50 years, the proportion of people over 65 will more than double, growing from 6.8 percent of the global population to 15.1 percent. In Western Europe, more than one person in four (27.5 percent) will be over 65 in 2050.

SOURCE: "Older Populations Expanding Rapidly" 1998

the worldwide prevalence of HIV was 30.6 million, meaning that a total of 30.6 million people worldwide were living with HIV infection in 1997 (World Health Organization and the United Nations Joint Programme on HIV/AIDS 1998).

Morbidity statistics include data on the incidence and prevalence of mental disorders. The American Psychiatric Association (1994) defines a **mental disorder** as a "behavioral or psychological syndrome or pattern that occurs in an individual and that is associated with present distress (e.g., painful symptoms) or disability (i.e., impairment in one or more important areas of functioning) or with a significantly increased risk of suffering, death, pain, disability, or an important loss of freedom" (p. xxi).

As we discuss later in this chapter, patterns of morbidity vary according to social factors such as poverty, education, sex, and race. Morbidity patterns also vary according to the level of development of a society and the age structure of the population. In the industrialized world, infectious and parasitic diseases have been largely controlled by advances in sanitation and immunizations. Noninfectious diseases such as cancer, circulatory diseases, mental disorders, respiratory diseases, and musculoskeletal diseases pose the greatest health threat to the industrialized world (World Health Organization 1997c). In developing nations, such as China and Mexico, infectious and parasitic diseases are more common than in industrialized countries, but chronic degenerative diseases are increasing. In the less developed countries, malnutrition, pneumonia, and infectious and parasitic diseases such as HIV disease, malaria (transmitted by mosquitos), and measles are major health concerns.

In many countries, birthrates have declined over the last few decades (see Chapter 15). At the same time, improvements in sanitation and the development of medical technologies (such as vaccinations) have contributed to increased longevity. Declining birthrates and increased longevity have resulted in the aging of the world's population.

The aging of the world's population means that the most common health problems are becoming those of adults rather than those of children. Diseases

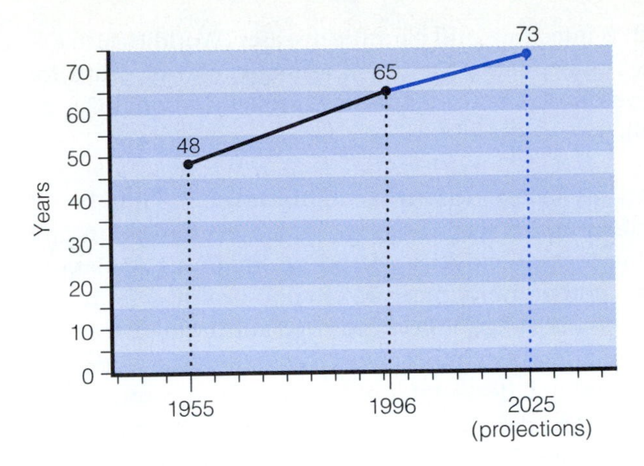

Figure 2.1 Global Life Expectancy, 1955–2025

SOURCE: World Health Organization. 1998. "Fifty Facts from the World Health Report 1998." **http://www.who.int/whr/1998/factse.htm** (8/8/98).

that need time to develop, such as cancer, heart disease, Alzheimer's disease, arthritis, and osteoporosis are becoming more common, and childhood illnesses, typically caused by infectious and parasitic diseases, are becoming less common. The shift from a society characterized by low life expectancy and parasitic and infectious diseases to one characterized by high life expectancy and chronic and degenerative diseases is called the **epidemiological transition.**

Patterns of Longevity

One indicator of the health of a population is the average number of years individuals born in a given year can expect to live, referred to as **life expectancy.** Worldwide, life expectancy has increased dramatically over the last half century (see Figure 2.1). However, wide disparities exist in life expectancy for different populations between and within societies. In 1998, Japan had the longest life expectancy: 80 years. In the same year, life expectancy was less than 40 in 3 countries (Malawi, Zambia, and Zimbabwe) (*Statistical Abstract of the United States: 1998,* Table 1345). In addition, more than 50 million people live in countries with a life expectancy of less than 45 years, and about 300 million people live in 16 countries where life expectancy actually decreased between 1975 and 1995 (World Health Organization 1998).

National and International Data

In 1998, life expectancy in the United States was 76.1. In the same year, 16 countries had life expectancies that exceeded that of the United States.

SOURCE: *Statistical Abstract of the United States: 1998,* Table 1345

CONSIDERATION

A longer life does not necessarily mean a healthier life. Many older individuals spend their later years suffering from pain and disability associated with chronic and debilitating diseases such as heart disease, cancer, diabetes, and mental disorders. The World Health Organization (1997c) suggests that **health expectancy**—defined as life expectancy in good health—is at least as important as life expectancy. Health expectancy refers to the average number of years an individual can expect to live in "good health."

International Data

In 1996, 28 countries had under-5 mortality rates that were lower than that of the United States.

SOURCE: UNICEF 1998

Patterns of Mortality

Rates of **mortality,** or death—especially those of infants, children, and women—provide sensitive indicators of the health of a population. Worldwide,

National Data

In 1996, 7.4 percent of live U.S. births were low-birthweight babies.

SOURCE: *Statistical Abstract of the United States: 1998,* Table 93

International Data

In 1995, 27 percent of all children under 5 were underweight.

SOURCE: World Health Organization 1998

the leading cause of death is infectious and parasitic diseases (World Health Organization 1998). In the United States, the three leading causes of death for both women and men are heart disease, cancer, and stroke (National Center for Health Statistics 1998).

Infant and Childhood Mortality Rates

Infant mortality rates, the number of deaths of live-born infants under 1 year of age per 1,000 live births (in any given year), provide an important measure of the health of a population. **Under-5 mortality rates,** another useful measure of child health, refer to the number of deaths of children under age 5. Between 1960 and 1996, there have been significant decreases in under-5 mortality rates throughout the world (see Table 2.1). However, the childhood death rates of less developed countries are still extremely high compared to the childhood death rates in the developed world. In less developed countries, 7 out of 10 childhood deaths can be attributed to five main causes: pneumonia, diarrhea, measles, malaria, and malnutrition (World Health Organization 1997b). Finally, many countries have lower childhood death rates than the United States.

Infant and child health is assessed not only by death rates, but by weight as well. Low-birthweight babies (weighing less than 5.5 lbs.) and underweight children are more likely to die early, and those who survive are more likely to suffer illness, stunted growth, and other health problems into adult life.

For a woman to die from pregnancy and childbirth is a social injustice. Such deaths are rooted in women's powerlessness and unequal access to employment, finances, education, basic health care, and other resources.

—Safe Motherhood Initiative

Maternal Mortality Rates

Maternal mortality rates, a measure of deaths that result from complications associated with pregnancy, childbirth, and unsafe abortion, also provide a sensitive indicator of the health status of a population. Maternal deaths are the leading cause of death for reproductive-age women in less developed countries.

The highest rates of maternal death are among African women. One of every 16 African women, compared with only 1 in 3,700 North American women and

Table 2.1 Under-5 Mortality Rates for Selected Countries, 1960 and 1996

Country	1960	1996
Canada	33	7
China	209	47
Ethiopia	280	177
France	34	6
India	236	111
Iran	233	37
Japan	40	6
Kenya	202	90
Mali	500	220
Mexico	148	32
Niger	320	320
Russian Federation	65	25
Sweden	20	4
United Kingdom	27	7
United States	30	8

SOURCE: UNICEF. 1998. *The State of the World's Children, 1998.* New York: UNICEF

1 in 1,400 European women are at risk of dying from pregnancy, childbirth, or unsafe abortion ("Maternal Mortality: A Preventable Tragedy" 1998).

Several factors contribute to high maternal mortality rates in less developed countries. Good health care and adequate nutrition and sanitation are far less available in poorer countries. Also, women in less developed countries experience higher rates of pregnancy and childbearing and begin childbearing at earlier ages. Thus, they face the risk of maternal death more often and before their bodies are fully developed (see also Chapter 15). Women in many countries also lack access to family planning services and/or do not have the support of their male partners to use contraceptive methods such as condoms. Consequently, many women resort to abortion to limit their childbearing, even in countries where abortion is illegal (see Chapter 14).

Illegal abortions in less developed countries have an estimated mortality risk of 100 to 1,000 per 100,000 procedures (Miller and Rosenfield 1996). In contrast, the U.S. mortality risk for legal abortion is very low: 0.6 per 100,000. Unsafe abortion represents a serious threat to the health and lives of women. Between 50,000 and 100,000 deaths result from unsafe abortions each year (Miller and Rosenfield 1996).

International Data

One-quarter to one-half of deaths among women in developing countries are attributed to pregnancy-related complications, compared with less than 1 percent in the United States.

SOURCE: "Maternal Mortality: A Preventable Tragedy" 1998

International Data

Out of an estimated 50 million pregnancies terminated worldwide every year, about 20 million are performed under unsafe and unsanitary conditions.

SOURCE: Abeysinghe 1998

Patterns of Burden of Disease

Researchers have developed a new approach to measuring the health status of a population. This new approach provides an indicator of the overall **burden of disease** on a population through a single unit of measurement that combines not only the number of deaths but also the *impact* of premature death and disability on a population (Murray and Lopez 1996). This comprehensive unit of measurement, called the **disability-adjusted life year (DALY)**, reflects years of life lost to premature death and years lived with a disability. More simply, 1 DALY is equal to 1 year of healthy life. A "premature" death is defined as one that occurs before the age to which the person could have expected to live if he or she were a member of one of the world's longest-surviving populations (82.5 years for women; 80 years for men). To calculate DALYs incurred through traffic road accidents in the United States in 1998, add the total years of life lost in fatal road accidents and the total years of life lived with disabilities by survivors of such accidents.

The Global Burden of Disease Study (Murray and Lopez 1996) calculated the burden of disease for various diseases and injuries. The study found that the burdens of mental illnesses, including depression, alcohol dependence, and schizophrenia, have been seriously underestimated by traditional approaches that focus on death and not disability. The study also concluded that tobacco is a more serious threat to human health than any single disease, including HIV (see also Chapter 3).

International Data

Although mental illnesses are responsible for only about 1 percent of deaths, they account for almost 11 percent of disease burden worldwide.

SOURCE: Murray and Lopez 1996

SOCIOLOGICAL THEORIES OF ILLNESS AND HEALTH CARE

The sociological approach to the study of illness, health, and health care differs from medical, biological, and psychological approaches to these topics. Next, we discuss how three major sociological theories—structural-functionalism, conflict theory, and symbolic interactionism—contribute to our understanding of illness and health care.

> Recent years have seen an increasing tendency to blame individuals for their own health problems . . . Yet . . . patterns of disease reflect social conditions as much as, if not more than, individual behaviors or biological characteristics.
>
> –Rose Weitz
> Sociologist
> Arizona State University

Structural-Functionalist Perspective

The structural-functionalist perspective is concerned with how illness, health, and health care affect and are affected by changes in other aspects of social life. For example, the women's movement and changes in societal gender roles have led to more women smoking and drinking, and experiencing the negative health effects of these behaviors. Increased modernization and industrialization throughout the world has resulted in environmental pollution—a major health concern. The emergence of HIV and AIDS in the U.S. gay male population was a force that helped unite and mobilize gay rights activists.

According to the structural-functionalist perspective, health care is a social institution that functions to maintain the well-being of societal members and, consequently, of the social system as a whole. Illness is dysfunctional in that it interferes with people performing needed social roles. To cope with nonfunctioning members and to control the negative effects of illness, society assigns a temporary and unique role to those who are ill—the sick role (Parsons 1951). This role assures that societal members receive needed care and compassion, yet at the same time, it carries with it an expectation that the person who is ill will seek competent medical advice, adhere to the prescribed regimen, and return as soon as possible to normal role obligations.

Structural-functionalists explain the high cost of medical care by arguing that society must entice people into the medical profession by offering high salaries. Without such an incentive, individuals would not be motivated to endure the rigors of medical training or the stress of being a physician.

Conflict Perspective

The conflict perspective focuses on how wealth, status, and power, or the lack thereof, influence illness and health care. Worldwide, the have-nots not only experience the adverse health effects of poverty, they also have less access to medical insurance and quality medical care. In societies where women have little status and power, their life expectancy is lower than in industrialized countries because of several social factors: eating last and eating less, complications of frequent childbearing and sexually transmitted diseases (because they have no power to demand abstinence or condom use), infections and hemorrhages following genital mutilation (which is practiced in about 28 countries), and restricted access to modern health care (Lorber 1997).

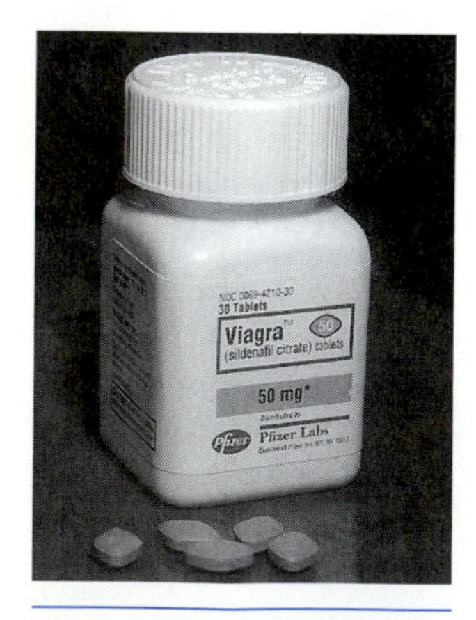

The emergence of Viagra outraged many women who discovered that some insurance companies covered Viagra while female contraceptives were not covered.

Medical research agendas are also shaped by wealth, status, and power. Although malaria kills twice as many people annually as does AIDS, malaria research receives less than one-tenth as much public funding as AIDS research (Morse 1998). Similarly, pneumonia and diarrheal diseases constitute 15.4 percent of the total global disease burden, but only 0.2 percent of the total global spending on research (Visschedijk and Simeant 1998). This is because northern developed countries (such as the United States), who provide most of the funding for world health-related research, do not feel threatened by malaria, pneumonia, and diarrheal diseases, which primarily affect less developed countries in Africa and Asia.

The male-dominated medical research community has also neglected women's health issues and has excluded women from medical research. When the male impotence drug Viagra made its debut in 1998, women across the United States were outraged by the fact that some insurance policies covered

Viagra (or were considering covering it), although female contraceptives were not covered. Women have also been excluded from participating in major health research studies, such as the Harvard Physicians Health Study (which looked at the relationship between aspirin use and heart disease) and the Multiple Risk Factor Intervention Trials (which examined how cholesterol levels, blood pressure, and smoking affect heart disease) (Johnson and Fee 1997).

The conflict perspective also focuses on how the profit motive influences health, illness, and health care. The profit motive underlies much of the illness, injury, and death that occurs from hazardous working conditions and dangerous consumer products (see Chapter 11). Corporations may influence health-related policies and laws through contributions to politicians and political candidates. In a study on the influence of tobacco industry campaign contributions on state legislators in six states, researchers found that legislators who received higher tobacco industry campaign contributions had more protobacco policy positions (Monardi and Glantz 1998).

National Data

The insurance industry was the top contributor to the Republican party in 1997–1998.

SOURCE: Common Cause 1998

Conflict theorists argue that the high costs of medical care in the United States are a result of a capitalistic system in which health care is a commodity, rather than a right. The conflict perspective views power and concern for profits as the primary obstacles to U.S. health care reform. Insurance companies realize that health care reform translates into federal regulation of the insurance industry. In an effort to buy political influence to maintain profits, the insurance industry has contributed millions of dollars to congressional candidates.

Symbolic Interactionist Perspective

Symbolic interactionists focus on (1) how meanings, definitions, and labels influence health, illness, and health care and (2) how such meanings are learned through interaction with others and through media messages and portrayals. According to the social construction of illness, "there are no illness or diseases in nature. There are only conditions that society, or groups within it, have come to define as illness or disease" (Goldstein 1999, 31). Psychiatrist Thomas Szasz (1970) argued that what we call "mental illness" is no more than a label conferred on those individuals who are "different," that is, who don't conform to society's definitions of appropriate behavior.

Those who dance are considered insane by those who can't hear the music.

—George Carlin
Comedian

Definitions of health and illness vary over time and from society to society. In some countries, being fat is a sign of health and wellness; in others it is an indication of mental illness or a lack of self-control. Before medical research documented the health hazards of tobacco, our society defined cigarette smoking as fashionable. Cigarette advertisements still attempt to associate positive meanings (such as youth, sex, and romance) with smoking to entice people to smoke. A study of top-grossing American films from 1985 to 1995 revealed that 98 percent had references that supported tobacco use and 96 percent had references that supported alcohol use (Everett, Schnuth, and Tribble 1998).

A growing number of behaviors and conditions are being defined as medical problems—a trend known as **medicalization**. Hyperactivity, insomnia, anxiety, and learning disabilities are examples of phenomena that some view as medical conditions in need of medical intervention. Increasingly, "normal" aspects of life, such as birth, aging, sexual development, menopause, and death, have come to be seen as medical events (Goldstein 1999).

In the past, men created witches; now they create mental patients.

—Thomas Szasz
Psychiatrist

Symbolic interactionists also focus on the stigmatizing effects of being labeled "ill." Individuals with mental illnesses, drug addictions, physical

Figure 2.2 **Leading Causes of Death in the United States, ages 15 to 24**

SOURCE: *Statistical Abstract of the United States: 1998,* 117th ed. U.S. Bureau of the Census. Washington, D.C.: U.S. Government Printing Office, Table 141.

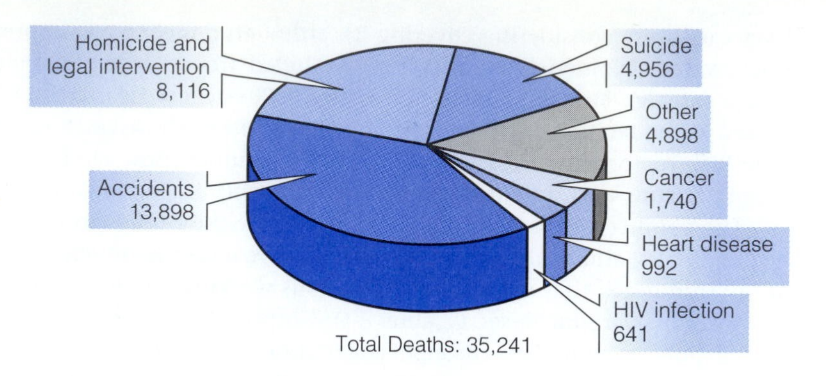

National Data

During the 30 days preceding the survey more than one-third (35 percent) of college students nationwide had ridden with a driver who had been drinking alcohol. About one-third of college men and one-fourth of college women had driven a vehicle after drinking alcohol.

SOURCE: Centers for Disease Control and Prevention 1997

deformities and disabilities, and HIV and AIDS are particularly prone to being labeled in negative ways. Having a stigmatized illness or condition often becomes a master status, obscuring other aspects of a person's social identity. One wheelchair-bound individual commented: "When I am in my chair I am invisible to some people; they see only the chair" (Ostrof 1998, 36).

HEALTH CONCERNS OF U.S. COLLEGE STUDENTS

Next, we look at health problems that are primary health concerns among typical-age U.S. college students. These include automobile accidents, HIV and other sexually transmitted infections, eating disorders, and suicide. Use of alcohol, tobacco, and other drugs is also a major health concern that is discussed in Chapter 3, and homicide is discussed in Chapter 4.

Motor Vehicle Accidents

In the United States, accidents are the leading cause of death among people aged 15 to 24 (see Figure 2.2). Men have a much higher risk of death or injury in accidents than do women. Although many types of accidents cause death, the most common accidents causing death are motor vehicle accidents. Alcohol consumption is frequently involved in motor vehicle accidents (see Chapter 3).

National Data

In the United States, at least one person in four will contract an STI at some point in his or her life.

SOURCE: SIECUS Fact Sheet 1998

HIV/AIDS and Other Sexually Transmitted Infections

Sexually transmitted infections (STIs), also known as sexually transmitted diseases (STDs), and their consequences represent a major public health concern, especially for developing countries that lack resources for preventing and treating STIs. Many infected individuals—particularly women—experience no symptoms.

Chlamydia, a sexually transmissible bacterial infection that may affect the genitals, eyes, and lungs, is the most common STI in the United States, on college campuses, and in the world. Other STIs include gonorrhea, human papilloma virus (HPV), herpes, syphilis, and HIV/AIDS. Untreated STIs can result in severe consequences, including uterine and fallopian tube infections, infertility, ectopic pregnancy (a potentially fatal pregnancy in which the fertilized egg begins to develop outside the uterus), chronic pain, cancer, and death.

National Data

Half of all new HIV infections in the United States occur among people under the age of 25.

SOURCE: National AIDS Clearinghouse 1998

Young adults are the age group at greatest risk of acquiring an STI because they are more likely to engage in intercourse without using condoms and to have multiple sex partners. About one in four U.S. teenagers acquires an STI annually (SIECUS Fact Sheet 1998). In a nationwide survey of U.S. college students, only 30 percent who had had sexual intercourse in the past 3 months reported using a condom at last intercourse (Centers for Disease Control and Prevention 1997).

One of the most urgent public health concerns around the globe is the spread of the human immunodeficiency virus (HIV), which causes acquired immunodeficiency syndrome (AIDS). Africa suffers the highest incidence of HIV/AIDS. This chapter's *Self & Society* allows you to assess your knowledge of AIDS.

In the United States HIV and AIDS rates are highest among gay and bisexual men. This is due to the high rate of anal intercourse—a high risk behavior—among gay and bisexual men. However, worldwide, the predominant mode of HIV transmission is through heterosexual contact (Inciardi and Harrison 1997). In the United States, the first-ever annual decrease in new AIDS cases occurred in 1996. The biggest improvement—a drop of 11 percent—was in homosexual men. However, among U.S. blacks and Hispanics, new AIDS cases rose among heterosexual men and women (World Health Organization and United Nations Joint Programme on HIV/AIDS 1998). In the United States, injection of illegal drugs is the risk behavior most frequently associated with heterosexual transmission of HIV.

Some students have a hard time understanding that the consequences of one unprotected sexual encounter may not be reversible.

—American College Health Association

National Data

Nationwide, 39 percent of college students have ever had their blood tested for HIV infection.

SOURCE: Centers for Disease Control and Prevention 1997

Eating Disorders

Eating disorders are characterized by a persistent pattern of abnormal eating or dieting behavior. These patterns are associated with significant emotional and physical disturbances. Two common eating disorders are anorexia nervosa and bulimia nervosa. **Anorexia nervosa** is characterized by weight loss, excessive exercise, food aversion, distorted body image, and an intense and irrational fear of body fat and weight gain. **Bulimia nervosa** is characterized by cycles of binge-eating and purging (self-induced vomiting, use of laxatives, and/or use of diuretics). Eating disorders are much more common among women than they are among men. Unrealistic media portrayals of female attractiveness that emphasize thinness contribute to the prevalence of eating disorders among U.S. women.

Anorexia can cause a variety of physical problems, including anemia, kidney dysfunction, cardiovascular problems, and osteoporosis (inadequate bone calcium), and can lead to death. Bulimia can cause electrolyte and mineral imbalances, dental enamel erosion, disruption of normal bowel functions, tearing of the esophagus, rupturing of the stomach, and irregularities in heart rhythm that can be fatal. Treatment for eating disorders may involve psychotherapy, medications, nutritional counseling, medical treatment, and support groups.

If you think that a friend or a roommate may have an eating disorder, what should you do? The Academy for Eating Disorders (1997) recommends that you approach the person in private and express in a caring but straightforward way what behaviors or symptoms you have observed and what your concerns are. Say you are worried and want to help, and give the person time to talk and express feelings. Listen without being judgmental. Provide information about resources for treatment and offer to go with the person and wait during the first appointment with a health care provider. If the person denies having a problem or refuses to get help, let the person know that you are still concerned and that you may bring the topic up again in the future.

National Data

At any given time, 10 percent or more of U.S. college-aged women report symptoms of eating disorders.

SOURCE: Academy for Eating Disorders 1997

AIDS Knowledge Scale

Indicate whether you think the following items are true or false.

	True	False
1. Hemophiliacs can get AIDS.		
2. AIDS is an epidemic.		
3. Only homosexuals get AIDS.		
4. The virus that causes AIDS is called human immunodeficiency virus (HIV).		
5. The AIDS virus can remain infectious outside the body for up to 10 days if it is at room temperature.		
6. One can get AIDS by sharing a meal with a person who has AIDS.		
7. People who have AIDS do not develop cancer.		
8. Today the blood supply in hospitals and blood donation centers is screened for the AIDS virus.		
9. Impaired memory and concentration and motor deficits may occur in some AIDS patients.		
10. One can get AIDS by sharing drug needles.		
11. The AIDS virus may live in the human body for years before symptoms appear.		
12. One can get AIDS from receiving blood or sperm from a donor who has AIDS.		
13. By using a condom when having sex, one is always safe from contracting AIDS.		
14. The HIV test is a blood test that can tell if a person has AIDS.		
15. There is a cure for AIDS.		
16. AIDS victims may show extreme tiredness, night sweats, fever, weight loss, diarrhea, etc.		
17. One can get AIDS by having sexual intercourse with an infected person.		
18. AIDS is spread by sneezing, coughing, or touching.		
19. AZT is the only drug approved by the U.S. Food and Drug Administration for the treatment of AIDS.		
20. One can get AIDS by having sex with someone who uses intravenous drugs.		
21. AIDS can be spread by having contact with towels or bed linens used by a person with AIDS.		
22. An infected mother can give the AIDS virus to the baby during pregnancy and/or through breast-feeding.		
23. About 400,000 people in the United States are infected with the HIV virus.		
24. In the United States, blacks and Hispanics show higher incidence rates of AIDS than other ethnic groups.		
25. More U.S. women than U.S. men have been infected by the AIDS virus.		

SCORING: The following items are true: 1, 2, 4, 8, 9, 10, 11, 12, 16, 17, 20, 22, and 24. The following items are false: 3, 5, 6, 7, 13, 14, 15, 18, 19, 21, 23, 25. The scale is scored by totaling the number of items answered correctly. Possible scores range from 0 to 25. The higher the score, the higher the degree of knowledge of HIV/AIDS.

COMPARISON: The average score of 68 undergraduate men at a large urban public university on the East Coast was 17.41; the average score of 98 undergraduate women at the same university was 17.87.

SOURCE: David S. Goh. 1993. "The Development and Reliability of the Attitudes toward AIDS Scale." *College Student Journal* 27:208–14. The scale is on p. 214. Reprinted by permission of *College Student Journal*.

Table 2.2 Suicide Warning Signs

Verbal Signs	
(Direct statements about suicide, or indirect statements or subtle hints indicating a wish to die)	
"I am going to commit suicide."	"I don't think I can take it much longer."
"I don't want to live anymore."	"I wish I could go to sleep and never wake up."
"You won't have to put up with me much longer."	
Behavioral Signs	
Sadness and crying	Abusing alcohol or other drugs
Changes in sleep or eating patterns	Obtaining guns, ropes, or pills
Withdrawal from social interaction	Prior suicide attempts
Giving away personal possessions	Drop in grades
Neglect of personal hygiene	Taking risks, frequent accidents
Situational Signs	
Loss of a relationship	Conflictual relationship with parents
Trouble with the law or at school	Unwanted pregnancy
Presence of mental or serious physical illness	Disliked or stigmatized by peers
Recent suicide attempt by a friend/family member	

Suicide

Suicide is the third leading cause of death in the United States among young persons aged 15 to 24. Although men are more likely than women to complete suicide, women are at higher risk for suicide attempts. Most individuals who attempt or complete suicide give warning signs that they are thinking about killing themselves (see Table 2.2).

If you suspect that someone you know is contemplating suicide, there are three steps you can take ("Youth Suicide Prevention" 1998):

1. **Show You Care.** Say something like "I'm concerned about you." "You mean a lot to me and I want to help." "Tell me about your pain." "I don't want you to kill yourself." "I'm on your side . . . we'll get through this."

2. **Ask Questions.** In a direct, but caring way, ask questions such as, "Are you thinking about suicide?" "Are you thinking about harming yourself or ending your life?" "How long have you been thinking about suicide?" "Have you thought about how you would do it?" "Do you have the _____?" (insert the lethal means they have mentioned). "Do you really want to die? Or do you want the pain to go away?"

3. **Get Help.** Call a local crisis hot line, psychiatrist or psychologist, or other mental health professional. If the person has expressed an immediate plan or has access to a gun or other potentially deadly means, do not leave him or her alone; get help immediately by calling the police (or dial 911). Things you can say to the person at risk include the following: "You are not alone. Let me help you." "Together I know we can figure something out to make you feel better." "I will stay with you . . . I can go with you to where we can get some help." "Let's talk to someone who can help . . . let's call the crisis line, now."

National Data

Nationwide, one in ten college students had seriously considered attempting suicide during the past year; 3.4 percent of black students and 1 percent of white students had attempted suicide.

SOURCE: Centers for Disease Control and Prevention 1997

SOCIAL FACTORS ASSOCIATED WITH HEALTH AND ILLNESS

Public health education campaigns, articles in popular magazines, college-level health courses, and health professionals emphasize that to be healthy, we must

It is tragic enough that 1.5 million children died as a result of wars over the past decade. But it is far more unforgivable that, in the same period, 40 million children died from diseases completely preventable with simple vaccines or medicine.

—Bill Clinton
U.S. President

adopt a healthy lifestyle. In response, many people have at least attempted to quit smoking, eat a healthier diet, and include exercise in their daily or weekly routine. However, health and illness are affected by more than personal lifestyle choices. In the following section, we examine how social factors such as poverty, education, race, and gender affect health and illness. Health problems related to environmental problems are discussed in Chapter 15.

Poverty

Poverty has been identified as the world's leading health problem by an international group of physicians ("Poverty Threatens Crisis" 1998). Poverty is associated with unsanitary living conditions, hazardous working conditions, lack of access to medical care, and inadequate nutrition (see also Chapter 10).

In the United States, socioeconomic status is related to numerous aspects of health and illness (National Center for Health Statistics 1998). People with lower family income tend to die at younger ages than those with higher income, and adults with low incomes are four to seven times as likely (depending on race, ethnicity, and sex) than those with higher incomes to report fair or poor health. Further, children from low-income families are less likely to be fully vaccinated against childhood diseases and are less likely to have medical insurance. Poor persons are also more likely to report an unmet need for health care. Also, research has found that low socioeconomic status is associated with increased risk of a broad range of psychiatric conditions (Williams and Collins 1999). Rates of depression and substance abuse, for example, are higher in the lower socioeconomic classes (Kessler et al. 1994). Why do poor people have higher rates of mental illness? One explanation suggests that lower-class individuals experience greater stress due to their deprived and difficult living conditions. Others argue that members of the lower class are simply more likely to have their behaviors identified and treated as mental illness.

International Data

In developing countries, malnutrition is associated with 54 percent of all deaths of children under age 5.

SOURCE: Visschedijk and Simeant 1998

CONSIDERATION

Poor health—either physical or mental—can cause, as well as result from, poverty. For example, individuals with poor mental or physical health have problems in achieving high levels of education and difficulty obtaining and keeping employment. The leading cause of personal bankruptcy is experiencing a medical emergency or health crisis.

International Data

Worldwide, 840 million people suffer from hunger; 86 countries are classified by the United Nations as food-deficient.

SOURCE: Thirunarayanapuram 1998

Lower socioeconomic groups have higher rates of mortality, in part, because they have higher rates of health risk behaviors such as smoking, alcohol drinking, being overweight, and being physically inactive. Other factors that explain the relationship between socioeconomic status and mortality include exposure to environmental health hazards and inequalities in access to and use of preventive and therapeutic medical care (Lantz et al. 1998). In addition, the lower class tends to experience high levels of stress, while having few resources to cope with it (Cockerham 1998). Stress has been linked to a variety of physical and mental health problems, including high blood pressure, cancer, chronic fatigue, and substance abuse.

Education

In general, low levels of education are associated with higher rates of health problems and mortality (National Center for Health Statistics 1998). For example, less educated women and men have higher rates of suicide.

Low birthweight and high infant mortality are also more common among the children of less educated mothers than among children of more educated mothers. This is partly due to the fact that women with less education are less likely to seek prenatal care and are more likely to smoke during pregnancy.

One reason that lower education levels are associated with higher mortality rates is that individuals with low levels of education are more likely to engage in health risk behaviors such as smoking and heavy drinking. The well educated, in contrast, are less likely to smoke and drink heavily and are more likely to exercise. However, research findings suggest that educational differences in mortality are best explained by the strong association between education and income (Lantz et al. 1998).

National Data

In 1994–95, suicide rates for women and men 25 to 44 years old with less than 13 years of education were about twice the rates for those with more education.

SOURCE: National Center for Health Statistics 1998

Gender

As noted earlier, women in developing countries suffer high rates of mortality and morbidity due to the high rates of complications associated with pregnancy and childbirth. Women in all countries also suffer more severe and frequent complications of STI's, in part because women are biologically more susceptible than men to becoming infected if exposed to an STI. Also, STIs are less likely to produce symptoms in women and are therefore more difficult to diagnose until serious problems develop. The low status of women in many less developed countries results in their being nutritionally deprived and having less access to medical care than do men.

Prior to the twentieth century, the life expectancy of U.S. women was shorter than that of men due to the high rate of maternal mortality that resulted from complications of pregnancy and childbirth. Currently, however, U.S. women have a higher life expectancy than U.S. men (see Figure 2.3).

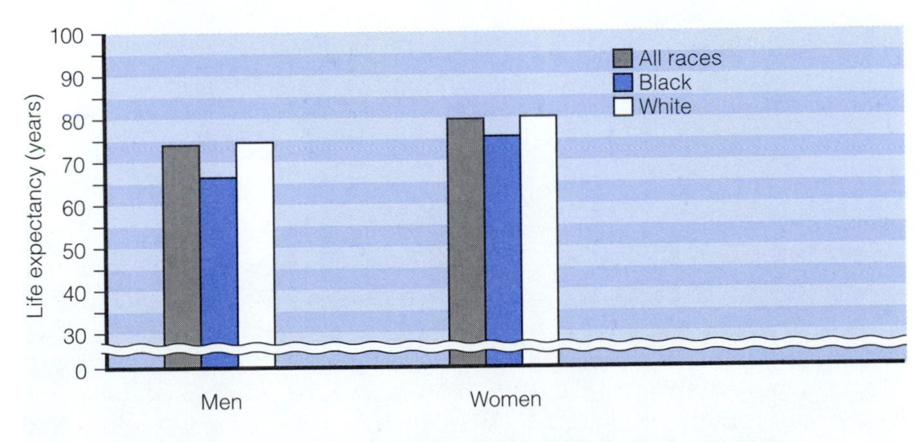

Figure 2.3 Life Expectancy at Birth, by Race and Sex: United States

SOURCE: National Center for Health Statistics. 1998. *Health, United States, 1998: With Socioeconomic Status and Health Chartbook.* Hyattsville, Md.: *U.S. Government Printing Office.*

Although U.S. women tend to live longer than men, they have higher rates of illness and disability than do U.S. men (Verbrugge 1999). Prevalence rates for nonfatal chronic conditions (such as arthritis, thyroid disease, and migraine headache) are typically higher for women. However, men tend to have higher rates of fatal chronic conditions (such as high blood pressure, heart disease, and diabetes). Women also tend to experience a higher incidence of acute conditions, such as colds and influenza, infections, and digestive conditions. "In sum, women live longer than men but experience more illness, whereas men experience relatively little illness but die quickly when illness strikes" (Weitz 1996, 56). Regarding mental health, men are more likely to abuse drugs and have higher rates of personality disorders, whereas women are more likely to suffer from mood disorders such as depression and anxiety (Cockerham 1998).

Women tend to have more non–life-threatening illnesses because of the stresses of routinized jobs, child care, elder care, and housework. Men are more prone to chronic and life-threatening diseases, such as coronary disease, because they are more likely than women to smoke, use alcohol and illegal drugs, and work under hazardous conditions. U.S. culture socializes men to be aggressive and competitive and to engage in risky behaviors, which contributes to their higher risk of death from injuries and accidents. Although women are more likely to attempt suicide, men are more likely to succeed at it because they use deadlier methods.

CONSIDERATION

Although men are more likely to be victims of violence outside the home, in the home women are more likely to be victims (see also Chapter 5). Rose Weitz (1996) notes that "although neither health care workers nor the general public typically thinks of it as a health problem, women battering is a major cause of injury, disability, and death among American women, as among women worldwide." (p. 56)

Men in our society are more likely than women to be heavy drinkers. As a result, men are nearly twice as likely as women to die of liver disease.

Racial and Ethnic Minority Status

In the United States, Asian Americans typically enjoy high levels of health, while other racial and ethnic minorities tend to suffer higher rates of mortality and morbidity. For example, the infant mortality rate among U.S. blacks is alarmingly high, exceeding infant mortality rates found in many poor countries such as Cuba, Czechoslovakia, and Chile.

Among U.S. adults, blacks, especially black men, have a lower life expectancy compared to whites (see Figure 2.3). Compared with U.S. whites, U.S. blacks have higher rates of death due to AIDS, heart disease, cancer, diabetes, liver disease and cirrhosis, and homicide. Compared with non-Hispanic whites, Hispanics have more diabetes, high blood pressure, tuberculosis, lung cancer, STIs, alcoholism, and homicide. Health problems of American Indians include high rates of diabetes, alcoholism (and death from automobile accidents), and suicide (Cockerham 1998). Because Asian Americans have the highest levels of income and education of any racial/ethnic U.S. minority group, they typically have high levels of health.

Socioeconomic differences between racial and ethnic groups are largely responsible for racial and ethnic differences in health status (Williams and Collins 1999). In addition, discrimination contributes to poorer health among oppressed racial and ethnic populations by restricting access to the quantity and quality of public education, housing, and health care. Racial and ethnic minorities are also less likely to have insurance coverage for health care. Blacks are more likely than whites to work in occupations where they are exposed to occupational hazards such as toxic chemicals, dust, and fumes. **Environmental racism** (see also Chapter 15)—the tendency for U.S. hazardous waste sites and polluting industries to be located in areas where the surrounding residential population is black—may also contribute to poorer health among U.S. blacks. Finally, racial and ethnic prejudice and discrimination may induce psychological distress that may adversely affect physical and mental health status and increase the likelihood of violence and substance abuse.

National Data

In 1995, the infant mortality rate for U.S. blacks was 14.6, compared with 6.3 for whites, 9.0 for American Indians and Alaska Natives, and 5.3 for Asians and Pacific Islanders.

SOURCE: National Center for Health Statistics 1998

CONSIDERATION

Cultural beliefs and values sometimes contribute to the poorer health of U.S. minorities. For example, about 14 percent of all Latinos living in the United States suffer from adult-onset diabetes, and they die twice as often from the disease as non-Hispanic whites. For Latinos with diabetes, fatalism ("I will get well if God wishes it"), cultural reluctance to place individual medical needs over the needs of the family, and faith in and use of folk remedies such as garlic, cactus, and aloe create barriers to adequate treatment (Lipton et al. 1998).

PROBLEMS IN U.S. HEALTH CARE

The United States boasts of having the best physicians, hospitals, and advanced medical technology in the world, yet problems in U.S. health care remain a major concern on the national agenda. After presenting a brief overview of U.S. health care, we address some of the major health care problems in the

National Data

In 1997, 15 percent of a national sample of U.S. adults said that health care costs/insurance was the most urgent health problem facing the United States.

SOURCE: Moore 1997

United States—the high cost of medical care and insurance, unequal access to medical care, and the managed care crisis.

National Data

A 1997 Harris Poll of 1,000 Americans found that 50 percent think the health care system is getting worse (compared to 45 percent who think it is getting better). However, only 4 percent think it is getting "a lot better" compared with 22 percent who see it as getting "a lot worse."

SOURCE: PNHP Data Updates 1997b

U.S. Health Care: An Overview

Various types of health insurance exist in the United States. In traditional health insurance plans, the insured choose their health care provider, who is reimbursed by the insurance company on a fee-for-service basis. The insured individual typically must pay out-of-pocket a certain amount called a "deductible" (perhaps $250.00 per year per person or per family) and then is often required to pay a percentage of medical expenses (e.g., 20 percent) until a maximum out-of-pocket expense amount is reached (after which insurance will cover 100 percent of medical costs). **Health maintenance organizations (HMOs)** are prepaid group plans in which a person pays a monthly premium for comprehensive health care services. HMOs attempt to minimize hospitalization costs by emphasizing preventive health care. **Preferred provider organizations (PPOs)** are health care organizations in which employers who purchase group health insurance agree to send their employees to certain health care providers or hospitals in return for cost discounts. In this arrangement, health care providers obtain more patients, but charge lower fees to buyers of group insurance.

Medicare and Medicaid are government-subsidized health programs established in 1965. **Medicare**, available to people 65 years and over, younger people who are receiving Social Security disability benefits, and persons who need dialysis or kidney transplants, consists of two separate programs: a hospital insurance program and a supplementary medical insurance program. The hospital insurance program is free, but enrollees may pay a deductible and a copayment, and there is limited coverage of home health nursing and hospice care. Medicare's medical insurance program care is not free; enrollees must pay a monthly premium as well as a copayment for services. Medicare does not cover prescription drugs, long-term nursing home care, and other types of services, which is why many individuals who receive Medicare also purchase supplementary private insurance known as **medigap policies** (see also Chapter 6).

National Data

In 1997, one-quarter of the U.S. population was enrolled in health maintenance organizations, double the enrollment of 1991.

SOURCE: National Center for Health Statistics 1998

Medicaid, which provides health care coverage for the poor, is administered by the states with matching funds from the federal government (see also Chapter 10). Eligibility rules and benefits vary from state to state and in many states Medicaid provides health care only for the very poor who are well below the federal poverty level.

Managed care refers to any medical insurance plan that controls costs through monitoring and controlling the decisions of health care providers. In many plans, doctors must call a **utilization review** office to receive approval before they can hospitalize a patient, perform surgery, or order an expensive diagnostic test.

The High Cost of Health Care and Insurance

National Data

More than half of all Americans are enrolled in managed care organizations.

SOURCE: Hellinger 1998

The United States spends more on health care per person than any other country in the world. Health care costs have risen faster than the rate of inflation. In 1950, health care expenditures accounted for 4.4 percent of the gross domestic product; by 1980, they rose to 8.9 percent, and in 1995, 14.2 percent (Cherner 1995, Cockerham 1998, National Center for Health Statistics 1998).

Almost 9 million U.S. families with health insurance spent more than 10 percent of their annual income for health care in 1997. On average, Medicare beneficiaries spend 19 percent of their annual income on medical care, and the 2 million elderly who are poor but not covered by Medicaid spend an average of 54 percent of income on out-of-pocket health care costs (PNHP Data Updates 1998). Half of these health care costs were spent on insurance (see Figure 2.4).

Several factors have contributed to escalating medical costs. First, the U.S. population is aging. Because people over age 65 use medical services more than younger individuals, the growing segment of the elderly population means more money is spent on medical care. Conrad and Brown (1999) forewarn that "we have yet to feel the full effect of medical costs on our aging population; as the so called 'baby boom' generation reaches older ages in the next century, health costs are expected to rise even more steeply" (p. 582).

National Data

In 1996, CEO pay at the nation's 35 largest for-profit HMOs averaged $1.3 million; 13 managed care CEOs also received nearly $70 million in stock options.

SOURCE: PNHP Data Updates 1997a

Health care administrative expenses, which are higher in the United States than in any other nation, are another reason for high medical care costs. About 26¢ of every health care dollar is spent on administrative overhead (Health Care Financing Administration 1997).

High costs of Medicaid and Medicare are partly due to fraud and abuse. A 1997 audit of 5,300 Medicare claims for 600 patients revealed errors in 30 percent of the claims. The most common problem was billing for medically unnecessary treatments (PNHP Data Updates 1997b). The study suggests Medicare overpayments of as much as $23 billion annually.

High salaries and benefits of for-profit managed care corporations also contribute to health care costs to the consumer. The use of expensive medical technology, unavailable just decades ago, also contributes to high medical bills.

High doctors' fees and hospital costs are major factors in the rising costs of health care (Weitz 1996). For example, an operation to extract cataracts and implant artificial lenses in the eye costs around $2,000, even though the procedure requires only about one hour to perform. Increasingly, hospitals are purchasing more expensive equipment than is needed. And new expensive hospitals are being built in areas where there is already a surplus of hospital beds (PNHP Data Updates 1998).

National Data

Fewer than half (43 percent) of Americans have health insurance paid for by a private employer.

SOURCE: PNHP Press Release 1999

High costs of public and private insurance have also contributed to escalating health care expenditures. Together, Medicaid and Medicare financed $351 billion in health care services in 1996—more than one-third of the nation's total health care bill (Health Care Financing Administration 1998c). In 1996, the self-employed paid between $5,000 and $7,000 for health insurance coverage for a healthy family (National Coalition on Health Care 1998). Businesses that provide health insurance benefits bear much of the expense of the insurance for their employees. With the rising cost of medical insurance, fewer companies today fully finance medical care coverage for their employees than in the past.

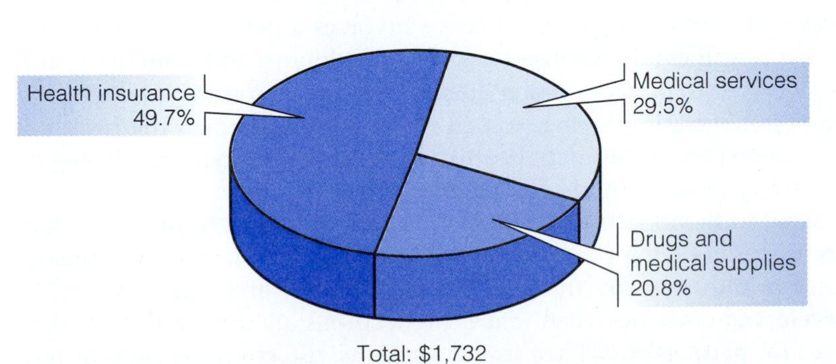

Figure 2.4 Average Annual Health Care Expenditures per U.S. Consumer, 1996

SOURCE: *Statistical Abstract of the United States: 1998,* 118th ed. U.S. Bureau of the Census. Washington, D.C.: U.S. Government Printing Office, Table 180.

National Data

In 1999, 16 percent of the U.S. population had no medical insurance coverage.
SOURCE: PNHP Press Release 1999

Unequal Access to Quality Health Care

The availability and quality of health care services are not equally distributed in the United States. Due to the high cost of insurance, many individuals in the United States do not have insurance.

The poor and near-poor who do not have insurance or who cannot meet their deductible or copayment may choose not to seek health care. Among adults ages 18 to 64, about one-third of poor persons reported an unmet need for care in 1994–95 compared to about 7 percent of high income persons (National Center for Health Statistics 1998). The rural poor have problems of access to health care as medical facilities and practitioners may not be available locally. Many doctors do not want to practice in rural areas, leaving a doctor shortage in many rural communities.

The Managed Care Crisis

In a National Coalition on Health Care survey, 8 in 10 Americans believed that the quality of medical care was being compromised in the interest of profit (Miller 1998). They believed that hospitals have cut corners to save money and that quality care is compromised by the health insurance companies' efforts to save money. In a survey of physicians' views on the effects of managed care, the majority responded that managed care has negative effects on the quality of patient care due to limitations in diagnostic tests, length of hospital stay, and choice of specialists (Feldman, Novack, and Gracely 1998).

STRATEGIES FOR ACTION: IMPROVING HEALTH AND HEALTH CARE

Because poverty underlies many of the world's health problems, improving the world's health requires strategies that reduce poverty—a topic covered in Chapter 10. Chapter 3 addresses problems associated with tobacco and illegal drugs, Chapter 11 addresses health hazards in the workplace, and Chapter 15 focuses on environmental health problems. Here, we discuss other strategies for improving health. These include global strategies to improve maternal and infant health, the use of computer technology in health care, and U.S. health care reform.

Improving Maternal and Infant Health

As discussed earlier, maternal deaths are a major cause of death among women of reproductive age in the developing world. In 1987, the Safe Motherhood Initiative was launched. This global initiative involves a partnership of governments, nongovernmental organizations, agencies, donors, and women's health advocates working to protect women's health and lives, especially during pregnancy and childbirth. Improving women's health also improves the health of infants; 30 to 40 percent of infant deaths are the result of poor care during labor and delivery (Safe Motherhood Initiative 1998).

The Safe Motherhood Initiative advocates improving maternal and infant health by first identifying the powerlessness that women face as an injustice that countries must remedy through political, health, and legal systems. In many developing countries, men make the decisions about whether or when their wives (or partners) will have sexual relations, use contraception, or bear

THE HUMAN SIDE

MANAGED CARE HORROR STORIES

1. A mother in Atlanta called her HMO at 3:30 A.M. to report that her 6-month-old boy had a fever of 104 degrees and was panting and limp. The hot line nurse told the woman to take her child to the HMO's network hospital 42 miles away, bypassing several closer hospitals. By the time the baby reached the hospital, he was in cardiac arrest and had already suffered severe damage to his limbs from an acute and often fatal disease, meningococcemia. Both his hands and legs had to be amputated. A court subsequently found the HMO at fault.
2. After conducting tests, a woman's physician suspected that his patient had a kidney dysfunction, but the doctor's repeated requests to make a referral to a urologist were denied by the patient's HMO. It took two years before the HMO finally allowed the woman to go to a specialist, at which time it was discovered that she was in the later stages of renal failure. An arbitrator ordered the HMO to pay the woman $1.1 million, saying the case presents "a compelling picture of the problems and pitfalls" of managed care.
3. A Virginia woman underwent a double mastectomy. Her doctor told her that she would spend several days in the hospital after the surgery. Instead, she was discharged at her HMO's insistence 20 hours later. She left the hospital groggy from anesthesia, in pain, and with four drainage tubes attached to her skin. She was given instructions on what to do for bleeding and blood clotting, prescriptions for painkillers and antibiotics, and exercise instructions. She went home, where she lived by herself. Once home, she was unable to get out of bed without great pain or to cook for herself.
4. A Chicago Medicaid recipient was visited at home by an HMO representative who claimed to be from the Illinois Department of Public Aid. The representative told the woman that she was required to sign up for the managed care plan, that the HMO would cover transportation, and that if she was dissatisfied, she could quickly disenroll. When the woman tried to use the HMO's services, she found that the promised free transportation was not provided, and she had to wait 2 to 3 hours for scheduled appointments. Moreover, she had to pay for childbirth related items that the HMO marketing agent had said were covered. When the woman tried to disenroll, every phone call she made to the HMO's toll-free line resulted in either a busy signal or a recording.
5. A 48-year-old Arizona woman diagnosed with multiple sclerosis obtained a referral from her primary care physician to go to a special treatment center. She visited the center three times. Her HMO denied her claims for the visits. A request to cover the cost of her catheters was also denied. The reason: her primary care doctor did not follow the proper procedure in granting the referral.
6. A 6-year-old boy suffering severe hearing loss was fitted for hearing aids at a cost of $600 apiece. The child's HMO denied reimbursement, telling the parents that hearing loss is a natural process of aging and therefore not covered under their plan.

SOURCE: "Fight Managed Care!" 1998. **http://www.his.com/~pico/1-25.htm**

children. Improving the status and power of women involves ensuring that they have the right to make decisions about their health and reproductive lives.

Improving maternal and infant health requires educating women and communities about the importance of maternal health care and the health benefits of breast-feeding. A major priority is the provision of free or affordable maternal and infant health services with facilities located close to where women live. Such facilities should have an adequate number of trained staff and a continuing supply of drugs and equipment.

Young adolescent girls are more likely to die during pregnancy and delivery, and children born to young adolescent girls are more likely to die during their first year of life. Thus, it is important to discourage pregnancy and childbearing at younger ages. One way to do this is to increase the minimum age for legal marriage. Delaying marriage often delays first births and can reduce the total number of pregnancies a woman has, which reduces her lifetime risk of maternal mortality and morbidity.

Promoting women's education, another health strategy, increases the status and power of women to control their reproductive lives, exposes women to information about health issues, and also delays marriage and childbearing. In 23 developing countries, women with a secondary education marry on average 4 years later than those with no education ("Economic Growth Unnecessary for Reducing Fertility" 1998).

Finally, women must have access to family planning services, affordable methods of contraception, and safe abortion services where legal. The Safe Motherhood Initiative recommends reforming laws and policies to support women's reproductive health and improve access to family planning services. This implies removing legal barriers to abortion—a highly controversial issue in many countries.

In the United States, a public health priority is to ensure that pregnant women receive prenatal care. One aspect of prenatal care involves ensuring that pregnant women receive adequate nutrition. In the United States, the Special Supplemental Food Program for Women, Infants, and Children (WIC) is a federally funded health promotion program aimed at improving the nutritional status of low-income pregnant, breast-feeding, and postpartum women and their infants and children up to age 5.

International Data

At least 35 percent of women in developing countries receive no prenatal care, almost 50 percent give birth without a trained attendant, and 70 percent receive no care in the 6 weeks following delivery.

SOURCE: Safe Motherhood Initiative 1998

International Data

More than 50 countries allow marriage at age 16 or below, and 7 allow marriage as early as age 12.

SOURCE: Safe Motherhood Initiative 1998

U.S. Health Care Reform

The United States is the only country in the industrialized world without national health insurance. Other countries such as Canada, Great Britain, Sweden, Germany, and Italy, have national health insurance systems, also referred to as **socialized medicine.** Despite differences in how socialized medicine works in various countries, what is common to all systems of socialized medicine is that the government (1) directly controls the financing and organization of health services, (2) directly pays providers, (3) owns most of the medical facilities (Canada is an exception), (4) guarantees equal access to health care, and (5) allows some private care for individuals who are willing to pay for their medical expenses (Cockerham 1998).

Most developed countries make health insurance a right, not a job benefit. We should too.

—Dr. David Himmelstein
Associate Professor of Medicine
Harvard Medical School

Since the early 1900s, many politicians have advocated establishing a national health insurance program in the United States. The Clinton administration's 1994 Health Security Act was a proposal that would provide health insurance to every American in a system financed by employers, employees, savings from Medicare and Medicaid, and new taxes on items such as cigarettes. The states would manage the plans by negotiating prices with health care networks and other providers that would be organized and administered by insurance companies. Although the Clinton plan was not passed by Congress, public dissatisfaction with the current state of U.S. health care will continue to pressure policy makers to implement health care reform to control costs and provide universal access to health care.

Advocates of a national health insurance system continue to lobby for health care reform that would provide health care benefits to all Americans. The primary goal of Physicians for a National Health Program, an organization with over 8,000 U.S. physicians, is to promote universal health care, eliminating for-profit health insurance and substituting a single-payer system funded through a payroll deduction. In a **single-payer system**, a single tax-financed public insurance program replaces private insurance companies. Massive numbers of administrative personnel needed to handle itemized billing to 1,500 private insurance companies would no longer be needed (Single Payer Fact Sheet 1999). In testimony before the National Medicare Commission, Douglas Robins (1998) stated:

> **As a result of the enormous savings that could be realized by eliminating the present insurance system . . . we could cover all of the 43 million Americans who now lack health care coverage. . . The GAO [Government Accounting Office] estimated a 10 percent administrative savings with a single payer system which would amount to more than 100 billion dollars a year in savings. (p. 1)**

The insurance industry, not surprisingly, opposes the adoption of such a system because the private health insurance industry would be virtually eliminated. The health insurance industry's opposition to a single-payer universal health plan is matched only by the persistent efforts of those who advocate such a plan. As these opposing forces continue their battle, other health care reform measures have taken place. A few reform measures are described below:

The Children's Health Insurance Program (CHIP) In 1998, over 10 million U.S. children—1 in 7—had no health insurance. Many of these children come from families with incomes too high to qualify for Medicaid but too low to afford private health insurance. CHIP was created to expand health coverage to uninsured children (Health Care Financing Administration 1998a).

Medicare Reform Congress has created the National Bipartisan Commission on the Future of Medicare to make recommendations on how to ensure the financial integrity of the Medicare program through 2020. This commission is debating increasing eligibility from age 65 to 67 by 2020. Other reform efforts are targeting Medicare fraud and abuse. The first annual report of the Health Care Fraud and Abuse Control Program revealed that nearly $1 billion was returned to the Medicare Trust Fund and convictions for health care fraud–related crimes increased by nearly 20 percent (Health Care Financing Administration 1998b).

Managed Care Reform At least 9 states have passed legislation requiring managed care plans to disclose financial incentives they offer participating physicians, 34 states have passed bills that restrict techniques that managed care plans can use to reduce the use of health services, and 32 states have passed

National Data

A poll conducted for the National Coalition on Health Care found that 84 percent of respondents agreed that the federal government should play a role in "making sure everyone has access to quality medical care regardless of their ability to pay."

SOURCE: PNHP Data Updates 1997a

laws banning "gag rules" (Hellinger 1998). These are clauses sometimes found in contracts between health plans and physicians that prohibit physicians from discussing with patients treatment options not covered under the plan. Because the number of Medicare and Medicaid beneficiaries enrolled in managed care plans has grown, the Health Care Financing Administration (HCFA), which administers these two programs, is the largest purchaser of managed care in the country. As a result, HCFA has also taken steps to protect beneficiaries in managed care. These steps include banning "gag clauses," providing response within 72 hours to appeals of care denials, and developing a new survey tool (Consumer Assessment of Health Plans) that will enable consumers to rate their health plan and then use the survey information in choosing among plans (Health Care Financing Administration 1998d).

The Health Insurance Portability and Accountability Act of 1996 This act is designed to protect the health insurance coverage for workers and their families when they change or lose their jobs. This act also provides funding for the adoption of uniform national standards for electronic processing of insurance claims and related transactions, which is estimated to result in a savings of as much as $9 billion per year on administrative overhead (Health Care Financing Administration 1997).

Computer Technology in Health Care

Computer technology offers numerous ways to reduce costs associated with health care delivery and to improve patient care. For example, public sexual health clinics of the future are expected to implement computer technology that will decrease the number of support staff, eliminate paperwork related to record keeping, supplement risk assessment of clients, and deliver interventions such as counseling and referrals (Conlon 1997). At one hospital, a computer information system used to store information and give treatment advice to physicians was successful in detecting 60 times as many adverse drug reactions in patients as traditional methods ("Using Computers to Advance Health Care" 1996). (This chapter's *Focus on Technology* describes additional uses of computers in health care.)

UNDERSTANDING ILLNESS AND HEALTH CARE

As we have seen, patterns of health and disease vary widely between developed countries such as the United States and less developed countries. The different types of health problems in developed versus less developed countries reflect their different socioeconomic statuses. Health problems are affected not only by economic resources, but also by other social factors such as aging of the population, gender, education, and race.

U.S. cultural values and beliefs emphasize the ability of individuals to control their lives through the choices they make. Thus, Americans and other westerners view health and illness as a result of individual behavior and lifestyle choices, rather than as a result of social, economic, and political forces. We agree that an individual's health is affected by the choices he or she makes—choices such as whether or not to smoke, exercise, engage in sexual activity, use condoms, wear a seatbelt, etc. However, the choices individuals make are influenced by social, economic, and political forces that must be taken into account if the

FOCUS ON TECHNOLOGY

THE EMERGING FIELD OF TELEMEDICINE

In December 1994, a Chinese college student named Zhu Ling became ill, experiencing such symptoms as dizziness, facial paralysis, and loss of hair. Doctors did not know what was wrong with her. After she lapsed into a coma, her friends at Beijing University posted a plea for help on the Internet, appealing to any medical professional anywhere for help in diagnosing Zhu Ling's mysterious illness. Shortly after the plea was posted (which included Zhu Ling's symptoms and photographs), an American doctor made the correct diagnosis of thallium poisoning. Doctors from all over the world consulted via the Internet on the best treatments for Zhu Ling, who eventually came out of her coma.

Zhu Ling's story illustrates just one of the numerous applications of telemedicine, which literally means "medicine at a distance." **Telemedicine** involves using information and communication technologies to deliver a wide range of health care services, including diagnosis, treatment, prevention, health support and information, and education of health care workers.

TYPES OF TELEMEDICINE SERVICES

Telemedicine can involve the transmission of three main types of information: data, audio, and images. A patient's medical records or vital signs (such as heart rate and blood pressure) can be transmitted from one location to another. Many hospitals and clinics store their medical records electronically, allowing doctors to access information about their patients very quickly and to update patient data from a distance. Specialized medical databases, such as MEDLINE, can be accessed via the Internet and offer a valuable resource for health care practitioners and researchers. The public may also utilize the Internet to gain health information and support. The simplest telemedicine service is telephone consultation. Transmitting medical images—either still or moving pictures—is another type of telemedicine. **Teleradiology,** or the transmission of radiological images (such as x-ray and ultrasound images) from one location to another for the purpose of interpretation or consultation, has become one of the most commonly used telemedicine services. **Telepathology** involves transmitting images of tissue samples to a pathologist in another location, who can look at the image on a monitor and offer an interpretation.

BENEFITS OF TELEMEDICINE

Telemedicine has the potential to improve public health by making health care available in rural and remote areas and by providing health information to health care workers and to the general population. In addition, "telemedicine allows the scarce resources of specialists and expensive equipment to be shared by a much greater number of patients. Doctors are no longer restricted by geographical boundaries; international specialists are able to spread their skills across continents, without leaving their own hospitals" (LaPorte 1997, 38).

Telemedicine can also be used in training and educating health care professionals and providing health care workers with up-to-date health information. For example, East Carolina University School of Medicine has a family practice training program where medical trainees live and practice in rural areas and are supervised over the state telemedicine network.

Telemedicine can also reduce health care costs by reducing the cost of travel to major health centers or to specialists and by reducing the length of hospitalization, since patients can be monitored at a distance.

Another benefit of telemedicine is the provision of health information and support services on the Internet, which helps empower individuals in managing their health concerns. Through e-mail, bulletin boards, and chat rooms, individuals with specific health problems can network with other similarly affected individuals. This social support assists in patient recovery, reduces the number of visits to physicians and clinics, and "provides disabled

(continued)

FOCUS ON TECHNOLOGY

FIELD OF TELEMEDICINE *(continued)*

individuals with an opportunity to achieve levels of social integration that were simply not possible before" (LaPorte 1997, 33).

Telemedicine also offers indirect benefits to the larger society. Availability of health care in remote areas helps to (1) slow population migration and attract people back to previously abandoned areas; (2) attract skilled personnel to remote and rural areas with a positive impact on local and national economies; and (3) improve health indicators of a country, which improves the overall image of a country and attracts investment (LaPorte 1997).

THE FUTURE OF TELEMEDICINE

In the twentieth century, advances in public health have been due largely to improvements in sanitation and immunization. Advocates of telemedicine have forecasted that in the twenty-first century, improvements in public health will result from the increased uses of information technology (LaPorte 1997).

One of the unresolved issues in telemedicine concerns how to pay health care professionals for their telemedicine services. In 1996, Norway became the first country to introduce a telemedicine fee schedule, making it reimbursable by the national health service (Sethov 1997). Leaders of the telemedicine movement are working to establish payment guidelines, develop standard consent forms, and pass legislation and medical licensing policies that allow professionals to practice medicine across state and national boundaries.

Telemedicine holds the promise of improving the health of individuals, families, communities, and nations. But whether or not telemedicine achieves its promise depends, in part, on whether resources are allocated to provide the technology and the training to use it.

SOURCES:

LaPorte, Ronald E. 1997. "Improving Public Health via the Information Superhighway." **http://www.the-scientist.library.upenn.edu/yr1997/august/opin_97018.html** (8/6/98).

Sethov, Inger. 1997. "The Wonders of Telemedicine." **http://www.techserver.com/newsroom/ntn/info/111097/info10_14458_noframes.html** (8/4/98).

goal is to improve the health of not only individuals, but also entire populations. Further, by focusing on individual behaviors that affect health and illness, we often overlook not only social *causes* of health problems, but social *solutions* as well. For example, at an individual level, the public has been advised to rinse and cook meat, poultry, and eggs thoroughly and to carefully wash hands, knives, cutting boards, and so on in order to avoid illness caused by *Escherichia coli* and salmonella bacteria. But whether or not one becomes ill from contaminated meat, eggs, or poultry is affected by more than individual behaviors in the kitchen. Governmental action in the 1980s reduced the number of government food inspectors and deregulated the meat-processing industry (Link and Phelan 1998). Just as governmental actions created the need for individuals to use caution in food preparation, governmental actions can also offer solutions by providing for more food inspectors and stricter regulations on food industries.

While certain changes in medical practices and policies may help to improve world health, "the health sector should be seen as an important, but not the sole, force in the movement toward global health" (Lerer et al. 1998, 18). Improving the health of a society requires addressing diverse issues, including poverty and economic inequality, gender inequality, population growth, environmental issues, education, housing, energy, water and sanitation, agriculture, and workplace safety. Health promotion is important not only in the hospital, clinic, or doctor's office—it must also occur in the various settings where people live, work, play, and learn (Antezana, Chollat-Traquet, and Yach 1998).

CRITICAL THINKING

1. An analysis of 161 countries found that, in general, countries with high levels of literacy have low levels of HIV (World Health Organization and United Nations Joint Programme on HIV/AIDS 1998). However, in the region of the world worst affected by HIV, sub-Saharan Africa, there is also a relationship between literacy rates and HIV, but the direction of the relationship is reversed. In this region, the countries with the highest levels of HIV infection are also those whose men and women are most literate. What are some possible explanations for this?
2. The Centers for Disease Control and Prevention (CDC) and the American College of Sports Medicine (ACSM) recommend that people aged 6 and older engage regularly, preferably daily, in light to moderate physical activity for at least 30 minutes per day. Experts agree that "if Americans who lead sedentary lives would adopt a more active lifestyle, there would be enormous benefit to the public's health and to individual well-being" (Pate et al. 1995, 406). Yet in a telephone survey of over 87,000 U.S. adults, only about 22 percent reported being active at the recommended level; 24 percent reported that they led a completely sedentary lifestyle (that is, they reported no leisure-time physical activity in the past month) (Pate et al. 1995). What social and cultural factors contribute to the sedentary lifestyle of many Americans?
3. In many countries, drug injecting accounts for more HIV infections than sex. Research evidence clearly suggests that needle exchange programs that provide sterile needles to intravenous drug users result in lower rates of HIV transmission (World Health Organization and United Nations Joint Programme on HIV/AIDS 1998). Why do you think the U.S. government does not support such programs? Should it?

KEY TERMS

acute condition
anorexia nervosa
bulimia nervosa
burden of disease
chronic condition
disability-adjusted life years (DALY)
environmental racism
epidemiological transition
epidemiologist
epidemiology
health
health expectancy
health maintenance organizations (HMOs)
incidence
infant mortality rate
life expectancy
managed care
maternal mortality rate
Medicaid
medicalization
Medicare
medigap policies
mental disorder
morbidity
mortality
preferred provider organizations (PPOs)
prevalence
single-payer system
socialized medicine
telemedicine
telepathology
teleradiology
under-5 mortality rate
utilization review

You can find more information on Medicaid, Medicare, the American Medical Association, and the National Health Care Plan at the *Understanding Social Problems* web site at **http://sociology.wadsworth.com.**

Either from the Wadsworth sociology resource center at **http://sociology.wadsworth.com** or directly from your web browser, you may access InfoTrac College Edition, an online university library that includes over 700 popular and scholarly journals in which you can find articles related to the topics in this chapter. Suggested articles and questions relating to these articles are listed below.

Bhopal, Raj. 1998. "Spectre of Racism in Health and Health Care." *British Medical Journal* 316(7149):1970–74.

1. What was the Tuskegee experiment and what is its legacy?
2. What differences can be seen in the way physicians treat white patients as opposed to black patients?
3. Why is seeking biological differences between races dangerous?

Kaveny, M. Cathleen. 1998. "Older Women and Health Care." *America* 179(6):15–16.

1. In what age group is there the most dramatic increase in the elderly population?
2. How does aging create social isolation?
3. What four systems of social support are needed to boost the well-being of widows in society?

Troiano, Richard P. and Katherine M. Fiegal. 1998. "Overweight Children and Adolescents: Description, Epidemiology, and Demographics." *Pediatrics* 101(3):497–505.

1. What measures do the authors use to classify obesity?
2. What do the authors see in the future for overweight children?
3. What factors do the authors identify to account for such trends?

CHAPTER THREE

Alcohol and Other Drugs

Is It True?

Is It True?

1. Illicit drug use is higher today than it has been in 30 years.
2. In 1995, the Supreme Court ruled that random drug testing of student athletes in public school is unconstitutional.
3. More Americans believe that environmental factors cause drug problems than biological factors.
4. The most commonly used illicit drug in the United States is marijuana.
5. The Dutch have decriminalized small quantities of heroin and have one of the lowest addiction rates in Europe.

Answers: 1 = T, 2 = F, 3 = T, 4 = T, 5 = T

Substance abuse, the nation's number-one preventable health problem, places an enormous burden on American society, harming health, family life, the economy, and public safety, and threatening many other aspects of life.

Robert Wood Johnson Foundation, Institute for Health Policy, Brandeis University

Scott Krueger was athletic, intelligent, handsome, and what you'd call an all-around "nice guy." A freshman at Massachusetts Institute of Technology (MIT) from Buffalo, New York, he was a three-letter athlete and one of the top 10 students in his high school graduating class of over 300. He was a "giver" not a "taker," tutoring other students in math after school while studying second-year calculus so he could pursue his own career in engineering. While at MIT he rushed a fraternity and celebrated his official acceptance into the brotherhood. The night he celebrated he was found in his room, unconscious, and after 3 days in an alcoholic coma he died. He was 18 years old (Moore 1997).

Drug-induced death is just one of many negative consequences that can result from alcohol and drug abuse. The abuse of alcohol and other drugs is a social problem when it interferes with the well-being of individuals and/or the societies in which they live—when it jeopardizes health, safety, work and academic success, family, and friends. But managing the drug problem is a difficult undertaking. In dealing with drugs, a society must balance individual rights and civil liberties against the personal and social harm that drugs promote—crack babies, suicide, drunk driving, industrial accidents, mental illness, unemployment, and teenage addiction. When to regulate, what to regulate, and who should regulate are complex social issues. Our discussion begins by looking at how drugs are used and regulated in other societies.

THE GLOBAL CONTEXT: DRUG USE AND ABUSE

Pharmacologically, a **drug** is any substance other than food that alters the structure or functioning of a living organism when it enters the bloodstream. Using this definition, everything from vitamins to aspirin constitutes a drug. Sociologically, the term drug refers to any chemical substance that (1) has a direct effect on the user's physical, psychological, and/or intellectual functioning, (2) has the potential to be abused, and (3) has adverse consequences for the individual and/or society. Societies vary in how they define and respond to drug use. Thus, drug use is influenced by the social context of the particular society in which it occurs.

Drug Use and Abuse around the World

Globally, drug use and abuse have increased over the last several decades. The production of opium has more than tripled since 1985; nine times the quantity of synthetic stimulants were confiscated in 1993 compared to 1978; and in the last 10 years seizures of most major drugs have increased dramatically. For example, 1996 cocaine seizures resulted in 251 tons of cocaine being confiscated worldwide (*World Drug Report* 1997). Further, individual indicators such as "emergency room visits, substance abuse related mortality cases, arrests of drug abusers, [and] number of countries reporting rising consumption levels—make clear that consumption has become a truly global phenomenon" (*World Drug Report* 1997, 29).

Italy decriminalized personal drug use and possession of small amounts of drugs in 1993 (*World Drug Report* 1997). The Netherlands, however, has had an

U.S. citizens visiting the Netherlands may be shocked or surprised to find people smoking marijuana and hashish openly in public places.

official government policy of treating the use of such drugs as marijuana, hashish, and heroin as a health issue rather than a crime issue since the mid-1970s. Overall, this "decriminalization" policy has had positive results. Although marijuana and heroin are readily available in Dutch cities, the use of both has declined over the years as have drug overdoses and HIV infections from intravenous drug use. The decline in the number of drug users may have occurred, in part, because addicts are free to seek help without fear of criminal reprisals.

Great Britain has also adopted a "medical model," particularly in regard to heroin and cocaine. As early as the 1960s, English doctors prescribed opiates and cocaine for their patients who were unlikely to quit using drugs on their own and for the treatment of withdrawal symptoms. By the 1970s, however, British laws had become more restrictive, making it difficult for either physicians or users to obtain drugs legally. Today, British government policy provides for limited distribution of drugs to addicts who might otherwise resort to crime to support their habits.

In stark contrast to such health-based policies, other countries execute drug users and/or dealers, or subject them to corporal punishment. The latter may include whipping, stoning, beating, and torture. Such policies are found primarily in less developed nations such as Malaysia, where religious and cultural prohibitions condemn any type of drug use, including alcohol and tobacco.

International Data

According to the 1997 *World Drug Report,* less than 10 percent of the global population consumes illicit drugs—about 3.3 to 4.1 percent of the world's population.

SOURCE: *World Drug Report* 1997

Drug Use and Abuse in the United States

According to officials and media campaigns (e.g., "just say no") there is a drug crisis in the United States—a crisis so serious that it warrants a multibillion-dollar-a-year "war on drugs." Americans' concern with drugs, however, has varied over the years. Ironically, in the 1970s when drug use was at its highest, concern over drugs was relatively low. Today, when a sample of Americans were asked, "What do you think is the most important problem facing this country . . . ?" , drugs ranked fourth behind crime, a moral deficit, and problems in education (Gallup Poll 1998).

As Table 3.1 indicates, use of alcohol is much more widespread than use of illicit drugs, such as marijuana and cocaine. In the United States, cultural

Table 3.1 Drug Use, by Type of Drug and Year, 1985 and 1996 (in percents)

	Ever Used		Current User[a]	
	1985	1996	1985	1996
Marijuana	29.4	32.0	9.7	4.7
Cocaine	11.7	10.3	3.0	.8
Inhalants	7.9	5.6	.6	.4
Hallucinogens	6.9	9.7	1.2	.6
Heroin	.9	1.1	.1	.1
Stimulants[b]	7.3	4.7	1.8	.4
Sedatives[b]	4.8	2.3	.5	.1
Tranquilizers[b]	7.6	3.6	2.2	.4
Analgesics[b]	7.6	5.5	1.4	.9
Alcohol	84.9	82.6	60.2	51.0

[a]A current user is someone who used the drug at least once within a month prior to the study.
[b]Nonmedical use; does not include over-the-counter drugs.

SOURCE: *Statistical Abstract of the United States: 1998,* 118th ed. U. S. Bureau of the Census. Washington, D.C.: U.S. Government Printing Office, Table 220.

definitions of drug use are contradictory—condemning it on the one hand (e.g., heroin), yet encouraging and tolerating it on the other (e.g., alcohol). At various times in U.S. history, many drugs that are illegal today were legal and readily available. In the 1800s and the early 1900s, opium was routinely used in medicines as a pain reliever, and morphine was taken as a treatment for dysentery and fatigue. Amphetamine-based inhalers were legally available until 1949, and cocaine was an active ingredient in Coca-Cola until 1906, when it was replaced with another drug—caffeine (Witters, Venturelli, and Hanson 1992).

SOCIOLOGICAL THEORIES OF DRUG USE AND ABUSE

Most theories of drug use and abuse concentrate on what are called psychoactive drugs. These drugs alter the functioning of the brain, affecting the moods, emotions, and perceptions of the user. Such drugs include alcohol, cocaine, heroin, and marijuana. **Drug abuse** occurs when acceptable social standards of drug use are violated, resulting in adverse physiological, psychological, and/or social consequences. For example, when an individual's drug use leads to hospitalization, arrest, or divorce, such use is usually considered abusive. Drug abuse, however, does not always entail chemical dependency. **Chemical dependency** refers to a condition in which drug use is compulsive—users are unable to stop because of their dependency. The dependency may be psychological, in that the individual needs the drug to achieve a feeling of well-being, and/or physical, in that withdrawal symptoms occur when the individual stops taking the drug.

Various theories provide explanations for why some people use and abuse drugs. Drug use is not simply a matter of individual choice. Theories of drug use explain how structural and cultural forces, as well as biological factors, influence drug use and society's responses to it.

Structural-Functionalist Perspective

Functionalists argue that drug abuse is a response to the weakening of norms in society. As society becomes more complex and rapid social change occurs, norms and values become unclear and ambiguous, resulting in **anomie**—a state of normlessness. Anomie may exist at the societal level, resulting in social strains and inconsistencies that lead to drug use. For example, research indicates that increased alcohol consumption in the 1830s and the 1960s was a response to rapid social change and the resulting stress (Rorabaugh 1979). Anomie produces inconsistencies in cultural norms regarding drug use. For example, while public health officials and health care professionals warn of the dangers of alcohol and tobacco use, advertisers glorify the use of alcohol and tobacco and the U.S. government subsidizes alcohol and tobacco industries. Further, cultural traditions, such as giving away cigars to celebrate the birth of a child and toasting a bride and groom with champagne, persist.

Anomie may also exist at the individual level as when a person suffers feelings of estrangement, isolation, and turmoil over appropriate and inappropriate behavior. An adolescent whose parents are experiencing a divorce, who is separated from friends and family as a consequence of moving, or who lacks parental supervision and discipline may be more vulnerable to drug use because of such conditions. Thus, from a structural-functionalist perspective, drug use is a response to the absence of a perceived bond between the individual and society, and to the weakening of a consensus regarding what is considered acceptable. Consistent with this perspective, Nylander, Tung, and Xu (1996) found that adolescents who reported that religion was important in their lives were less likely to use drugs than those who didn't.

Conflict Perspective

Conflict perspectives emphasize the importance of power differentials in influencing drug use behavior and societal values concerning drug use. From a conflict perspective, drug use occurs as a response to the inequality perpetuated by a capitalist system. Societal members, alienated from work, friends, and family, as well as from society and its institutions, turn to drugs as a means of escaping the oppression and frustration caused by the inequality they experience. Further, conflict theorists emphasize that the most powerful members of society influence the definitions of which drugs are illegal and the penalties associated with illegal drug production, sales, and use.

> There are but three ways for the populace to escape its wretched lot. The first two are by route of the wine-shop or the church; the third is by that of the social revolution.
>
> –Mikhail A. Bakunin
> Anarchist and revolutionary

For example, alcohol is legal because it is often consumed by those who have the power and influence to define its acceptability—white males (*Statistical Abstract of the United States: 1997,* Table 228). This group also disproportionately profits from the sale and distribution of alcohol and can afford powerful lobbying groups in Washington to guard the alcohol industry's interests. Since tobacco and caffeine are also commonly used by this group, societal definitions of these substances are also relatively accepting.

Conversely, crack cocaine and heroin are disproportionately used by minority group members, specifically, blacks and Hispanics (Reiman 1998; HHS 1997). Consequently, the stigma and criminal consequences associated with the use of these drugs are severe. The use of opium by Chinese immigrants in the 1800s provides a historic example. The Chinese, who had been brought to the United States to work on the railroads, regularly smoked opium as part of their cultural tradition. As unemployment among white workers increased, however, so did resentment of Chinese laborers. Attacking the use of opium became a convenient means of attacking the Chinese, and in 1877 Nevada became the first of many states to prohibit opium use. As Morgan (1978) observes:

> **The first opium laws in California were not the result of a moral crusade against the drug itself. Instead, it represented a coercive action directed against a vice that was merely an appendage of the real menace—the Chinese—and not the Chinese per se, but the laboring "Chinamen" who threatened the economic security of the white working class. (p. 59)**

The criminalization of other drugs, including cocaine, heroin, and marijuana, follows similar patterns of social control of the powerless, political opponents, and/or minorities. In the 1940s, marijuana was used primarily by minority group members and carried with it severe criminal penalties. But after white middle-class college students began to use marijuana in the 1970s, the government reduced the penalties associated with its use. Though the nature and pharmacological properties of the drug had not changed, the population of users was now connected to power and influence. Thus, conflict theorists regard the regulation of certain drugs, as well as drug use itself, as a reflection of differences in the political, economic, and social power of various interest groups.

Symbolic Interactionist Perspective

> Drug use by friends is consistently the strongest predictor of a person's involvement in drug use.
>
> —U.S. Department of Justice

Symbolic interactionism, emphasizing the importance of definitions and labeling, concentrates on the social meanings associated with drug use. If the initial drug use experience is defined as pleasurable, it is likely to recur, and over time, the individual may earn the label of "drug user." If this definition is internalized so that the individual assumes an identity of a drug user, the behavior will likely continue and may even escalate.

Drug use is also learned through symbolic interaction in small groups. First-time users learn not only the motivations for drug use and its techniques, but also what to experience. Becker (1966) explains how marijuana users learn to ingest the drug. A novice being coached by a regular user reports the experience:

> **I was smoking like I did an ordinary cigarette. He said, "No, don't do it like that." He said, "Suck it, you know, draw in and hold it in your lungs . . . for a period of time." I said, "Is there any limit of time to hold it?" He said, "No, just till you feel that you want to let it out, let it out." So I did that three or four times. (Becker 1966, 47)**

Marijuana users not only learn how to ingest the smoke, but also learn to label the experience positively. When certain drugs, behaviors, and experiences are defined by peers as not only acceptable but pleasurable, drug use is likely to continue.

> **Because they [first-time users] think they're going to keep going up, up, up till they lose their minds or begin doing weird things or something. You have to like reassure them, explain to them that they're not really flipping or anything, that they're gonna be all right. You have to just talk them out of being afraid. (Becker 1966, 55)**

Interactionists also emphasize that symbols may be manipulated and used for political and economic agendas. The popular DARE (Drug Abuse Resistance Education) program, with its antidrug emphasis fostered by local schools and police, carries a powerful symbolic value that politicians want the public to identify with. "Thus, ameliorative programs which are imbued with these potent symbolic qualities (like DARE's links to schools and police) are virtually assured wide-spread public acceptance (regardless of actual effectiveness) which in turn advances the interests of political leaders who benefit from being associated with highly visible, popular symbolic programs" (Wysong, Aniskiewicz, and Wright 1994, 461).

Biological Theories

Biological research has primarily concentrated on the role of genetics in predisposing an individual to alcohol abuse (Pickens and Svikis 1988). Research indicates that children of alcoholics have a 50 percent chance of becoming alcoholics themselves. At a recent International Conference on Genetics, one researcher stated that by looking at inherited traits, ". . . we can predict [in] childhood with 80 percent accuracy who is going to develop alcoholism later in life" (AAP 1998). At the same time, many alcoholics do not have parents who abuse alcohol, and many alcoholic parents have offspring who do not abuse alcohol.

Biological theories of drug use hypothesize that some individuals are physiologically predisposed to experience more pleasure from drugs than others and, consequently, are more likely to be drug users. According to these theories, the central nervous system, which is composed primarily of the brain and spinal cord, processes drugs through neurotransmitters in a way that produces an unusually euphoric experience. Individuals not so physiologically inclined report less pleasant experiences and are less likely to continue use (Jarvik 1990).

National Data

More Americans believe that environmental factors lead to drug abuse than biological factors. When asked about the cause of alcoholism, 33 percent responded it is determined "completely or mostly by heredity or genes," followed by "somewhat by heredity or genes" (44 percent), and "not at all by heredity or genes" (22 percent). In reference to drug addiction, the percentages are 15, 44, and 38, respectively.

SOURCE: U.S. News and Bozell Poll 1997

Psychological Theories

Psychological explanations focus on the tendency of certain personality types to be more susceptible to drug use. Individuals who are particularly prone to anxiety may be more likely to use drugs as a way to relax, gain self-confidence, and ease tension. Those who have dependent personalities and have a compulsive need for love may be more inclined to use drugs to numb the frustration when other needs are not being met.

Psychological theories of drug abuse also emphasize that drug use is maintained by positive and negative reinforcement. Positive reinforcement occurs when drug use results in desirable experiences, such as excitement, pleasure, and peer approval. Negative reinforcement occurs when drug use results in the temporary alleviation of undesirable experiences, such as pain, anxiety, boredom, and loneliness.

FREQUENTLY USED LEGAL AND ILLEGAL DRUGS

Social definitions regarding which drugs are legal or illegal have varied over time, circumstance, and societal forces. In the United States, two of the most dangerous and widely abused drugs, alcohol and tobacco, are legal.

> Drunkenness is the ruin of reason. It is premature old age. It is temporary death.
>
> –St. Basil
> Bishop of Caesarea

Alcohol

Americans' attitudes toward alcohol have had a long and varied history (this chapter's *Self & Society* deals with attitudes toward alcohol). Although alcohol

Alcohol Attitude Test

If you strongly agree with the following statements, write in 1. If you agree, but not strongly, write in 2. If you neither agree nor disagree, write in 3. If you disagree, but not strongly, write in 4. If you strongly disagree, write in 5.

SET 1

_____ 1. If a person concentrates hard enough, he or she can overcome any effect that drinking may have on driving.
_____ 2. If you drive home from a party late at night when most roads are deserted, there is not much danger in driving after drinking.
_____ 3. It's all right for a person who has been drinking to drive, as long as he or she shows no signs of being drunk.
_____ 4. If you're going to have an accident, you'll have one anyhow, regardless of drinking.
_____ 5. A drink or two helps people drive better because it relaxes them.
_____ Total score for questions 1 through 5

SET 2

_____ 6. If I tried to stop someone from driving after drinking, the person would probably think I was butting in where I shouldn't.
_____ 7. Even if I wanted to, I would probably not be able to stop someone from driving after drinking.
_____ 8. If people want to kill themselves, that's their business.
_____ 9. I wouldn't like someone to try to stop me from driving after drinking.
_____ 10. Usually, if you try to help someone else out of a dangerous situation, you risk getting yourself into one.
_____ Total score for questions 6 through 10

SET 3

_____ 11. My friends would not disapprove of me for driving after drinking.
_____ 12. Getting into trouble with my parents would not keep me from driving after drinking.
_____ 13. The thought that I might get into trouble with the police would not keep me from driving after drinking.
_____ 14. I am not scared by the thought that I might seriously injure myself or someone else by driving after drinking.
_____ 15. The fear of damaging the car would not keep me from driving after drinking.
_____ Total score for questions 11 through 15

SET 4

_____ 16. The 55 mph speed limit on the open roads spoils the pleasure of driving for most teenagers.
_____ 17. Many teenagers use driving to let off steam.
_____ 18. Being able to drive a car makes teenagers feel more confident in their relations with others their age.
_____ 19. An evening with friends is not much fun unless one of them has a car.
_____ 20. There is something about being behind the wheel of a car that makes one feel more adult.
_____ Total score for questions 16 through 20

SCORING

Set 1. 13–25 points: realistic in avoiding drinking-driving situations; 5–6 points: tends to make up excuses to combine drinking and driving.
Set 2. 15–25 points: takes responsibility to keep others from driving when drunk; 5–9 points: wouldn't take steps to stop a drunk friend from driving.
Set 3. 12–25 points: hesitates to drive after drinking; 5–7 points: is not deterred by the consequences of drinking and driving.
Set 4. 19–25 points: perceives auto as means of transportation; 5–14 points: uses car to satisfy psychological needs, not just transportation.

SOURCE: Courtesy of National Highway Traffic Safety Administration. National Center for Statistics and Analysis, from *Drug Driving Facts*. Washington, D.C.: NHTSA, 1988.

was a common beverage in early America, by 1920 the federal government had prohibited its manufacture, sale, and distribution through the passage of the Eighteenth Amendment to the Constitution. Many have argued that Prohibition, like the opium regulations of the late 1800s, was in fact a "moral crusade" (Gusfield 1963) against immigrant groups who were more likely to use alcohol. The amendment had little popular support and was repealed in 1933. Today, the United States is experiencing a resurgence of concern about alcohol. What has been called a "new temperance" has manifested itself in federally mandated 21-year-old drinking age laws, warning labels on alcohol bottles, increased concern over fetal alcohol syndrome and teenage drinking, and stricter enforcement of drinking and driving regulations.

National Data

In 1996, over half (51 percent) of Americans aged 12 and older reported being current users of alcohol.

SOURCE: HHS 1997

Alcohol is the most widely used and abused drug in America. Although most people who drink alcohol do so moderately and experience few negative effects, alcoholics are psychologically and physically addicted to alcohol and suffer various degrees of physical, economic, psychological, and personal harm.

International Data

In England, Sweden, and much of Europe, ALCOPOPS, soft drinks containing as much as 5 percent alcohol, are being marketed for and illicitly sold to children.

SOURCE: Crime and Justice International 1997

The 1996 National Household Survey on Drug Abuse conducted by the Department of Health and Human Services reports that 109 million Americans 12 and older used alcohol in the survey month—about 51 percent of the population. Of these, 11 million were heavy drinkers (defined as drinking 5 or more drinks per occasion on 5 or more days in the survey month) and 32 million were binge drinkers (defined as drinking 5 or more drinks on at least one occasion during the survey month). Even more troubling were the 9 million current users of alcohol who were 12 to 20 years old, 50 percent of whom were binge drinkers (HHS 1997).

A recent study by the Harvard School of Public Health (Wechsler, Dowdall, Davenport, and Dejong 1998) of over 17,000 college students at 140 four-year colleges and universities indicates that:

- In the 2 weeks prior to the survey 44 percent of college students had engaged in binge drinking
- 73 percent of men and 68 percent of women reported that getting drunk was an important reason for drinking
- Students who were white, members of a fraternity or sorority, or athletes were the most likely to binge
- 61 percent of men and 39 percent of women drank alcohol on 10 or more occasions in the past 30 days
- Over 80 percent of women and men reported having a hangover, over 50 percent doing something they regretted later, over 40 percent missing class due to drinking, and over 30 percent getting behind in schoolwork because of drinking

Not only were binge drinkers more likely to report using other controlled substances, but also the more frequently a student binged, the higher the probability of reporting other drug use. The most commonly reported other drugs used by frequent binge drinkers were, in order, cigarettes, marijuana, hallucinogens, and chewing tobacco (HHS 1998). Concerns over binge drinking have led some colleges to turn down alcohol advertisements at athletic events (MADD 1998a).

Although technology was developed years ago to remove nicotine from cigarettes and to control with precision the amount of nicotine in cigarettes, they are still marketing cigarettes with levels of nicotine that are sufficient to produce and sustain addiction.

—David Kessler
Head of Federal Drug Administration

Tobacco

Although nicotine is an addictive psychoactive drug and tobacco smoke has been classified by the Environmental Protection Agency as a Group A carcinogen, tobacco continues to be among the most widely used drugs in the United

States. According to the U.S. Department of Health and Human Services survey (1997), 62 million Americans continue to smoke cigarettes—29 percent of the population—including 4.1 million 12-to-17-year-olds.

International Data

According to the World Health Organization, if smoking among young people worldwide continues at the present rate of growth, more than 200 million children and teenagers will die from tobacco use between 2025 and 2050.

SOURCE: American Heart Association 1997c

Much of the concern about smoking surrounds the use of tobacco by young people (see Figure 3.1). Overall smoking rates have increased among high school students by almost a third in the last 6 years. A 1997 Youth Risk Behavior Survey indicates that among 16,000 students in grades 9 through 12, almost half of male students and more than a third of female students report using cigarettes, cigars, or smokeless tobacco in the last month. African-American student smoking, once thought to be declining, is also on the rise (HHS 1998).

Given the above, it is not surprising that the tobacco industry is under attack for targeting youth, minorities, and women in their tobacco advertising. Advertising campaigns such as "Joe Camel" directed toward youth, the placement of billboards advertising cigarettes at a rate four times greater in black communities than white communities, and the packaging of cigarettes in pastel perfumed boxes are but a few of the techniques used by the tobacco industry (American Heart Association 1997c).

Tobacco was first cultivated by Native Americans, who introduced it to the European settlers in the 1500s. The Europeans believed it had medicinal properties, and its use spread throughout Europe, assuring the economic success of the colonies in the New World. Tobacco was initially used primarily through chewing and snuffing, but in time smoking became more popular even though scientific evidence that linked tobacco smoking to lung cancer existed as early as 1859 (Feagin and Feagin 1994). Today, the health hazards of tobacco use are well documented and have resulted in the passage of federal laws that require warning labels on cigarette packages and prohibit cigarette advertising on radio and television. Smoking is associated with lung cancer, cardiovascular disease, strokes, emphysema, spontaneous abortion, premature birth, and neonatal death. Over a million new cases of lung cancer were diagnosed in 1998 (*Statistical Abstract of the United States: 1998,* Table 239).

Despite increased awareness of the dangers of cigarettes, the tobacco industry continues to enjoy enormous revenues—$48 billion annually—in part because of its ability to influence public opinion and governmental policy. For example, tobacco corporations such as Philip Morris and R. J. Reynolds donated more than $4.8 million to the two major political parties within 18 months of the 1996 elections (American Heart Association 1997b). Such donations are designed

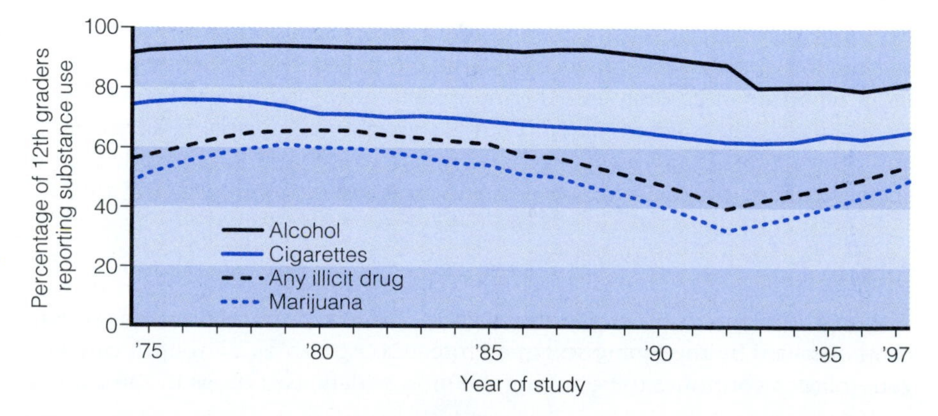

Figure 3.1 Lifetime Prevalence of Drug Use by Twelfth Graders: Monitoring the Future Study

SOURCE: Ann Crowe. 1998. "Drug Identification and Testing in the Juvenile Justice System." Office of Juvenile Justice and Delinquency Prevention. Washington, D.C.: U.S. Government Printing Office.

to influence favorable tobacco legislation and to elect tobacco-friendly officials. In 1998, a "tobacco-control" bill, which would have raised the price of cigarettes and allowed the government to regulate nicotine, failed to pass Congress (Koch 1998).

At the same time, a smoking prohibition movement is gaining momentum. Organizations such as the American Cancer Society, the American Heart Association, and the American Lung Association have sought to expose the fatal effects of tobacco and to encourage legislation banning its use. Many states now prohibit smoking in public places such as airports, restaurants, bars, elevators, schools, hospitals, libraries, and retail stores. Recent polls show that over 80 percent of Americans support government efforts to prevent children from using tobacco products (American Heart Association 1997b).

Although the tobacco companies have organized smoking rights organizations, the trend to restrict and/or ban tobacco use continues. An important effect of the antismoking campaign is that millions of Americans are giving up the habit annually. Since 1965, smoking has declined 40 percent (American Heart Association 1997a).

Everybody knew it's addictive. Everybody knew it causes cancer. We were all in it for the money.

—Victor Crawford
Former tobacco lobbyist and smoker who developed lung cancer

National Data

People who smoke are more likely to drink heavily and are slightly more likely to be male; they are more likely to be white rather than African-American or Hispanic, and to live in the South rather than in other parts of the country. Smoking is also correlated with education —the higher one's education, the lower the probability an individual smokes.

SOURCE: HHS 1997

Marijuana

Although drug abuse ranks among the top concerns of many Americans, surveys indicate that illegal drug use of all kinds is far less common than alcohol abuse. Marijuana remains the most commonly used and most heavily trafficked illicit drug in the world (*World Drug Report* 1997). According to the Health and Human Services survey, 77 percent of all illicit drug users use marijuana; 54 percent of illicit drug users use marijuana only (HHS 1997).

Marijuana's active ingredient is THC which, in varying amounts, may act as a sedative or as a hallucinogen. There are an estimated 200 to 250 million marijuana users worldwide, predominantly in Africa and Asia. Marijuana use dates back to 2737 B.C. in China and has a long tradition of use in India, the Middle East, and Europe. In North America, hemp, as it was then called, was used for making rope and as a treatment for various ailments. Nevertheless, in 1937 Congress passed the Marijuana Tax Act, which restricted its use; the law was passed as a result of a media campaign that portrayed marijuana users as "dope fiends" and, as conflict theorists note, was enacted at a time of growing sentiment against Mexican immigrants (Witters, Venturelli, and Hanson 1992, 357–59).

Marijuana use is most common among 18- to 25-year-olds (*Statistical Abstract of the United States: 1998,* Table 237). Over the last 10 years there has been a steady decline in the number of young people who believe marijuana is dangerous and, thus, a steady increase in the use of marijuana among young people. However, recent evidence suggests that use may be leveling off and for some age groups actually decreasing (ERS 1998).

While the effects of alcohol and tobacco are, in large part, indisputable, there is less agreement about the effects of marijuana. Although the carcinogenic effects of marijuana are as lethal as nicotine's, other long-term physiological effects are unknown. An important concern is that marijuana may be a **gateway drug** that causes progression to other drugs such as cocaine and heroin. More likely, however, is that persons who experiment with one drug are more likely to experiment with another. Indeed, most drug users are polydrug users with the most common combination being alcohol, tobacco, and marijuana.

National Data

Of all current illicit drug users in the United States, 74 percent are white, 14 percent African-American, and 8 percent Hispanic; use is higher in the Western and North Central States than in the South and Northeast, and in metropolitan areas compared to nonmetropolitan areas; the ratio of men to women users is 2 to 1.

SOURCE: HHS 1997

CONSIDERATION

In 1996 both Arizona and California passed acts known as "marijuana medical bills," that is, they made the use and cultivation of marijuana, under a physician's orders, legal. While proponents of the bills argue marijuana should be available to those who need it either to treat a disease or relieve pain and suffering, opponents argue the legislation was unnecessary and will further victimize those it's purported to help. According to a U.S. Department of Justice document (DEA 1997), marijuana (1) weakens the immune system, further jeopardizing those with AIDS, (2)contains carcinogens leading to and exacerbating cancer, and (3) does not prevent blindness due to glaucoma. Further, THC, the active ingredient in marijuana, is already available by prescription for those benefitting from it. Government officials contend that both the California and Arizona initiatives were funded by marijuana prolegalization organizations.

Cocaine

Cocaine is classified as a stimulant and, as such, produces feelings of excitation, alertness, and euphoria. Although such prescription stimulants as methamphetamine and dextroamphetamine are commonly abused, over the last 10 to 20 years societal concern over drug abuse has focused on cocaine. Such concerns have been fueled by its increased use, addictive qualities, physiological effects, and worldwide distribution. More than any other single substance, cocaine has led to the present "war on drugs."

Cocaine, which is made from the coca plant, has been used for thousands of years, but anticocaine sentiment in the United States did not emerge until the early 1900s, when it was primarily a response to cocaine's heavy use among urban blacks (Witters, Venturelli, and Hanson 1992, 260). Cocaine was outlawed in 1914 by the Harrison Narcotics Act, but its use and effects continued to be misunderstood. For example, a 1982 *Scientific American* article suggested that cocaine was no more habit forming than potato chips (Van Dyck and Byck 1982). As demand and then supply increased, prices fell from $100 a dose to $10 a dose, and "from 1978 to 1987 the United States experienced the largest cocaine epidemic in history" (Witters, Venturelli, and Hanson 1992, 256, 261).

International Data

Most of the world's coca leaf cultivation occurs in Peru, Columbia, and Bolivia, which combined account for 98 percent of the world's cocaine supply. World production of cocaine doubled between 1985 and 1995.

SOURCE: *World Drug Report* 1997

The current number of cocaine users is about 1.7 million, down from a high of 5.7 million in 1985 (HHS 1997). Whether the decrease is a function of increased government surveillance, anti-cocaine media messages and celebrity deaths, demographic changes, or a greater awareness of the substance's dangers are unknown.

Crack Cocaine

Crack is a crystallized product made by boiling a mixture of baking soda, water, and cocaine. The result, also called rock, base, and gravel, is relatively inexpensive and was not popular until the mid-1980s. Crack is one of the most dangerous drugs to surface in recent years. Crack dealers often give drug users their first few "hits" free, knowing the drug's intense high and addictive qualities are likely to lead to returning customers. Recent data, however, suggest that the number of new users may be decreasing as young people begin to associate crack use with "burnouts" and "junkies" (ONDCP 1998d). This chapter's *The Human Side* graphically describes conditions in a crack house and associated criminal behaviors through the eyes of sociologist-ethnographer Terry Williams.

> Crack is a drug peddler's dream: it is cheap, easily concealed and provides a short-duration high that invariably leaves the user craving more.
>
> —Tom Morganthau
> Journalist

THE HUMAN SIDE

AN EXCERPT FROM *THE COCAINE KIDS*

The door opens a crack before I can knock, a tall African-American man brusquely thrusts his palm toward me and asks, "You got three dollars?" He motions excitedly, "If you ain't got three dollars you can't come in here." The entrance fee. I pay and walk in.

The establishment is desolate, uninviting, dank and smoky. The carpet in the first room is shit-brown and heavily stained, pockmarked by so many smoke burns that it looks like an abstract design. In the dim light, all the people on the scene seem to be in repose, almost inanimate, for a moment.

As my eyes adjust to the smoke, several bodies emerge. I see jaws moving, hear voices barking hoarsely into walkie-talkies—something about money; their talk is jagged, nasal and female. One woman takes out an aluminum foil packet, snorts some of its contents, passes it to her partner then disappears into another room. In a corner near the window, a shadowy figure moans. One woman sits with her skirt over her head, while a bobbing head writhes underneath her. In an adjacent alcove, I see another couple copulating. Somewhere in the corridor a man and woman argue loudly in Spanish. Staccato rap music sneaks over the grunts and hollers.

The smell is a nauseating mix of semen, crack, sweat, other human body odors, funk and filth. Two men dicker about who took the last "hit" (puff); two others are on their hands and knees looking for crack particles they claim they have lost in the carpet.

In the crack houses, the sharing rituals associated with snorting are being supplanted by more individualistic, detached arrangements where people come together for erotic stimulation, sexual activity, and cocaine smoking. They may be total strangers, seeking only brief and superficial physical contact, encounters designed to heighten sensations; the smoking act is a narcissistic fix—there is little thought for the other person. The emotional content is largely due to the momentary excitation of the setting and the cocaine. Much of the sexual behavior is performed to acquire more cocaine.

Nothing better exemplifies the new attitude than the act of *Sancocho* (a word meaning to cut up in little pieces and stew). To sancocho is to steal crack, drugs or money from a friend or other person who is not alert, a regular practice in the crack houses. Another example is the "hit kiss" ritual: after inhaling deeply, basers literally "kiss"—put their lips together and exhale the smoke into each other's mouths. This not only saves all the valuable smoke, but also stimulates the other sexually. Other versions of the kiss extend to other orifices.

SOURCE: Terry Williams. 1989. From *The Cocaine Kids: The Inside Story of a Teenage Drug Ring*, by Terry Williams. Copyright © 1989, by Terry Williams. Reprinted by permission of Perseus Books Publishers, a member of Perseus Books, L.L.C.

Other Drugs

Other drugs abused in the United States include hallucinogens (e.g., lysergic acid diethylamide [LSD], MDMA), narcotics (e.g., heroin), prescription drugs (e.g. tranquilizers, amphetamines), and inhalants (e.g., glue).

Hallucinogens Hallucinogens may be classified as either natural, such as peyote, or synthetic, such as LSD. In either case users report an altering of perceptions and thought patterns. Although the use of LSD has decreased in recent years, the use of so-called designer drugs such as MDMA and ketamine has increased. These drugs, marketed illegally under such names as "ecstasy," "China White," and "Special K," are not only dangerous but expensive. Ketamine, an anesthetic for human and animal use, costs veterinarians $7 a vial, drug dealers $30–$45 a vial, and drug users $100–$200 a vial (U.S. Department of Justice 1997).

International Data

Almost 90 percent of the world's opium production comes from two areas—the "Golden Crescent" (Afghanistan, Iran, and Pakistan) and the "Golden Triangle" (Lao PDR, Myanmar, Thailand). In the 1990s more than 300 tons of heroin were produced from these areas.

SOURCE: *World Drug Report* 1997

Heroin Heroin use has risen in recent years with the number of heroin-related deaths increasing by 134 percent between 1990 and 1995. There is some indication that the increase in heroin use is a consequence of the decreased use of crack cocaine. While crack cocaine has become less fashionable among youthful offenders, heroin, an opium-based narcotic, has increased in acceptability to the point of being glamorized in recent motion pictures and song lyrics (Heroin Drug Conference 1997).

Prescription Drugs Tranquilizers and antidepressants are often used in the treatment of psychiatric disorders. For example, Prozac, Valium, and Halcion are often prescribed for the treatment of insomnia, anxiety, and depression. Despite most often being obtained through prescription, these drugs continue to be abused, and an illegal market for them persists. They remain, however, the top selling prescription drugs in the United States.

Amphetamines are stimulants rather than "downers" and are also legal when prescribed by a physician. Illegal use of amphetamines, made in clandestine laboratories, has increased in recent years. In 1996, the number of persons who had ever tried methamphetamine, for example, was 4.7 million, or 2.2 percent of the U.S. population (U.S. Department of Justice 1997). Worldwide, the production of amphetamines is increasing dramatically with nine times the quantity seized by government officials in 1993 than in 1978 (*World Drug Report* 1997).

Inhalants Common inhalants include lighter fluid, air fresheners, hair spray, glue, paint, and correction fluid, although there are over 1,000 other household products that are currently abused. Inhalants are the third most common substance abused by 12- to 14-year-olds, behind cigarettes and alcohol, and are extremely dangerous due to their toxicity. Young people often use inhalants believing they are harmless, or that any harm caused requires prolonged use. Sudden Sniff Death Syndrome, however, occurs instantly when the toxic fumes inhaled exceed certain levels of concentration (Join Together 1998).

SOCIETAL CONSEQUENCES OF DRUG USE AND ABUSE

Drugs are a social problem not only because of their adverse effects on individuals, but also as a result of the negative consequences their use has for society as a whole. Everyone is a victim of drug abuse. Drugs contribute to problems within the family and to crime rates, and the economic costs of drug abuse are enormous. Drug abuse also has serious consequences for health on both an individual and societal level.

When one or both parents use or abuse drugs, family funds may be diverted to the purchasing of drugs rather than necessities. Children raised in such homes have a higher probability of suffering from neglect.

Family Costs

The cost to families of drug use is incalculable. When one or both parents use or abuse drugs, needed family funds may be diverted to purchasing drugs rather than necessities. Children raised in such homes have a higher probability of neglect, behavioral disorders, and absenteeism from school as well as lower self-concepts (Tubman 1993; Easley and Epstein 1991). Further, lower intelligence quotients (IQs) in toddlers have been linked to prenatal exposure to marijuana (Perkins 1997). Drug abuse is also associated with family disintegration. For example, alcoholics are seven times more likely to separate or divorce than nonalcoholics, and as much as 40 percent of family court problems are alcohol related (Sullivan and Thompson 1994, 347). In a recent Gallup Poll (1997), one-third of all Americans reported alcohol-related family problems in the last year.

Abuse between intimates is also linked to drug use (Foster, Forsyth, and Herbert 1994; U.S. Department of Justice 1998). In a study of 320 men who were married or living with someone, twice as many reported hitting their partner only after they had been drinking compared with those who reported the same behavior while sober (Leonard and Blane 1992). Additionally, Straus and Sweet (1992) found that alcohol consumption and drug use were associated with higher levels of verbal abuse among spouses.

Crime Costs

The drug behavior of persons arrested, those incarcerated, and persons in drug treatment programs provides evidence of the link between drugs and crime. Drug users commit a disproportionate number of crimes. A study of chronic drug users found that 50 percent of both male and female addicts had engaged in illegal behavior in the 30 days prior to the survey. Homicide rates of those 18 and under have more than doubled since 1985, a statistic attributable in part to the increase in drug-related gang violence in the United States (ONDCP 1998d). Further, a recent Columbia University report concludes that as many as 80 percent of our nation's prisoners are drug involved (Califano 1998).

International Data

In Great Britain, alcohol is a factor in 60 to 70 percent of all homicides, 75 percent of all stabbings, 70 percent of all beatings, and 50 percent of all fights and domestic assaults.

SOURCE: Institute of Alcohol Studies 1997

The only livin' thing that counts is the fix . . . Like I would steal off anybody—anybody, at all, my own mother gladly included.

—Heroin addict

CONSIDERATION

The relationship between crime and drug use is a complex one. Researchers disagree as to whether drugs actually "cause" crime or whether, instead, criminal activity leads to drug involvement (Gentry 1995). Further, since both crime and drug use are associated with low socioeconomic status, poverty may actually be the more powerful explanatory variable. Studies on cocaine use among white, middle-class, educated users indicate drugs and crime are not necessarily related (Waldorf, Reinarman, and Murphy 1991). After extensive study of the assumed drug-crime link, Gentry (1995) concludes that "the assumption that drugs and crime are causally related weakens when more representative or affluent subjects are considered." (p. 491)

In addition to the hypothesized crime–drug use link, some criminal offenses are drug defined: possession, cultivation, production, and sale of controlled substances; public intoxication; drunk and disorderly conduct; and driving while intoxicated. Between 1996 and 1997 arrests for drug abuse violations increased for both adults and juveniles (Bureau of Justice Statistics 1999). Driving while intoxicated is one of the most common drug-related crimes. Interestingly, legal limits of intoxication vary from state to state, and some states have lower blood alcohol requirements for younger drivers (see Figure 3.2). Whereas 4 in 10 violent victimizations, the most common being simple assault, are alcohol related, over 40 percent of all traffic fatalities are alcohol related (U.S. Department of Justice 1998).

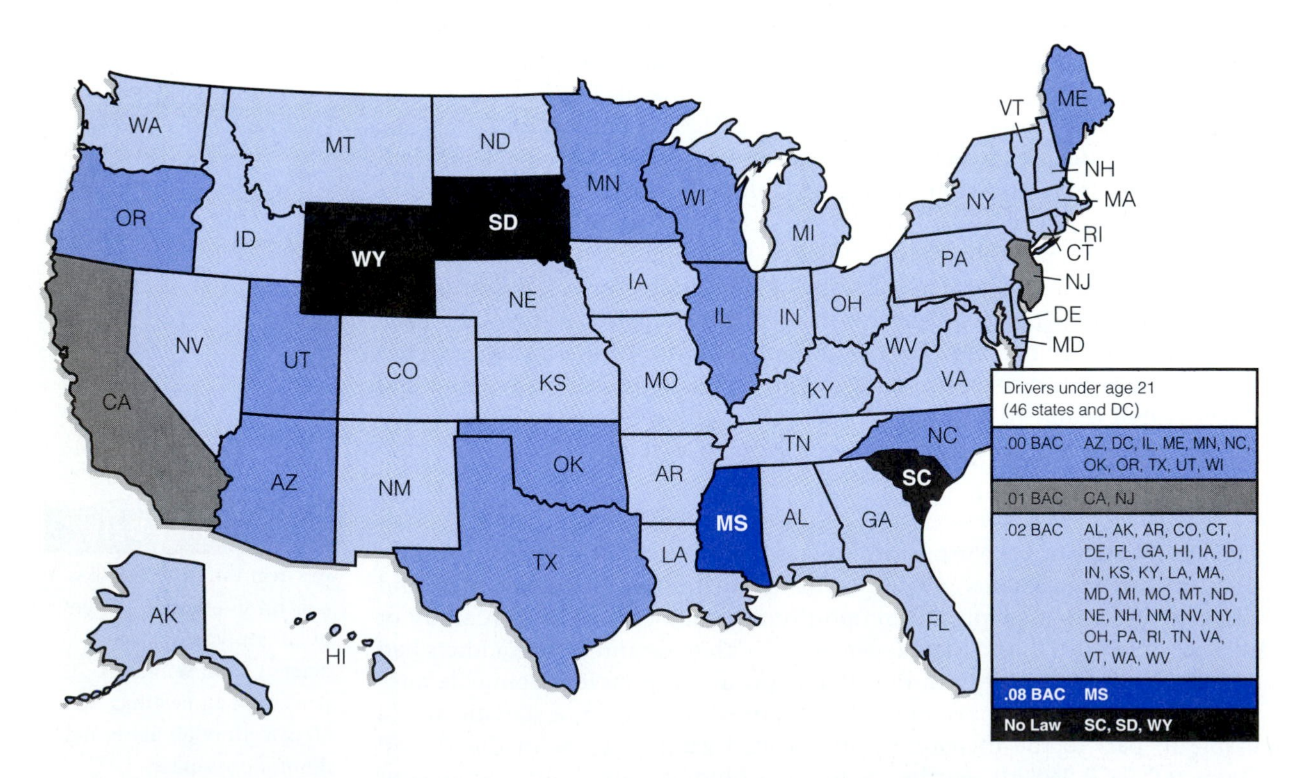

Figure 3.2 **States with Lower Blood Alcohol Levels for Young Drivers (as of January 1998)**
SOURCE: National Highway Traffic Safety Administration, 1998

Economic Costs

The economic costs of drug use are high, over $65 billion annually, 70 percent of which is attributable to the cost of drug-related crime. Federal, state, and local governments spend an estimated $25 billion on drug enforcement alone—50¢ for every dollar spent on illicit drugs (ONDCP 1998a). Also lost are billions of corporate dollars due to reduced worker productivity, absenteeism, premature deaths, and insurance and health care costs (Perspective 1996). Concern that on-the-job drug use may impair performance and/or cause fatal accidents has led to drug testing. For many employees, such tests are routine both as a condition for employment and as a requirement for keeping their job. This chapter's *Focus on Technology* reviews some of the issues related to drug testing and narcotics detection.

Other economic costs of drug abuse include the cost of homelessness, the cost of implementing and maintaining educational and rehabilitation programs, and the cost of health care. The cost of fighting the "war on drugs" is also enormous and likely to increase as organized crime develops new patterns of involvement in the illicit drug trade.

National Data

A 1998 survey found that a majority of Americans (53 percent) fear drunk drivers more than any other highway problem. The survey also found that 70 percent favor lowering the illegal blood alcohol level to .08, 97 percent said they would stop a drunk friend from driving, and 33 percent said that the laws are ineffective in addressing the problem.

SOURCE: MADD 1997

FOCUS ON TECHNOLOGY

DRUG TESTING

The technology available to detect whether a person has taken drugs was used during the 1970s by crime laboratories, drug treatment centers, and the military. Today, employers in private industry have turned to chemical laboratories for help in making decisions on employment and retention, and parents and school officials use commercial testing devices to detect the presence of drugs. An individual's drug use can be assessed through the analysis of hair, blood, or urine. New technologies include portable breath (or saliva) alcohol testers, THC detection strips, passive alcohol sensors, interlock vehicle ignition systems, and fingerprint screening devices. Countertechnologies have even been developed, for example, shampoos that rid hair of toxins and "Urine Luck," a urine additive that is advertised to speed the breakdown of unwanted chemicals.

In 1986, the President's Commission on Organized Crime recommended that all employees of private companies contracting with the federal government be regularly subjected to urine testing for drugs as a condition of employment. This recommendation was based on the belief that if employees such as air traffic controllers, airline pilots, and railroad operators are using drugs, human lives may be in jeopardy due to impaired job performance.

In 1987, an Amtrack passenger train crashed outside Baltimore, killing 16 and injuring hundreds. There was evidence of drug use by those responsible for the train's safety. As a result, the Supreme Court ruled in 1989 (by a vote of 7–2), that it is constitutional for the Federal Railroad Administration to administer a drug test to railroad crews if they are involved in an accident. Testing those in "sensitive" jobs for drug use may save lives. In 1995 the Supreme Court ruled that random drug testing of student athletes in public schools is not unconstitutional. Today, drug testing is not only routine but, according to a recent Gallup Poll, supported by those it most affects with 97 percent of employees responding that workplace drug testing is appropriate under certain circumstances (National Clearinghouse of Alcohol and Drug Information 1998).

An alternative perspective is that drug testing may be harmful (O'Keefe 1987; "This is a Test" 1998; O'Connor 1995). First, the tests can be inaccurate because they may result in either "false positives" (the person who does not use drugs is identified as doing so) or

continued

FOCUS ON TECHNOLOGY

DRUG TESTING–cont'd

"false negatives" (the person who uses drugs is identified as not doing so). The false-positives problem is serious: an innocent person could lose his or her job because of faulty technology. Second, urinalysis may also reveal that a person is pregnant, is being treated for heart disease, or has epilepsy. The person may want these aspects of his or her private life to remain private. Third, such drug testing may violate basic constitutional rights. The Fourth Amendment states that "the right of the people to be secure in their persons . . . against unreasonable searches and seizures, shall not be violated." However, where the welfare of the public is at stake, the government has taken the position that it has the right to administer drug tests (e.g., to airline pilots).

Drug testing has enabled employers to detect drug use among employees and applicants. When combined with drug treatment programs, it has contributed to the rehabilitation of drug abusers. However, Ackerman (1995) warns:

> To the individual drug user denied a job because of a positive urine test, and to the society on which he or she must remain (unemployed and) dependent, drug testing may compound the problems caused by drug addiction. (p. 487)

As a society, we are moving toward a policy of enforced drug testing for more and more segments of our society. Counterdrug technologies, such as high energy x-ray systems that are 50 to 70 times as powerful as a typical airport x-ray machine are also being developed (NCJRS 1997), and there is little doubt that the production of drug-fighting technologies will continue. The question in a complex and increasingly diverse society is how to balance the rights of an individual with the needs of society as a whole.

SOURCES: Deborah L. Ackerman. 1995. "Drug Testing." In *Handbook on Drug Abuse and Prevention,* ed. Robert H. Coombs and Douglas Ziedonis, pp. 473–89. Boston: Allyn and Bacon; National Clearinghouse of Alcohol and Drug Information. 1998. "Drug Testing in the Workplace." *Prevention Primer.* **http://health.org;** NCJRS (National Criminal Justice Resource Service). 1997. "Letter to The Honorable Benjamin A. Gilman." April 15. **http://www.ncjrs.org;** *Vernonia School District v Wayne Acton et ux, Guardians ad Litem for James Acton,* 115 S. Ct. 2386, 132 L. ed. 2d 564 (1995); Anne Marie O'Keefe. 1987. "The Case against Drug Testing." *Psychology Today,* June, 34–38; "This Is a Test: The Dilemmas of Drug Testing." 1998. *Issues in Ethics* 1: 1–3.

Health Costs

The physical health consequences of drug use for the individual are tremendous: shortened life expectancy; higher morbidity (e.g., cirrhosis of the liver, lung cancer); exposure to HIV infection, hepatitis, and other diseases through shared needles; a weakened immune system; birth defects such as fetal alcohol syndrome; drug addiction in children; and higher death rates. Death rates from drug-induced and alcohol-induced causes are significantly higher for males and minorities than for females and whites (*Statistical Abstract of the United States: 1998,* Tables 153,154). However, the incidence of lung cancer in women has increased dramatically as their smoking rates have increased. In 1964 the male-female ratio of lung cancer victims was 6 to 1; it is predicted that within the decade the ratio will be 1 to 1 (Brody 1998).

National Data

Tobacco use is responsible for 1 in 5 deaths in the United States. Of those who continue to smoke, 50 percent will die prematurely from smoking, and of these, 50 percent will die between the ages of 35 and 69—a loss of between 20 and 25 years of life.

SOURCE: American Cancer Society 1998

Heavy alcohol and drug use are also associated with negative consequences for an individual's mental health. Longitudinal data on both male and female adults have shown that drug users are more likely to suffer from anxiety disorders (e.g., phobias), depression, and antisocial personalities (White and Labouvie 1994). Other data confirm that drug users, particularly in adolescence, have a higher incidence of suicide (Bureau of Justice Statistics 1992). Marijuana, the drug most commonly used by adolescents, is also linked to short-term memory loss, learning disabilities, motivational deficits, and retarded emotional development.

The societal costs of drug-induced health concerns are also extraordinary. An Institute of Health and Aging study estimates that the cost for medical services for drug users is $3.2 billion (cited in ONDCP 1998a). Health costs also include the cost of disability insurance, the effects of secondhand smoke, the spread of AIDS, and the medical costs of accident and crime victims, as well as unhealthy infants and children. For example, infants are five times as likely to die from sudden infant death syndrome if adults smoke in their rooms (Klonoff-Cohen et al. 1995), drug use by pregnant women increases the probability of complications for both mother and child, and 1 in 500 babies is born with fetal alcohol syndrome (ONDCP 1998a).

Physicians should advise their patients during prenatal care and after delivery not to smoke around an infant.

—Sandra Klonoff-Cohen
Physician/Researcher

TREATMENT ALTERNATIVES

Helping others to overcome chemical dependency is expensive. In 1998, the federal budget included over $3 billion for substance abuse treatment (ONDCP 1998c), and the 1999 budget includes a recommended increase of $491 million for treatment and prevention programs (ONDCP 1998b). Persons who are interested in overcoming chemical dependency have a number of treatment alternatives from which to choose. Their options include hospitalization, family therapy, drug counseling, private and state treatment facilities, behavior modification, community care programs, drug maintenance programs, and employee assistance programs. Two other commonly used rehabilitative techniques are 12-step programs and therapeutic communities.

People who smoke are more likely to be heavy drinkers and current illicit drug users. Some evidence suggests that giving up smoking leads to a reduction in alcohol consumption.

12-Step Programs

Both Alcoholics Anonymous (AA) and Narcotics Anonymous (NA) are voluntary associations whose only membership requirement is the desire to stop drinking or taking drugs. AA and NA are self-help groups in that they are operated by nonprofessionals, offer "sponsors" to each new member, and proceed along a continuum of 12 steps to recovery. These include an acknowledgment of one's helplessness over addiction and a recognition of a higher power as a source of help. In addition to the 12 steps, which require abstinence, humility, penance, and commitment, AA and NA members are immediately immersed in a fellowship of caring individuals with whom they meet daily or weekly to affirm their commitment. Some have argued that AA and NA members trade their addiction to drugs for feelings of interpersonal connectedness by bonding with other group members.

Symbolic interactionists emphasize that AA and NA provide social contexts in which people develop new meanings. Abusers are surrounded by others who convey positive labels, encouragement, and social support for sobriety. Sponsors tell the new members that they can be successful in controlling alcohol and/or drugs "one day at a time" and provide regular interpersonal reinforcement for doing so. Some AA members may also choose to take **Antabuse**, a prescribed medication, which when combined with alcohol produces severe nausea. Some alcoholics regard Antabuse as a "crutch," but others feel that anything that helps the alcoholic to remain sober should be used.

AA members essentially trade addiction to the bottle to a network of friends who share a common bond.

—Barry Lubetkin
Psychologist

AA and NA have reputations as two of the most successful drug rehabilitation programs. But because membership is voluntary and anonymous, accurate evaluation of these programs is difficult. Further, since the desire to change is a prerequisite for joining AA or NA, their success rates are artificially elevated, making other programs appear less successful.

Therapeutic Communities

In **therapeutic communities**, which house between 35 and 500 people for up to 15 months, participants abstain from drugs, develop marketable skills, and receive counseling. Synanon, which was established in 1958, was the first therapeutic community for alcoholics and was later expanded to include other drug users. More than 400 residential treatment centers are now in existence, including Daytop Village and Phoenix House. The longer a person stays at such a facility, the greater the chance of overcoming his or her dependency. Symbolic interactionists argue that behavioral changes appear to be a consequence of revised self-definition and the positive expectations of others.

Stay'N Out, a therapeutic community for the treatment of incarcerated drug offenders, has demonstrated its success in reducing recidivism. When participating in the program for at least nine months, only 23 percent of the inmates, compared with 50 percent of those receiving no treatment, reoffended. The Cornerstone Program has also been successful with drug abusers in prison. Both programs include a holistic treatment approach that focuses on the social and psychological difficulties of returning to acceptable social roles (Lipton 1994, 336).

STRATEGIES FOR ACTION: AMERICA RESPONDS

> The goal of a drug-free America is an unrealistic one; it is realistic, however, to strive to reduce use to the pre-1960 era when drug use was a small problem in America.
>
> —Herbert D. Kleber, M.D. Former deputy director, White House Office of National Drug Control Policy

Drug use is a complex social issue exacerbated by the structural and cultural forces of society that contribute to its existence. While the structure of society perpetuates a system of inequality creating in some the need to escape, the culture of society, through the media and normative contradictions, sends mixed messages about the acceptability of drug use. Thus, developing programs, laws, or initiatives that are likely to end drug use may be unrealistic. Nevertheless, numerous social policies have been implemented or proposed to help control drug use and its negative consequences.

Government Regulations

The largest social policy attempt to control drug use in the United States was Prohibition. Although this effort was a failure by most indicators, the government continues to impose drug regulations such as the 1988 Anti–Drug Abuse Act, the Crime Control Act of 1990, the Violent Crime Control and Law Enforcement Act of 1994, and the Drug-Free Communities Act of 1997 (see Table 3.2).

In the 1980s the federal government declared a "war on drugs." Yet, as recently as 1998, the Clinton administration revealed a 5-year initiative called the National Youth Anti-Drug Media Campaign, as well as a recommended 1999 drug-control budget of $17.1 billion—an increase of 6.8 percent from the previous year (see Figure 3.3) (McCaffrey 1998; ONDCP 1998b). Although there appears to be a movement toward treatment and prevention as indicated by funding specifics, there remains an emphasis on deterrence and "get tough" policies, particularly in dealing with large-scale traffickers and international offenders (Clinton 1998).

Table 3.2 **Four Major Federal Antidrug Laws from the 1980s and 1990s**

The 1988 Anti-Drug Abuse Act

- Increased penalties for offenses related to drug trafficking, created new federal offenses and regulatory requirements, and changed criminal procedures.
- Altered the organization and coordination of federal antidrug efforts.
- Increased treatment and prevention efforts aimed at reduction of drug demand.
- Endorsed the use of sanctions aimed at drug users to reduce the demand for drugs.
- Targeted for reduction of drug production abroad and international trafficking in drugs.

The Crime Control Act of 1990

- Doubled the appropriations authorized for drug law enforcement grants to states and localities.
- Expanded drug control and education programs aimed at the nation's schools.
- Expanded specific drug enforcement assistance to rural states.
- Expanded regulation of precursor chemicals used in the manufacture of illegal drugs.
- Provided additional measures aimed at seizure and forfeiture of drug traffickers' assets.
- Sanctioned anabolic steroids under the Controlled Substances Act.
- Included provisions on international money laundering, rural drug enforcement, drug-free school zones, drug paraphernalia, and drug enforcement grants.

The Violent Crime Control and Law Enforcement Act of 1994

- Expanded the federal death penalty to cover large-scale drug trafficking.
- Provided new and stiffer penalties for drug trafficking crimes committed by gang members.
- Required mandatory life imprisonment without possibility of parole for federal offenders with three or more convictions for serious violent felonies or drug trafficking crimes.
- Provided $383 million for prison drug treatment.

The Drug-Free Communities Act of 1997

- Provided needed funds for coalitions of community members in the fight against youthful substance abuse.
- Provided that such funds are under the supervision of the Director of the Office of National Drug Control Policy (ONDCP).
- Established an Advisory Commission on Drug-Free Communities to consult with the Director of the Office of National Drug Control Policy.

SOURCES: U.S. Department of Justice, Office of Justice Programs, Bureau of Justice Statistics. 1992. *Drugs, Crime, and the Justice System,* NCJ-1333652, p. 86. Washington, D.C.: U.S. Government Printing Office; *Violent Crime Control and Law Enforcement Act of 1994.* 1995; "Highlights of the Drug-Free Communities Act of 1997" **http://www.whitehousedrugpolicy.gov**

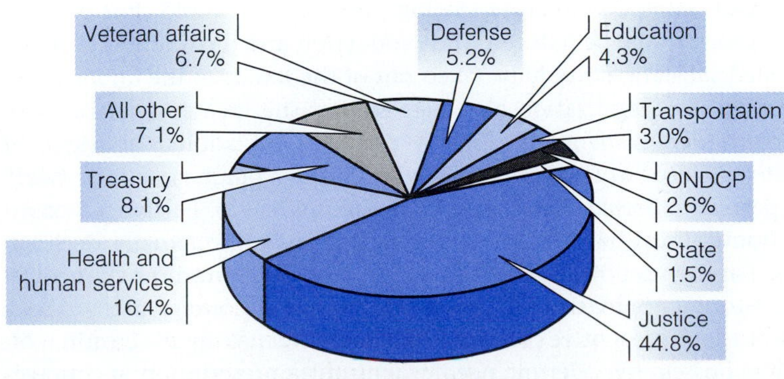

Figure 3.3 **Federal Drug Control Spending by Department, 1999**

SOURCE: ONDCP. 1998. "1999 Drug Budget Program Highlights: Overview." **http://www.whitehousedrugpolicy.gov/drugfacts/budget-1.html**

National Data

In 1995, the Drug Enforcement Administration (DEA) seized over 319 clandestine drug laboratories, and $500,000 in assets and made 23,000 arrests, which lead to 14,000 convictions.

SOURCE: *Statistical Abstract of the United States: 1997*, Table 334

Some would argue, however, that the war on drugs has done more harm than good. Yale law professor Steven Duke and coauthor Albert C. Gross in their book *America's Longest War* (1994) argue that the war on drugs, much like Prohibition, has only intensified other social problems: drug-related gang violence and turf wars, the creation of syndicate-controlled black markets, unemployment, the spread of AIDS, overcrowded prisons, corrupt law enforcement officials, and the diversion of police from other serious crimes. Consistent with conflict theory, still others argue that the "war on drugs is actually a war against African-American males, while the drug use of affluent whites goes largely ignored" (Duster 1995).

Further, U.S. drug policies have implications that extend beyond domestic concerns and affect international relations and the economies of foreign countries. Distributing drugs worldwide requires a complex network of social actors. As the *World Drug Report* (1997) concludes:

> **The most prominent trafficking organizations appear to be characterized by highly centralized management control at the upper echelons, with compartmentalization of functions and task specialization at the lower levels. A seemingly endless stream of willing recruits for the most menial tasks provides the industry with a wide range of specialized personnel including chemists, chemical engineers, pilots, communication specialists, money launderers, accountants, lawyers, security guards and "hit men."**

Many of the countries in which drug trafficking occurs are characterized by government corruption and crime, military coups, and political instability. Ironically, the United States provides foreign aid and military assistance to some of these countries. Some argue that trade sanctions should be imposed in addition to crop eradication programs and interdiction efforts. Others, however, noting the relative failure of such programs in reducing the supply of illegal drugs entering the United States, argue that the "war on drugs" should be abandoned and that legalization is preferable to the side effects of regulation.

National Data

A recent Discovery/ABC poll found that 22 percent of Americans favor the legalization of marijuana in small amounts for personal use; 75 percent are opposed. When asked whether or not doctors should be able to prescribe marijuana for medical purposes, 69 percent said "Yes," 27 percent "No," and 4 percent had "No Opinion."

SOURCE: Discovery/ABC Poll 1998

Legalization

Proponents for the **legalization** of drugs affirm the right of adults to make an informed choice. They also argue that the tremendous revenues realized from drug taxes could be used to benefit all citizens, that purity and safety controls could be implemented, and that legalization would expand the number of distributors, thereby increasing competition and reducing prices. Drugs would thus be safer, drug-related crimes would be reduced, and production and distribution of previously controlled substances would be taken out of the hands of the underworld.

Those in favor of legalization also suggest that the greater availability of drugs would not increase demand. When a sample of 600 adults was asked, "If cocaine were legalized, would you personally consider purchasing it or not?" less than 1 percent reported that they would (Dennis 1993). Further, in countries where some drugs have been decriminalized, use has actually declined (Kort 1994). Finally, **decriminalization** of drugs would promote a medical rather than criminal approach to drug use that would encourage users to seek treatment and adopt preventive practices. For example, making it a criminal offense to sell or possess hypodermic needles without a prescription encourages the use of nonsterile needles that spread infections such as HIV and hepatitis.

Opponents of legalization argue that it would be construed as government approval of drug use and, as a consequence, drug experimentation and abuse would increase. Further, while the legalization of drugs would result in substantial revenues for the government, since all drugs would not be decriminalized (e.g., crack), drug trafficking and black markets would still flourish (Bennett 1993). Legalization would also require an extensive and costly bureaucracy to regulate the manufacture, sale, and distribution of drugs. Finally, the position that drug use is an individual's right cannot guarantee that others will not be harmed. It is illogical to assume that a greater availability of drugs will translate into a safer society.

Deregulation

The government has also approached the drug problem through **deregulation** or the reduction of government control of certain drugs. While all states now require that individuals be 21 years old to purchase alcohol and 18 to purchase cigarettes, both substances are legal and purchased freely. Further, in some states, possession of marijuana in small amounts is now a misdemeanor rather than a felony. Deregulation of marijuana permits the criminal justice system to concentrate on fighting more serious drugs such as crack.

U.S. law also allows distribution of methadone in treatment centers for heroin addicts. Methadone, a synthetic opiate, is taken orally and inhibits the euphoric effect of heroin, thus blocking the motivation for its use. The drug produces no "high," and the recovering addict can begin to lead a normal life. Persons in methadone maintenance programs also participate in family counseling and job training programs, both designed to help the addict make a successful return to society. Today, there are over 100,000 methadone patients in the United States (Lindesmith Center 1998). A recent Drug Abuse Treatment Outcome Study conducted by the National Institute on Drug Abuse concludes that outpatient methadone treatment reduced heroin use by 70 percent, cocaine use by 48 percent, crime by 57 percent, and increased full-time employment by 24 percent (DATOS 1998).

Collective Action

Social action groups such as Mothers Against Drunk Driving **(MADD)** have successfully lobbied legislators to raise the drinking age to 21 and to provide harsher penalties for driving while impaired. In 1998 MADD pressured Congress to pass the Repeat Offender and Open Container Law. The new legislation "prohibits the possession or consumption of an open container of alcohol by the driver and all passengers in a motor vehicle" and establishes minimum penalties for repeat drunk drivers (MADD 1998b). States are required to adopt the new guidelines by the year 2001 or suffer the loss of federal highway funds (MADD 1998a).

> If some drunk gets out and kills your kid, you'd probably be a little crazy about it, too.
>
> —Kathy Prescott
> MADD president

MADD, with 3.5 million members and 60 chapters, has also put pressure on alcohol establishments to stop "two for one" offers and has pushed for laws that hold the bartender personally liable if a served person is later involved in an alcohol-related accident. Even hosts in private homes can now be held liable if they allow a guest to drive who became impaired while drinking at their house. Most importantly perhaps, MADD seeks to change the meaning of alcohol use by, for example, redefining drunk driving "accidents" as violent crimes.

Between 1980 and 1996 the DUI arrest rate for 16 to 20 year olds was cut in half. Reasons for such a dramatic decrease include an increase in the minimum drinking age, stricter punishments, and MADD campaigns that have successfully redefined drunk driving as a violent crime. Here, high school students view a vehicle involved in an alcohol-related fatal accident. Programs such as these are designed to make students think twice before drinking and driving.

Sensitized to the danger of driving while impaired, high school principals and school boards have encouraged students to become members of Students Against Drunk Driving **(SADD).** Members often sign a formal pledge and put an emblem on their car to signify a commitment against alcohol. To reduce the number of teenagers driving while drinking, local groups of parents have also organized parties at bowling alleys or school gyms as alternatives to high school graduation parties.

Collective action is also being taken against tobacco companies by smokers, ex-smokers, and the families of smoking victims. They charge that tobacco executives knew over 30 years ago that tobacco was addictive and concealed this fact from both the public and the government. Furthermore, they charge that tobacco companies manipulate nicotine levels in cigarettes with the intention of causing addiction. Recently, a Florida jury ordered a tobacco company to pay $1 million dollars—the largest settlement to date—to the family of a man who had died of lung cancer after smoking for 50 years (Mauro 1998).

Further, with the failure of the $350 billion federal comprehensive tobacco bill, individual states continue to try to negotiate health cost–related actions against the tobacco industry. In 1997, Mississippi, Florida, Texas, and Minnesota received over $36 billion in settlements. In 1998, the remaining 46 states and the tobacco industry reached a $206 billion settlement in which tobacco companies were assured legal protection from future suits (Tobacco Wars 1998). However, tobacco companies are also being sued by health insurance companies and labor union health funds (Torry 1998). In addition to these legal pressures, health lobbyists want to force the Food and Drug Administration to regulate nicotine as a drug.

In summarizing what we know about substance abuse, drugs and their use are socially defined. As the structure of society changes, the acceptability of one drug or another changes as well. As conflict theorists assert, the status of a drug as legal or illegal is intricately linked to those who have the power to define acceptable and unacceptable drug use. There is also little doubt that rapid social change, anomie, alienation, and inequality further drug use and abuse. Symbolic interactionism also plays a significant role in the process—if people are labeled as "drug users" and expected to behave accordingly, drug use is likely to continue. If there is positive reinforcement of such behaviors and/or a biological predisposition to use drugs, the probability of drug involvement is even higher. Thus, the theories of drug use complement rather than contradict one another.

Drug use must also be conceptualized within the social context in which it occurs. In a study of high-risk youths who had become drug involved, Dembo et al. (1994) suggest that many youths in their study had been "failed by society":

> **Many of them were born into economically-strained circumstances, often raised by families who neglected or abused them, or in other ways did not provide for their nurturance and wholesome development . . . few youths in our sample received the mental health and substance abuse treatment services they needed. (p. 25)**

However, many treatment alternatives, emanating from a clinical model of drug use, assume that the origin of the problem lies within the individual rather than in the structure and culture of society. Although admittedly the problem may lie within the individual at the time treatment occurs, policies that address the social causes of drug abuse provide a better means of dealing with the drug problem in the United States.

Prevention is preferable to intervention, and given the social portrait of hard drug users—young, male, minority—prevention must entail dealing with the social conditions that foster drug use. Some data suggest that inner city adolescents are particularly vulnerable to drug involvement due to their lack of legitimate alternatives (Van Kammen and Loeber 1994).

> **Illegal drug use may be a way to escape the strains of the severe urban conditions and dealing illegal drugs may be one of the few, if not the only, ways to provide for material needs. Intervention and treatment programs, therefore, should include efforts to find alternate ways to deal with the limiting circumstances of inner-city life, as well as create opportunities for youngsters to find more conventional ways of earning a living. (p. 22)**

Social policies dealing with drug use have been predominantly punitive rather than preventive. Recently, however, the Clinton administration pledged to concentrate on educating the public on the dangers of alcohol and drug use. DARE is one of the most popular school-based drug education programs in the United States. Uniformed police officers deliver 17 weekly lessons for 45 to 60 minutes with the goal of providing students with antidrug views and the social skills to resist using drugs. Short-term effects have been positive ("Community

National Data

The 1999 cost of drugs totaled more than 14,000 lives and $17 billion.

SOURCE: Yost 1999

Drug Prevention" 1990). However, Wysong, Aniskiewicz, and Wright (1994) compared 228 high school seniors exposed to DARE as seventh graders with 355 nonexposed seniors. No significant differences in drug use behaviors or attitudes were found between the two groups, a finding consistent with several other program evaluations (Society 1999). The authors argue that the popularity of the DARE program is part of a "symbolic crusade" against the "drug crisis," as defined by the Bush war on drugs campaign and increased media attention to illicit drugs as a major social problem.

In the United States and throughout the world, millions of people depend on legal drugs for the treatment of a variety of conditions, including pain, anxiety and nervousness, insomnia, overeating, and fatigue. While drugs for these purposes are relatively harmless, the cultural message "better living through chemistry" contributes to alcohol and drug use and its consequences. But these and other drugs are embedded in a political and economic context that determines who defines what drugs, in what amounts, as licit or illicit and what programs are developed in reference to them.

CRITICAL THINKING

1. Are alcoholism and other drug addictions a consequence of nature or nurture? If nurture, what environmental factors contribute to such problems? Which of the three sociological theories best explains drug addiction?
2. Measuring alcohol and drug use is often very difficult. This is particularly true given the tendency for respondents to acquiesce, that is, respond in a way they believe is socially desirable. Consider this and other problems in doing research on alcohol and other drugs, and how such problems would be remedied.
3. If, as symbolic interactionists argue, social problems are those conditions so defined, how might the manipulation of social definitions virtually eliminate many "drug" problems?
4. Look at the distribution of federal drug control spending in Figure 3.3. Give examples of the ways in which each of the departments are contributing to the "war on drugs."

KEY TERMS

anomie
Antabuse
chemical dependency
decriminalization
deregulation
drug
drug abuse
gateway drug
legalization
MADD
SADD
therapeutic communities

INTERNET

You can find more information on the effects of drugs, drug testing in the workplace, and the arguments for and against the legalization of marijuana at the *Understanding Social Problems* web site at **http://sociology.wadsworth.com.**

INFOTRAC COLLEGE EDITION

Either from the Wadsworth sociology resource center at **http://sociology.wadsworth.com** or directly from your web browser, you may access InfoTrac College Edition, an online university library that includes over 700 popular and scholarly journals in which you can find articles related to the topics in this chapter. Suggested articles and questions relating to these articles are listed below.

Field, Gary. 1998. "From the Institution to the Community." *Corrections Today* 60 (6): 94–99

1. How is the Oregon system of drug rehabilitation different from the California system?
2. What reason does the author give for the continuation of drug rehabilitation post-institutionalization?
3. What elements of the justice system inhibit the effectiveness of post-institution rehabilitation?

Manisses Communications Group Inc. 1998. "Promising Data from Criminal Justice Pilot Leads to Expansion." *Alcoholism & Drug Abuse Weekly.* Nov. 16: 1–3.

1. Describe the basic goals of the *Breaking the Cycle* program.
2. What are some of the results from the implementation of this program?
3. Name some of the expansions the creators of *Breaking the Cycle* hope to achieve in the near future.

Schneider, Cathy Lisa. 1998. "Racism, Drug Policy and AIDS." *Political Science Quarterly* 113 (3): 427–47.

1. What are several theoretical explanations for the prevalence of HIV infection among lower class blacks and Latinos?
2. When was drug use made illegal in America and what was the first drug the government declared illegal?
3. What are some of the myths associated with minority use of illicit narcotics?

CHAPTER FOUR

Crime and Violence

- The Global Context: International Crime and Violence
- Sources of Crime Statistics
- *Self & Society: Criminal Activities Survey*
- Sociological Theories of Crime and Violence
- *Focus on Technology: DNA Evidence*
- Types of Crime
- Demographic Patterns of Crime
- Costs of Crime and Violence
- Strategies for Action: Responding to Crime and Violence
- *The Human Side: Witness to an Execution*
- Understanding Crime and Violence

Is It True?

Is It True?

1. Most persons arrested for violent crime are between the ages of 35 and 45.
2. Although sociologists have different theories about the causes of crime, they agree that crime is always harmful to society.
3. Amnesty International identifies the United States as a country that violates human rights because capital punishment is legal here.
4. The role of criminal homicide is 14 times higher in the United States than in other industrialized countries.
5. The majority of U.S. homicide victims had some type of relationship with their murderer.

Answers: 1 = F, 2 = F, 3 = T, 4 = T, 5 = T

Unjust social arrangements are themselves a kind of extortion, even violence . . .

John Rawls, from A Theory of Justice

The recent rash of killings by elementary and middle school children have shocked the nation. On March 23, 1998, in Jonesboro, Arkansas, two boys, 13 and 11, allegedly pulled a fire alarm to ensure that the students and teachers at Westside Middle School would be in the school yard. After stealing guns from one boy's grandfather, the two boys, hiding in trees on the perimeter of the school grounds, shot and killed four young girls and a teacher, critically wounding three others.

This incident and others in Paducah, Kentucky, and Pearl, Mississippi, have focused national attention on the problems of crime and violence and, specifically, the problem of juvenile crime. Although incidents such as these are not representative of a typical homicidal event, in 1996 15 percent of those arrested for murder were under age 18 (*Statistical Abstract of the United States: 1998,* Table 355).

Crime and violence are among Americans' foremost concerns. In a 1998 Gallup Poll 20 out of every 100 Americans responded that crime ". . . is the most important problem facing this country today" (Gallup Poll 1998). Such apprehensions are warranted. In 1995, the chance of being a victim of a serious crime was 1 in 19 (Reiman 1998, 15).

This chapter examines crime rates as well as theories, types, and demographic patterns of criminal behavior. The economic, social, and psychological costs of crime and violence are also examined. The chapter concludes with a discussion of social policies and prevention programs designed to reduce crime and violence in America.

. . . crime is no longer bound by the contraints of borders. Such offenses as terrorism, nuclear smuggling, organized crime, computer crime and drug trafficking can spill over from other countries into the United States. Regardless of origin, these and other oversea crimes impact directly on our citizens and our economy.

–Louis Freeh
Director, FBI

THE GLOBAL CONTEXT: INTERNATIONAL CRIME AND VIOLENCE

Crime and violence rates vary dramatically by country. Data indicate that the United States has the highest rate of violent crime of all industrialized nations. The rate of criminal homicide, for example, is 14 times higher in the United States than in other industrialized countries (Zimring and Hawkins 1997). It is important to note, however, that the incidences of violent *but nondeadly* crimes such as robbery and burglary are *not* higher than in other Western nations. Such comparisons have led Zimring and Hawkins (1997) to conclude that "crime is not the problem" in the United States—lethal violence is.

International Data

Personal force without a weapon is used in 81 percent of assaults in London; assaults without a weapon occur in only 13 percent of New York City's reported cases.

SOURCE: Zimring and Hawkins 1997

The nature of crime and violence vary from country to country. Russian rubles, arms, and precious metals are smuggled out of that country daily; Chinese Triads operate in large cities worldwide netting billions of dollars a year from prostitution, drugs, and other organized crime activities; Columbian cocaine cartels flourish and spread to sub-Saharan countries with needy economies; and Nigerian crime syndicates continue to defraud victims of billions of dollars through the promise of high returns on bogus investments (United Nations 1997; INTERPOL 1998). Because of the unprecedented increase in international crime, in 1998 President Clinton introduced to Congress the International Crime Control Act of 1998. The bill, designed to "deter and punish international crime, to protect United States nationals and

United Nations data indicate that the United States has one of the highest rates of violent crime in the world. Ironically, such data also indicated that Americans are also the most likely to believe the police are doing a good job in fighting crime.

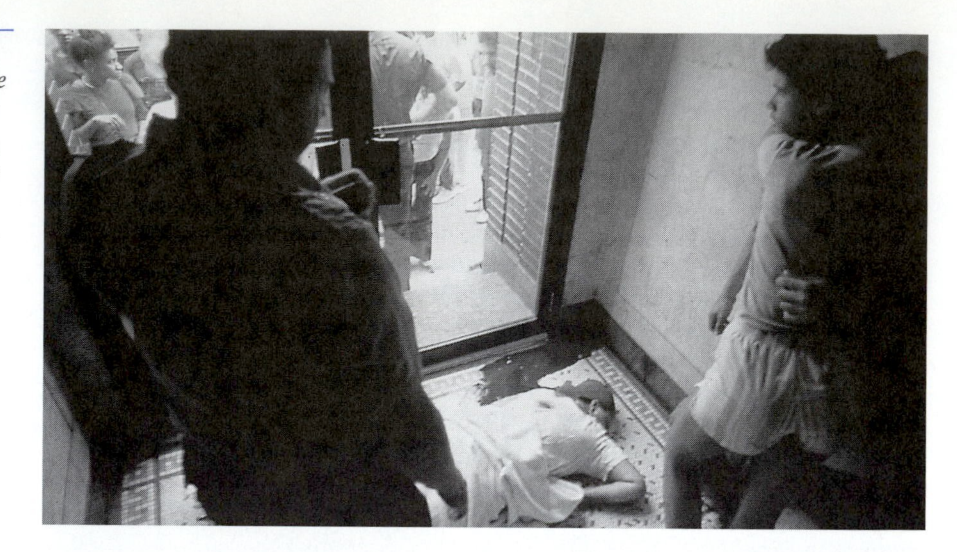

interests at home and abroad, and to promote global cooperation against international crime" (ICCA 1998, 1), contains eight objectives (The White House 1998, 12–13):

- Extending the first line of defense against international crime beyond U.S. borders
- Protecting U.S. borders by attacking smuggling and smuggling-related crimes
- Denying a "safe haven" for international criminals
- Fighting international financial crime (e.g. money laundering)
- Preventing criminal exploitation of international trade
- Responding to emerging (e.g. computer crime, trafficking in human beings) international crime threats
- Fostering a spirit of international cooperation and rule of law
- Optimizing international crime fighting efforts

International Data

Despite its small size, in just 9 months in 1998 drug control agents in Nepal arrested over 1,000 people, and seized 5,500 lbs. of hashish, 3,000 lbs. of cannabis, and 12 lbs. of heroin.

SOURCE: Shabad 1999

SOURCES OF CRIME STATISTICS

The U.S. government spends millions of dollars annually compiling and analyzing crime statistics. A **crime** is a violation of a federal, state, or local criminal law. For a violation to be a crime, however, the offender must have acted voluntarily and with intent and have no legally acceptable excuse (such as insanity) or justification (e.g., self-defense) for the behavior. The three major types of statistics used to measure crime are official statistics, victimization surveys, and self-report offender surveys.

Ultimately, any crime statistic is only as useful as the reader's understanding of the processes that generated it.

—Robert M. O'Brien
Sociologist, University of Oregon

Official Statistics

Local sheriffs' departments and police departments throughout the United States collect information on the number of reported crimes and arrests and voluntarily report them to the Federal Bureau of Investigation (FBI). The FBI then compiles these statistics annually and publishes them, in summary form, in the Uniform Crime Reports (UCR). The UCR lists **crime rates** or the number of crimes committed per 100,000 population, as well as the actual number of crimes and the percentage of change over time.

These statistics have several shortcomings (DiIulio 1999). Not only do many incidents of crime go unreported, not all crimes reported to the police are recorded. Alternatively, some rates may be exaggerated. Motivation for such distortions may come from the public (e.g., demanding that something be done) or from political (e.g., election of a sheriff) and/or organizational pressures (e.g., budget requests). For example, a police department may "crack down" on drug-related crimes in an election year. The result is an increase in the recorded number of these offenses. Such an increase reflects a change in the behavior of law enforcement personnel, not a change in the number of drug violations. Thus, official crime statistics may be a better indicator of what police are doing than what criminals are doing.

Victimization Surveys

Victimization surveys ask people if they have been victims of crime. The Department of Justice's National Crime Victimization Survey (NCVS), conducted annually, interviews nearly 83,000 people about their experiences as victims of crime. Interviewers collect a variety of information, including the victim's background (e.g., age, race and ethnicity, sex, marital status, education, and area of residence), relationship to offender (stranger or nonstranger), and the extent to which the victim was harmed. Although victimization surveys provide detailed information about crime victims, they provide less reliable data on offenders.

Self-Report Offender Surveys

Self-report surveys ask offenders about their criminal behavior. The sample may consist of a population with known police records, such as a prison population, or it may include respondents from the general population, such as college students. Self-report data compensate for many of the problems associated with official statistics but are still subject to exaggerations and concealment. The Criminal Activities Survey (see this chapter's *Self & Society*) asks you to indicate whether you have engaged in a variety of illegal activities.

CONSIDERATION

Self-report surveys reveal that virtually every adult has engaged in some type of criminal activity. Why then is only a fraction of the population labeled as criminal? Like a funnel, which is large at one end and small at the other, only a small proportion of the total population of law violators is ever convicted of a crime. For an individual to be officially labeled as a criminal, his or her behavior must first be observed or known to have occurred. Next, the crime must be reported to the police. The police must then respond to the report, find sufficient evidence to substantiate that a crime has taken place, file an official report, and have enough evidence to make an arrest. The arrestee must then go through a preliminary hearing, an arraignment, and a trial and may or may not be convicted. At every stage of the process, an offender may be "funneled" out. For example, for every 1,000 burglaries, only 495 are reported to the police, 72 lead to an arrest, 21 to a conviction, and 16 to an offender being incarcerated (Felson 1998, 7). As Figure 4.1 indicates, the measures of crime used at various points in time lead to different results.

Criminal Activities Survey

Read each of the following questions. If, since the age of 16, you have ever engaged in the behavior described, place a "1" in the space provided. If you have not engaged in the behavior, put a "0" in the space provided. After completing the survey, read the section on interpretation to see what your answers mean.

Questions	1 (Yes)	0 (No)
1. Have you ever been in possession of drug paraphernalia?	___	___
2. Have you ever failed to return a rented automobile on time?	___	___
3. Have you ever lied about your age, or about anything else when making application to rent an automobile?	___	___
4. Have you ever intentionally destroyed or erased someone else's phone messages?	___	___
5. Have you ever tampered with a coin-operated vending machine or parking meter?	___	___
6. Have you ever fired a gun in a public place or right-of-way?	___	___
7. Have you ever loaned or given away lewd or obscene materials?	___	___
8. Have you ever driven with a revoked, suspended, or canceled driver's licence?	___	___
9. Have you ever begun and/or participated in an office basketball or football pool?	___	___
10. Have you ever provided false information to a police officer?	___	___
11. Have you ever bought beer for someone under the age of 16?	___	___
12. Have you ever received cable services without paying for them?	___	___
13. Have you ever carried a concealed weapon?	___	___
14. Have you ever given a cigarette to someone under the age of 18?	___	___
15. Have you ever used "filthy, obscene, annoying, or offensive" language while on the telephone?	___	___
16. Have you ever driven on a beach on which it was prohibited?	___	___
17. Have you ever signed a fictitious name to a petition and/or signed it more than once?	___	___
18. Have you ever given or sold a beer to someone under the age of 21?	___	___
19. Have you ever been on someone else's property (land, house, boat, structure, etc.) without their permission?	___	___
20. Have you ever forwarded a chain letter with the intent to profit from it?	___	___
21. Have you ever recorded video images (for example, movies off of television) or audio sounds (for example, music off of the radio) without the permission of the copyright owner?	___	___
22. Have you ever improperly gained access to someone else's e- mail or other computer account?	___	___

INTERPRETATION

Each of the activities described in these questions represents criminal behavior that was subject to fines, imprisonment, or both under the laws of Florida in 1998. For each activity, the following table lists the maximum prison sentence and/or fine for a first-time offender. To calculate your "prison time" and/or fines, sum the numbers corresponding to each activity you have engaged in.

Criminal Activities Survey—cont'd

Maximum Prison Sentence	Maximum Fine	Offense
1. One year	$1000	Possession of drug paraphernalia
2. Five years	$5000	Unauthorized use of vehicle
3. Five years	$5000	Fraud
4. One year	$1000	Unlawful interference with telecommunications
5. Two months	$500	Fraud
6. One year	$1000	Unlawful discharge of a firearm
7. One year	$1000	Unlawful dissemination of obscene material
8. Two months	$500	Illegal use of driver's license
9. Five years	$5000	Illegal gambling
10. One year	$1000	Obstruction of justice
11. One year	$1000	Illegal distribution of alcohol
12. One year	$1000	Fraud
13. One year	$1000	Illegal possession of a firearm
14. Two months	$500	Illegal possession of tobacco
15. Two months	$500	Harassing/obscene telecommunications
16. Two months	$500	Environmental vandalism
17. One year	$1000	Fraud
18. Two months	$500	Illegal distribution of alcohol
19. One year	$1000	Trespassing
20. One year	$1000	Illegal gambling
21. One year	$25,000	Fraudulent use of copyrighted material
22. Five years	$5000	Illegal misappropriation of cybercommunication

SOURCE: Florida Criminal Code. 1997. **http://www.leg.state.fl.us./citizen/documents/statutes**

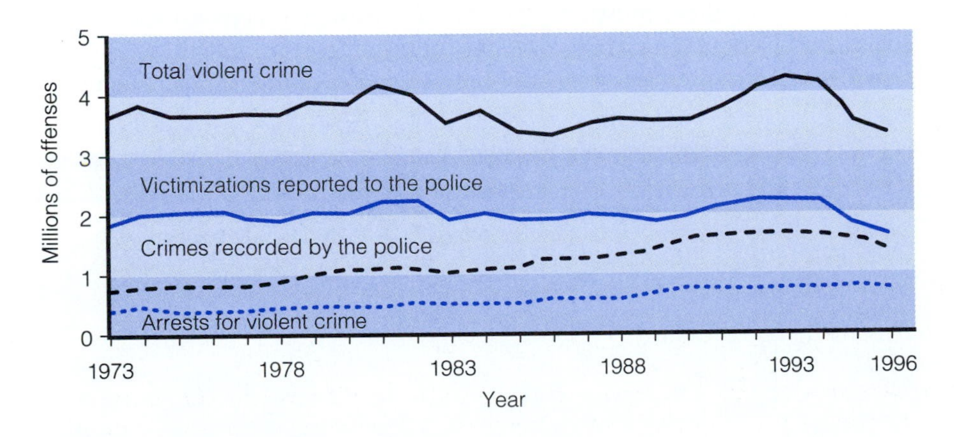

Figure 4.1 **Four Measures of Serious Violent Crime**

SOURCE: U.S. Department of Justice. 1997. Bureau of Justice Statistics. "Serious violent crime levels declined between 1995 and 1996." **http://www.ojp.usdo;.gov/bjs/glance/cv2.html**

SOCIOLOGICAL THEORIES OF CRIME AND VIOLENCE

Some explanations of crime and violence focus on psychological aspects of the offender such as psychopathic personalities, unhealthy relationships with parents, and mental illness. Other crime theories focus on the role of biological variables such as central nervous system malfunctioning, vitamin or mineral deficiencies, chromosomal abnormalities, and a genetic predisposition toward aggression (see this chapter's *Focus on Technology*). Sociological theories of crime and violence emphasize the role of social factors in criminal behavior and societal responses to it.

Structural-Functionalist Perspective

> Poverty is the mother of crime.
>
> —Magnus Aurelius Cassiodorus
> Roman historian

According to Durkheim and other structural-functionalists, crime is functional for society. One of the functions of crime and other deviant behavior is that it strengthens group cohesion:

> **The deviant individual violates rules of conduct which the rest of the community holds in high respect; and when these people come together to express their outrage over the offense . . . they develop a tighter bond of solidarity than existed earlier. (Erikson 1966, 4)**

Crime may also lead to social change. For example, an episode of local violence may "achieve broad improvements in city services . . . [and] be a catalyst for making public agencies more effective and responsive, for strengthening families and social institutions, and for creating public-private partnerships" (National Research Council 1994, 9–10).

While functionalism as a theoretical perspective deals directly with some aspects of crime and violence, it is not a theory of crime per se. Three major theories of crime and violence have developed from functionalism, however. The first, called **strain theory,** was developed by Robert Merton (1957) using Durkheim's concept of anomie, or normlessness. Merton argues that when legitimate means (for example, a job) of acquiring culturally defined goals (for example, money) are limited by the structure of society, the resulting strain may lead to crime.

Individuals, then, must adapt to the inconsistency between means and goals in a society that socializes everyone into wanting the same thing but only provides opportunities for some (see Table 4.1). Conformity occurs when indi-

Table 4.1 Merton's Five Types of Adaptation

	Culturally Defined Goals	Structurally Defined Means
1. Conformity	+	+
2. Innovation	+	−
3. Ritualism	−	+
4. Retreatism	−	−
5. Rebellion	−/+	−/+

Key: + = acceptance of/access to; − = rejection of/lack of access to; −/+ = rejection of culturally defined goals and structurally defined means and replacement with new goals and means.

SOURCE: Robert K. Merton. 1957. Reprinted with permission of The Free Press, a division of Simon & Schuster from *Social Theory and Social Structure* by Robert K. Merton. Copyright © 1957 by The Free Press: copyright renewed 1985 by Robert K. Merton.

> Every time you stop a school, you will have to build a jail. What you gain at one end you lose at the other. It's like feeding a dog on its own tail. It won't fatten the dog.
>
> —Mark Twain
> American author

FOCUS ON TECHNOLOGY

DNA EVIDENCE

In 1954, Sam Sheppard, a prominent Cleveland physician, was accused of killing his wife, Marilyn Sheppard. Over 40 years later this case was the basis for the movie *The Fugitive,* and the drama carries on today as Sam Sheppard Jr. continues to try and clear his father's name. This time, however, he is armed with DNA evidence—evidence that suggests that someone other than his father may have killed Marilyn Sheppard (Parker 1998).

Increasingly, law enforcement officers both in the United States and in Europe are using what is called DNA fingerprinting in the identification of criminal suspects. DNA stands for deoxyribonucleic acid, which is found in the nucleus of every cell and contains an individual's complete and unique genetic makeup. Developed in the mid-1980s, DNA fingerprinting is a general term used to describe the process of analyzing and comparing DNA from different sources, including evidence found at a crime scene, for example, blood, semen, hair, saliva, fibers, and skin tissue, and the DNA of a suspect.

Since 1988, the FBI crime lab's DNA analysis unit has tested over 12,000 suspects, of which 3,000, or one-fourth, have been exonerated by the DNA evidence (Willing 1997). Despite these success stories, however, there is tremendous concern about the use of DNA evidence and, specifically, questions about how donors would be selected and what methods of data collection would be used.

Libertarians fear possible violations of civil rights, particularly after a recent court decision by a French judge who ruled that all men in a village would be DNA tested in order to locate the killer of a 13-year-old girl who had been murdered. In Germany, DNA screening of U.S. soldiers led to a rape/murder conviction and in England police officials legally take samples of blood or skin tissue from every criminal suspect (Gleick 1997, 3). That is why, until recently, England had the only nationwide DNA data bank in the world (Goldberg 1998; Gleick 1997).

In the fall of 1998, the FBI initiated the Combined DNA Index System. This database contains the DNA fingerprints of 250,000 convicted felons and over 4,600 DNA samples from unsolved crime scenes. Despite the database being in operation less than a year, the "FBI claims that 200 outstanding cases have already been solved" (Kluger 1999, 1). Concerns over the Fourth Amendment's prohibition against unreasonable search and seizure, however, remain.

Regardless of the legal battles, the future of DNA fingerprinting is likely to be bright. It's less expensive than ever before, predicted to be as low as $10 a test within a few years compared to earlier costs of $200 to $300. And as technology has become more and more sophisticated, the time it takes to conduct the analysis has decreased from weeks to days (Goldberg 1998). Further, even if it doesn't survive the legal scrutiny it's likely to come under, it remains a valuable identification technique used in biology, archeology, medical diagnosis, paleontology, and forensics.

SOURCES: Gleick, Elizabeth. 1997. "The Killer Left a Trace." *Time,* September 1, 150.
Goldberg, Carey. 1998. "DNA Databanks Giving Police a Powerful Weapon, and Critics." *New York Times,* February 19, 1.
Kluger, Jeffery. 1999. "DNA Detectives." *Time Daily* 153(1):1. **http://www.time.com**
Parker, Laura. 1998. "Tests Point to Sheppard's Innocence." *USA Today,* March 5, A3.
Willing, Richard. 1997. "Tests Exonerate One in Four Suspects." *USA Today,* November 28, A3.

viduals accept the culturally defined goals and the socially legitimate means of achieving them. Merton suggests that most individuals, even those who do not have easy access to the means and goals, remain conformists. Innovation occurs when an individual accepts the goals of society, but rejects or lacks the socially legitimate means of achieving them. Innovation, the mode of adaptation most associated with criminal behavior, explains the high rate of crime committed by uneducated and poor individuals who do not have access to legitimate means of achieving the social goals of wealth and power.

Another adaptation is ritualism, in which the individual accepts a lifestyle of hard work, but rejects the cultural goal of monetary rewards. The ritualist goes through the motions of getting an education and working hard, yet is not committed to the goal of accumulating wealth or power. Retreatism involves rejecting both the cultural goal of success and the socially legitimate means of achieving it. The retreatist withdraws or retreats from society and may become an alcoholic, drug addict, or vagrant. Finally, rebellion occurs when an individual rejects both culturally defined goals and means and substitutes new goals and means. For example, rebels may use social or political activism to replace the goal of personal wealth with the goal of social justice and equality.

While strain theory explains criminal behavior as a result of blocked opportunities, **subcultural theories** argue that certain groups or subcultures in society have values and attitudes that are conducive to crime and violence. Members of these groups and subcultures, as well as other individuals who interact with them, may adopt the crime-promoting attitudes and values of the group. For example, subcultural norms and values contribute to street crime. Sociologist Elijah Anderson (1994) explains that many inner city African-American youths live by a survival code on the streets that emphasizes gaining the respect of others through violence—the tougher you are and the more others fear you, the more respect you have in the community.

National Data

Between 1986 and 1996 the juvenile population increased 10 percent while the number of rapes committed by juveniles increased 17 percent, murder 39 percent, and aggravated assault 64 percent.

SOURCE: DiIulio 1999

But, if blocked opportunities and subcultural values are responsible for crime, why don't all members of the affected groups become criminals? **Control theory** may answer that question. Hirschi (1969), consistent with Durkheim's emphasis on social solidarity, suggests that a strong social bond between individuals and the social order constrains some individuals from violating social norms. Hirschi identified four elements of the social bond: attachment to significant others, commitment to conventional goals, involvement in conventional activities, and belief in the moral standards of society. Several empirical tests of Hirschi's theory support the notion that the higher the attachment, commitment, involvement, and belief, the higher the social bond and the lower the probability of criminal behavior. For example, Laub, Nagin, and Sampson (1998) found that a good marriage contributes to the cessation of a criminal career. Further, Warner and Rountree (1997) report that local community ties, athough varying by neighborhood and offense, decrease the probability of crimes occurring.

Conflict Perspective

Conflict theories of crime suggest that deviance is inevitable whenever two groups have differing degrees of power; in addition, the more inequality in a society, the greater the crime rate in that society. Social inequality leads individuals to commit crimes such as larceny and burglary as a means of economic survival. Other individuals, who are angry and frustrated by their low position in the socioeconomic hierarchy, express their rage and frustration through crimes such as drug use, assault, and homicide.

> There are two criminal justice systems in this country. There is a whole different system for poor people. It's the same courthouse—it's not separate—but it's not equal.
>
> —Paul Petterson
> Public defender

According to the conflict perspective, those in power define what is criminal and what is not, and these definitions reflect the interests of the ruling class. Laws against vagrancy, for example, penalize individuals who do not contribute to the capitalist system of work and consumerism. Rather than viewing law as a mechanism that protects all members of society, conflict theorists focus on how laws are created by those in power to protect the ruling class. While wealthy corporations contribute money to campaigns to influence politicians to enact tax laws that serve corporate interests (Jacobs 1988), the "criminal justice system grows increasingly punitive as labor surplus increases," that is, as greater social control is needed (Hochstetler and Shover, 1997).

Furthermore, conflict theorists argue that law enforcement is applied differentially, penalizing those without power and benefitting those with power. For example, female prostitutes are more likely to be arrested than are the men who seek their services. Unlike street criminals, corporate criminals are often punished by fines rather than by lengthy prison terms, and rape laws originated to serve the interests of husbands and fathers who wanted to protect their property—wives and unmarried daughters.

Societal beliefs also reflect power differentials. For example, "rape myths" are perpetuated by the male-dominated culture to foster the belief that women are to blame for their own victimization, thereby, in the minds of many, exonerating the offender. Such myths include the notion that when a woman says "no" she means "yes," that "good girls" don't get raped, that appearance indicates willingness, and that women secretly want to be raped. Not surprisingly, in societies where women and men have greater equality, there is less rape (Sanday 1981).

Symbolic Interactionist Perspective

Two important theories of crime and violence emanate from the symbolic interactionist perspective. The first, **labeling theory**, focuses on two questions: How do crime and deviance come to be defined as such, and what are the effects of being labeled as criminal or deviant? According to Howard Becker (1963):

> **Social groups create deviance by making rules whose infractions constitute deviance, and by applying those rules to particular people and labeling them as outsiders. From this point of view, deviance is not a quality of the act a person commits, but rather a consequence of the application by others of rules and sanctions to an "offender." The deviant is one to whom the label has successfully been applied; deviant behavior is behavior that people so label. (p. 238)**

Labeling theorists make a distinction between primary deviance, which is deviant behavior committed before a person is caught and labeled as an offender, and secondary deviance, which is deviance that results from being caught and labeled. After a person violates the law and is apprehended, that person is stigmatized as a criminal. This deviant label often dominates the social identity of the person to whom it is applied and becomes the person's "master status," that is, the primary basis on which the person is defined by others.

Being labeled as deviant often leads to further deviant behavior because (1) the person who is labeled as deviant is often denied opportunities for engaging in nondeviant behavior, and (2) the labeled person internalizes the deviant label, adopts a deviant self-concept, and acts accordingly. For example, the teenager who is caught selling drugs at school may be expelled and thus denied opportunities to participate in nondeviant school activities (e.g., sports, clubs) and associate with nondeviant peer groups. The labeled and stigmatized teenager may also adopt the self-concept of a "druggie" or "pusher" and continue to pursue drug-related activities and membership in the drug culture.

The assignment of meaning and definitions learned from others are also central to the second symbolic interactionist theory of crime, **differential association.** Edwin Sutherland (1939) proposed that, through interaction with others, individuals learn the values and attitudes associated with crime as well as the techniques and motivations for criminal behavior. Individuals who are exposed to more definitions favorable to law violation (e.g., "crime pays") than unfavorable (e.g., "do the crime, you'll do the time") are more likely to engage in criminal behavior. Thus, children who see their parents benefit from crime, or who live in high crime neighborhoods where success is associated with illegal behavior, are more likely to engage in criminal behavior.

TYPES OF CRIME

The FBI identifies eight **index offenses** as the most serious crimes in the United States. The index offenses, or street crimes as they are often called, may be against a person (called violent or personal crimes) or against property (see Table 4.2). Other types of crime include vice crime, such as drug use, gambling, and prostitution, as well as organized crime, white-collar crime, computer crime, and juvenile delinquency.

Street Crime: Violent Offenses

In general, violent crime has increased in recent decades although, according to the NCVS, it has decreased since 1994 (NCVS 1997; SSBR 1999). Violent crime includes homicide, assault, rape, and robbery.

Homicide refers to the willful or nonnegligent killing of one human being by another individual or group of individuals. Although homicide is the most serious of the violent crimes, it is also the least common accounting for less than 2 percent of the violent index crimes in 1996 (*Statistical Abstract of the*

International Data

Homicide and other aggressive crimes are more common in countries with high beer consumption such as the United States, Canada, New Zealand, Poland, Germany, the Netherlands, Finland, and Australia.

SOURCE: Russell 1995

Table 4.2 Index Crimes Known to the Police

	Rate per 100,000		
	1986	1996	Percentage Change in Rate (1986–1996)
Violent crime			
Murder	8.6	7.4	−14.0%
Forcible rape	37.9	36.1	−4.7%
Robbery	225.1	202.4	−10.2%
Aggravated assault	346.1	388.2	+12.2%
Total	617.7	634.2	+2.7%
Property crime*			
Burglary	1,344.6	943.0	−29.9%
Larceny/theft	3,010.3	2,975.9	−1.1%
Motor vehicle theft	507.8	525.9	+3.6%
Total	4,862.6	4,444.8	−8.6%

*Sufficient data are not available to estimate totals for arson.

SOURCE: Adapted from *Statistical Abstract of the United States: 1998,* 118th ed. U.S. Bureau of the Census. Washington, D.C.: U.S. Government Printing Office, Table 335.

United States: 1998, Table 335). The majority of homicide victims had some type of relationship with their murderer and were killed over such issues as jealousy and money.

In most murders, the victim is the same race as the offender. Ninety-four percent of black victims are killed by other blacks, and 84 percent of white victims are killed by other whites (FBI 1995). The rate of victimization of young black men is over seven times higher than that of young white men. In 1996, the rate of victimization for African-American males was 56.3 compared to 7.8 for white males (*Statistical Abstract of the United States: 1998,* Table 142).

Another form of violent crime, aggravated assault, involves the attacking of another with the intent to cause serious bodily injury. Like homicide, aggravated assault occurs most often between members of the same race. Unlike homicide, however, aggravated assault is more likely to take place between strangers than between acquaintances and is a fairly common occurrence. In 1996, the assault rate was over 50 times greater than the murder rate—assaults comprised 61 percent of all violent crime.

Rape is also classified as a form of violent crime and is also intraracial, that is, the victim and offender are from the same racial group. The FBI definition of rape contains three elements: sexual penetration, force or the threat of force, and nonconsent of the victim. In 1996, more than 95,000 forcible rapes were reported in the United States, a 4 percent decline from the previous year (*Statistical Abstract of the United States: 1998,* Table 335). Given the problems with official statistics, however, this number may be less than 10 percent of the actual number of rapes that occurred that year.

Perhaps as many as 80 percent of all rapes are **acquaintance rapes**—rapes committed by someone the victim knows. While acquaintance rapes are the most likely to occur, they are the least likely to be reported and the most difficult to prosecute. Unless the rape is what Williams (1984) calls a "**classic rape**"—that is, the rapist was a stranger who used a weapon, and the attack resulted in serious bodily injury—women hesitate to report the crime out of fear of not being believed. The increased use of "rape drugs" such as Rohypnol, which renders its victims unconscious, may lower reporting levels even further. In 1996 a federal law was passed that added 20 years in prison to an offender using drugs to incapacitate a victim (AP 1996).

Robbery, although involving theft, also involves force or the threat of force, or putting a victim in fear, and is thus considered a violent crime. Officially, in 1996 more than 537,000 robberies took place in the United States (*Statistical Abstract of the United States: 1998,* Table 335). However, the National Crime Victimization Survey indicates that only around half of all robberies are reported (Perkins and Klaus 1996). Robberies are committed by young people, with over two-thirds of all robberies committed by those under the age of 25 (FBI 1995). As with rape, victims who resist a robbery are more likely to stop the crime, but are also more likely to be physically harmed.

National Data

The recent decrease in homicide rates is greatest in cities with a population of 1 million or more, falling from 35.5 to 20.3 per 100,000 between 1991 and 1997.

SOURCE: Price 1999

International Data

The results of interviews with a random sample of women in Toronto, Canada (N = 420), suggest that over 50 percent of women have been the victims of forced, or attempted forced, sexual intercourse.

SOURCE: Randall and Haskell 1995

Street Crime: Property Offenses

Property crimes are those in which someone's property is damaged, destroyed, or stolen; they include larceny, motor vehicle theft, burglary, and arson. Property crimes have declined in recent years with a 7 percent decrease between 1996 and 1997 (SSBR 1999). Larceny, or simple theft, is the most common property crime comprising over half of all property arrests in 1996 (see Table 4.2).

Larceny involving automobiles and auto accessories is the largest single category of thefts, but because of the cost involved, theft of a motor vehicle is considered a separate index offense. In 1996, nearly 1.4 million auto thefts took place in the United States (*Statistical Abstract of the United States: 1998,* Table 335). Because twice as many cars are stolen to be used temporarily as are sold for economic gain, most stolen vehicles are recovered (Clarke and Harris 1992).

Burglary, which is the second most common index offense, entails entering a structure, usually a house, with the intent to commit a crime while inside. Official statistics indicate that 2,502,000 burglaries were reported to the police in 1996 (*Statistical Abstract of the United States: 1998,* Table 335). Interestingly, whether or not the burglary is reported to the police depends on the type of entry. When forced entry occurs (breaking a window or a door), 77 percent of the burglaries are reported; when unlawful entry takes place (going through an open window or door), only 43 percent are reported (Barkan 1997).

Arson involves the malicious burning of the property of another. Estimating the frequency and nature of arson is difficult given the legal requirement of "maliciousness." Of the reported cases of arson, 53 percent involved structures (the majority of which were residential), 26 percent movable property (e.g., boat, car), and the remainder were miscellaneous personal items (FBI 1996).

International Data

In England, over 90 percent of the total number of crimes committed are considered "offenses against property," including burglary, theft, criminal damages, and fraud.

SOURCE: U.S. Department of Justice 1998

Vice Crimes

Vice crimes are illegal activities that have no complaining party and are therefore often called **victimless crimes.** Vice crimes include using illegal drugs, engaging in or soliciting prostitution (except for legalized prostitution, which exists in Nevada), illegal gambling, and pornography.

Most Americans view drug use as socially disruptive. In a 1998 poll, drugs were listed as one of the most significant problems facing America today. (Gallup Poll 1998). There is less consensus, nationally or internationally, that gambling and prostitution are problematic. In the Netherlands the Prostitution Information Centre in Amsterdam offers a 6-day course on "prostitution as a career-option," and the Australia Council of Trade Unions recently recognized women in prostitution as a labor sector (CATW 1997). In the United States, many states have legalized gambling, including casinos in Nevada, New Jersey, Connecticut, and many other states, as well as state lotteries, bingo parlors, horse and dog racing, and jai alai. Further, some have argued that there is little difference, other than societal definitions of acceptable and unacceptable behavior, between gambling and other risky ventures such as investing in the stock market. Conflict theorists are quick to note that the difference is who's making the wager.

National Data

Vice crime arrests in 1996 were most common for drug violations, followed by arrests for commercialized vice and prostitution, sex offenses, and gambling.

SOURCE: *Statistical Abstract of the United States: 1998,* Table 354

Organized crime refers to criminal activity conducted by members of a hierarchically arranged structure devoted primarily to making money through illegal means. Many of these activities include providing illegal goods and services such as drugs and gambling, as well as the coercion of legitimate businesses for profit. For example, organized crime groups may force legitimate businesses to pay "protection money" by threatening vandalism or violence.

The traditional notion of organized crime is the Mafia—a national band of interlocked Italian families—but members of many ethnic groups engage in organized crime. In recent years Asian organized crime groups have emerged as one of the most violent and economically successful groups in the United States dealing primarily in kidnaping, gambling, drug trafficking, counterfeiting, and prostitution (Lindberg et al., 1997).

International Data

It is estimated that 10 percent of the female population of Thailand are prostitutes; further, over 25,000 girls and women are imported annually from Burma, China, and the former Soviet Union to work in Bangkok's sex industries.

SOURCE: Leidholdt 1997

Organized crime also occurs at the international level, such as smuggling illegal drugs and arms. Since the fall of the Soviet Union, it is estimated that as much as 25 percent of the Russian gross national income is from organized crime activities generated by 5,600 separate crime groups. Today, organized crime is considered "... one the most important political problems in Russia ..." whereby a "... Russian citizen's personal safety" can no longer be guaranteed (Shabalin, Alibini, and Rogers 1995).

> ... organized crime in Russia threatens not only the safety of this country but the safety of the U.S.A.
>
> —Louis Freeh
> Director, FBI

White-Collar Crime

White-collar crime includes both occupational crime, where individuals commit crimes in the course of their employment, and corporate crime, where corporations violate the law in the interest of maximizing profit. Occupational crime is motivated by individual gain. Employee theft of merchandise (pilferage) or money (embezzlement) are examples of occupational crime. Price-fixing, antitrust violations, and fraud are examples of corporate crime, that is, crime that benefits the organization.

National Data

In 1992, Charles Keating was sentenced to 10 years for 18 counts of securities fraud; the savings and loan scandal cost American taxpayers over $2.6 billion in reimbursements to bank depositors.

SOURCE: Reiman 1998

Corporate violence, a form of corporate crime, refers to the production of unsafe products and the failure of corporations to provide a safe working environment for their employees. Corporate violence is the result of negligence, the pursuit of profit at any cost, and intentional violations of health, safety, and environmental regulations. General Motors (GM) C-K model pickup trucks built between 1973 and 1987 were fitted with side-mounted fuel tanks. Unlike other vehicles, the fuel tanks were not protected by the frame, resulting in what consumer groups have called the "the most lethal auto defect in U.S. history" (Zagaroli 1997). While it is estimated that over 800 people have burned to death in side-impact collisions as a result of the defect, over 4 million trucks remain on the road. The National Highway Traffic Safety Administration agreed not to recall the trucks after GM agreed to fund a $51 million auto safety program (AP 1998).

Residents of Love Canal were victimized by corporate violence when toxic waste, dumped by the Hooker Chemical Company, began to seep into the basements of homes and schools. Many residents of Love Canal moved out of the area; others complained of high rates of miscarriages, birth defects, and cancer. Although Love Canal was allegedly cleaned up, residents protested against resettling the area. After 20 years of litigation, the last suit was settled in 1998 with the City of Love Canal receiving $250,000 and the construction of a park near the site of the spill.

> Although "street crimes" such as robbery and murder may spark the greatest fear in the public, and gain the most attention from the media, the damage they do is dwarfed by that of "respectable" criminals who wear white collars and work for our most powerful organizations.
>
> —James W. Coleman
> Sociologist

Table 4.3 Types of White-Collar Crime

Crimes against Consumers	Crimes against Employees
Deceptive advertising	Health and safety violations
Antitrust violations	Wage and hour violations
Dangerous products	Discriminatory hiring practices
Manufacturer kickbacks	Illegal labor practices
Physician insurance fraud	Unlawful surveillance practices
Crimes against the Public	**Crimes against Employers**
Toxic waste disposal	Embezzlement
Pollution violations	Pilferage
Tax fraud	Misappropriation of government funds
Security violations	Counterfeit production of goods
Police brutality	Business credit fraud

The National Commission on Product Safety estimates that there are 110,000 permanently disabling injuries, 30,000 deaths, and millions of serious injuries each year as a result of unsafe consumer products (Barkan 1997, 336). In many cases, corporate executives are aware that the product is unsafe. One example is the Ford Pinto, which had a defective gas tank that exploded in rear-end collisions. Hundreds of burn deaths occurred before the problem was revealed to the public. Yet the Ford Motor Company knew of the dangers of the car and made a profit-motivated decision to do nothing:

> **The $11 repairs for all Pintos would cost $137 million but 180 burn deaths and 180 serious burn injuries and 2100 burned vehicles would cost only $49.5 million (each death was figured at $200,000, and each injury at $67,000). Therefore, the company could anticipate a savings or profit [by doing nothing] of $87.5 million. (Hills 1987, 425)**

Table 4.3 summarizes some of the major categories of white-collar crime.

Computer Crime

Computer crime refers to any violation of the law in which a computer is the target or means of criminal activity. Seventy to 90 percent of all cases of computer crime are committed by employees (Albanese and Pursley 1993; Lewis 1998). Conklin (1998) has identified several examples of computer crime:

- In 1996 two individuals were charged with theft of 80,000 cellular phone numbers. Using a device purchased from a catalogue, the thieves picked up radio waves from passing cars, determined private cellular codes, reprogrammed computer chips with the stolen codes and then, by inserting the new chips into their own cellular phones, charged calls to the original owners.
- A programmer made $300 a week by programming a computer to round off each employee's paycheck down to the nearest 10¢ and then to deposit the extra few pennies in the offender's account.

- An oil company illegally tapped into another oil company's computer to get information that allowed the offending company to underbid the other company for leasing rights.
- In 1995, Kevin Mitnick was arrested for breaking into an Internet provider computer system and stealing 20,000 credit card numbers.

CONSIDERATION

With the increased popularity and use of the Internet there has been an associated increase in the use of the "Net" to support criminal activity. In 1996 a 6-year-old California girl's accusations of sexual abuse led to an FBI investigation of the "Orchid Club," a chat room where people interested in child pornography shared pictures and other information. In July of 1996, 13 men were indicted, and arrest warrants were issued for several others in Finland, Canada, and Australia. Other crimes tied to the Internet include advertising of illegal products (e.g., prostitution, drugs, pornography), fraud, illegal distribution of copyrighted materials (e.g., books, tapes, compact discs (CDs), computer software), and illegal lotteries and gambling. (CATW 1997; Hughes 1997).

> In America today, no population poses a greater threat to public safety than juvenile criminals.
>
> —Bill McCollum
> U.S. Representative

Juvenile Delinquency

In general, children under the age of 18 are handled by the juvenile court either as status offenders or as delinquent offenders. A status offense is a violation that can only be committed by a juvenile, such as running away from home, truancy, and underage drinking. A delinquent offense is an offense that would be a crime if committed by an adult, such as the eight index offenses. The most common status offenses handled in juvenile court are underage drinking, truancy, and running away. In 1996, 146,000 juveniles were arrested for running away, the only status offense in which arrests of girls exceeded those of boys (*Statistical Abstract of the United States: 1998,* Table 355).

Juveniles commit more property offenses than violent offenses. Of all delinquency cases disposed by juvenile courts in 1996, over four times as many were for serious property crimes as for serious violent crimes (*Statistical Abstract of the United States: 1998,* Table 371). Murder is the least likely violent crime committed by a juvenile and robbery the most likely. Between 1985 and 1995 the rate of murder committed by teens increased 172 percent, and the rate of juvenile killings with a gun quadrupled (Fox 1997). Not surprisingly, several states have introduced laws to toughen the treatment of youthful offenders, and the federal government is in the process of reforming the federal juvenile justice system.

National Data

In 1996, of all juveniles arrested for drug possession, possession of marijuana was the most common, followed by possession of heroin and cocaine, and possession of synthetic narcotics.

SOURCE: *Statistical Abstract of the United States: 1998,* Table 353

National Data

There are an estimated 23,388 youth gangs in the U.S. with a combined membership of 664,906 in all 50 states.

SOURCE: NIJ 1998

DEMOGRAPHIC PATTERNS OF CRIME

Although virtually everyone violates a law at some time, persons with certain demographic characteristics are disproportionately represented in the crime statistics. Victims, for example, are disproportionately young, lower-class, minority males from urban areas. Similarly, the probability of being an offender varies by gender, age, race, social class, and region.

National Data

The ratio of males to females for all arrests in 1996 was 8 male offenders to 2 female offenders; for murder, the ratio was 10:1, for theft, 7:3; embezzlement, 5:5; and for prostitution, 4:6.

SOURCE: *Statistical Abstract of the United States: 1998*, Table 335

Gender and Crime

Both official statistics and self-report data indicate that males commit more crime than females. Why are males more likely to commit crime than females? One explanation is that society views female lawbreaking as less acceptable and thus places more constraints on female behavior: "women may need a higher level of provocation before turning to crime—especially serious crime. Females who choose criminality must traverse a greater moral and psychological distance than males making the same choice" (Steffensmeier and Allan 1995, 88).

Further, data suggest that males and females tend to commit different types of crimes. Men, partly because of more aggressive socialization experiences, are more likely than women to commit violent crimes. Men who kill outnumber women by 10 to 1 (*Statistical Abstract of the United States: 1998*, Table 355.). Females are less likely than males to commit serious offenses, and the monetary value of female involvement in theft, property damage, and illegal drugs is typically less than that for similar offenses committed by males. Nevertheless, a growing number of women have become involved in characteristically male criminal activities such as gang-related crime and drug use.

CONSIDERATION

Feminist criminology focuses on how the subordinate position of women in the social structure impacts female criminality. For example, violent female offenders, although few in number, are more likely to be married and less likely to have a history of offending than their male counterparts. This suggests the possibility that violent female offenders may be recruited into crime by their husbands (Barlow 1993). Medna Chesney-Lind (1996) reports that arrest rates for runaway juvenile females are higher than for males not only because they are more likely to run away as a consequence of sexual abuse in the home, but also because police with paternalistic attitudes are more likely to arrest female runaways than male runaways. Feminist criminology thus adds insights into understanding crime and violence often neglected by traditional theories by concentrating on gender inequality in society.

Females who join gangs often do so to win approval from their boyfriends who are gang members; increasingly, however, females are forming independent "girl gangs."

Age and Crime

In general, criminal activity is more prevalent among younger persons than older persons. The highest arrest rates are for individuals under the age of 25. Crimes committed by people in their teens or early 20s tend to involve property crimes like burglary, larceny, arson, vandalism, and liquor/drug violations. The median age of people who commit more serious crimes such as aggravated assault and homicide is in the late 20s (Steffensmeier and Allan 1995). However, given the increase of young people in the next decade, it is projected that by the year 2005, there will be a 25 percent increase in teen arrests for homicide (Heubusch 1997, 1).

Why is criminal activity more prevalent among individuals in their teens and early 20s? One reason is that juveniles are insulated from many of the legal penalties for criminal behavior. Younger individuals are also more likely to be unemployed or employed in low-wage jobs. Thus, as strain theorists argue, they have less access to legitimate means for acquiring material goods.

Some research suggests, however, that high school students who have jobs become more, rather than less, involved in crime (Felson 1998, 120). In earlier generations, teenagers who worked did so to support themselves and/or their families. Today, teenagers who work typically spend their earnings on recreation and "extras," including car payments and gasoline. The increased mobility associated with having a vehicle also increases opportunities for criminal behavior and reduces parental control.

National Data

While the total U.S. population will rise about 12 percent by 2005, the number of youths aged 15 to 19 will increase by 21 percent. Young African-American and Hispanic men—those with the highest violent crime rate—will increase 24 percent and 47 percent, respectively.

SOURCE: Gest and Friedman 1994

Race, Social Class, and Crime

Race is a factor in who gets arrested. Although whites represented two-thirds of those arrested for all criminal offenses in 1996, (*Statistical Abstract of the United States: 1998,* Table 354), minorities are overrepresented in official crime statistics. For example, although African-Americans represent about 13 percent of the population, they account for more than 30.9 percent of all arrests. Further, while minorities are underrepresented in white-collar crimes, often lacking the opportunity to commit such offenses, blacks are arrested for 35.1 percent of all index crimes (*Statistical Abstract of the United States: 1998,* Table 354).

Nevertheless, it is inaccurate to conclude that race and crime are causally related. First, official statistics reflect the behaviors and policies of criminal justice actors. Thus, the high rate of arrests, conviction, and incarceration of minorities may be a consequence of individual and institutional bias (e.g., police prejudice) not only against minorities, but against the lower class in general. Second, race and social class are closely related in that nonwhites are overrepresented in the lower classes. Since lower-class members lack legitimate means to acquire material goods, they may turn to instrumental, or economically motivated, crimes. Further, while the "haves" typically earn social respect through their socioeconomic status, educational achievement, and occupational role, the "have-nots" more often live in communities where respect is based on physical strength and violence, as subcultural theorists argue.

Thus, the apparent relationship between race and crime may, in part, be a consequence of the relationship between these variables and social class. Research indicates, however, that even when social class backgrounds remain the same between blacks and whites, blacks have higher rates of criminality, particularly for violent crime (Wolfgang, Figlio, and Sellin 1972; Elliot and Ageton 1980). Further, to avoid the bias inherent in official statistics, researchers have compared race, class, and criminality by examining self-report data and victim studies. Their findings indicate that while there are racial and class differences in criminal offenses, the differences are not as great as official data would indicate (U.S. Department of Justice 1993; Walker, Spohn, and Delone 1996; Hagan and Peterson, 1995).

National Data

In 1996, the violent crime rate in metropolitan areas was 715 per 100,000 compared with 222 per 100,000 in rural areas. The property crime rate in the same year was 4,798 in metropolitan areas compared to 1,828 in rural areas.

SOURCE: *Statistical Abstract of the United States: 1998,* Table 336

Region and Crime

Crime rates are higher in urban than suburban areas, although suburban crime is growing faster than crime in the cities (Conklin 1998). Higher crime rates in urban areas are due to several factors. First, social control is a function of small intimate groups socializing their members to engage in law-abiding behavior, expressing approval for their doing so and disapproval for their noncompliance. In large urban areas, people are less likely to know each other and thus are not influenced by the approval or disapproval of strangers. Demographic factors also explain why crime rates are higher in urban areas: cities have large concentrations of poor, unemployed, and minority individuals.

Property crime rates vary by region with the highest 1995 property crime rate in western states, followed by southern, midwestern, and northeastern states. The highest violent crime rate in 1995 was in the South, followed by the West, Northeast, and Midwest (*Statistical Abstract of the United States: 1997,* Table 315). The murder rate is particularly high in the South—32 percent higher than the rest of the country (Conklin 1998). The high rate of southern lethal violence has been linked to high rates of poverty and minority populations in the South, a southern "subculture of violence," higher rates of gun ownership in the South, and a warmer climate that facilitates victimization by increasing the frequency of social interaction.

COSTS OF CRIME AND VIOLENCE

Obviously crime pays, or there'd be no crime.

—G. Gordon Liddy
Radio personality

Crime and violence often result in physical injury and loss of life. In 1995, homicide was the tenth leading cause of death among Americans and was the second leading killer of those aged 15 to 24 (*Statistical Abstract of the United States: 1998,* Table 141). It is not surprising that the U.S. Public Health Service now defines "violence" as one of the top health concerns facing Americans. In addition to death, physical injury, and loss of property, crime also has economic, social, and psychological costs.

Economic Costs of Crime and Violence

We cannot expect to rein in the costs of our health-care system if emergency rooms are overflowing with victims of gun violence.

—Major Owens
Congressman

Conklin (1998, 71–72) suggests that the financial costs of crime can be classified into at least six categories. First are direct losses from crime such as the destruction of buildings through arson, of private property through vandalism, and of the environment by polluters. Second are costs associated with the transferring of property. Bank robbers, car thieves, and embezzlers have all taken property from its rightful owner at tremendous expense to the victim and society. For example, in 1993 computer thieves stole nearly $2 billion worth of copyrighted software programs over the Internet, and in 1995 an estimated $250 million worth of CDs and cassette tapes were illegally copied in China (Meyer and Underwood 1994; Faison 1996).

A third major cost of crime is that associated with criminal violence, such as the loss of productivity of injured workers and the medical expenses of victims. Fourth are the costs associated with the production and sale of illegal goods and services, that is, illegal expenditures. The expenditure of money on drugs, gambling, and prostitution diverts funds away from the legitimate economy and enterprises and lowers property values in high crime neighborhoods. Fifth is the cost of prevention and protection, that is, the millions of dollars spent on house alarms, security devices, weapons for protection, bars for windows, timers for lights, automobile security systems, and the like.

Finally, there is the cost of the criminal justice system. In 1996, more people were incarcerated in the United States than ever before—1.5 million people behind bars with a combined state and federal cost of $30 billion a year—an annual cost per inmate of $25,000 (*Statistical Abstract of the United States: 1998,* Table 376; ACLU 1996; Barkan 1997, 520). Reasons for such growth include increases in the rate of arrest and conviction, changes in the sentencing structure, public attitudes toward criminals, the growing number of young males, and the war on drugs. Regardless of the cause, however, the staggering cost of institutionalization has led to the "privatization" of prisons whereby the private sector increasingly supplies needed prison services (Carlson 1998).

What is the total economic cost of crime? One estimate suggests that the total cost of crime and violence in the United States is more than $450 billion a year (National Research Council 1994). While costs from "street crimes" are staggering, the costs from "crimes in the suites" such as tax evasion, fraud, false advertising, and antitrust violations may be as high as 50 times the cost of property crime (Rosoff, Pontell, and Tillman 1998).

National Data

The direct costs of robbery, rape, assault, auto theft, and personal and household theft is about $17.6 billion.

SOURCE: Barkan 1997

International Data

The Australian Institute of Criminology estimates that $3 billion a year is lost because of white-collar crime; the cost of murder is estimated to be $323 million and the cost of assaults, including sexual assaults, $980 million a year.

SOURCE: Lagan 1997

Social and Psychological Costs of Crime and Violence

Crime and violence entail social and psychological, as well as economic, costs. A *Newsweek* poll of children aged 10 to 17 found that over half of white children and two-thirds of minority children were worried about some family member becoming a victim of violent crime (Morganthau 1993). In the same poll, 34 percent of minority children and 16 percent of white children said they did not feel safe from violent crime in their neighborhood after dark. Fear of crime and violence also affects community life:

> **If frightened citizens remain locked in their homes instead of enjoying public spaces, there is a loss of public and community life, as well as a loss of "social capital"—the family and neighborhood channels that transmit positive social values from one generation to the next. (National Research Council 1994, 5–6)**

White-collar crimes also take a social and psychological toll at both the individual and the societal level. Moore and Mills (1990, 414) state that the effects of white-collar crime include "(a) diminished faith in a free economy and in business leaders, (b) loss of confidence in political institutions, processes and leaders, and (c) erosion of public morality." Crime also causes personal pain and suffering, the destruction of families, lowered self-esteem, and shortened life expectancy and disease.

National Data

Identity theft, that is, the use of your name and social security number, cost taxpayers over $745 million dollars in 1997; a major credit card company reports that 66 percent of all consumer inquiries relate to identity fraud.

SOURCE: Mannix 1998

The eight blunders that lead to violence in society:
-Wealth without work
-Pleasure without conscience
-Knowledge without character
-Commerce without morality
-Science without humanity
-Worship without sacrifice
-Politics without principle
-Rights without responsibilities

–Mohandas K. Gandhi
Indian nationalist leader and peace activist

STRATEGIES FOR ACTION: RESPONDING TO CRIME AND VIOLENCE

For the first time in history, the Attorney General of the United States announced a "Department of Justice Strategic Plan" for dealing with crime and violence, nationally and internationally, into the twenty-first century (Reno 1997). However, in addition to economic policies designed to reduce unemployment and poverty, numerous social policies and programs have already been initiated to alleviate the problem of crime and violence. These policies and programs are directed toward children at risk of being offenders, community crime prevention, media violence, and criminal justice policies.

Youth Programs

Early intervention programs acknowledge that it is better to prevent crime than to "cure" it once it has occurred. Preschool enrichment programs, such as the Perry Preschool Project, have been successful in reducing rates of aggression in young children. After random assignment of children to either a control or experimental group, experimental group members received academically oriented interventions for 1 to 2 years, frequent home visits, and weekly parent-teacher conferences. When control and experimental groups were compared, the experimental group had better grades, higher rates of high school graduation, lower rates of unemployment, and fewer arrests (Murray, Guerra, and Williams 1997).

Recognizing the link between juvenile delinquency and adult criminality, many anticrime programs are directed toward at-risk youths. These prevention strategies, including youth programs such as Boys and Girls Clubs, are designed to keep young people "off the streets," provide a safe and supportive environment, and offer activities that promote skill development and self-esteem. According to Gest and Friedman (1994), housing projects with such clubs report 13 percent fewer juvenile crimes and a 25 percent decrease in the use of crack.

Finally, many youth programs are designed to engage juveniles in noncriminal activities. While sports are probably the most common example, other recreation-based programs entail music, theater arts, dance, and the visual arts. One such program was developed in Chicago in 1991. Aimed at the city's thousands of inner city, unemployed, low-income youths between the ages of 14 and 21, "Gallery 37" now provides over 2,000 paid apprenticeship positions with local artists. The program is so successful that similar programs now exist in 16 U.S. cities as well as Adelaide, Australia, and London, England (OJP 1998).

Although we seem to go to extraordinarily expensive lengths to punish criminals, we are reluctant to spend money on programs that have proven effective at preventing youngsters from slipping into a life of crime in the first place.

–Sylvia Anne Hewlett
President of National Parenting Association

Community Programs

Neighborhood watch programs involve local residents in crime-prevention strategies. For example, MAD DADS in Omaha, Nebraska, patrol the streets in high crime areas of the city on weekend nights, providing positive adult role models for troubled children. Members also report crime and drug sales to police, paint over gang graffiti, organize gun buy-back programs, and counsel incarcerated fathers. Some communities also offer alternative dispute resolution (ADR) centers. These centers encourage community members who are involved in a conflict to meet with mediators to discuss their conflicts and try to find a mutually agreeable resolution.

Victim-offender dispute resolution programs are also increasing with over 300 such programs now in the United States. The growth of these programs is no doubt a reflection of their success rate: two-thirds of cases referred result in face-to-face meetings, over 90 percent of these cases result in a written restitution agreement, and 90 percent of the written restitution agreements are completed within 1 year (VORP 1998).

Media Violence

Violence is portrayed in music lyrics, music videos, video games, cartoons, television series, and movies. A 1996 report found that 85 percent of premium channel shows, 59 percent of basic cable shows, and 44 percent of broadcast channel shows included at least one act of violence (Farhi 1996). Such media images may desensitize individuals to violence and serve as models for violent behavior. Exposure to television violence is associated with increased aggressive behavior and decreased sensitivity to the pain and suffering of others (Comstock 1993; Murray 1993).

At the Harvard Center for Health Communications, director Jay Winsten collaborates with entertainment industry executives to promote antiviolence in popular music and television programming. The Center for Health Communications, whose goal is "to mobilize the immense power of mass communication to improve human health," also uses the media to address other social problems such as drug abuse, teen pregnancy, and drunken driving (Rubin 1995, A5). One of the Center's recent campaigns, called "Squash It," aims to teach kids that it is cool to walk away from a confrontation. "Squash it" is an African-American street term for "Back off, let's not fight" (Rubin 1995).

National Data

In a recent Harris Poll, more Americans believed that the media is more likely to report "the bad news about terrible, violent crime" (89.0 percent) rather than the "good news, that violent crime is decreasing" (6.6 percent).

SOURCE: Harris Poll 1996

Criminal Justice Policy

The criminal justice system is based on the principle of **deterrence**, that is, the use of harm or the threat of harm to prevent unwanted behaviors. It assumes that people rationally choose to commit crime, weighing the rewards and consequences of their actions. Thus, the recent emphasis on "get tough" measures holds that maximizing punishment will increase deterrence and cause crime rates to decrease. Research indicates, however, that the effectiveness of deterrence is a function of not only the severity of the punishment, but the certainty and swiftness of the punishment as well. Further, "get tough" policies create other criminal justice problems, including overcrowded prisons, and, consequently, the need for plea bargaining and early release programs.

Capital Punishment With capital punishment, the State (the federal government or a state) takes the life of a person as punishment for a crime. Only sixteen states, however, allow persons under the age of 18 to be executed. In 1996, 19 states executed 45 prisoners, 36 by lethal injection, 7 by electrocution, 1 by hanging, and 1 by firing squad. The average stay in 1996 on death row was 10 years and 5 months, a reduction of 9 months from the previous year (U.S. Department of Justice 1997b). In this chapter's *The Human Side,* Robert Johnson, a professor at American University, describes his reaction as "witness to an execution."

National Data

In a 1997 survey of persons 18 and older in the United States, when asked "Do you believe in capital punishment?", 75.3 percent were in favor of the death penalty and 21.9 percent were opposed to it.

SOURCE: Harris Poll 1997

Proponents of capital punishment argue that executions of convicted murderers are necessary to convey the public disapproval and intolerance for such heinous crimes. Those against capital punishment believe that no one, including

The legal system is greatly overestimated in its ability to sort out innocent from guilty.

—Stephen Bright
Visiting lecturer,
Yale Law School

To do justice, to break the cycle of violence, to make Americans safer, prisons need to offer inmates a chance to heal like a human, not merely heel like a dog.

—Richard Stratton
Former prison inmate

THE HUMAN SIDE

WITNESS TO AN EXECUTION

At 10:58 the prisoner entered the death chamber. He was, I knew from my research, a man with a checkered, tragic past. He had been grossly abused as a child, and went on to become grossly abusive of others. I was told he could not describe his life, from childhood on, without talking about confrontations in defense of a precarious sense of self—at home, in school, on the streets, in the prison yard. Belittled by life and choking with rage, he was hungry to be noticed. Paradoxically, he had found his moment in the spotlight. . . .

En route to the chair, the prisoner stumbled slightly, as if the momentum of the event had overtaken him. Were he not held securely by two officers, one at each elbow, he might have fallen. . . . Once the prisoner was seated, again with help, the officers strapped him into the chair.

Arms, legs, stomach, chest, and head were secured in a matter of seconds. Electrodes were attached to the cap holding his head and to the strap holding his exposed right leg. A leather mask was placed over his face. The last officer mopped the prisoner's brow, then touched his hand in a gesture of farewell. . . .

The strapped and masked figure sat before us, utterly alone, waiting to be killed . . . waiting for a blast of electricity that would extinguish his life. Endless seconds passed. His last act was to swallow, nervously, pathetically, with his Adam's apple bobbing. I was struck by the simple movement then, and can't forget it even now. It told me, as nothing else did, that in the prisoner's restrained body, behind that mask, lurked a fellow human being who, at some level, however primitive, knew or sensed himself to be moments from death.

. . . Finally, the electricity hit him. His body stiffened spasmodically, though only briefly. A thin swirl of smoke trailed away from his head and then dissipated quickly. The body remained taut, with the right foot raised slightly at the heel, seemingly frozen there. A brief pause, then another minute of shock. When it was over, the body was flaccid and inert.

Three minutes passed while the officials let the body cool. (Immediately after the execution, I'm told, the body would be too hot to touch and would blister anyone who did.) All eyes were riveted to the chair; I felt trapped in my witness seat, at once transfixed and yet eager for release. I can't recall any clear thoughts from that moment. One of the death watch officers later volunteered that he shared this experience of staring blankly at the execution scene. Had the prisoner's mind been mercifully blank before the end? I hope so.

The physician listened for a heartbeat. Hearing none, he turned to the warden and said, "This man has expired." The warden, speaking to the Director, solemnly intoned: "Mr. Director, the court order has been fulfilled." . . .

SOURCE: Robert Johnson. 1989. "'This Man Has Expired': Witness to an Execution." *Commonwealth,* January 13, 9–15. Copyright by the Commonwealth Foundation. Reprinted with permission.

the State, has the right to take another person's life and that putting convicted murderers behind bars for life is a "social death" that conveys the necessary societal disapproval.

Proponents of capital punishment also argue that it deters individuals from committing murder. Critics of capital punishment hold, however, that since most homicides are situational, and are not planned, offenders do not consider the consequences of their actions before they commit the offense. Critics also point out that the United States has a much higher murder rate than Western European nations that do not practice capital punishment and that death sentences are racially discriminatory.

Capital punishment advocates suggest that executing a convicted murderer relieves the taxpayer of the costs involved in housing, feeding, guarding, and providing medical care for inmates. Opponents of capital punishment argue that the principles that decide life and death issues should not be determined by financial considerations. In addition, taking care of convicted murderers for life may actually be less costly than sentencing them to death, due to the lengthy and costly appeal process for capital punishment cases (Garey 1985; Myths 1998).

Nevertheless, those in favor of capital punishment argue that it protects society by preventing convicted individuals from committing another crime, including the murder of another inmate or prison official. Opponents contend, however, that capital punishment may result in innocent people being sentenced to death. According to Radelet, Bedeau, and Putnam (1992), at least 139 innocent people were sentenced to death in the United States between 1985 and 1990. Twenty-three of these people were executed.

> I feel morally and intellectually obligated to concede that the death penalty experiment has failed.
>
> –Justice Harry Blackmun
> U.S. Supreme Court

National Data

In 1996, 56.5 percent of the inmates on death row were racial or ethnic minorities.

SOURCE: *Statistical Abstract of the United States: 1998,* Table 380

Rehabilitation versus Incapacitation Another important debate focuses on the primary purpose of the criminal justice system: Is it to rehabilitate offenders or to incapacitate them through incarceration? Both **rehabilitation** and **incapacitation** are concerned with recidivism rates, or the extent to which criminals commit another crime. Advocates of rehabilitation believe recidivism can be reduced by changing the criminal, whereas proponents of incapacitation think it can best be reduced by placing the offender in prison so that he or she is unable to commit further crimes against the general public.

Societal fear of crime has lead to a public emphasis on incapacitation and a demand for tougher mandatory sentences, a reduction in the use of probation and parole, support of a "three strikes and you're out" policy, and truth–in–sentencing laws (DiIulio 1999). While incapacitation is clearly enhanced by longer prison sentences, rehabilitation may not be. Rehabilitation assumes that criminal behavior is caused by sociological, psychological, and/or biological forces rather than being solely a product of free will. If such forces can be identified, the necessary change can be instituted. Rehabilitation programs include education and job training, individual and group therapy, substance abuse counseling, and behavior modification. While the evaluation of rehabilitation programs is difficult and results are mixed, incapacitation must of necessity be a temporary measure. Unless all criminals are sentenced to life, at some point about 98 percent will be returned to society.

International Data

Among the 59 nations surveyed by the Sentencing Project, Russia had the highest incarceration rate per 100,000 population (690), followed by the United States (600). The U.S. rate is 6 to 10 times higher than that of other industrialized nations; Canada's rate is 115, France's 100, Germany's 95, Sweden's 65, Japan's 37, and India's 24.

SOURCE: Mauer 1997

Gun Control One of the most passionately debated issues in America is gun control. Those against gun control argue that citizens have the right to own guns and that the nearly 20,000 firearm regulations thus far enacted have not reduced the incidence of violence. After reviewing the research literature, Wright (1995) suggests that "a compelling case for stricter gun controls is difficult to make on empirical grounds and that solutions to the problems of crime and violence in this nation will have to be found elsewhere" (p. 496).

Advocates of gun control insist that the nearly 200 million privately owned guns in the United States, 65 million of which are handguns, significantly contribute to the violent crime rate in the United States. In 1996, 68 percent of all homicides were the result of a gunshot wound (Cook and Ludwig 1997; *Statistical Abstract of the United States: 1998,* Table 339). In addition, guns contribute to accidental injury and death, often of friends and family members of the gun owner. In 1996, over 1,000 Americans died as a result of firearm accidents (*Statistical Abstract of the United States: 1998,* Table 148).

After a seven-year battle with the National Rifle Association (NRA), gun control advocates achieved a small victory in 1993 when Congress passed the **Brady Bill.** This law requires a 5-day waiting period on handgun purchases so that sellers can screen buyers for criminal records or mental instability. The extent to which the Brady Bill has been successful is still being assessed.

National Data

In 1996, 1,225 U.S. residents lost their lives to accidents involving firearms.

SOURCE: *Statistical Abstract of the United States: 1998,* Table 148

National Data

Firearm ownership is more common among arrestees than the general population. Of those arrested, 37 percent report owning a handgun and 14 percent report carrying a handgun all or most of the time.

SOURCE: Decker, Pennell, and Caldwell 1997

Law Enforcement Agencies Police policies and practices can also impact crime rates. The Crime Control Act of 1994 established the Office of Community Oriented Policing Services (COPS). Community oriented policing involves collaborative efforts among the police, the citizens of a community, and local leaders. As part of community policing efforts, officers speak to citizens groups, consult with social agencies, and enlist the aid of corporate and political leaders in the fight against neighborhood crime (COPS 1998; Lehrur 1999).

Officers using community policing techniques often employ "practical approaches" to crime intervention. Such solutions may include what Felson (1998) calls "situational crime prevention." Felson argues that much of crime could be prevented simply by minimizing the opportunity for its occurrence. For example, cars could be outfitted with unbreakable glass, flush-sill lock buttons, an audible reminder to remove keys, and a high-security lock for steering columns (Felson 1998, 168).

Finally, police departments frequently initiate crime prevention programs. One such program is Operation Cease Fire, implemented in Boston in 1994. The Boston Police Department, with the help of clergy, social workers, educators, and probation officers, identified hot spots—locations in the city where gang violence is consistently high. Once an area was identified, a team of negotiators met with gang members to develop a nonviolent plan of action in which elimination of all firearms was an intricate part. Failure to agree and comply with the pact led to severe penalties. Between 1990 and 1994 there were 155 killings of children and teenagers. In 1995, there were no homicides in Boston of juveniles under 18 (Operation Cease Fire 1997).

National Data

One in five residents have face-to-face contact with police officers annually. The most common reason for such contact is reporting a crime, followed by receiving a ticket, asking for help, a casual encounter, reporting a problem, and victimization.

SOURCE: U.S. Department of Justice 1997a

UNDERSTANDING CRIME AND VIOLENCE

What can we conclude from the information presented in this chapter? Research on crime and violence supports the contentions of both functionalists and conflict theorists. The inequality in society, along with the emphasis on material well-being and corporate profit, produces societal strains and individual frustrations. Poverty, unemployment, urban decay, and substandard schools, the symptoms of social inequality, in turn lead to the development of criminal subcultures and definitions favorable to law violation. Further, the continued weakening of social bonds between members of society and individuals and society as a whole, the labeling of some acts and actors as "deviant," and the differential treatment of minority groups at the hands of the criminal justice system encourage criminal behavior.

While crime and violence constitute major social problems in American society, they are also symptoms of other social problems such as poverty and economic inequality, racial discrimination, drug addiction, an overburdened educational system, and troubled families. The criminal justice system continues to struggle to find effective and just measures to deal with crime and criminal offenders. Many citizens and politicians have embraced the idea that society should "get tough on crime." Get-tough measures include building more prisons and imposing lengthier mandatory prison sentences on criminal offenders. Advocates of harsher prison sentences argue that "getting tough on crime" makes society safer by keeping criminals off the streets and deterring potential criminals from committing crime. But skeptics are not convinced. As one argues:

> **Prison . . . has a minimal impact on crime because it is a response after the fact, a mop-up operation. It doesn't work. The idea of punishing the few to deter the many is counterfeit because potential criminals either think they're not going to get caught or they're so emotionally desperate or psychologically distressed that they don't care about the consequences of their actions (Rideau 1994, 80).**

Prison sentences may not only be ineffective in preventing crime, they may also promote it by creating an environment in which prisoners learn criminal behavior, values, and attitudes from each other.

National Data

According to a recent report, 80 percent of all American inmates are drug or alcohol involved. Of those in state prisons for committing violent crimes, 21 percent were high on alcohol, 3 percent on crack, and 1 percent on heroin at the time the act was committed.

SOURCE: Califano 1998

Salaries and other benefits for custodial staff constitute the largest proportion of prison expenses.

Rather than getting tough on crime after the fact, some advocate getting serious about prevention:

> It is better to prevent crimes than to punish them.
>
> —Cesare Beccaria
> Italian economist

> **The only effective way to curb crime is for society to work to prevent the criminal act in the first place. . . . Our youngsters must be taught to respect the humanity of others and to handle disputes without violence. It is essential to educate and equip them with the skills to pursue their life ambitions in a meaningful way. As a community, we must address the adverse life circumstances that spawn criminality. (Rideau 1994, 80)**

Reemphasizing the values of honesty, responsibility, and civic virtue is a basic line of prevention with which most agree. In a 1997 poll, 16 percent of Americans responded that the lack of "ethics, morality and a decline in family values" is the second most important problem facing Americans—surpassed only by crime and violence (Gallup Poll 1998). Community of Caring is a nationwide program of value socialization in which public school teachers emphasize caring, trust, respect, and responsibility (Henry 1995, D1). As a nation, our focus should be on the moral deficit as well as on the budget deficit. Sociologist Amitai Etzioni observed, "No society can survive moral anarchy" (Potok and Stone 1995, A1).

William DeJong (1994) of the Harvard School of Public Health echoes a similar sentiment concerning violence prevention:

> We have to do something to try to change or we will be destroyed by violence.
>
> —Arun Gandhi
> Peace activist

> **If we are truly committed to preventing violence, we must do more than teach individual children how to survive in the dysfunctional environments in which they are growing up. We must also strive to change those environments by addressing the broader social, cultural, institutional, and physical forces that are at work. (p. 9)**

CRITICAL THINKING

1. Crime statistics are sensitive to demographic changes. Explain why crime rates in the United States began to rise in the 1960s as baby-boom teenagers entered high school, and why they may increase again as we move into the 21st century.
2. Some countries have high gun ownership and low crime rates. Other have low gun ownership and low crime rates. What do you think accounts for the differences between these countries?
3. One of the criticisms of crime theories is that they don't explain all crime, all of the time. Identify the theories of crime that are most useful in explaining *categories* of crime, for example, white-collar crime, violent crime, sex crimes, etc. Explain your choices.
4. The use of technology in crime-related matters is likely to increase dramatically over the next several decades. DNA testing, and the use of heat sensors, blood detecting chemicals, and computer surveillance are just some of the ways science will help fight crime. As with all technological innovations, however, there is the question, "Who benefits?". Are there gender, race, and/or class implications of these new technologies?

KEY TERMS

acquaintance rape
Brady Bill
classic rape
computer crime
control theory
corporate violence
crime
crime rate
deterrence
differential association
incapacitation
index offenses
labeling theory
organized crime
rehabilitation
strain theory
subcultural theory
victimless crimes
white-collar crime

INTERNET

You can find more information on crime statistics and the measurement of crime, DNA testing, gun control, and capital punishment at the *Understanding Social Problems* web site at **http://sociology.wadsworth.com**.

INFOTRAC COLLEGE EDITION

Either from the Wadsworth sociology resource center at **http://sociology.wadsworth.com** or directly from your web browser, you may access InfoTrac College Edition, an online university library that includes over 700 popular and scholarly journals in which you can find articles related to the topics in this chapter. Suggested articles and questions relating to these articles are listed below.

Hannon, Lance, and Lynn Resnick Dufour. 1998. "Still Just the Study of Men and Crime?" *Sex Roles* 38(1–2):63–72.

1. What is androcentricity and, according to feminist theorists, what effects does it have upon crime statistics?
2. What differences did the authors find between criminology articles written in the 1970's and those written in the 1990's?
3. What is over generalization and how does it affect criminology?

Hannon, Lance and James Defronzo. 1998. "The Truly Disadvantaged, Public Assistance and Crime." *Social Problems* 45(3):383–93.

1. What is the anomie theory of crime?
2. What have previous studies postulated needs to be done to reduce crime in society?
3. What three factors have led to the creation of an underclass?

Longshore, Douglas. 1998. "Self-control and Criminal Opportunity." *Social Problems*. 45(1):102–14.

1. What are the six components or traits of self-control?
2. What have previous studies discovered about the relationship between self-control and crime?
3. What were the two limitations of the study detailed in the article?

CHAPTER FIVE

Family Problems

Is It True?

Is It True?

1. Due to mandatory sex education, the United States has one of the lowest rates of teenage childbirth of any industrialized nation.
2. Nearly one in three single-parent families headed by a woman lives in poverty.
3. The United States has the highest divorce rate in the world.
4. About one-third of all U.S. births are to unmarried women.
5. Under U.S. law, a husband forcing his wife to have sex is not considered to be an act of rape.

Answers: 1 = F; 2 = T; 3 = T; 4 = T; 5 = F

An unhealthy culture cannot have healthy families.

Mary Pipher, from *The Shelter of Each Other: Rebuilding Our Families*

Politicians often point to the "breakdown" of the family as one of the primary social problems in the world today—a problem of such magnitude that it leads to such secondary social problems as crime, poverty, and substance abuse. Although strengthening family relationships and values is frequently offered as the principal solution to such problems, sociologists argue that family problems and their solutions are rooted in the structure and culture of society. For example, the structure of the economic institution requires many parents to work long hours, often without benefits. Without a national child care system, such as those in France and Sweden, employed parents of young children must often leave their offspring with inadequate or no supervision. Parents and spouses who are stressed by long work hours, poor working conditions, and little pay often bring their stress home, contributing to child abuse, spouse abuse, and divorce.

Families in the United States are also influenced by cultural factors. For example, the American value of **individualism**, which stresses the importance of individual happiness, contributes to divorce as spouses leave marriages to pursue their individual goals. In other countries, such as India, the cultural value of **familism** encourages spouses to put their family's welfare above their individual and personal needs.

The United States has long been known as the world's most individualistic society.

–David Popenoe
Sociologist

In this chapter we look at some of the major social problems facing American families—domestic violence, divorce, unmarried childbirth, and teenage parenthood. Prior and subsequent chapters in this text discuss other problems facing families, including problems in balancing work and family demands, poverty, lack of affordable housing, illness and inadequate health care, drug abuse, inequality in education, and prejudice and discrimination.

THE GLOBAL CONTEXT: FAMILIES OF THE WORLD

Family is a central aspect of every society throughout the world. But family forms are diverse. Although the only legal form of marriage in the United States is between one woman and one man, other societies practice *polygamy*—a form of marriage in which there are more than two spouses. Other societies legally recognize same-sex unions.

Family values, roles, and norms are also highly variable across societies. For example, in most Western countries, fathers are expected to be the child's primary adult male figure in terms of nurturance, socialization, and economic support. Yet in some African societies, it is the mother's brother—the child's uncle—who fulfills such a role. Unlike in Western societies, in Asian countries some marriages are arranged by the parents who select mates for their children. Another traditional Asian practice is for the eldest son and his wife to move in with the son's parents, where his wife takes care of her husband's parents. In some societies, it is normative for married couples to view each other as equal partners in the marriage, whereas in other societies social values dictate that wives be subservient to their husbands. The role of children also varies across societies. In some societies, children are expected to work full-time to help support the family (see discussion of child labor in Chapter 11).

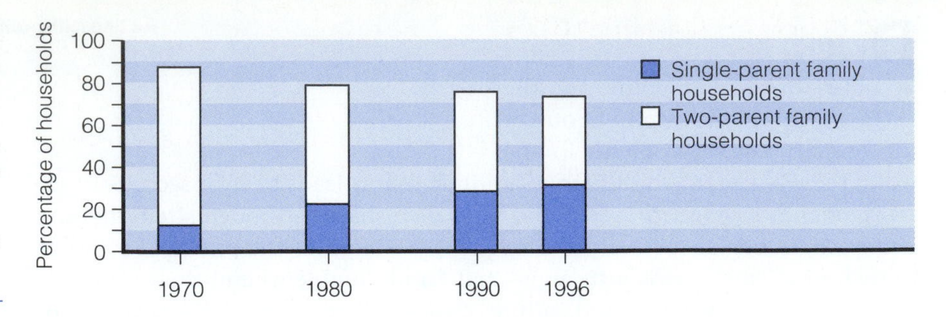

Figure 5.1 Two-Parent and Single-Parent Family Households with Children Under 18 Since 1970

SOURCE: *Statistical Abstract of the United States: 1998,* 118th ed. U.S. Bureau of the Census. Washington, D.C.: U.S. Government Printing Office, Table 78; and *Statistical Abstract of the United States: 1995,* 115th ed. U.S. Bureau of the Census. Washington, D.C.: U.S. Government Printing Office, Table 75.

National Data

In 1997, 23 percent of U.S. families with children under 18 were headed by single mothers; 5 percent were headed by single fathers.

SOURCE: *Statistical Abstract of the United States: 1998,* Table 79

National Data

In 1997, 22 percent of U.S. white families with children under 18 were headed by a single parent. The percentages for black and Hispanic families were 60 percent and 31 percent, respectively.

SOURCE: *Statistical Abstract of the United States: 1998,* Table 79

National Data

About one in five married couples with children under age 18 are a blended family. More than one in three children born in the United States will be in a blended family before reaching age 18.

SOURCE: Ganong and Coleman 1994

It is clear from the previous discussion that families are shaped by the social and cultural context in which they exist. Next we look at variations in U.S. families.

Variations in U.S. Families and Households

For the U.S. Census, a **family household** consists of two or more persons related by birth, marriage, or adoption who reside together. As Figure 5.1 shows, over the last few decades, two-parent family households have declined, while single-parent family households have increased.

Another family variation that has increased in recent decades is the blended, or step, family. A **blended family** consists of remarried spouses with at least one of the spouses having a child from a previous relationship.

Nonfamily households may consist of one person who lives alone, two or more people as roommates, and cohabiting heterosexual or homosexual couples involved in a committed relationship. As "nonfamily" (according to the U.S. Census) households and variations in family forms have become more common, many family scholars are advocating new cultural and legal definitions of the family. Geile (1996) suggests that "recognition as 'family' should . . . not be confined to the traditional two-parent unit connected by blood, marriage, or adoption, but should be extended to include kin of a divorced spouse . . . same-sex partners, congregate households of retired persons, group living arrangements, and so on" (pp. 103–104).

Increasingly, social and legal policy and court decisions are expanding the concept of family to include unmarried living-together couples and same-sex couples who live together and view themselves as a family. In some U.S. cities and counties, heterosexual cohabiting couples as well as gay and lesbian couples who apply for a **domestic partnership** designation are granted legal entitlements such as health insurance benefits and inheritance rights that have traditionally been reserved for married couples. Some employers offer health insurance benefits to cohabiting partners of employees and grant family leave to employees whose cohabiting partner becomes ill, dies, or gives birth or adopts.

Although marriage and family forms have become diversified, marriage and family are still important to most Americans. An examination of the sociological theories of the family will provide a framework for our discussion of various problems related to the family.

SOCIOLOGICAL THEORIES OF THE FAMILY

Three major sociological theories of the family—structural-functionalism, conflict theory, and symbolic interactionism—explain different aspects of the family.

Structural-Functionalist Perspective

The structural-functionalist perspective emphasizes that families perform functions that are important for the survival of society. Families replenish the population through reproduction and contribute to the socialization and education of youth. Well-functioning families also provide emotional and physical care for their members.

The structural-functionalist perspective is also concerned with how other institutions affect families. The educational, health care, religious, economic, and political insitutions each affect families and their members.

National Data

Between 1980 and 1997, the number of unmarried couples (two unrelated adults of the opposite sex living together) in the United States increased more than two and one-half times.

SOURCE: *Statistical Abstract of the United States: 1998,* Table 66

Conflict Perspective

Conflict theory focuses on how wealth and power influence marriages and families. Within families, the unequal distribution of power among women, men, and children may contribute to spouse and child abuse. The unequal distribution of wealth between men and women, with men traditionally earning more money than women, contributes to inequities in power and fosters economic dependence of wives on husbands. The term **patriarchy** refers to male-dominated families. Families have traditionally been dominated by men, with the wife taking her husband's name, family residence defined by the husband's place of work, and the standard of living dictated by the male's income.

Conflict theorists emphasize that social programs and policies that affect families are largely shaped by powerful and wealthy segments of society. The interests of corporations and businesses are often in conflict with the needs of families. Corporations and businesses strenuously fought the passage of the 1993 Family and Medical Leave Act that guarantees employees of firms with more than 50 workers unpaid time off for parenting leave, illness or death of a family member, and elder care. Hewlett and West (1998) note that corporate interests undermine family life "by exerting enormous downward pressure on wage levels for young, child-raising adults" (p. 32). Government, which is largely influenced by corporate interests through lobbying and political financial contributions, enacts policies and laws that serve the interests of for-profit corporations, rather than families.

National Data

Over 70 percent of first-year U.S. college and university students regard raising a family as an "essential" or "very important" personal objective.

SOURCE: "This Year's Freshmen: A Statistical Profile" 1999

Our leaders talk as though they value families, but act as though families were a last priority.

—Sylvia Hewlett
Cornell West
Family advocates

Symbolic Interactionist Perspective

Symbolic interactionism emphasizes that human behavior is largely dependent upon the meanings and definitions that emerge out of small group interaction. Divorce, for example, was once highly stigmatized and informally sanctioned through the criticism and rejection of divorced friends and relatives. As societal definitions of divorce became less negative, however, the divorce rate increased. The social meanings surrounding single parenthood, cohabitation, and delayed childbearing and marriage have changed in similar ways. As the definitions of each of these family variations became less negative, the behaviors became more common.

Symbolic interactionists also point to the effects of labeling on one's self-concept and the way the self-fulfilling prophecy can affect family members' behavior toward one another. The **self-fulfilling prophecy** implies that we behave according to the expectations of others. When family members label children as "bad" or "stupid," children may internalize these labels and view themselves as "bad" or "stupid." For some children, a negative self-concept and

Family Functioning Scale

INSTRUCTIONS:

Every family has strengths and capabilities, although different families have different ways of using their abilities. This questionnaire asks you to indicate whether or not your family is characterized by 26 different qualities. Please read each statement, then *circle* the response that is most true for your family (people living in your home). Please give your honest opinions and feelings. Remember that your family will not be like all the statements.

How is your family like the following statements:	Not at All Like My Family	A Little Like My Family	Sometimes Like My Family	Usually Like My Family	Almost Always Like My Family
1. We make personal sacrifices if they help our family.	1	2	3	4	5
2. We usually agree about how family members should behave.	1	2	3	4	5
3. We believe that something good always comes out of even the worst situations.	1	2	3	4	5
4. We take pride in even the smallest accomplishments of family members.	1	2	3	4	5
5. We share our concerns and feelings in useful ways.	1	2	3	4	5
6. Our family sticks together no matter how difficult things get.	1	2	3	4	5
7. We usually ask for help from persons outside our family if we cannot do things ourselves.	1	2	3	4	5
8. We usually agree about the things that are important to our family.	1	2	3	4	5
9. We are always willing to pitch in and help each other.	1	2	3	4	5
10. We find things to do that keep our minds off our worries when something upsetting is beyond our control.	1	2	3	4	5
11. We try to look at the bright side of things no matter what happens in our family.	1	2	3	4	5
12. We find time to be together even with our busy schedules.	1	2	3	4	5
13. Everyone in our family understands the rules about acceptable ways to act.	1	2	3	4	5
14. Friends and relatives are always willing to help whenever we have a problem or crisis.	1	2	3	4	5

Family Functioning Scale *(continued)*

How is your family like the following statements:	Not at All Like My Family	A Little Like My Family	Sometimes Like My Family	Usually Like My Family	Almost Always Like My Family
15. Our family is able to make decisions about what to do when we have problems or concerns.	1	2	3	4	5
16. We enjoy time together even if it is doing household chores.	1	2	3	4	5
17. We try to forget our problems or concerns for a while when they seem overwhelming.	1	2	3	4	5
18. Family members listen to both sides of the story during a disagreement.	1	2	3	4	5
19. We make time to get things done that we all agree are important.	1	2	3	4	5
20. We can depend on the support of each other whenever something goes wrong.	1	2	3	4	5
21. We usually talk about the different ways we deal with problems and concerns.	1	2	3	4	5
22. Our family's relationships will outlast our material possessions.	1	2	3	4	5
23. We make decisions like moving or changing jobs for the good of all family members.	1	2	3	4	5
24. We can depend upon each other to help out when something unexpected happens.	1	2	3	4	5
25. We try not to take each other for granted.	1	2	3	4	5
26. We try to solve our problems first before asking others to help.	1	2	3	4	5

(May be duplicated without permission with proper acknowledgment and citation.)

SCORING:
Circling a 5 represents the most optimum family functioning response in terms of family strengths. Circling a 1 represents the least optimum family functioning response. The scale was administered to 206 mothers and 35 fathers of preschool children. The majority of items had average ratings between 3.00 and 4.00.

SOURCE: Carol M. Trivette, Carl J. Dunst, Angela G. Deal, Deborah W. Hamby, and David Sexton. 1994. "Family Functioning Style Scale" in Chapter 10, (Assessing Family Strengths and Capabilities.) *Supporting and Strengthening Families: Volume 1—Methods, Strategies and Practices,* ed. Carl J. Dunst, Carol M. Trivette, and Angela G. Deal, p. 139. Cambridge, Mass.: Brookline Books. Permission to reproduce granted in text.

the self-fulfilling prophecy contribute to violence, crime and delinquency, school failure, and mental health problems such as depression and anxiety. As the following section explains, painful criticism of spouses, children, and elders may constitute a form of domestic abuse.

VIOLENCE AND ABUSE IN THE FAMILY

Unless we end the cult of silence and complacency toward the violence at the domestic core of our civilizations, we can't hope to end the biased crime and violence against races, religious groups, and nations.

—Daniela Gioseffi
Poet, novelist, activist

Although marriage and family relationships provide many individuals with a sense of intimacy and well-being, for others, these relationships involve violence and abuse causing terror, pain, injury, and, for some, death. Domestic abuse can take various forms besides physical violence, including verbal and emotional abuse, and sexual abuse. **Neglect**, another form of abuse, includes failure to provide adequate attention and supervision, food and nutrition, hygiene, medical care, and a safe and clean living environment.

Violence and Abuse in Marriage and Cohabiting Relationships

Although both men and women may be victims of abuse by intimate partners, domestic violence primarily involves female victims. According to the U.S. Department of Justice (1994), women experience over 10 times as many incidents of violence by an intimate as men do. In 1996, three in every four victims of intimate murder were female (U.S. Department of Justice 1998).

Rape is another form of abuse that occurs in marital and cohabiting relationships. An estimated 7 to 14% of married women have been raped by their husbands (Monson, Byrd, and Langhinrichsen-Rohling 1996).

Physical and emotional abuse is, no doubt, a factor in many divorces. In addition to affecting the happiness and stability of relationships, abuse affects the physical and psychological well-being of the victim. Battering is the single major cause of injury to women in the United States ("Domestic Violence Fact Sheet" 1999).

National Data

Women aged 16 to 24 are at highest risk for being victims of intimate violence.

SOURCE: U.S. Department of Justice 1998

Violence between intimate partners or ex-partners may also include unintentional death and intentional murder. Most (65 percent) of intimate murders in 1996 were committed with firearms (U. S. Department of Justice 1998).

Domestic violence is also a primary cause of homelessness. One study found that half of homeless women and children were fleeing from abuse ("Domestic Violence and Homelessness" 1998).

About 40 percent of battered women are abused during their pregnancy, resulting in a high rate of miscarriage and birth defects (North Carolina Coalition Against Domestic Violence 1991). Witnessing marital violence is related to emotional and behavioral problems in children and subsequent violence in their own relationships (Busby 1991). Children may also commit violent acts against a parent's abusing partner.

Breast bruised, brains battered
Skin scarred, soul shattered.
Can't scream—neighbors stare,
Cry for help—no one's there.

—Stanza from a poem by Nenna Nehru, a battered Indian woman

Child Abuse

Child abuse refers to the "physical or mental injury, sexual abuse, negligent treatment, or maltreatment of a child under the age of eighteen by a person who is responsible for the child's welfare . . ." (Willis, Holden, and Rosenberg 1992, 2). As noted in Figure 5.2, the most common type of reported child abuse is neglect.

Reviews of research on the effects of child abuse suggest that abused children tend to exhibit aggression, low self-esteem, depression, and low academic

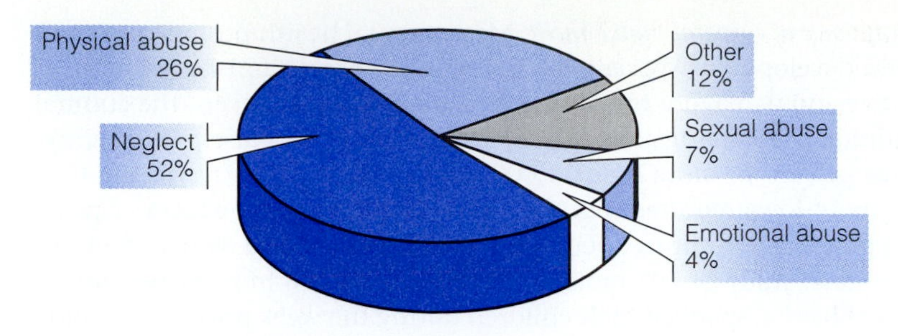

Figure 5.2 Types of Reported Child Abuse, 1997

SOURCE: Ching-Tung Wang and Deborah Daro. 1998. "Current Trends in Child Abuse Reporting and Fatalities: The Results of the 1997 Annual Fifty State Survey." Chicago, Ill.: National Committee to Prevent Child Abuse.

achievement (Gelles and Conte 1991; Lloyd and Emery 1993). Adults who were physically abused as children may exhibit low self-esteem, depression, unhappiness, anxiety, an increased risk of alcohol abuse, and suicidal tendencies. Physical injuries sustained by child abuse cause pain, disfigurement, scarring, physical disability, and death.

Among adolescent females, sexual abuse is associated with lower self-esteem, higher levels of depression, antisocial behavior (e.g., running away from home, illegal drug use), and suicide attempts (Morrow and Sorrell 1989). Sexually abused girls are also more likely to experience teenage pregnancy, have higher numbers of sexual partners in adulthood and are more likely to acquire sexually transmitted infections and experience forced sex (Browning and Laumann 1997; Stock et al., 1997).

Adult males who were sexually abused as children tend to exhibit depression, substance abuse, and difficulty establishing intimate relationships (Krug 1989). Sexually abused males also have a higher risk of anxiety disorders, sleep and eating disorders, and sexual dysfunctions (Elliott and Briere 1992).

National Data

In 1997, the number of child abuse reports exceeded 3 million; over 1 million of these reports were substantiated. In 1997, 15 out of every 1,000 U.S. children were substantiated as victims of child maltreatment.

SOURCE: Wang and Daro 1998

National Data

In a national survey of adults, 27 percent of the women and 16 percent of the men reported being victims of child sexual abuse.

SOURCE: Finkelhor et al. 1990

Elder Abuse

Community-based surveys suggest that 3 to 6 percent of elders in the United States (over age 65) report having experienced elder abuse (Lachs et al. 1997). **Elder abuse** includes physical abuse, psychological abuse, financial abuse (such as improper use of the elder's financial resources), and neglect. Elder neglect includes failure to provide basic health and hygiene needs such as clean clothes, doctor visits, medication, and adequate nutrition. Neglect also involves unreasonable confinement, isolation of elderly family members, lack of supervision, and abandonment. Most abuses of the elderly are committed by adult children, followed by spouses (Lachs et al. 1997).

Factors Contributing to Domestic Violence and Abuse

Research suggests that cultural, community, and individual and family factors contribute to domestic violence and abuse (Willis, Holden, and Rosenberg 1992).

Cultural Factors

In many ways, American culture tolerates and even promotes violence. Violence in the family stems from our society's acceptance of violence as a legitimate means of enforcing compliance and solving conflicts at personal, national, and international levels (Viano 1992). Violence and abuse in the family may be linked to cultural factors such as violence in the media (see Chapter 4), acceptance of corporal punishment, gender inequality, and the view of women and children as property:

1. *Acceptance of corporal punishment.* Many mental health professionals and child development specialists argue that corporal punishment is ineffective and damaging to children. Yet many parents accept the cultural tradition of spanking as an appropriate form of child discipline. Eighty-three percent of more than 11,000 undergraduate students at the University of Iowa reported that they had experienced some form of physical punishment during their childhood (Knutson and Selner 1994). In a national study of 807 mothers of children aged 6 to 9, 44 percent reported having spanked their children during the week prior to the study and that they had spanked them an average of 2.1 times that week (Straus, Sugarman, and Giles-Sims 1997). Another national survey of parental discipline practices found that in 1998, 45 percent of parents reported that they had spanked or hit their child in the past month (compared to 62 percent in 1988) (Daro 1998). Although not everyone agrees that all instances of corporal punishment constitute abuse, undoubtedly, some episodes of parental "discipline" are abusive.
2. *Gender inequality.* Traditional male gender roles have taught men to be aggressive. Traditionally, men have also been taught that they are superior to women and that they may use their aggression toward women because "women need to be put in their place." Traditional female gender roles have also taught women to be submissive to their male partner's control.

   Anderson (1997) found that men who earn less money than their partners are more likely to be violent toward them. "Disenfranchised men then must rely on other social practices to construct a masculine image. Because it is so clearly associated with masculinity in American culture, violence is a social practice that enables men to express a masculine identity" (p. 667).

> Our culture promotes sexual victimization of women and children when it encourages males to believe that they have overpowering sexual needs that must be met by whatever means available.
>
> —Edwin Schur
> Sociologist

3. *View of women and children as property.* Prior to the late nineteenth century, a married woman was considered to be the property of her husband. A husband had a legal right and marital obligation to discipline and control his wife through the use of physical force. The expression "rule of thumb" can be traced to an old English law that permitted a husband to beat his wife with a rod no thicker than his thumb. This "rule of thumb" was originally intended as a humane measure to limit how harshly men could beat their wives. This traditional view of women as property may contribute to men doing with their "property" as they wish.

   The view of women and children as property also explains marital rape and father-daughter incest. Historically, the penalties for rape were based on property right laws designed to protect a man's property—his wife or daughter—from rape by other men; a husband or father "taking" his own property was not considered rape (Russell 1990). In the past, a married woman who was raped by her husband could not have her husband arrested because marital rape was not considered a crime.

National Data

In 1978, only 3 states recognized marital rape as a law violation. Today, all 50 states recognize marital rape as a crime.

CONSIDERATION

In some countries, husbands are permitted by law to use physical force against their wives. In some societies, **"honor murder"** of wives is practiced when a man suspects his wife of being unfaithful. **Dowry deaths** in other countries may occur when the wife brings an insufficient dowry into the marriage.

Community Factors Community factors that contribute to violence and abuse in the family include social isolation and inaccessible or unaffordable health care, day care, elder care, and respite care facilities:

1. *Social isolation.* Living in social isolation from extended family and community members increases a family's risk for abuse. Isolated families are removed from material benefits, care-giving assistance, and emotional support from extended family and community members.
2. *Inaccessible or unaffordable community services.* Failure to provide medical care to children and elderly family members (a form of neglect) is sometimes due to the lack of accessible or affordable health care services in the community. Failure to provide supervision for children and adults may result from inaccessible day care and elder care services. Without elder care and respite care facilities, socially isolated families may not have any help with the stresses of caring for elderly family members and children with special needs.

Individual and Family Factors Individual and family factors that are associated with domestic violence and abuse include a family history of violence, drug and alcohol abuse, poverty, and fatherless homes:

1. *Family history of abuse.* Most parents who abuse or neglect their children were themselves abused or neglected as children (Gelles and Conte 1991). But, many adults who were abused as children do not continue the cycle with their own children.
2. *Drug and alcohol abuse.* More than half of prison and jail inmates serving time for violence against an intimate had been using alcohol, drugs, or both at the time of the incident for which they were incarcerated (U.S. Department of Justice 1998). O'Keefe (1997) found that higher use of alcohol and other drugs was also associated with violence in high school dating relationships. Alcohol increases aggression in some individuals and enables the offender to avoid responsibility by blaming his or her violent behavior on alcohol.
3. *Poverty.* Abuse in adult relationships occurs among all socioeconomic groups. However, Kaufman and Zigler (1992) point to a relationship between poverty and child abuse:

    > **Although most poor people do not maltreat their children, and poverty, per se, does not cause abuse and neglect, the correlates of poverty, including stress, drug abuse, and inadequate resources for food and medical care, increase the likelihood of maltreatment. (p. 284)**

4. *Fatherless homes.* Numerous studies show that children are more likely to be sexually abused by a stepfather or mother's boyfriend than by their biological father (Blankenhorn 1995). This is largely because stepfathers and mothers' boyfriends are not constrained by the cultural incest taboo that prohibits fathers from having sex with their children.

STRATEGIES FOR ACTION: PREVENTING AND RESPONDING TO FAMILY VIOLENCE AND ABUSE

Strategies to prevent family violence and abuse can be applied at three different levels (Gelles 1993; Harrington and Dubowitz 1993). **Primary prevention**

Public service campaigns against child abuse use sports figures such as Tom Gugliotta to promote their message of non-violence.

strategies target the general population, whereas **secondary prevention** strategies target groups thought to be at high risk for family violence and abuse. **Tertiary prevention** strategies target families who have experienced abuse; these strategies are designed to reduce the adverse effects of abuse and stop the abuse from happening again.

Primary Prevention Strategies Public education and media campaigns that target the general population may help reduce domestic violence by conveying the criminal nature of domestic assault and offering ways to prevent abuse ("When you are angry at your child, count to 10 and call a friend . . ."). Ultimately, though, to prevent or reduce family violence, elements of American culture that contribute to such violence must change. Parents and educators must be taught and encouraged to use methods of child discipline that do not involve physical punishment. Parent training classes are a high school graduation requirement in some states (Shapiro and Schrof 1995). Violence in the media must be curbed or eliminated, and traditional gender roles and views of women and children as property must be replaced with egalitarian gender roles and respect for women and children.

Another strategy involves reducing violence-provoking stress by reducing poverty and unemployment and providing adequate housing, nutrition, medical care, and educational opportunities. However, rather than strengthening the supports for poor families with children, welfare reform legislation enacted in 1996 limits cash assistance to poor single parents to two consecutive years with a five-year lifetime limit (some exceptions are granted). Although the previous welfare program Aid to Families with Dependent Children (AFDC) entitled single mothers to public financial aid, the new welfare-to-work program, Temporary Aid to Needy Families (TANF) pushes women into the labor force. As discussed in Chapter 10, many women going from welfare to work will experience greater hardships as a result of a loss of food stamp benefits, increases in federal housing rent, loss of Medicaid benefits, cost of transportation to work, and child care costs (Edin and Lein 1997). Many women forced to go to work and unable to afford child care will leave their children unattended, increasing child neglect. The cumulative stresses and hardships that the welfare-to-work legislation will have on single parents may very well contribute to increases in child neglect and abuse.

Secondary Prevention Strategies Families who are at risk of experiencing violence and abuse include low-income families, parents with a history of depression or psychiatric care, single parents, teenage mothers, parents with few social and family contacts, individuals who experienced abuse in their own childhood, and parents or spouses who abuse drugs or alcohol. Secondary prevention strategies, designed to prevent abuse from occurring in high-risk families, include parent education programs, parent support groups, individual counseling, substance abuse treatment, and home visiting programs.

> **National Data**
>
> As of May 1998, Healthy Families America had resulted in 270 pilot programs in 40 states and the District of Columbia.
>
> SOURCE: Prevention Programs 1998

A nationwide program called **Healthy Families America** (HFA) offers one or two home visits to all new parents. During these visits, service providers share information about baby care and development and parenting skills and explain how parents can utilize other organizations and agencies that provide family support. Families most at risk of child maltreatment are offered intensive home visitation services (at least once a week) for 3 to 5 years.

CONSIDERATION

Programs such as Healthy Families America cost money, but in the long run, they save money. For every $3 spent on child abuse prevention, we save at least $6 that might otherwise have been spent on child welfare services, special education services, medical care, foster care, counseling, and housing juvenile offenders. (Healthy Families America Fact Sheet 1994)

> **National Data**
>
> In 1996, President Clinton launched the National Domestic Violence Hotline, a 24-hour, toll-free service that provides crisis assistance and local shelter referral for callers across the country. The number is (800) 787-SAFE; for hearing impaired callers, the number is (800) 787-3224.

Tertiary Prevention Strategies What social interventions are available to families that are experiencing abuse or neglect? Abused women and children may seek relief at a shelter or "safe house" for abused women. Shelters provide abused women and their children with housing, food, and counseling services. Safe houses are private homes of individuals who volunteer to provide temporary housing to abused women who decide to leave their violent home. Battered men are not allowed to stay at women's shelters, but many shelters help abused men by providing money for a motel room, counseling, and support services. Some communities have abuse shelters for victims of elder abuse.

Abused or neglected children may be removed from the home. All states have laws that permit abused or neglected children to be placed in out-of-home care, such as foster care (Stein 1993). However, federal law requires that states have programs to prevent family breakup when desirable and possible without jeopardizing the welfare of children in the home. Family preservation programs are in-home interventions for families who are at risk of having a child removed from the home due to abuse or neglect.

Alternatively, a court may order an abusing spouse or parent to leave the home. Abused spouses or cohabiting partners may obtain a restraining order prohibiting the perpetrator from going near the abused partner. About half of the states and Washington, D.C., now have mandatory arrest policies that require police to arrest abusers, even if the victim does not want to press charges. However, legal action does not always protect victims of family violence. Nearly 4 in 100 violent offenders sentenced to local jail for intimate violence were on probation or under a restraining order at the time they committed the offense (U.S. Department of Justice 1998).

DIVORCE

National Data

Half of all marriages entered into in recent years will end in divorce or separation if recent marital dissolution rates continue.

SOURCE: Glenn 1997

The United States has the highest divorce rate in the world (United Nations 1992). Individual and relationship factors that contribute to divorce include incompatibility in values or goals, poor communication, lack of conflict resolution skills, sexual incompatibility, extramarital relationships, substance abuse, emotional or physical abuse or neglect, boredom, jealously, and difficulty coping with change or stress related to parenting, employment, finances, in-laws, and illness. Other demographic and life course factors that are predictive of divorce include marriage order (second and subsequent marriages are more prone to divorce than first marriages), cohabitation (couples who live together before marriage are more prone to divorce), teenage marriage, premarital pregnancy, and low socioeconomic status.

Various social factors also contribute to the high rate of divorce in U.S. society. These include the following structural and cultural forces.

The model of the two-parent family, based on a lasting, monogamous marriage, is both possible and desirable. This form of the family, in fact, is by far the most efficacious one for child rearing and for the long-term well-being of individuals and society.

—David Popenoe
Sociologist

1. *Changing Family Functions* Prior to the Industrial Revolution, the family constituted a unit of economic production and consumption, provided care and protection to its members, and was responsible for socializing and educating children. During industrialization, other institutions took over these functions. For example, the educational institution has virtually taken over the systematic teaching and socialization of children. Today, the primary function of marriage and the family is the provision of emotional support, intimacy, affection, and love. When marital partners no longer derive these emotional benefits from their marriage, they may consider divorce with the hope of finding a new marriage partner to fulfill these affectional needs.
2. *Increased Economic Autonomy of Women* Before 1940, most wives were not employed outside the home and depended on their husband's income. During World War II, the United States needed women in

the labor force. Today, the employment rate among married women is higher than in previous decades.

Wives who are unhappy in their marriage are more likely to leave the marriage if they have the economic means to support themselves. Unhappy husbands may also be more likely to leave a marriage if their wives are self-sufficient and can contribute to the support of the children.

3. *Increased Work Demands* Another factor influencing divorce is increased work demands and the stresses of balancing work and family roles. Workers are putting in longer hours, often working overtime or taking second jobs. Many employed parents—particularly mothers—come home to work a "second shift"—the work involved in caring for children and household chores. Balancing work and family roles can lead to marital tension and resentment, as reflected in the following excerpt:

> **My husband's a great help watching our baby. But as far as doing housework or even taking the baby when I'm at home, no. He figures he works five days a week; *he's* not going to come home and clean. But he doesn't stop to think that I work *seven* days a week. Why should I have to come home and do the housework without help from anybody else? My husband and I have been through this over and over again. Even if he would just pick up from the kitchen table and stack the dishes for me, that would make a big difference. He does nothing. On his weekends off, I have to provide a sitter for the baby so he can go fishing. When I have a day off, I have the baby all day long without a break. He'll help out if I'm not here, but the minute I am, all the work at home is mine. (quoted in Hochschild 1997, pp. 37–38)**

4. *Liberalized Divorce Laws* Prior to 1970, the law required a couple who wanted a divorce to prove that one of the spouses was at fault and had committed an act defined by the state as grounds for divorce—adultery, cruelty, or desertion. In 1969, California became the first state to initiate **no-fault divorce,** which permitted a divorce based on the claim that there were "irreconcilable differences" in the marriage. No-fault divorce law has contributed to the U.S. divorce rate by making divorce easier to obtain. Although U.S. divorce rates started climbing before California instituted the first no-fault divorce law, the widespread adoption of such laws has probably contributed to their continued escalation. Today, all 50 states recognize some form of no-fault divorce.
5. *Cultural Values* The U.S. value of individualism has contributed to divorce. In the pursuit of individual happiness, spouses and children may be left behind. Familism is still prevalent among Asian Americans and Mexican Americans, however, which helps to explain why the divorce rate is lower among these groups than among whites and African Americans (Mindel, Habenstein, and Wright 1998).

   The value of marriage has also changed in American culture. The increased social acceptance of nonmarital sexuality, nonmarital childbearing, cohabitation, and singlehood reflect the view that marriage is an option, rather than an inevitability. The view of marriage as an option, rather than as an imperative, is mirrored by the view that

National Data

In 1975, only 41 percent of married women were employed. By 1997, 62 percent of U.S. wives were employed.

SOURCE: *Statistical Abstract of the United States: 1998,* Table 652

In this new model of family and work life, a tired parent flees a world of unresolved quarrels and unwashed laundry for the reliable orderliness, harmony, and managed cheer of work.

–Arlie Hochschild
Sociologist

It is now widely accepted that men and women have the right to expect a happy marriage, and that if a marriage does not work out, no one has to stay trapped.

–Sylvia Ann Hewlett
President, National Parenting Association

divorce is also an acceptable option. Divorce today has less social stigma than in previous generations.

Consequences of Divorce

When parents have bitter and unresolvable conflict, and/or if one parent is abusing the child or the other parent, divorce may offer a solution to family problems. But divorce often has negative effects for ex-spouses and their children and also contributes to problems that affect society as a whole.

Health Consequences Divorce tends to have deleterious physical and psychological health consequences on ex-spouses. Divorce is associated with anxiety and psychological distress, unhappiness, loneliness, depression, illness, and suicide (Waite 1995; Kitson and Morgan 1991; Song 1991; Heim and Snyder 1991).

Economic Consequences While both women and men experience a drop in income following divorce, women may suffer more. Forty percent of divorcing women lose half of their family income, whereas fewer than 17 percent of men experience this degree of loss (Arendell 1995).

Effects on Children Although divorce, following high conflict, may actually improve the emotional well-being of children relative to staying in a conflicted home environment (Jekielek 1998), for many children parental divorce has detrimental effects. With only one parent in the home, children of divorce, as well as children of never-married mothers, tend to have less adult supervision compared with children in two-parent homes. Lack of adult supervision is related to higher rates of juvenile delinquency, school failure, and teenage pregnancy (Popenoe 1993). A survey of 90,000 teenagers found that the mere physical presence of a parent in the home after school, at dinner, and at bedtime significantly reduces the risk of teenage suicide, violence, and drug use (Resnick et al., 1997). Based on a review of 37 studies involving more than 81,000 adults who experienced the divorce of their parents, researchers found that adult children of divorced parents experienced lower levels of psychological well-being (depression, low life satisfaction), family well-being (low marital quality, divorce), socioeconomic well-being (low educational attainment, income, and occupational prestige), and poorer physical health (Amato and Keith 1991).

CONSIDERATION

Concern about the effects of fatherlessness on children has overshadowed the issue of how child absence affects fathers. Many divorced fathers are overwhelmed with feelings of failure, guilt, anger, and sadness over the separation from their children (Knox 1998). If separation from their children causes anguish to fathers, why do some divorced fathers sever all ties with their children? Hewlett and West (1998) explain that "visiting their children only serves to remind these men of their painful loss, and they respond to this feeling by withdrawing completely" (p. 169).

THE HUMAN SIDE

COLLEGE STUDENTS' EXPERIENCES OF PARENTAL DIVORCE

Untitled
(by Rachel Crenshaw, 1997)

Mom and Dad are getting a divorce,
I guess time has taken its course.
They were great lovers,
And now who knows how long it will take Mom to recover.
Dad says he will still be around for us kids
but Mom knows that won't be enough
For times ahead are going to be rough.
So she gets another job to make ends meet,
And Dad becomes a deadbeat.
No holidays, phone calls, or child support checks like he promised. So much for being honest.
Now that we are out of his life, he has found himself a new wife
And a kid to call him "Dad."
Why does he give her what I never had?
No one knows where my dad is now,
But I'm determined to find him somehow.
It's been seven years since I've seen his face
And now . . . not even a trace.
It is just as well, cause I've turned out just fine,
And my life is in line.
But I can't help but wonder about all those years
. . . when I find him, I will point my finger,
Because he is the cause of so much anger.

What Did I Do Now?
(by Lakeisha Austin, 1998)

What did I do now?
They've had another spat.
I could've sworn I put away my ball and bat.
As he raises his hand to take a swing,
All I can do is close my eyes and wish this was all a dream.
But I know that it's not, it's my reality.
And I feel that this is all happening because of me.
I promise I'll be good and they'll never have to worry.
There has to be some way to make them believe,
But now it's too late and Mommy wants to leave.
But how will we make it without Daddy in our lives?
Mommy doesn't know, but she says that we'll survive.
I feel terrible and I shoulder all the blame.
It's years later now and I truly understand
It wasn't my fault, my Daddy just wasn't a real man.

I Remember a Time
(by Jillian Senn, 1998)

I remember a time a long, long time ago
When my mom and dad were happy, very much so.
They would hold hands and smile when they looked into each
Other's eyes,
And to each other they never told any lies.
We were all a big family, always so happy and free
And someone was always there to take care of "little me."
Then one day something happened and Mommy got very upset
And in my dad's eyes there was a look of regret.

(continued)

National Data

One-fourth of all first-year college students in the United States have parents who are divorced or separated.

SOURCE: "This Year's Freshmen: A Statistical Profile" 1999

THE HUMAN SIDE

COLLEGE STUDENTS' EXPERIENCES OF PARENTAL DIVORCE–(continued)

Daddy moved out and things were never the same,
I couldn't get them together, not even for my basketball games.
I began to see Daddy less and less every year
And my mom began to lose that happy glow and cheer.
Soon Daddy had a new family and I didn't see him anymore,
And Mommy and me were left lonely and poor.
Things are better now and we're doing pretty well,
But something is missing and you can definitely tell.
I remember that time, a long, long time ago
When we were all so happy, very much so.
I will always remember that time a long, long time ago,
But times are different now and I must learn to let go.

If Only It Would Stop
(by Megan Moye, 1998)

Reflecting on her childhood, she remembers the fights
The yelling and screaming, she only wished it would stop.
She would scream at the top of her lungs
"I am going to call Grandma"
Thinking Grandma would be able to help
But Grandma was not capable of fixing her family.
As she got older, the fighting got worse
And she got a little bit wiser.
She vowed "I won't have a marriage like this."
She still wished it would stop.
The little girl grew up, she's a freshman in college now.
She's lived her whole life with two parents.
One day she gets a phone call.
"I am leaving your Father," her Mother says.
She knew it was coming, it had been for a long time.
She finally got her wish.
Now the fighting is over
The family is broken
The children are torn
Yet this is what she wished for.

I'm Not Giving Up!!!
(by Leslie Liles, 1998)

Sitting in my room, my father is coming soon.
Will it make my mom mad, or will she just get sad?
It really breaks my heart to see them so far apart
Will they really try or will they say goodbye?
Wouldn't it be nice for us to be together again
More as a family and not just friends.
It has been a year, and they still shed tears.
I hear a knock, I open the door and go into shock.
I see my father standing there with another woman, and he seems not to care.
My mother comes up and asks how they've been,
As if she always knew that it was really the end.
I guess it's something I'll have to accept,
But I'm not going to give up quite just yet.

NOTE: These poems were written by students in the authors' sociology classes and are reproduced with their permission.

STRATEGIES FOR ACTION: RESPONDING TO THE PROBLEMS OF DIVORCE

Two general strategies for responding to the problems of divorce are (1) strategies to prevent divorce and strengthen marriages and (2) strategies to strengthen postdivorce families.

Divorce Prevention Strategies

One strategy for preventing divorce is to require or encourage couples to participate in premarital education before getting married. Several states have proposed legislation requiring or encouraging premarital education (Clark 1996; Peterson 1997). Proposed policies include mandating premarital education, lowering marriage license fees for those who participate, and imposing delays on issuing marriage licenses for those who refuse premarital education.

Researchers have found that couples who participated in a widely used couples' education program called PREP (the Prevention and Relationship Enhancement Program), had a lower divorce/separation rate 5 years after completing the program compared with couples who did not participate (Stanley et al. 1995). PREP couples also showed significant improvement in conflict management, maintained higher levels of marital satisfaction, and reported significantly lower levels of aggression than did the controls.

Some family scholars and policy makers advocate strengthening marriage by reforming divorce laws to make divorce harder to obtain. At least a dozen states have considered, and in some cases implemented, **divorce law reform** (Brienza 1996). In most cases, these measures are designed to make breaking up harder to do by requiring proof of fault (e.g., adultery, abuse) or extending the waiting period required before divorce is granted.

Opponents argue that divorce law reform measures would increase acrimony between divorcing spouses (which harms the children as well as the adults involved), increase the legal costs of getting a divorce (which leaves less money to support any children), and delay court decisions on child support and custody and distribution of assets. Efforts in many state legislatures to repeal no-fault divorce laws have largely failed.

However, in June 1997, the Louisiana legislature became the first in the nation to pass a law creating a new kind of marriage contract. Under the new law, couples can voluntarily choose between the standard marriage contract that allows a no-fault divorce (after a 6-month separation) or a **"covenant marriage"** that permits divorce only under condition of fault or after a marital separation of more than 2 years. Couples who choose a covenant marriage must also get premarital counseling. Representative Tony Perkins, who sponsored the bill, believes the new law will prevent potentially weak marriages. Perkins explains:

> **When a man says he wants a no-fault marriage and a woman says she wants a covenant marriage, that's going to raise some red flags. She's going to say "What? You're not willing to have a lifelong commitment to me?" (Louisiana Divorce 1997, 1)**

It is too early to assess the effect of Lousiana's new covenant marriage law. Nevertheless, some states are considering similar legislation.

Strengthening Postdivorce Families

Negative consequences of divorce for children may be minimized if both parents continue to spend time with their children on a regular and consistent basis and communicate to their children that they love them and are interested in their lives. Parental conflict, either in intact families or divorced families, negatively influences the psychological well-being of children (Demo 1992; Demo 1993). By maintaining a civil coparenting relationship during a separation and after divorce, parents can minimize the negative effects of divorce on their children. However, about 20 percent of separated and divorced spouses have an angry, hostile relationship (Masheter 1991).

What can society do to promote cooperative parenting by ex-spouses? One answer is to encourage, or even mandate, divorcing couples to participate in **divorce mediation.** In divorce mediation divorcing couples meet with a neutral third party, a mediator, who helps them resolve issues of property division, child custody, child support, and spousal support in a way that minimizes conflict and encourages cooperation. Children of mediated divorces adjust better to the divorce than children of litigated divorces (Marlow and Sauber 1990). An increasing number of jurisdictions and states have mandatory child custody mediation programs, whereby parents in a custody or visitation dispute must attempt to resolve their dispute through mediation before a court will hear their case.

Another trend aimed at strengthening postdivorce families is the establishment of parenting programs for divorcing parents (Shapiro and Schrof 1995). Programs such as "Sensible Approach to Divorce" (Wyandotte County, Kansas) and "Parenting after Divorce" (Orange County, North Carolina) emphasize that it is important for both parents to remain involved in their children's lives. Such programs also teach parents skills to help their children adjust to the divorce and advise parents to avoid expressing hostility or criticism toward the ex-spouse in front of the children. Connecticut and Utah require all divorcing parents to attend a parenting program as do more than 100 courts throughout the United States. Florida goes beyond these programs by offering a course for children of divorcing parents.

National Data

The U.S. teenage birthrate in 1997 was 52.9 births per 1,000 women aged 15 to 19 years—a 15 percent decrease from the rate of 62.1 reported for 1991.

SOURCE: Ventura, Mathews, and Curtin 1998

National Data

In 1996, 11 percent of white births, 23 percent of black births, and 17 percent of Hispanic births were to teenagers. In the same year, 26 percent of white births, 70 percent of black births, and 41 percent of Hispanic births were to unmarried women.

SOURCE: *Statistical Abstract of the United States:1998,* Table 93

UNMARRIED AND TEENAGE PARENTHOOD

In addition to domestic violence and the high rate of divorce, another pressing social problem related to families is the high rate of unmarried and teenage parents. The birthrate for U.S. teenagers has fallen steadily in the 1990s. The rate of teenage mothers who have a second child has also decreased. The current decline in teenage birthrates has been accompanied by decreases in abortion rates as well. Thus, teenage pregnancy rates have fallen in the 1990s (Ventura, Mathews, and Curtin 1998).

Most births to teenagers are nonmarital. However, teenagers do not account for the majority of births to unmarried women. In 1975, more than half of all births to unmarried women were to teenagers, but by 1997, only 31 percent of U.S. births to unmarried women were to teens (Ventura, Mathews, and Curtin 1998).

Black, Hispanic, and Native American women, who are disproportionately poor, are more likely than white women to give birth without being married and to experience teenage pregnancy.

> Marriage has become an almost forgotten institution among black teens. In whole sections of the black community, children are being raised almost exclusively by very young mothers without male role models.
>
> —Daniel Patrick Moynihan
> U.S. Senator

Social Factors That Encourage Unmarried Parenthood

The teenage birthrate in the United States, as measured by births per 1,000 teenagers, is significantly higher than in other industrialized countries. Teenagers in other countries are as sexually active as U.S. teenagers, but they grow up in societies that promote responsible contraceptive use.

In the United States, perceived lack of future occupational opportunities also contributes to teenage parenthood (Luker 1996). Teenage females who do poorly in school may have little hope of success and achievement in pursuing educational and occupational goals. They may think that their only remaining option for a meaningful role in life is to become a parent. In addition, some teenagers feel lonely and unloved and have a baby to create a sense of feeling needed and wanted.

The most rapid rise in single parenthood is among older, educated, professional women, although they still represent only a small fraction of all single mothers (Ingrassia 1993). Most would prefer to be married before having a child, but feel that their biological clock won't wait (Sapiro 1990).

Increased social acceptance of cohabitation also contributes to unmarried parenthood. Births to cohabiting couples are technically considered unwed births, even though the parents are living together.

International Data

The social acceptance of unmarried couples having children varies widely throughout the world. Acceptance of this lifestyle ranges from 90 percent or more in parts of Western Europe to under 15 percent in Singapore and India.

SOURCE: Global Study of Family Values 1998

Social Problems Related to Teenage and Unmarried Parenthood

Teenage and unmarried childbirth are considered social problems because of the consequences typically associated with them. These include the following:

1. *Poverty for single mothers and children.* Many unmarried mothers, especially teenagers, have no means of economic support or have limited earning capacity. Single mothers and their children often live in substandard housing and have inadequate nutrition and medical care. The public bears some of the economic burden of supporting unmarried mothers and their children, but even with public assistance, many unwed and teenage parents often struggle to survive economically.

   The high rate of divorce and out-of-wedlock births means that 60 percent of all children (and 90 percent of black children) will spend some time in a single-parent household (Levitan, Mangum, and Mangum 1998). The 1997 poverty rate of children under age six who lived in female-headed families with no spouse present was 59 percent (Center on Budget and Policy Priorities 1998).
2. *Poor health outcomes.* Teenage and unmarried women are less likely to seek prenatal care and more likely than older and married women to smoke, drink alcohol, and take drugs. These factors have adverse effects on the health of the baby. Indeed, babies born to unmarried or teenage mothers are more likely to have low birth weights and to be born prematurely. Children of teenage and unmarried mothers are also more likely to be developmentally delayed. These outcomes are largely a result of the association between teenage and unmarried childbearing and poverty.
3. *Low academic achievement.* Low academic achievement is both a contributing factor and a potential outcome of teenage parenthood. Teenage mothers are at risk for dropping out of school and, as a consequence, have a much higher probability of remaining poor throughout

> For many disadvantaged teen-agers, childbearing reflects—rather than causes—the limitations of their lives.
>
> —Ellen W. Freeman
> Karl Rickels

National Data

The poverty rate of U.S. married couple families was 5.2 in 1997, compared to 31.6 for families headed by a woman with no husband present.

SOURCE: U.S. Census Bureau 1998

The social consequences of family change over the past few decades have been largely negative, especially for children.

–David Popenoe
Sociologist

Fatherlessness is the most destructive trend of our generation.

–David Blankenhorn
President, American Institute for Family Values

Dad is destiny. More than virtually any other factor, a biological father's presence in the family will determine a child's success and happiness.

–Joseph Shapiro
Joannie Schrof

their lives. Since poverty is linked to unmarried parenthood, a cycle of successive generations of teenage pregnancy may develop.

4. *Children without fathers.* About one-third of U.S. children who live in households without their fathers are the product of unmarried childbirth (Hewlett and West 1998). Shapiro and Schrof (1995) report that children who grow up without fathers are more likely to drop out of school, be unemployed, abuse drugs, experience mental illness, and be a target of child sexual abuse. They also note that "a missing father is a better predictor of criminal activity than race or poverty" (p. 39). Popenoe (1996) believes that fatherlessness is a major cause of the degenerating conditions of our young.

STRATEGIES FOR ACTION: INTERVENTIONS IN TEENAGE PREGNANCY

Some interventions regarding teenage childbearing aim at prevention, while others attempt to minimize its negative effects. One preventive intervention is to provide sex education and family planning programs before unwanted or unintended pregnancy occurs. Schools, churches, family planning clinics, and public health departments may offer sex education programs. Research on the effectiveness of such programs in preventing pregnancy is disheartening, however. Stout and Rivara (1989) reviewed five studies and concluded that school-based sex education programs have little or no effect on reducing teenage pregnancy. However, programs that include sex education curricula with easily accessible health clinics (Jacobs and Wolf 1995) have resulted in decreases in adolescent pregnancy rates.

One obstacle to sex education in the schools is the fear that teaching youth about sex and contraception actually encourages sexual activity. However, the World Health Organization reviewed 35 controlled studies of sex education programs in the United States and Europe and found that students who participated in the programs did not initiate intercourse at an earlier age than students in the control group that were exposed to no programs (Berne and Huberman 1996).

Other programs aim at both preventing teenage and unmarried childbearing and minimizing its negative effects by increasing the life options of teenagers and unmarried mothers. Such programs include educational programs, job training, and skill-building programs. Other programs designed to help teenage and unwed mothers and their children include public welfare (such as WIC), prenatal programs to help ensure the health of the mother and baby, and parenting classes for both teenage fathers and unmarried mothers.

Efforts to prevent teenage pregnancy and minimize its negative effects on teen parents and their children are also built into the 1996 welfare-to-work legislation (Personal Work and Responsibility Act of 1996). Under this law, unmarried minor parents are required to live with a responsible adult or in an adult-supervised setting and participate in educational and training activities in order to receive public assistance. Welfare reform legislation also provides funds for abstinence education to prevent teen births.

In this chapter we have focused on only a few topics related to family concerns—violence and abuse, divorce, and teenage and unmarried parenthood. But families are affected by numerous other problems, discussed elsewhere in this text—problems such as crime; disease, illness, and inadequate health care; alcohol and drugs; gender inequality; prejudices against racial, ethnic, and sexual orientation minorities; poverty; war; and problems at the workplace. This chapter's *Focus on Social Problems Research* feature reveals that issues concerning the workplace and economic concerns are at the top of U.S. parents' political priorities.

FOCUS ON SOCIAL PROBLEMS RESEARCH

Parents' Political Priorities

In 1996, independent pollsters Penn & Schoen conducted a nationwide poll of parents' political priorities for the National Parenting Association (NPA). This research represents an important contribution to policy makers concerned with supporting and strengthening families in the United States. According to Hewlett and West (1998):

> Despite all the political posturing on family values, up until this survey there were virtually no hard data on what kinds of support parents want and need, from government and from the community at large. With all the hand-wringing in Washington over family disintegration, it is astonishing that no one has seen fit to ask the views of those most deeply affected–moms and dads. This survey allowed us to listen to the voices of parents and tease out their most desperate needs, desires, and yearnings (Hewlett and West, 1998, p. xv).

Sample and Method

The sample consisted of 500 parents with children living at home or as their dependents. All respondents were 18 years or older, and welfare recipients were excluded. Respondents were selected at random through computerized random digit dialing and interviewed by trained interviewers.

Findings

The poll results show that a central concern of parents across the nation is a "parental time famine." "Mothers and fathers . . . strive to balance the competing demands of work and family in a world newly dominated by shrinking paychecks, lengthening work weeks, and single parenthood" (Hewlett and West 1998, p. 216). At the top of parents' political "wish list" is a set of policies that allow parents more time for and with their families and that ease the economic burden of raising children. Specific policies and the percentages of parents who favored them are described below.

Time-Enhancing Workplace Policies	Percentage of Parents Favoring
• Government tax incentives to encourage employers to adopt family-friendly policies that offer compressed work weeks, flextime, job-sharing, and benefits for part-time work.	90%
• A law guaranteeing three days of paid leave annually for child-related responsibilities such as attending a parent-teacher conference or taking a child to the dentist.	87%
• Legislation allowing workers to take time off instead of extra pay for overtime.	79%
• Legislation allowing workers to trade two weeks' pay for an extra two weeks of vacation time per year.	71%
• Legislation requiring companies to offer up to twelve weeks of paid job-protected parenting leave following childbirth or adoption. (Companies with fewer than 25 employees would be exempt.)	76%

Relief From the Economic Burdens of Child-Rearing	Percentage of Parents Favoring
• Federal tax deductions/credits to help pay for preschool and college education for families earning less than $100,000.	94%
• Tax breaks on children's necessities, such as diapers, car seats, school supplies, and learning tools. This could be done by eliminating state and local sales taxes on these items.	82%
• A tripling of the dependent deduction for children in families with annual incomes under $100,000, bringing the deduction up to $6,500 per child (almost the level it was at in the early 1950s).	82%
• The raising of wages to lift all full-time working parents above the poverty line.	81%

(continued)

A Longer School Day and School Year	**Percentage of Parents Favoring**
• Keeping schools open longer for classes, supervised homework, or extracurricular activities (so that school schedules better match the working day).	72%

Interviewers also asked parents to indicate how helpful they thought various ideas were in helping parents and families (see Table). The items in Group 2 reflect issues often stressed by advocates of "family values." However, these "family values" issues were not viewed as important to most parents in the survey.

> Parents make it abundantly clear that a focus on practical issues such as income support, tax breaks, and the creation of workplace policies that would enable them to give more and better time to their children would be most helpful. The priorities of the family-values advocates—banning abortion, outlawing same-sex marriage—simply do not register with American parents when they are asked how government and the wider community can best help families with children. (Hewlett and West, 1998, p. 224)

Table: Parents' Priorities

Question: In general, which set of ideas do you think would be more helpful to parents and families, Group 1 or Group 2?

Group 1 (selected by 77% of parents as more helpful to parents and families)

Laws to keep guns away from children.
Letting parents work flexible hours in their jobs.
Raising wages so that all full-time workers are above the poverty level.
Increasing tax deductions/credits for parents with children.
Making the school day and year more in sync with the work day and year.

Group 2 (selected by 16% of parents as more helpful to parents and families)

Banning marriages among people of the same sex.
Making divorce laws punish adultery in awarding support and child custody.
Using tax dollars to help parochial schools.
Cutting off welfare payments after two years to mothers who can't find a job.
Making abortion illegal.

A Collective Voice

Finally, the survey revealed that "parents want a collective voice, an organization that reflects their urgent concerns and can produce new clout for moms and dads in Congress, in state assemblies, on school boards, and in the private sector" (Hewlett and West 1998, p. 225). Eighty-nine percent of parents in the survey favor the creation of a parents' organization similar to the American Association for Retired Persons (AARP). About half say they are strongly in favor. "Millions of moms and dads across this nation are facing an urgent set of unmet needs and an unresponsive political establishment. If they can band together and find collective strength as senior citizens have done, they stand a much better chance of persuading a politician—or a party—to provide the serious support they so desperately need" (Hewlett and West, 1998, p. 228).

SOURCES: National Parenting Association. 1996. *What Will Parents Vote For? : Findings of the First National Survey of Parent Priorities.* New York: NPA.
Hewlett, Sylvia Ann and Cornel West. 1998. *The War Against Parents: What We Can Do for Beleaguered Moms and Dads.* Boston: Houghton Mifflin Company.

UNDERSTANDING **FAMILY PROBLEMS**

Societal definitions of the family have changed with increased acceptance of many previously stigmatized behaviors, such as cohabitation, divorce, and nonmarital parenthood. As these behaviors have become more socially acceptable, they have also become more common.

The impact of family problems, including divorce, abuse, and nonmarital childbearing, is felt not only by family members, but by the larger society as well. Family members experience such life difficulties as poverty, school failure, low self-esteem, and mental and physical health problems. Each of these difficulties contributes to a cycle of family problems in the next generation. The impact on society includes large public expenditures to assist single-parent families and victims of domestic violence and neglect, increased rates of juvenile delinquency, and low academic achievement of children who are struggling to cope with family problems.

> It is absolutely necessary that we create that extended family that we grew up with.
>
> —Yolanda King
> Daughter of Martin Luther King, Jr.

Family problems can best be understood within the context of the society and culture in which they occur. Although domestic violence, divorce, and teenage pregnancy and unmarried parenthood may appear to result from individual decisions, these decisions are influenced by a myriad of social and cultural forces.

Given the social context of family problems, it is important that we look to social intervention for solutions. This chapter's *Focus on Social Problems Research* feature shows that parents want economic help in raising their children and work policies that help them balance their work and family roles. Perhaps with more economic and social supports and more "family-friendly" workplaces, families would be less likely to experience violence and abuse and divorce. Politicians have enormous power in shaping policies that affect families. Recognizing this, Blankenhorn (1995) suggests:

> **The U.S. Congress should pass, and the President should support, a resolution stating that the first question of policy makers regarding all proposed domestic legislation is whether it will strengthen or weaken the institution of marriage. Not the sole question, of course, but always the first. (p. 231)**

CRITICAL THINKING

1. Some scholars and politicians argue that "stable families are the bedrock of stable communities." Others argue that "stable communities and economies are the bedrock of stable families." Which of these two positions would you take and why?
2. Is individualism necessarily incompatible with familism? Why or why not?
3. Research has suggested that secondhand smoke from cigarettes represents a health hazard for those who are exposed to it. Consequently, smoking is now banned in many public places and workplaces. Do you think that parents should be banned from smoking in enclosed areas (home or car) to protect their children from secondhand smoke? Do you think that smoking in enclosed areas with one's children present should be considered a form of child abuse? Why or why not?
4. Some judges are imposing "shame sentences" on convicted abusers. For example, Texas Judge Ted Poe ordered an abusive husband to publicly apologize to his wife on the steps of City Hall and ordered another batterer to go to a local mall carrying a sign that read, "I went to jail for assaulting my wife. This could be you" ("In the News" 1998). What do you think about these types of "shame sentences" for batterers?

5. When divorcing parents participate in mediation, the mediator does not speak of "child custody" and "visitation." Rather, the mediator uses phrases such as "parenting plan" and "plan for spending time with your children." Why do you think mediators use terms such as "parenting plan" and "plan for spending time with your children" rather than the terms "child custody" and "visitation?"

KEY TERMS

blended family
child abuse
covenant marriage
divorce law reform
divorce mediation
domestic partnership
dowry death
elder abuse
emotional abuse
familism
family household
Healthy Families America
"honor murder"
individualism
neglect
no-fault divorce
patriarchy
primary prevention
secondary prevention
self-fulfilling prophecy
tertiary prevention

INTERNET

You can find more information on teenage and unwed parenthood, family violence and abuse, divorce, and fathers' rights organizations at the *Understanding Social Problems* web site at **http://www.sociology.wadsworth.com.**

INFOTRAC COLLEGE EDITION

Either from the Wadsworth sociology resource center at **http://sociology.wadsworth.com** or directly from your web browser, you may access InfoTrac College Edition, an online university library that includes over 700 popular and scholarly journals in which you can find articles related to the topics in this chapter. Suggested articles and questions relating to these articles are listed below.

Mazur, Allan and Joel Michalek. 1998. "Marriage, Divorce and Testosterone." *Social Forces* 77(1):315–25.

1. What method of research was used in this study?
2. What do the authors argue is the relationship between testosterone levels and marital status? Between testosterone levels and criminal behavior?

Griffin, Heidi. 1998. "Courts Order Divorcing Parents to Parenting Classes." *Trial* 34(10):100–03.

1. According to the article, what rights do children have whose parents are going through a divorce?
2. What are some of the topics covered in parenting classes?
3. What does *CCWD* stand for and how does it operate?

Jacobson, Neil S. and John M. Gottman. 1998 (March–April). "Anatomy of a Violent Relationship." *Psychology Today* 31(2):60–69.

1. List some of the findings associated with the researchers' eight year study of abusers.
2. Create a scenario that, according to the article, would typically lead to an abusing situation.
3. Describe the therapy program (HEALS) discussed at the end of the article.

Section 2

Problems of Human Diversity

People are diverse. They vary on a number of dimensions, including age, gender, sexual orientation, and race and ethnicity. In most societies, including the United States, these characteristics are imbued with social significances and are used to make judgments about an individual's worth, intelligence, skills, and personality. Such labeling creates categories of people who are perceived as "different" by others as well as themselves and, as a result, are often treated differently.

A **minority** is defined as a category of people who have unequal access to positions of power, prestige, and wealth in a society. In effect, minorities have unequal opportunities and are disadvantaged in their attempt to gain societal resources. Even though they may be a majority in terms of numbers, they may still be a minority sociologically. Before Nelson Mandela was elected president of South Africa, South African blacks suffered the disadvantages of a minority, even though they were a numerical majority of the population.

Terms that apply to all minorities include stereotyping, prejudice, and discrimination. A **stereotype** is an exaggerated and overgeneralized truth (e.g., "all Puerto Ricans like spicy food"). **Prejudice** is an attitude, often negative, that prejudges an individual. **Discrimination** is differential treatment by members of the majority group against members of the minority that has a harmful impact on members of the subordinate group. The groups we will discuss in this section are all victims of stereotyping, prejudice, and discrimination.

Minority groups usually have certain characteristics in common. In general, members of a minority group know that they are members of a minority, stay within their own group, have relatively low levels of self-esteem, are disproportionately in the lower socioeco-

Nine Characteristics of Four Minorities

	Old/Young	Women	Racial and Ethnic Minorities	Homosexuals
1. Status ascribed based on:	Age	Sex	Race/ethnicity	Sexual orientation
2. Visibility:	High	High	High	Low
3. Attribution of minority status (correctly or incorrectly) based on:	Hair Skin elasticity/Size Posture	Anatomy Shape	Skin color Facial features Hair	Mannerisms Style of dress
4. Summary image:	Dependent	Weak	Inferior	Sick or immoral
5. Derogatory and offensive terms:	Old codger/Brats Battle-Ax/Punks	Bitch Whore	Nigger Spic	Faggot Dyke
6. Control through feigning various characteristics:*	Frailty/Helplessness	Weakness	Ignorance	Heterosexuality
7. Discrimination:	Yes	Yes	Yes	Yes
8. Victims of violence:	Yes	Yes	Yes	Yes
9. Segregation:	Yes	Yes	Yes	Yes

*Being aware of their lack of power, minority group members may try to exert control by feigning certain characteristics. For example, a slave couldn't tell his or her owner, "I'm not going to plow the fields–do it yourself!" But by acting incompetent, the slave may avoid the work. Gays pass as straight by bragging about heterosexual conquests. The elderly in nursing homes whose family members might otherwise not visit act ill in the hope of eliciting a visit. The problem with these acts is that they contribute to and stabilize the summary images.

nomic strata, and are viewed as having negative traits. Other characteristics of specific minority groups are identified in the accompanying table.

In the following chapters, we discuss categories of minorities based upon age (Chapter 6), gender (Chapter 7), race and ethnicity (Chapter 8), and sexual orientation (Chapter 9). Although other categories of minorities exist (e.g., disabled/handicapped, religious minorities), we have chosen to concentrate on these four because each is surrounded by issues and policies that have far-reaching social, political, and economic implications.

CHAPTER SIX

The Young and the Old

- The Global Context: The Young and the Old around the World
- Youth and Aging
- Sociological Theories of Age Inequality
- Problems of Youth in America
- *The Human Side: Growing up in the Other America*
- Demographics: The "Graying of America"
- Problems of the Elderly
- *Self & Society: Alzheimer's Quiz*
- *Focus on Technology: Physician-Assisted Suicide and Social Policy*
- Strategies for Action: The Young and the Old Respond
- Understanding the Young and the Old

Is It True?

Is It True?

1. The more primitive a society, the more likely its members are to practice senilicide—the killing of the elderly.
2. The older a person, the less likely that person is to define a particular age as old.
3. A 1989 bill proposed that teenagers be paid less than mimimum wage during training periods.
4. In the United States, individuals age 65 and older are more likely to vote than any other age group.
5. The U.S. poverty rate for children is more than double that of every other major industrialized country.

Answers: 1 = T; 2 = T; 3 = T; 4 = T; 5 = T

> **For age is opportunity, no less than youth itself; although in another dress.**
> **And as the evening twilight fades away**
> **The sky is filled with stars Invisible by day.**
>
> —Henry Wadsworth Longfellow, Poet

One paradox in life is that the young want to be older and the old want to be younger. Toddlers want to be "old enough to go to school" with their brothers and sisters, children want to be "old enough to go to the mall by themselves," teenagers speak of "being old enough to drive," and the elderly reminisce about the "good old days" and complain of "being too old." Sociologically, at what age are people "old"? What does being "old enough" mean? And how much time is there between being "not old enough" and being "too old"?

While age diversity is often celebrated (e.g., birthdays, high school graduation, retirement), there are countless examples of the way in which social life oppresses children and devalues the elderly. The young and the old represent major population segments of American society. In this chapter, we examine the problems and potential solutions associated with youth and aging. We begin by looking at age in a cross-cultural context.

> I . . . remember when young people had respect for their elders and for the law. Those were the good old days.
>
> —Messages to the next generation AARP On-line

THE GLOBAL CONTEXT: THE YOUNG AND THE OLD AROUND THE WORLD

The young and the old receive different treatment in different societies. Differences in the treatment of the dependent young and old have traditionally been associated with whether the country is developed or less developed. Although there are proportionately more elderly in developed countries than in less developed ones, these societies have fewer statuses for the elderly to occupy. Their positions as caretakers, homeowners, employees, and producers are often usurped by those aged 18 to 65. Paradoxically, the more primitive the society the more likely that society is to practice senilicide—the killing of the elderly. In some societies the elderly are considered a burden and left to die or, in some cases, actually killed.

Not all societies treat the elderly as a burden. Scandinavian countries provide government support for in-home care workers for elderly who can no longer perform such tasks as cooking and cleaning. Eastern cultures such as Japan revere their elderly, in part, because of their presumed proximity to honored ancestors. This is in stark contrast to U.S. youth who view the elderly as "culturally irrelevant" (Kolland 1994). Japanese elders sit at the head of the table, enter a room and bathe first, and are considered the head of the family. While the United States has Mother's Day and Father's Day, Japan has "Respect for Elders Day."

Unlike the United States and other Western countries, Eastern cultures, such as Japan's, revere their elderly partly because of their presumed proximity to honored ancestors.

Societies also differ in the way they treat children. In less developed societies, children work as adults, marry at a young age, and pass from childhood directly to adulthood with no recognized period of adolescence. In contrast, in industrialized nations, children are often expected to attend school for 12 to 16 years and, during this time, to remain financially and emotionally dependent on their families.

Because of this extended period of dependence, the United States treats "minors" differently than adults. There is a separate justice system for juveniles and age limits for driving, drinking alcohol, joining the military, entering into a contract, marrying, dropping out of school, and voting. These limitations would not be tolerated if placed on individuals on the basis of sex or race.

There were only two periods in the life of a Chinese male when he possessed maximum security and minimal responsibility—infancy and old age. Of the two, old age was the better because one was conscious of the pleasure to be derived from such an almost perfect period.

–Paul T. Welty
Historian

International Data

One-fifth of the world's elderly population is Chinese. By 2020, there will be over 230 million elderly Chinese, about 15.6 percent of their total population.

SOURCE: DHHS 1998b

Hence **ageism**, the belief that age is associated with certain psychological, behavioral and/or intellectual traits, at least in reference to children, is significantly more tolerated than sexism or racism in the United States.

Despite this differential treatment, people in the United States are fascinated with youth and being young. This was not always the case. The elderly were once highly valued in the United States—particularly older men who headed families and businesses. Younger men even powdered their hair, wore wigs, and dressed in a way that made them look older. It should be remembered, however, that in 1900 only 4.1 percent of the population was 65 or older; over 40 percent of the population was under 18 (Harris 1990; DHHS 1998b). Being old was rare and respected; to some it was a sign that God looked upon the individual favorably.

One theory argues that the shift from valuing the old to valuing the young took place during the transition from an agriculturally based society to an industrial one. Land, which was often owned by elders, became less important as did their knowledge and skills about land-based economies. With industrialization, technological skills, training, and education became more important than land ownership. Called **modernization theory**, this position argues that as a society becomes more technologically advanced, the position of the elderly declines (Cowgill and Holmes 1972).

CONSIDERATION

Using participant observation techniques, Cahill (1993) recorded the treatment of children in public places for nearly 300 hours over a 2-year period. He observed that children are restricted in the places they go, for example, often discouraged from eating in some restaurants, and excluded from some apartment complexes. Further, public accommodations don't adequately meet the needs of children—pay telephones, sinks, and water fountains are out of children's reach. Cahill concludes that such limitations, as well as the segregation of children in schools, limits communication between children and adults, creating differences in their respective social worlds. The result is ". . . a distinct [youth] subculture of which adults are only vaguely aware . . ." (p. 400).

YOUTH AND AGING

Age is largely socially defined. Cultural definitions of "old" and "young" vary from society to society, from time to time, and from person to person. For example, the older a person, the less likely that person is to define a particular age as old. In a national study of more than 2,503 men and women aged 18 to 75, 30 percent of those under 25 responded that "old" is between 40 and 64 years of age. But only 8 percent of those over the age of 65 reported that 65 was old (Clements 1993). In ancient Greece or Rome where the average life expectancy was 20 years, one was old at 18; similarly, one was old at 30 in medieval Europe and at 40 in the United States in 1850.

Age is also a variable that has a dramatic impact on one's life (Matras 1990 identified 1–4):

1. Age determines one's life experiences since the date of birth determines the historical period in which a person lives. As Harrigan notes (1992), 50 years ago Americans couldn't have imagined metal detectors at school entrances.

2. Different ages are associated with different developmental stages (physiological, psychological, and social) and abilities. Ben Franklin observed, "[A]t 20 years of age the will reigns; at 30 the wit; at 40 judgment."
3. Age defines roles and expectations of behavior. The expression "act your age" implies that some behaviors are not considered appropriate for people of certain ages.
4. Age influences the social groups to which one belongs. Whether one is part of a sixth grade class, a labor union, or a seniors' bridge club depends on one's age.
5. Age defines one's legal status. Sixteen-year-olds can get a driver's license, 18-year-olds can vote and get married without their parents' permission, and 65-year-olds are eligible for Social Security benefits.

That's exactly how it goes. You're young and the whole world is ahead of you, and then one day you're not young anymore, and the whole world is behind you, and it's all over.

—David Letterman
Comedian

Childhood, Adulthood, and Elderhood

Every society assigns different social roles to different age groups. **Age grading** is the assignment of social roles to given chronological ages (Matras 1990). Although the number of age grades varies by society, most societies make at least three distinctions—childhood, adulthood, and elderhood.

National Data

In 1997, the median age was 35 years: in 2025 the median age is projected to be 38 years.

SOURCE: *Statistical Abstract of the United States: 1998*, Table 13

Childhood The period of childhood in our society is from birth through age 17 and is often subdivided into infancy, childhood, and adolescence. Infancy has always been recognized as a stage of life, but the social category of childhood only developed after industrialization, urbanization, and modernization took place. Prior to industrialization, infant mortality was high due to the lack of adequate health care and proper nutrition. Once infants could be expected to survive infancy, the concept of childhood emerged, and society began to develop norms in reference to children. In the U.S., child labor laws prohibit children from being used as cheap labor, educational mandates require that children attend school until the age of 16, and federal child pornography laws impose severe penalties for the sexual exploitation of children.

Adulthood The period from age 18 through 64 is generally subdivided into young adulthood, adulthood, and middle age. Each of these statuses involves dramatic role changes related to entering the workforce, getting married, and having children. The concept of "middle age" is a relatively recent one that has developed as life expectancy has been extended. Some people in this phase are known as members of the "**sandwich generation**" since they are often emotionally and economically responsible for both their young children and their aging parents.

National Data

Approximately 46 million Americans are over the age of 60; of that number, Americans over the age of 100, that is, the old-old, are the fastest growing age group.

SOURCE: DHHS 1998b

Elderhood At age 65 one is likely to be considered elderly, a category that is often subdivided into the young-old, old, and old-old. Membership in one of these categories does not necessarily depend on chronological age. The growing number of healthy, active, independent elderly are often considered to be the young-old, whereas the old-old are less healthy, less active, and more dependent.

SOCIOLOGICAL THEORIES OF AGE INEQUALITY

Three sociological theories help explain age inequality and the continued existence of ageism in the United States. These theories—structural-functionalism, conflict theory, and symbolic interactionism—are discussed in the following sections.

Structural-Functionalist Perspective

Structural-functionalism emphasizes the interdependence of society—how one part of a social system interacts with other parts to benefit the whole. From a functionalist perspective, the elderly must gradually relinquish their roles to younger members of society. This transition is viewed as natural and necessary to maintain the integrity of the social system. The elderly gradually withdraw as they prepare for death, and society withdraws from the elderly by segregating them in housing such as retirement villages and nursing homes. In the interim, the young have learned through the educational institution how to function in the roles surrendered by the elderly. In essence, a balance in society is achieved whereby the various age groups perform their respective functions: the young go to school, adults fill occupational roles, and the elderly, with obsolete skills and knowledge, disengage. As this process continues, each new group moves up and replaces another, benefitting society and all of its members.

This theory is known as **disengagement theory** (Cummings and Henry 1961). Some researchers no longer accept this position as valid, however, given the increased number of elderly who remain active throughout life (Riley 1987). In contrast to disengagement theory, **activity theory** emphasizes that the elderly disengage in part because they are structurally segregated and isolated, not because they have a natural tendency to do so. For those elderly who remain active, role loss may be minimal. In studying 1,720 respondents who reported using a senior center in the previous year, Miner, Logan, and Spitze (1993) found that those who attended were less disengaged and more socially active than those who did not.

Although some elderly people may like to skateboard and some young people may like to play bridge, age tends to influence the range of behavior each group engages in.

Conflict Perspective

The conflict perspective focuses on age grading as another form of inequality as both the young and the old occupy subordinate statuses. Some conflict theorists emphasize that individuals at both ends of the age continuum are superfluous to a capitalist economy. Children are untrained, inexperienced, and neither actively producing nor consuming in an economy that requires both. Similarly, the elderly, although once working, are no longer productive and often lack required skills and levels of education. Both young and old are considered part of what is called the dependent population; that is, they are an economic drain on society. Hence, children are required to go to school in preparation for entry into a capitalist economy, and the elderly are forced to retire.

Other conflict theorists focus on how different age strata represent different interest groups that compete with one another for scarce resources. Debates about funding for public schools, child health programs, Social Security, and Medicare largely represent conflicting interests of the young versus the old.

Symbolic Interactionist Perspective

The symbolic interactionist perspective emphasizes the importance of examining the social meaning and definitions associated with age. The elderly are often defined in a number of stereotypical ways contributing to a host of myths surrounding the inevitability of physical and mental decline. Table 6.1 identifies some of these myths.

Media portrayals of the elderly contribute to the negative image of the elderly. The young are typically portrayed in active, vital roles and are often overrepresented in commercials. In contrast, the elderly are portrayed as difficult, complaining, and burdensome and are often underrepresented in commercials. A recent study of the elderly in popular 1940s through 1980s films concluded that "[O]lder individuals of both genders were portrayed as less friendly, having less romantic activity, and enjoying fewer positive outcomes than younger characters at a movie's conclusion" (Brazzini et al. 1997, 541).

The elderly are also portrayed as childlike in terms of clothes, facial expressions, temperament, and activities—a phenomenon known as **infantilizing elders** (Arluke and Levin 1990). For example, young and old are often paired together. A promotional advertisement for the movie *Just You and Me, Kid* with Brooke Shields and George Burns described it as "the story of two juvenile delinquents." Jack Lemmon and Walter Matthau in *Grumpy Old Men* get "cranky" when they get tired, and the media focus on images of Santa visiting nursing homes and local elementary school children teaching residents arts and crafts. Finally, the elderly are often depicted in role reversal, cared for by their adult children as in the situation comedies *Golden Girls* and *Frazier.*

Negative stereotypes and media images of the elderly engender **gerontophobia**—a shared fear or dread of the elderly, which may create a self-fulfilling prophecy. For example, an elderly person forgets something and attributes his or her behavior to age. A younger person, however, engaging in the same behavior, is unlikely to attribute forgetfulness to age given cultural definitions surrounding the onset of senility. Thus, the elderly, having learned the social meaning associated with being old, may themselves perpetuate the negative stereotypes.

Table 6.1 Myths and Facts about the Elderly

Health

Myth The elderly are always sick; most are in nursing homes.

Fact Over 80 percent of the elderly are healthy enough to engage in their normal activities. Only 6 percent are confined to a nursing home.

Automobile Accidents

Myth The elderly are dangerous drivers and have a lot of wrecks.

Fact Drivers ages 45 to 74 have only 12 accidents per 100 drivers per year. Drivers under age 20 have 37 accidents per 100 drivers per year. Up to age 75, older drivers tend to drive fewer miles than younger drivers and compensate for their reduced reaction time by driving more carefully. However, drivers 75 and older have 25 accidents per 100 drivers per year. The increased incidence is "rooted in the normal processes of aging: diminishing vision and hearing and decreasing attention spans" (Carney 1989).

Mental Status

Myth The elderly are senile.

Fact Although some of the elderly learn more slowly and forget more quickly, most remain oriented and mentally intact. Only 20–25 percent develop Alzheimer's disease or some other incurable form of brain disease. Senility is not inevitable as one ages.

Employment

Myth The elderly are inefficient employees.

Fact Although only about 17 percent of men and 9 percent of women 65 years old and over are still employed, those who continue to work are efficient workers. When compared to younger workers, the elderly have lower job turnover, fewer accidents, and less absenteeism. Older workers also report higher satisfaction in their work.

Politics

Myth The elderly are not politically active.

Fact In 1986, 1988, 1990, 1992, 1994, and 1996 individuals age 65 and older were more likely to be registered to vote and/or to vote than any other age group.

Sexuality

Myth Sexual satisfaction disappears with age.

Fact Many elderly persons report sexual satisfaction. For example, in a study of 61 elderly men (average age = 71) both with and without sexual partners, sexual satisfaction was rated at an average of 6.3 on a scale where 1 = no satisfaction and 10 = extremely high satisfaction (Mulligan and Pagluta 1991).

Adaptability

Myth The elderly cannot adapt to new working conditions.

Fact A high proportion of the elderly are flexible in accepting change in their occupations and earnings. Adaptability depends on the individual: many young are set in their ways, and many older people adapt to change readily.

SOURCES: Robert H. Binstock. 1986. "Public Policy and the Elderly." *Journal of Geriatric Psychiatry* 19: 115–43; Teresa E. Seeman and Nancy Adler. 1998. "Older Americans: Who Will They Be?" *National Forum,* Spring, 22–25; James Carney. 1989. "Can a Driver Be Too Old?" *Time,* January 16, 28; Administration on Aging. 1998. "Employment and Older Americans: A Winning Partnership." *Fact Sheet.* Department of Health and Human Services. Washington, D.C.; T. Mulligan, and R. F. Palguta Jr. 1991. "Sexual Interest, Activity and Satisfaction among Male Nursing Home Residents." *Archives of Sexual Behavior* 20: 199–204; Erdman B. Palmore. 1984. "The Retired." In *Handbook on the Aged in the United States,* ed. Erdman B. Palmore, pp. 63–76, Westport, Conn.: Greenwood Press; and *Statistical Abstract of the United States: 1997,* 117th ed. U.S. Bureau of the Census. Washington, D.C.: U.S. Government Printing Offices.

PROBLEMS OF YOUTH IN AMERICA

> For in its innermost depths, youth is lonelier than old age.
>
> —Unknown

By the year 2000, there will be an estimated 17.9 million children under the age of 5, 35.7 million between the ages of 5 and 13, and 15.6 million between the ages of 14 and 17 for a total of 69.2 million "youth" in America (*Statistical Abstract of the United States: 1998,* Table 17). In spite of the presumed benefits of being young, numerous problems are associated with childhood. Further, some of our most pressing social problems can be traced to early childhood experiences and adolescent behavioral problems (Weissberg and Kuster 1997).

A 1998 Children's Defense Fund report documents the condition of children in America (1998a). Not surprising, what happens to children is increasingly defined as a social problem (see Table 6.2).

Children and the Law

> American children now stand in a transitional state between chattel and constitutionally protected child-citizen.
>
> —Charles Gill
> Attorney

Historically, children have had little control over their lives. They have been "double dependent" on both their parents and the state. Indeed, colonists in America regarded children as property. Beginning in the 1960s, however, the view that children should have more autonomy became popular and was codified in several legal decisions.

The dominant view of children today, as expressed by the courts as well as the public, involves taking "a protective stance toward children rather than empowering children to care for themselves" ("Children's Legal Rights" 1993, 342; Regoli and Hewitt 1997). Examples of protective legislation include requiring child abuse prevention and treatment programs, ensuring education for disabled children, and setting the constitutional threshold for executions at age 16. Alternatively, empowering children includes such cases as Gregory Kingsley's "divorce" from his mother and more recent court decisions that provide that children aged 15 or older should be permitted to testify as to their custody and visitation preferences.

In spite of these gains, children in the United States have received only limited empowerment. The United States has not ratified the United Nations'

Table 6.2 Every Day in America . . .

3	people under 25 die from HIV
6	children commit suicide
13	children are murdered
14	children die from firearms
81	babies die
443	babies are born to mothers with late or no prenatal care
781	low-birth-weight babies are born
1,403	babies are born to teenage mothers
1,827	babies are born without health insurance
2,430	babies are born into poverty
2,756	children drop out of school
3,436	babies are born to unwed mothers
8,470	children are reported abused or neglected

SOURCE: Children's Defense Fund 1998a.

Treaty on Children's Rights although 132 countries have (UNICEF 1998). One reason the United States has not signed the pact may be Article 11, which requires that children be assured "the highest attainable standard of health." Some people are concerned that accepting such a position would result in cases being brought to court that they are simply unwilling, at present, to hear.

Children are both discriminated against and granted special protections under the law. Although legal mandates require that children go to school until age 16, other laws provide a separate justice system whereby children have limited legal responsibility based upon their age status. As violence by minors increases, many Americans have begun to question the wisdom of a separate legal structure for minors. The laws are changing: most states have lowered the age of accountability, capital punishment of 16-year-olds is allowed in a number of states, and the number of parent-liability laws has increased. While children's rights may expand in some areas such as self-determination, recent legal changes may cost minors their protected status in the courts.

Economic Discrimination

Children are discriminated against in terms of employment, age restrictions, wages, training programs, and health benefits. Traditionally, children worked on farms and in factories but were displaced by the Industrial Revolution. In 1938, Congress passed the Fair Standards Labor Act, which required factory workers to be at least 16. Although the law was designed to protect children, it was also discriminatory in that it prohibited minors from having free access to jobs and economic independence. More recently, a 1989 federal bill proposed that employers pay teenagers less than the minimum wage during their initial training period. No such law exists for any other age group.

National Data

In 1997, the median hourly wage was $8.84; the median hourly wage for those over 25 was $9.91; for those 16–24, $6.22, and for those 16–19, $5.84.

SOURCE: *Statistical Abstract of the United States: 1998,* Table 700

Further, although the federal government gives the American Association of Retired Persons $75 million annually to provide training and placement programs for senior citizens, the 1996 unemployment rate for persons 16–19 was 16.0 percent, the highest for any age category (George 1992; *Statistical Abstract of the United States: 1998,* Table 677). Regarding health benefits, the federal government spends more on health care for the elderly during their last year of life than on all the health care needs of children—11 times as much money on those 65 and over as on those under the age of 18 (Hewlett 1992; Peterson 1997).

Worldwide, there is also evidence that when global economic decisions are made, children suffer. Bradshaw et al. (1992) observed that the global debt crisis has led the world's business community to pressure developing countries into financial strategies that directly affect the well-being of children. For example, funding for immunization programs for children is cut out of economic necessity. "This potential relationship between international debt and children's quality of life is one of the most important sociological issues in poor countries today" (p. 630).

National Data

Minority children are disproportionately poor. In 1997 while 11 percent of white children were poor, 39.9 percent of African-American children, 40.3 percent of Hispanic-American children, and 19.5 percent of Asian/Pacific Islander children were poor.

SOURCE: Children's Defense Fund 1998c

Because of economic discrimination at both the individual and the institutional level, many children live in poverty. The United States does not have a universal government policy to protect children. According to a recent Columbia University study, children are the poorest age group in America—twice as poor as those over the age of 65 (NCCP 1998). The Social Security system keeps many elderly out of poverty and, since 1972, has been indexed to keep up with inflation. But Aid to Families with Dependent Children (now called TANF or Temporary Assistance to Needy Families) has been cut 42 percent in

Officially, about one in five children in the United States lives in poverty. The United States has the highest child poverty rate of 18 industrialized nations and the lowest government benefits to families with poor children.

real dollars since 1970. Social Security keeps 8 out of 10 elderly from being poor; AFDC kept less than 1 in 3 children out of poverty (Taylor 1993), and recent welfare reform is likely to reduce that number even further—two-thirds of welfare recipients are children (Children Now 1998).

By the end of the 20th century, a child in America is almost twice as likely to be poor as an adult. This is a condition that has never before existed in our history.

–Danial Patrick Moynihan
U.S. Senator and social scientist.

CONSIDERATION

The United States has the highest child poverty rate of the 18 industrialized nations. Based on a recent worldwide study of children, the U.S. poverty rate for children is more than double that of every other major industrialized country. In 1999, 22 percent of U.S. children lived below the poverty line, up from 15 percent in 1960 (UNICEF 1994; Bergmann 1999). Childhood poverty is related to school failure (Fields and Smith 1998), negative involvement with parents (Harris and Marmer 1996), stunted growth, reduced cognitive abilities, limited emotional development (Brooks-Gunn and Duncan 1997), and a higher likelihood of dropping out of school (Duncan et al. 1998).

So long as little children are allowed to suffer, there is no true love in this world.

–Isadora Duncan
Dancer

Children and Violence

Today, gang violence, suicide, child abuse, and crime are all-too-common childhood experiences (see this chapter's *The Human Side*). In 1996, almost half of all high school males and one-third of high school females reported being involved in a physical fight during the previous year, and 20 percent reported carrying a weapon to school (DHHS 1998a). In the same year, there were 3.1 million cases of child abuse or neglect, over 4,700 children killed by firearms, and some 2,000 youth suicides (*Yearbook* 1998). Between 1950 and 1993 homicide and suicide rates among children under 15 years of age tripled and quadrupled respectively (AP 1997). Reasons for such increases include poverty, the portrayal of violence in the media, increased substance abuse, and greater access to guns (Children's Defense Fund 1998b).

International Data

The victimization of children is a worldwide phenomenon. In the Philippines there are an estimated 60,000 to 100,000 child prostitutes; in Brazil there are over 100,000 "street children"; and in Africa there will be an estimated 5 million AIDS orphans by the year 2000.

SOURCE: World Congress 1998

Kids in Crisis

Childhood is a stage of life that is socially constructed by structural and cultural forces of the past and present. The old roles for children as laborers and

> The job of parenting is being devalued, and with it the quality of children's lives and society's future.
>
> –"The Progress of Nations" (UNICEF)

THE HUMAN SIDE

GROWING UP IN THE OTHER AMERICA

Author Alex Kotlowitz conducted a 2-year participant observation research study of children in a Chicago housing project. The following description captures the horrific living conditions endured by Lafeyette, Pharoah, and Dede—three children living in the "jects."

The children called home "Hornets" or, more frequently, "the projects" or, simply, the "jects" (pronounced jets). Pharoah called it "the graveyard." But they never referred to it by its full name: the Governor Henry Horner Homes.

Nothing here, the children would tell you, was as it should be. Lafeyette and Pharoah lived at 1920 West Washington Boulevard, even though their high-rise sat on Lake Street. Their building had no enclosed lobby; a dark tunnel cut through the middle of the building, and the wind and strangers passed freely along it. Those tenants who received public aid had their checks sent to the local currency exchange, since the building's first-floor mailboxes had all been broken into. And since darkness engulfed the building's corridors, even in the daytime, the residents always carried flashlights, some of which had been handed out by a local politician during her campaign.

Summer, too, was never as it should be. It had become a season of duplicity.

On June 13, a couple of weeks after their peaceful afternoon on the railroad tracks, Lafeyette celebrated his twelfth birthday. Under the gentle afternoon sun, yellow daisies poked through the cracks in the sidewalk as children's bright faces peered out from behind their windows. Green leaves clothed the cottonwoods, and pastel cotton shirts and shorts, which had sat for months in layaway, clothed the children. And like the fresh buds on the crabapple trees, the children's spirits blossomed with the onset of summer.

Lafeyette and his nine-year-old cousin Dede danced across the worn lawn outside their building, singing the lyrics of an L. L. Cool J rap, their small hips and spindly legs moving in rhythm. The boy and girl were on their way to a nearby shopping strip, where Lafeyette planned to buy radio headphones with $8.00 he had received as a birthday gift.

Suddenly, gunfire erupted. The frightened children fell to the ground. "Hold your head down!" Lafeyette snapped, as he covered Dede's head with her pink nylon jacket. If he hadn't physically restrained her, she might have sprinted for home, a dangerous action when the gangs started warring. "Stay down," he ordered the trembling girl.

The two lay pressed to the beaten grass for half a minute, until the shooting subsided. Lafeyette held Dede's hand as they cautiously crawled through the dirt toward home. When they finally made it inside, all but fifty cents of Lafeyette's birthday money had trickled from his pockets.

SOURCE: Alex Kotlowitz. 1991. *There Are No Children Here: The Story of Two*...by Alex Kotlowitz.

farm helpers are disappearing, yet no new roles have emerged. While being bombarded by the media, children must face the challenges of an uncertain economic future, peer culture, music videos, divorce, incidents of abuse, poverty, and crime. Parents and public alike fear children are becoming increasingly involved with sex, drugs, alcohol, and violence. Some even argue that childhood as a stage of life is disappearing (Adler 1994).

In a poll commissioned by *Newsweek* and the Children's Defense Fund, 758 children between the ages of 10 and 17 reported their perspectives on life:

> **What emerges is a portrait of a generation living in fear. The security of their parents' generation, and the optimistic view of the future, is no longer taken for granted by young people today. For them, the poll found, the American Dream may be dying. (Ingrassia 1993, 52)**

Many reported that they feared being victims of violent crime (56 percent), not being able to afford a doctor (51 percent), and that their parents would get a divorce (p. 52). A study by the Fordham Institute on the social health of children also found a general decline in the well-being of American youth (Miringoff 1989).

> Knowledge by definition puts an end to innocence.
>
> —Philip Adler
> Writer

National Data

In a 1998 survey in which parents were asked whether or not they felt good about their children's future, 62.5 percent responded they felt good, 16.6 percent responded that they did not feel good, and 19.3 percent didn't know.

SOURCE: Harris Poll 1998a

DEMOGRAPHICS: THE "GRAYING OF AMERICA"

The population of America, as in many other countries around the world is "graying," that is, getting older. Here, the definition of "older" is age 65 or beyond. The origin of this arbitrary age is the Social Security Act of 1935, which established 65 as the age of retirement.

The number of elderly is increasing for three reasons (see Figure 6.1). First, 76 million baby boomers born between 1946 and 1964 are getting older. Second, life expectancy has increased as a result of better medical care; sanitation, nutrition, and housing improvements; and a general trend toward modernization. Finally, lowered birthrates contribute to a higher percentage of the elderly. Since the proportion of the elderly is influenced by such variables as the number of baby boomers, life expectancy, and birthrates, different countries have different proportions of the elderly. Western Europe and the Scandinavian countries have a higher percentage of the elderly than the United States; Canada, Japan, Israel, and the former Soviet Union have lower percentages.

National Data

By the year 2030 there will be 70 million elderly in the United States, making up 20 percent of the total population and more than two times the number of elderly in 1997.

SOURCE: AOA 1999

Age Pyramids

Age pyramids are a way of showing in graph form the percentage of a population in various age groups. In 1900 the U.S. age pyramid looked very much like a true pyramid: the base of the pyramid was large, indicating that most people were in their younger years, and the top of the pyramid was much smaller, showing that only a small percentage of the population was elderly. By the year 2030, the "pyramid" will look more like a pillar with the exception of the very top, which will reflect the large proportion of elderly people in the population (see Figure 6.1).

The number of people at various ages in a society is important because the demand for housing, education, health care, and jobs varies as different age groups, particularly baby boomers, move through the pyramid. For example, as

> This is the century of old age, or, as it has been called, the "Age of Aging."
>
> —Robert Butler
> Gerontologist

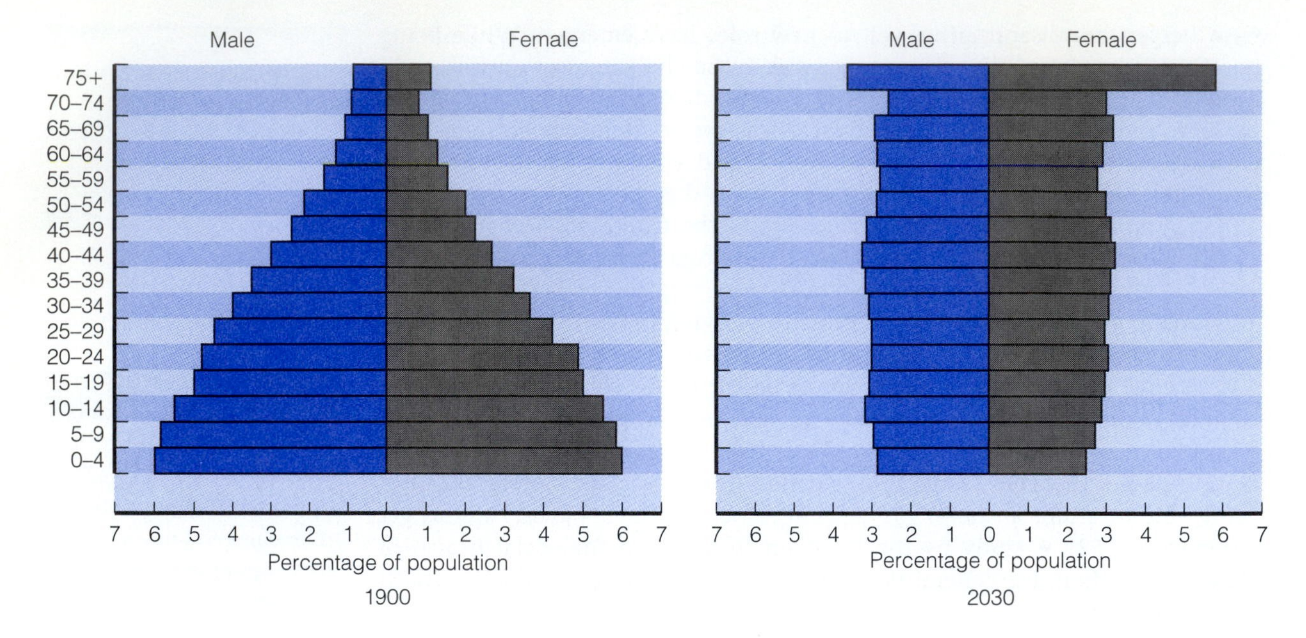

Figure 6.1 U.S. Population Pyramids

SOURCE: *Statistical Abstract of the United States: 1993,* 113th ed. U.S. Bureau of Census. Washington D.C.: U.S. Government Printing Office, Table 17 for 2030 projections.

America "grays," colleges will recruit older students, advertisements will be directed toward older consumers, and elderly housing and medical care needs will increase.

Age and Race and Ethnicity

In 1997, about 15 percent of the elderly were racial/ethnic minorities characterized by "multiple jeopardy," that is, they had more than one minority status. Minority populations are projected to represent 25 percent of the U.S. elderly population by 2030, up from 13 percent in 1990. While the elderly white population is projected to increase 79 percent between 1997 and 2030, the elderly minority population is expected to increase 238 percent in the same time period. The largest increases are for Hispanic Americans, followed by Asian Americans and Pacific Islanders, American Indians, Eskimos and Aleuts, and African Americans. Although racial and ethnic minorities have a shorter life expectancy than whites, their increased numbers in the general population are responsible for their higher growth rates of the elderly (AOA 1999).

National Data

Elderly women are five times more likely to be widows than elderly men are to be widowers; older women also have a higher poverty rate than older men, and older black women living alone are the poorest of the elderly.

SOURCE: AOA 1998c

Age and Gender

In the United States, elderly women outnumber elderly men. The 1996 sex ratio for ages 65 and over was 145 women for every 100 men. Because differences increase with advanced age, at age 85 and over, there are 257 females for every 100 males (AOA 1998c). Men die at an earlier age than women for both biological and sociological reasons—heart disease, stress, and occupational risk (see Chapter 2). The fact that women live longer results in a sizable number of elderly women who are poor. Not only do women, in general, make less money than men, but older women may have spent their savings on their husband's illness, and as homemakers, they are not eligible for Social Security benefits. Further, retirement benefits and other major sources of income may be lost with a husband's death. Almost three-quarters of the elderly living in poverty are women. Poverty among elderly minority women is especially high—three out of five African-American and two out of five Hispanic elderly women are living in poverty (AOA 1998c).

Age and Social Class

How long a person lives is influenced by his or her social class. In general, the higher the social class, the longer the person lives, the fewer the debilitating illnesses, the greater the number of social contacts and friends, the less likely to define oneself as "old," and the greater the likelihood of success in adapting to retirement. Higher social class is also related to fewer residential moves, higher life satisfaction, more leisure time, and more positive self-rated health. Of those 65 and older, 26 percent of those with annual incomes of $35,000 or higher reported that their health was "excellent," whereas only 10 percent of those with incomes under $10,000 rated their health as such. Functional limitations such as problems with walking, dressing, and bathing were also lower among higher income groups (Seeman and Adler 1998). In short, the higher one's socioeconomic status, the longer, happier, and healthier one's life.

PROBLEMS OF THE ELDERLY

Many of the problems of the elderly can be traced to the ideology of ageism. In response to this ideology, Betty Friedan (1993) advocates abandoning the cultural emphasis on youth and moving toward "the fountain of age." She suggests looking

> **into this new period of life with openness, with change, with vitality . . . It's time to look at age on its own terms, and put names on its values and strengths, breaking through the definition of age solely as deterioration or decline from youth. The problem is how to break through the cocoon of our illusory youth and risk a new stage in life, where there are no prescribed roles, no models, nor rigid rules or visible rewards—how to step out into the true existential unknown of these years of life now open to us and find our own terms for living them. (p. 69)**

Friedan's book, *The Fountain of Age* (1993), describes how most Americans view the aging process with fear and anxiety over what is believed to be inevitable physical and mental decline. She argues that the problem is not age per se, but the inability to look beyond youth to the possibilities of creative aging. She warns, however, that the elderly cannot reach their full potential if society's barriers, both structural and cultural, remain.

National Data

In 1970, about 27 percent of men and 10 percent of women aged 65 and over participated in the labor force. By 1997, these rates had decreased to 17 percent and 9 percent, respectively.

SOURCE: AOA 1999

Lack of Employment

What one does (occupation), for how long (work history), and for how much (wages), are important determinants of retirement income. Indeed, employment is important because it provides the foundation for economic resources later in life. Yet, for the elderly who want to work, entering and remaining in the labor force may be difficult because of negative stereotypes, lower levels of education, reduced geographical mobility, fewer employable skills, and discrimination. The likelihood of being employed has decreased for elderly men and women over the last few decades.

In 1967, Congress passed the Age Discrimination in Employment Act (ADEA), which was designed to ensure continued employment for people between the ages of 40 and 65. In 1986, the upper limit was removed, making mandatory retirement illegal in most occupations. Nevertheless, thousands of cases of age discrimination occur annually. While employers cannot advertise a position by age, they can state that the position is an "entry level" one or that "2–3 years' experience" is required. However, research on job-training opportunities at 688 organizations and businesses in the United States found that the elderly, in general, were unlikely to be discriminated against (Knoke and Kalleberg 1994).

If displaced, older workers have difficulty finding a new job. Some research indicates that older workers remain unemployed longer than younger workers, often have to accept lower salaries, and may be more likely to give up looking for work (Moone and Hushbeck 1989; AOA 1998a).

Retirement

Retirement is a relatively recent phenomenon. Prior to Social Security, individuals continued to work into old age. Today, although the government no longer requires a person to retire at a specific age, it limits Social Security payments to those over 65 years of age—age 67 by the year 2022. If recent proposals in Congress are passed, age limits will be extended even further—age 68 by the year 2017 and 70 by the year 2065 (AARP Bulletin 1998). Such policies are certainly an inducement to retire after 65. Many people, however, retire before age 65; 50 percent of all men and an even higher percentage of women retire before they are eligible to receive Social Security benefits.

Retirement is difficult in that in the U.S. "work" is often equated with "worth." A job structures one's life and provides an identity; the end of a job culturally signifies the end of one's productivity. Retirement may also involve a dramatic decrease in personal income. Consequently, the desire to remain financially independent and a lack of confidence in the Social Security system have lead many workers to remain in the labor force longer than they would have expected (Simon-Rusinowitz et al. 1996).

In spite of the potential problems with retirement, most people retire willingly. In a study of Finnish adults 35 and over, 55 percent of the workers said they would stop work immediately upon reaching retirement age (Huuhtanen and Piispa 1996). Most retirees report enjoying retirement and having a greater sense of well-being than younger people who are still working (Russell 1989). Some workers benefit from **phased retirement**, which permits them to withdraw gradually. As the number of elderly and, thus, the number of retirees increase, it will become easier to retire—there will be more retirement communities, senior citizen discounts, and products for the elderly. Negative stereotypes of the retired elderly may also change.

National Data

Recent data indicate that less than 30 percent of Americans have long-term savings plans; half of all American families have less than $1,000 in financial assets; and Americans in their late 50s have a median savings of less than $10,000.

SOURCE: Peterson 1998

Poverty

A considerable proportion of the elderly report being concerned about money. When a national sample of U.S. citizens was asked what would provide the greatest feeling of security in old age, 40 percent reported that money was at the top of their list compared to spouses (29 percent), children (12 percent), friends (5 percent), career and job (4 percent), or home (4 percent) (Clements 1993). As Table 6.3 indicates, such concern is justified.

Poverty among the elderly also varies dramatically by such characteristics as gender, race, ethnicity, marital status, and age: women, minorities, those who are single or widowed, and the "old-old" are most likely to be poor (AOA 1999; *Statistical Abstract of the United States: 1997,* Table 739). There is also a great deal of diversity among the elderly. The poorest fifth receive only 6 percent of the total resources for the aged, whereas the richest fifth receive 46 percent (Crystal and Shea 1990).

National Data

A 1998 Harris Poll of a national sample of persons 18 and older asked about government spending on the elderly; 8 percent responded "too much" was being spent, 58.6 percent responded "too little," and 32 percent responded that government spending on the elderly was "about right."

SOURCE: Harris Poll 1998b

Social Security Actually titled "Old Age, Survivors, Disability, and Health Insurance," Social Security is a major source of income for the elderly. When Social Security was established in 1935, however, it was never intended to be a person's sole economic support in old age; rather, it was to supplement other savings and assets.

Spending on the elderly has increased over time as have Social Security benefits. Today, 92 percent of Americans aged 65 and over receive Social Security, and another 3 percent will be eligible once they retire. Social Security is the major source of income for 63 percent of the elderly population and is the sole source of income for about 25 percent of the elderly. Social Security benefits keep 15 million older Americans out of poverty, and millions more from near poverty (Sherman 1998).

Although the number of poor elderly has been reduced by as much as 50 percent, the Social Security system has been criticized for being based on number of years worked and preretirement earnings. Hence, women and minorities, who often earn less during their employment years, receive less in retirement benefits. Another concern is whether funding for the Social Security system will be adequate to provide benefits for the increased numbers of elderly in the next decades.

We are not a special interest group. We are simply your mothers, fathers, and grandparents. . . . We are simply the generation or two that preceded you. When we are gone you will move up to the vanguard and another generation will wonder what to do with you short of pushing you off a cliff.

—Irene Paull and Bulbul
Authors/Activists

Table 6.3 Persons 65 Years Old and Older below Poverty Level

	Percent Below Poverty Level		
	1970	1990	1996
Black	48.0	33.8	25.3
Hispanic	n/a	22.5	24.4
White	22.6	10.1	9.4
All	24.6	12.2	10.8
Families	14.8	5.8	5.6
Unrelated individuals	47.2	24.7	20.9

SOURCES: *Statistical Abstract of the United States: 1997,* 117th ed. U.S. Bureau of the Census. Washington, D.C.: U.S. Goverment Printing Office, Table 740; *Statistical Abstract of the United States: 1998,* 118th ed. U.S. Bureau of the Census. Washington, D.C.: U.S. Government Printing Office, Table 760.

CONSIDERATION

The term **"new ageism"** refers to viewing the elderly as a burden on the economy and, specifically, on the youth of America. Younger workers are concerned that the size and number of benefits given to the elderly, particularly as the baby boomers move through the ranks, will drain the Social Security system. In 1984, some people, concerned that children and the poor were suffering at the hands of the elderly, formed Americans for Generational Equality (AGE). AGE members argue that as the number of claimants and the size of Social Security benefits grow, present tax levels will be unable to meet future expenditures. If that is the case, the government will have to increase taxes, decrease benefits, or find new sources of revenues. Concern in other countries over "equity between the generations" has already impacted public policy. For example, in Japan government benefits can be taken back if redistribution is necessary, in Australia employee pension plans are mandatory, and in South Korea workers must save 35 percent of their income for retirement in anticipation of government shortfalls (Peterson 1998).

Health Issues

The biology of aging is called **senescence.** It follows a universal pattern but does not have universal consequences. "Massive research evidence demonstrates that the aging process is neither fixed nor immutable. Biologists are now showing that many symptoms that were formerly attributed to aging—for example, certain disturbances in cardiac function or in glucose metabolism in the brain—are instead produced by disease" (Riley and Riley 1992, 221). Biological functioning is also intricately related to social variables. Altering lifestyles, activities, and social contacts affects mortality and morbidity. For example, a longitudinal study of men and women between 70 and 79 found that regular physical activity, higher levels of ongoing positive social relationships, and a sense of self-efficacy enhanced physical and cognitive functioning (Seeman and Adler 1998).

National Data

When a national sample of U.S. adults was asked to identify their greatest fear about getting old, 35 percent expressed concerns about illness and failing health.

SOURCE: Clements 1993

Biological changes are consequences of either **primary aging** due to physiological variables such as cellular and/or molecular variation (e.g., gray hair) or **secondary aging.** Secondary aging entails changes attributable to poor diet, lack of exercise, increased stress, and the like. Secondary aging exacerbates and accelerates primary aging.

Alzheimer's disease received national attention when former President Ronald Reagan announced that he had the disease. Named for German neurologist Alois Alzheimer, the debilitating disease affects both the mental and physical condition of some 4 million Americans—a projected 14 million by 2050 (AA Fact Sheet 1998). This chapter's *Self & Society* assesses your knowledge about this often misunderstood disease.

Many of the elderly are healthy, but as age increases, health tends to decline. Older people account for 49 percent of all days in the hospital with an average length of stay of 7.1 days compared with 5.4 days for people under 65 (AOA 1999). Health is a major quality-of-life issue for the elderly, especially because they face higher medical bills with reduced incomes. While those under 25 years of age spend 2.5 percent of their total expenditures on health care, those 65 to 74 and 75 and over spend 10.3 and 14 percent respectively (*Statistical Abstract of the United States: 1998,* Table 180). The poor elderly, often women and/or minorities spend an even higher proportion of their resources on health care.

Medicare was established in 1966 to provide medical coverage for those over the age of 65; in 1995, the program cost more than $181 billion (*Statistical*

Alzheimer's Quiz

	True	False	Don't Know
1. Alzheimer's disease can be contagious.	___	___	___
2. A person will almost certainly get Alzheimer's if they just live long enough.	___	___	___
3. Alzheimer's disease is a form of insanity.	___	___	___
4. Alzheimer's disease is a normal part of getting older, like gray hair or wrinkles.	___	___	___
5. There is no cure for Alzheimer's disease at present.	___	___	___
6. A person who has Alzheimer's disease will experience both mental and physical decline.	___	___	___
7. The primary symptom of Alzheimer's disease is memory loss.	___	___	___
8. Among persons older than age 75, forgetfulness most likely indicates the beginning of Alzheimer's disease.	___	___	___
9. When the husband or wife of an older person dies, the surviving spouse may suffer from a kind of depression that looks like Alzheimer's disease.	___	___	___
10. Stuttering is an inevitable part of Alzheimer's disease.	___	___	___
11. An older man is more likely to develop Alzheimer's disease than an older woman.	___	___	___
12. Alzheimer's disease is usually fatal.	___	___	___
13. The vast majority of persons suffering from Alzheimer's disease live in nursing homes.	___	___	___
14. Aluminum has been identified as a significant cause of Alzheimer's disease.	___	___	___
15. Alzheimer's disease can be diagnosed by a blood test.	___	___	___
16. Nursing-home expenses for Alzheimer's disease patients are covered by Medicare.	___	___	___
17. Medicine taken for high blood pressure can cause symptoms that look like Alzheimer's disease.	___	___	___

ANSWERS:

1–4, 8, 10, 11, 13–16 = False; remaining items = True.

SOURCE: Copyright by Neal E. Cutler, Boettner Institute of Financial Gerontology, University of Pennsylvania. Originally published in *Psychology Today,* 20th Anniversary Issue, "Life Flow: A Special Report—The Alzheimer's Quiz," 1987, vol. 21, no. 5, pp. 89, 93. Used by permission of Neal E. Cutler.

Abstract of the United States: 1997, Table 161). Although it is widely assumed that the medical bills of the elderly are paid by the government, the elderly are responsible for as much as 25 to 45 percent of their total health costs. Medicare, for example, pays about half the cost of a visit to the doctor, but does not pay for prescriptions, most nursing home care, dental care, glasses, and hearing aids. The difference between Medicare benefits and the actual cost of medical care is called the **medigap.**

Since health is associated with income, the poorest old are often the most ill: they receive less preventive medicine, have less knowledge about health care issues, and have limited access to health care delivery systems. Medicaid is a federally and state funded program for those who cannot afford to pay for medical care. Over 35 percent of Medicaid expenditures are for people 65 and over (*Statistical Abstract of the United States: 1997,* Table 168). However, eligibility requirements often disqualify many of the aged poor, often minorities and women.

Quality of Life

It's just a very hard process to watch yourself age. It's sad sometimes.

—Jessica Lange
Actress

While some elderly do suffer from declining mental and physical functioning, many others do not. Being old does not mean being depressed, poor, and sick. Interestingly, research indicates that the elderly may be less depressed than the young, with several studies concluding that the elderly may have the lowest depression rates of all age groups. Other research suggests that depression is "curvilinear" with age, that is, highest at the extremes of the age continuum (DeAngelis 1997).

Among the elderly who are depressed, two social factors tend to be in operation. One is society's negative attitude toward the elderly. Words and phrases such as "old," "useless," and "a has-been" reflect cultural connotations of the aged that influence feelings of self-worth. The roles of the elderly also lose their clarity. How is a retiree suppose to feel or act? What does a retiree do? As a result, the elderly become dependent on external validation that may be weak or absent.

National Data

There are an estimated 2.5 million elderly suffering from alcohol-related problems.

SOURCE: Public Health Reports 1999

The second factor contributing to depression among the elderly is the process of "growing old." This process carries with it a barrage of stressful life events all converging in a relatively short time period. These include health concerns, retirement, economic instability, loss of significant other(s), physical isolation, job displacement, and increased salience of the inevitability of death due to physiological decline. All of these events converge on the elderly and increase the incidence of depression and anxiety.

Living Arrangements

The elderly live in a variety of contexts, depending on their health and financial status (see Table 6.4). Most elderly do not want to be institutionalized but prefer to remain in their own homes or in other private households with friends and relatives. Of the noninstitutionalized elderly population, 7 out of 10 live in a family setting (AOA 1998b). Homes of the elderly, however, are usu-

Table 6.4 Living Arrangements of Noninstitutionalized Women and Men 65 Years Old and Over, 1997

Women	
Living with spouse	40%
Living with other relative	17%
Living alone or with nonrelative	43%
Men	
Living with spouse	72%
Living with other relative	8%
Living alone or with nonrelative	20%

SOURCE: AOA (Administration on Aging). 1999. "Profile of Older Americans: 1998." Washington, D.C.: Department of Health and Human Services.

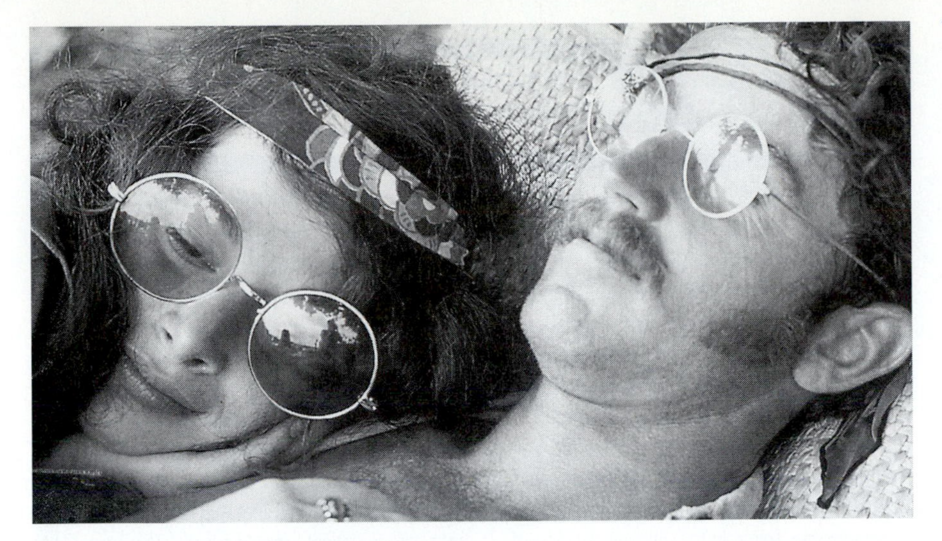

Little did members of this age cohort know that they were to become the middle-aged "Sandwich Generation," emotionally and economically responsible for both their young children and their elderly parents.

ally older, located in inner city neighborhoods, in need of repair, and often too large to be cared for easily.

Although many of the elderly poor live in government housing or apartments with subsidized monthly payments, the wealthier aged often live in retirement communities. These are often planned communities, located in states with warmer climate, and are often very expensive. These communities offer various amenities and activities, have special security, and are restricted by age. One criticism of these communities is that they segregate the elderly from the young and discriminate against younger people by prohibiting them from living in certain areas.

Those who cannot afford retirement communities or may not be eligible for subsidized housing often live with relatives in their own home or in the homes of others. It is estimated that more than 5 million people provide care for aging family members; 90 percent are adult daughters who bear most of the responsibility of caring for their elderly parents as well as for their children (Brody 1990; Dychtwald 1990; AOA 1998b).

Other living arrangements include shared housing, modified independent living arrangements, and nursing homes. With shared housing, people of different ages live together in the same house or apartment; they have separate bedrooms but share a common kitchen and dining area. They share chores and financial responsibilities. The advantage of this pattern is that it integrates age groups and utilizes skills, talents, and strengths of both the young and the old.

In modified independent living arrangements, the elderly live in their own house, apartment, or condominium within a planned community where special services such as meals, transportation, and home repairs are provided. Skilled or semiskilled health care professionals are available on the premises, and call buttons are installed so help can be summoned in case of an emergency. The advantage of this arrangement is that it provides both autonomy and support for people who are too ill or disabled to live alone. The individual can still maintain some independence even when ill.

Nursing homes are residential facilities that provide full-time nursing care for residents. Nursing homes may be private or public. Private facilities are very expensive and are operated for profit by an individual or a corporation. Public facilities are nonprofit and are operated by a governmental agency, religious

International Data

By the year 2020, there will be over 230 million elderly Chinese with fewer young people to support them because of the one-child policy begun in the 1970s. The Chinese constitution mandates that "adult children have a responsibility to support their elderly parents."

SOURCE: DHHS 1998b

FOCUS ON TECHNOLOGY

PHYSICIAN-ASSISTED SUICIDE AND SOCIAL POLICY

Everyone's parents may experience a significant drop in quality of life and a total loss of independence due to illness, accident, or, more typically, aging. Adult children or spouses are often asked their recommendations about withdrawing life support (food, water, or mechanical ventilation), starting medications to end life (intravenous vasopressors), or withholding certain procedures to prolong life (cardiopulmonary resuscitation). The recommendation of family members is sometimes given considerable weight in the ultimate decision of the caregiver/attending physician. Seventy-six percent of the caregivers in one study said that "family preference" was the most important factor influencing the restriction of technological interventions (Randolph et al. 1997). However, most family recommendations are based on the physician's recommendations to limit care or request such limitations (Luce 1997). When the physician and family member are in conflict about what to do, the physician usually defers to the preference of the family member (Prendergast and Luce 1997).

Though all 50 states now have living-will statutes permitting individuals who are not terminally ill (or family members) to refuse artificial nutrition (food) and hydration (water), there is no constitutional right of a family member or physician to end the life of another. **Euthanasia** is defined as the deliberate taking of an individual's life at her or his request (Zalcberg and Buchanan 1997). The issue has received nationwide attention in reference to Dr. Jack Kevorkian, who has been involved in over 30 physician-assisted suicides (PAS) at the patient's request.

The Supreme Court has ruled that state law will apply in regard to physician-assisted suicide. As of 1998, only Oregon recognizes the right of PAS with its Death with Dignity Act. Two physicians must agree that the patient is terminally ill and is expected to die within 6 months, the patient must ask three times for death both orally and in writing, and the patients must swallow the barbiturates themselves rather than be injected with a drug by the physician. However, the Drug Enforcement Administration has said that it will revoke the federal narcotics licenses of doctors who prescribe drugs to help terminally ill patients take their own lives.

Although 6 percent of a national sample of physicians reported that they had complied with the request of a patient for assisted suicide (Meier et al. 1998), the official position of the American Medical Association is that physicians must respect the patient's decision to forgo life-sustaining treatment but that they should not participate in patient-assisted suicide. The Association disagrees with Dr. Kevorkian. Arguments against PAS focus on who has the right to decide that a person may die. The practice is subject to abuse: for example, one spouse may encourage a physician to kill the other when he or she may thereby be relieved of a burden and inherit a lot of life insurance money. When euthanasia is defined as the deliberate ending of a life by a second person based on that second person's judgment about the quality of that life, ethical issues are raised. An objection to formalizing and legalizing the current practice of medical euthanasia is that this recognition may lead to toleration of deliberately ending the lives of members of minority, ethnic, or psychiatrically disturbed individuals, including the elderly—the "slippery slope" argument (Allen 1998).

Euthanasia remains controversial. Between 1935 and 1994, 144 people have been charged with mercy killing. One of these was Roswell Gilbert, who shot his wife of 51 years twice in the head, allegedly to end her suffering. He was convicted of first-degree murder but after 5 years in prison was paroled.

A team of researchers identified the attitudes of elderly individuals (patients) and their family members toward PAS (Koenig, Wildman-Hanlon, and Schmader 1996). While 40 percent of the elderly individuals had positive attitudes toward PAS, 59 percent of the relatives expressed favorable attitudes. Patients who opposed PAS were more likely to be women, black individuals, and those with less education.

PHYSICIAN-ASSISTED SUICIDE AND SOCIAL POLICY *(continued)*

Physician-assisted suicide has been legal in Holland for 10 years. But there have been some abuses. Though physicians are required by law to report their involvement in PAS, 60 percent of physicians in one study did not do so. Also, more than half had suggested euthanasia to patients who were not necessarily terminally ill. One-fourth of the physicians did not have the consent of the patient even though many of the patients were competent to make such a decision (Hendin, Ruthenfrans, and Zylicz 1997). Were PAS to become legal in the U.S., 36 percent of a national sample of physicians reported that they would be willing to hasten the death of a patient by prescribing medication (Meier et al. 1998).

SOURCES:
F. C. L. Allen. 1998. Euthanasia: Why Torture Dying People When We Have Sick Animals Put Down? *Australian Psychologist* 33: 12–15; Herbert Hendin, C. Rutenfrans, and Z. Zylicz. 1997. "Physician-Assisted Suicide and Euthanasia in the Netherlands: Lessons from the Dutch." *Journal of the American Medical Association* 277: 1720–23; H. G. Koenig, D. Wildman-Hanlon, and K. Schmader. 1996. "Attitudes of Elderly Patients and Their Families toward Physician-Assisted Suicide." *Archives of Internal Medicine* 156:2240–48; J. M. Luce. 1997. "Withholding and Withdrawal of Life Support: Ethical, Legal, and Clinical Aspects." *New Horizons* 5: 30–37; D. E. Meier, C. A. Emmons, S. Wallenstein, T. Quill, S. Morrison, and C. Cassel. "A National Survey of Physician-Assisted Suicide and Euthanasia in the United States." *New England Journal of Medicine* 338:1193–1201; T. J. Prendergast and J. M. Luce. 1997. "Increasing Incidence of Withholding and Withdrawal of Life Support from the Critically Ill." *American Journal of Respiratory and Critical Care Medicine* 155:15–20; A. G. Randolph, M. B. Zollo, R. S. Wigton, and T. S. Yeh. 1997. "Factors Explaining Variability among Caregivers in the Intent to Restrict Life-Support Interventions in a Pediatric Intensive Care Unit." *Critical Care Medicine* 25:435–39; J. R. Zalcberg and J. D. Buchanan. 1997. "Clinical Issues in Euthanasia." *Medical Journal of Australia* 166:150–52.

organization, or the like. The probability of being in such an extended care facility is associated with race, age, and sex: whites, the old-old, and women are more likely to be in residence. The elderly with chronic health problems are also more likely to be admitted to nursing homes. Nursing homes vary dramatically in cost, services provided, and quality of care.

Victimization and Abuse

Although abuse may take place in private homes by family members, the elderly, like children, are particularly vulnerable to abuse when they are institutionalized. A 1989 federal report entitled "Board and Care Homes in America: A National Tragedy" details the investigation of facilities for the elderly. Researchers interviewed residents, visited facilities, and collected data on homes in nine states. The result was a list of 18 categories of victimization, including preventable deaths, negligence, overmedication and sedation, poor sanitary conditions, lack of medical care, mail censorship, restriction of movement, inadequate staff, safety violations, inadequate diets, theft of personal funds, and life insurance fraud. The study also cited evidence of sexual and physical abuse and the use of restraining straps.

National Data

The National Council on Elder Abuse reports a 150 percent increase in cases between 1986 and 1996. In 1997, 2.1 million elderly were abused, neglected, or otherwise victimized.

SOURCE: AOA 1998a

Media exposure and heightened social awareness of such conditions and practices led to the definition of abused institutionalized elderly as a social problem. The result was the passage of the 1990 Nursing Home Reform Act, which established various rights for nursing care residents: the right to be free of mental and physical abuse, the right not to be restrained unless necessary as a safety precaution, the right to choose one's physician, and the right to receive mail and telephone communication (Harris 1990, 362). In 1998, in response to continued reports of nursing home abuses, President Clinton instructed state agencies to make random and frequent inspections of such facilities, and to immediately institute fines where violations of federal law were found (AARP 1998).

Whether the abuse occurs within the home or in an institution, the victim is most likely to be female, widowed, white, frail, and over 75. The abuser tends to be an adult child or spouse of the victim, who misuses alcohol (Anetzberger, Korbin, and Austin 1994). Some research suggests that the perpetrator of the abuse is more often an adult child who is financially dependent on the elderly victim (Boudreau 1993). Whether the abuser is an adult child or a spouse may simply depend on who the elder victim lives with.

Many of the problems of the elderly are compounded by their lack of interaction with others, loneliness, and inactivity. This is particularly true for the old-old. The elderly are also segregated in nursing homes and retirement communities, separated from family and friends, and isolated from the flow of work and school. As with most problems of the elderly, the problems of isolation, loneliness, and inactivity are not randomly distributed. They are higher among the elderly poor, women, and minorities. A cycle is perpetuated—being poor and old results in being isolated and engaging in fewer activities. Such withdrawal affects health, which makes the individual less able to establish relationships or participate in activities.

STRATEGIES FOR ACTION: THE YOUNG AND THE OLD RESPOND

> It would be much better, wouldn't it, to be a force to be reckoned with rather than always reckoning with other forces?
>
> —Raphael Sonensheim
> Children's rights advocate

Activism by or on behalf of children or the elderly has been increasing in recent years and, as their numbers grow, such activism is likely to escalate and to be increasingly successful. For example, global attention to the elderly led to 1999 being declared the "International Year of Older Persons" (DHHS 1998b), and "the first nearly universally ratified human rights treaty in history" deals with children's rights (UNICEF 1998). Such activism takes several forms, including collective action through established organizations and the exercise of political and economic power.

Collective Action

There are countless organizations working on behalf of children, including the Children's Defense Fund, UNICEF, Children's Partnership, Children Now, and the Children's Action Network. Many successes take place at the local level where parents, teachers, corporate officials, politicians, and citizens join together in the interest of children. In Kentucky, AmeriCorp volunteers raised the reading competency of underachieving youths 116 percent in just 6 months; in Dayton, Ohio, behavioral problems of students at an elementary school were significantly reduced once "character education" was introduced; and in Los Angeles's Crenshaw High School, student entrepreneurs established "Food from the Hood," a garden project that is projected to earn $50,000 in profit this year.

Similarly, more than a thousand organizations are directed toward realizing political power, economic security, and better living conditions for the elderly. These organizations include the National Council of Senior Citizens, the Older Women's League, the National Council on Aging, the National Retired Teachers Association, the National Committee to Preserve Social Security and

Medicare, the Gray Panthers, the Senior Action in a Gay Environment, and the American Association of Retired Persons.

One of the earliest and most radical groups is the Gray Panthers, founded in 1970 by Margaret Kuhn. The Gray Panthers were responsible for revealing the unscrupulous practices of the hearing aid industry, persuading the National Association of Broadcasters to add "age" to "sex" and "race" in the Television Code of Ethics statement on media images, and eliminating the mandatory retirement age. In view of these successes, it is interesting to note that the Gray Panthers, with only 50,000 to 70,000 members, is a relatively small organization when compared with the American Association of Retired Persons (AARP).

The AARP has more than 33 million members, age 50 and above, over 90 percent of whom are white. Services of the AARP include discounted mail-order drugs, investment opportunities, travel information, volunteer opportunities, and health insurance. The AARP is the largest volunteer organization in the United States with the exception of the Roman Catholic church.

National Data

There are an estimated 14,000 illegal telemarketing businesses in the United States defrauding Americans of $40 billion annually; half of the victims are over 50.

SOURCE: Elder Fraud 1999

CONSIDERATION

Critics of the AARP argue that the organization is geared more toward making money than toward helping the elderly. "In 1990, for example, the AARP spent about as much on office furniture and equipment as it did on programs to help its 33 million elderly members . . . $43 million was spent on salaries for the 1,110 headquarters employees" (George 1992). Nevertheless, the major stated goal of the AARP is the wielding of political power.

Political Power

Children are unable to hold office, to vote, or to lobby political leaders. Child advocates, however, acting on behalf of children, have wielded considerable political influence in such areas as child care, education, health care reform, and crime prevention. Further, funding of such programs is supported by most Americans. In a 1998 Harris Poll, only 8 percent of a national sample of adults responded that "too much" money was being spent on children—62.8 percent responded "too little" and 26.2 percent responded that government spending on children was "about right" (Harris Poll 1998a).

Although the elderly represent about 13 percent of the U.S. population, they are 16 percent of the voting public. As conflict theorists emphasize, the elderly compete with the young for limited resources. They have more political power than the young and more political power in some states than in others. In Florida, for example, there is concern that the elderly may eventually wield too much political power and act as a voting bloc demanding excessive services at the cost of other needy groups. For example, if the elderly were concentrated in a particular district, they could block tax increases for local schools. To the extent that future political issues are age based and the elderly are able to band together, their political power may increase as their numbers grow over time (Matras 1990; Thurow 1996).

In 1997, representatives of the United States, France, Germany, Italy, Canada, England, Japan, and Russia met for the first time to discuss the sociopolitical implications of the world's aging population. The summit stressed the importance of nations' learning from one another in order "to promote active aging through information exchanges and cross-national research" (DHHS 1998b).

One of the first priorities of any civilized society is to take care of its children, to prevent needless suffering in the ranks of the vulnerable and the blameless.

—Sylvia Ann Hewlett
President, National Parenting Association

Economic Power

While children have little economic power, the elderly's economic power has grown considerably in recent years (see, for example, Table 6.3) leading one economist to refer to the elderly as a "revolutionary class" (Thurow 1996). In 1996, the median income for persons over the age of 65 was $16,684 for men and $9,626 for women. Households headed by persons over 65 had a median income of $28,983. Although these incomes are significantly lower than for men and women between 45 and 54 years of age, income is only one source of economic power:

> **Some three quarters of those 65 and over own their own homes (compared to a total U.S. home ownership rate of 65%), and of the total financial assets held by U.S. families, 40% are accounted for by the 12% of the population aged 65 and above . . . . Of even greater significance from the point of view of corporate America is the fact that people 55 and over control about one-third of the discretionary income in the U.S. and spend 30% of it in the marketplace, roughly twice that of households headed by persons under 35. (Minkler 1989, 18)**

All would live long, but none would be old.

—Ben Franklin
Inventor

Advertisers actively seek the discretionary income of the elderly. Minkler (1989) observed that one favorable outcome of such a marketing focus is a more positive commercial image of the elderly. Better products and more services for older Americans are also benefits.

National Data

Of the 31.5 million persons over the age of 65 reporting an income in 1996, 40 percent reported incomes of less than $10,000 and 18 percent reported incomes of $25,000 a year or more.

SOURCE: AOA 1998c

Health Care Reform

Many of the organizations mentioned in the section on collective action have had considerable impact on health care reform in the United States. In 1988, Congress passed the Medicare Catastrophic Coverage Act (MCCA), which was the most significant change in Medicare since its establishment in 1966. The new benefits included unlimited hospitalization, an upper limit on the amount of money recipients would pay for physician's services, home health care and nursing home services, and unlimited hospice care. These changes were particularly significant since many of the illnesses of the elderly are chronic in nature.

The reforms were financed by increasing monthly medical premiums $4 a month and imposing an annual fee based on a person's federal income tax bracket. The maximum premium paid was $800 a person and $1,600 a couple (Harris 1990; Torres-Gil 1990). The AARP initially supported the reforms but later withdrew its support as did many other organizations. The additional monies paid were simply not worth the new benefits, they contended. The AARP also argued that the elderly should not have to bear the burden of reforms necessary for the general public. In 1989, under pressure from the AARP and other organizations of the elderly, Congress repealed the MCCA. As discussed in Chapter 2, Congress is currently in the process of reforming health care policy, including the strengthening of government regulation of managed health care organizations (AARP 1998).

In 1997 Congress passed the Children's Health Care Insurance Program (CHIP) which "gives hope to millions of American families who work hard and play by the rules, but don't make enough money to give their children the health care they need" (Shalala 1998). The CHIP, at a cost of $24 billion, will provide health care to some of the 10.6 million children who do not have health care coverage (see Chapter 2).

National Data

Children's health is improving; in 1999, 77 percent of pre-school children had up-to-date immunizations, and infant mortality was at an all time low.

SOURCE: Child and Family Statistics 1999

UNDERSTANDING THE YOUNG AND THE OLD

What can we conclude about youth and aging in American society? Age is an ascribed status and, as such, is culturally defined by role expectations and implied personality traits. Society regards both the young and the old as dependent and in need of the care and protection of others. Society also defines the young and old as physically, emotionally, and intellectually inferior. As a consequence of these and other attributions, both age groups are sociologically a minority with limited opportunity to obtain some or all of society's resources.

Although both the young and the old are treated as minority groups, different meanings are assigned to each group. In the United States, in general, the young are more highly valued than the old. Functionalists argue that this priority on youth reflects the fact that the young are preparing to take over important statuses while the elderly are relinquishing them. Conflict theorists emphasize that in a capitalist society, both the young and the old are less valued than more productive members of society. Conflict theorists also point out the importance of propagation, that is, the reproduction of workers, which may account for the greater value placed on the young than the old. Finally, symbolic interactionists describe the way images of the young and the old intersect and are socially constructed.

> Working to make a better life for the next generation.
>
> —South Korean company motto

The collective concern for the elderly and the significance of defining ageism as a social problem have resulted in improved economic conditions for the elderly. Currently, they are one of society's more powerful minorities. Research indicates, however, that despite their increased economic status, the elderly are still subject to discrimination in such areas as housing, employment, and medical care and are victimized by systematic patterns of stereotyping, abuse, and prejudice.

In contrast, the position of children in the United States has steadily declined as evidenced by a general increase in poverty, homelessness, and unemployment (see Chapters 10 and 11). Wherever there are poor families, there are poor children who are educated in inner city schools, live in dangerous environments, and lack basic nutrition and medical care. Further, age-based restrictions limit their entry into certain roles (e.g., employee) and demand others (e.g., student). While most of society's members would agree that children require special protections, concerns regarding quality-of-life issues and rights of self-determination are only recently being debated.

> All the little ones of the time are collectively the children of adults of the time, and are entitled to our general care.
>
> —Thomas Hardy
> Novelist

Age-based decisions are potentially harmful. For example, "by 2003 expenditures on the elderly (plus interest payments) will take 75 percent, and by 2013. . . 100 percent" of tax revenues if present laws remain unchanged (Thurow 1996). If budget allocations were based on indigence rather than age, more resources would be available for those truly in need. Further, age-based decisions may encourage intergenerational conflict. Government assistance is a zero-sum relationship—the more resources one group gets, the fewer resources another group receives.

Social policies that allocate resources on the basis of need rather than age would shift the attention of policy makers to remedying social problems rather than serving the needs of special interest groups. Age should not be used to impact negatively on an individual's life any more than race, ethnicity, gender, or sexual orientation. While eliminating all age barriers or requirements is unrealistic, a movement toward assessing the needs of individuals and their abilities would be more consistent with the American ideal of equal opportunity for all.

CRITICAL THINKING

1. In many ways, American society discriminates against children. Children are segregated in schools, in a separate justice system, and in the workplace. Identify everyday examples of the ways in which children are treated like "second-class" citizens in the United States.
2. Age pyramids pictorially display the distribution of people by age. How do different age pyramids influence the treatment of the elderly?
3. Regarding children and the elderly, what public policies or programs from other countries might be beneficial to incorporate in the United States? Do you think policies from other countries would necessarily be successful here?
4. How might the "fountain of age" described by Friedan be accomplished in American society? In what ways does gender impact definitions of the "fountain of age"?

KEY TERMS

activity theory
age grading
age pyramids
ageism
discrimination
disengagement theory
euthanasia
gerontophobia
infantilizing elders
medigap
minority
modernization theory
new ageism
phased retirement
prejudice
primary aging
sandwich generation
secondary aging
senescence
stereotype

INTERNET

You can find more information on youth suicide, teenage unemployment, social security, and Alzheimer's disease at the *Understanding Social Problems* web site at **http://sociology.wadsworth.com**

INFOTRAC COLLEGE EDITION

Either from the Wadsworth sociology resource center at **http://sociology.wadsworth.com** or directly from your web browser, you may access InfoTrac College Edition, an online university library that includes over 700 popular and scholarly journals in which you can find articles related to the topics in this chapter. Suggested articles and questions relating to these articles are listed below.

Executive Health Report Society. 1998. "Myths of Aging. *Executive Health's Good Health Report."* 34(10):5–7.

1. What positive effects does one receive by quitting smoking?
2. What is wrong with the myth that older people have no desire for sexual intimacy?
3. To what extent has life expectancy in the United States increased since 1900?

Hillyer, Barbara. 1998. "The Embodiment of Old Women: Silences." *Frontiers* 19(1):48–61.

1. According to the author, what problems does the literature on older women's bodies create?
2. What stereotypes are associated with an elderly woman, and what steps do women take to avoid such stereotypes as they age?
3. What reasons does the author offer as to why personal stories of older women are not more prevalent in the literature on aging?

Schwab, Michael. 1997. "Children's Rights and the Building Blocks of Democracy." *Social Justice* 24(3):177–92.

1. What does Hart see as the fallacy of democratic elections in U.S. public schools?
2. What was the result of childrens' involvement in the workings of the Brazilian government?
3. What four reasons does Hart give for childrens' involvement in British community development in the 1970's?

CHAPTER SEVEN

Gender Inequality

- The Global Context: The Status of Women and Men
- Sociological Theories of Gender Inequality
- Gender Stratification: Structural Sexism
- The Social Construction of Gender Roles: Cultural Sexism
- *Focus on Technology: Women, Men, and Computers*
- Social Problems and Traditional Gender Role Socialization
- Strategies for Action: Toward Gender Equality
- *Self & Society: Attitudes Toward Feminism Scale*
- *The Human Side: Declaration of the Father's Fundamental Pre-natal Rights*
- Understanding Gender Inequality

Is It True?

Is It True?

1. Researchers have observed that elementary and secondary school teachers pay more attention to girls than they do to boys.
2. Less than 5 percent of senior managers, vice presidents, or higher-level executives in Fortune 500 companies are women.
3. The more industrialized the society, the more modern its gender role ideology.
4. About half of all degrees in medicine, dentistry, and law are awarded to women.
5. Women in the United States occupy a greater percentage of legislative posi-

Answers: 1 = F; 2 = T; 3 = T; 4 = F; 5 =

> **Only a radical transformation of the relationship between women and men to one of full and equal partnership will enable the world to meet the challenges of the 21st century.**
>
> Beijing Declaration and Platform for Action

The term "gender inequality" begs the question, "Unequal in what way?" Depending on the issue, both women and men are victims of inequality. When income, career advancement, and sexual harassment are the focus, women are most often disadvantaged. But when life expectancy, mental and physical illness, and access to one's children following divorce are considered, it is often men who are disadvantaged. In this chapter, we seek to understand inequalities in both genders.

In the previous chapter, we discussed the social consequences of youth and aging. This chapter looks at **sexism**—the belief that there are innate psychological, behavioral, and/or intellectual differences between women and men and that these differences connote the superiority of one group and the inferiority of the other. As with age, such attitudes often result in prejudice and discrimination at both the individual and institutional levels. Individual discrimination is reflected by the physician who will not hire a male nurse because he or she believes that women are more nurturing and empathetic and are, therefore, better nurses. Institutional discrimination, that is, discrimination built into the fabric of society, is exemplified by the difficulty many women experience in finding employment; they may have no work history and few job skills as a consequence of living in traditionally defined marriage roles.

> We live in a state of gender warfare . . . there is a growing public awareness that many women are abused, discriminated against, and hindered in their personal development.
>
> –Daniel J. Levinson
> Author

Discerning the basis for discrimination is often difficult because gender, age, sexual orientation, and race intersect. For example, elderly African-American and Hispanic women are more likely to receive lower wages and work in less prestigious jobs than younger white women. They may also experience discrimination if they are "out" as homosexuals. Such **double** or **triple jeopardy** occurs when a person is a member of two or more minority groups. In this chapter, however, we emphasize the impact of gender inequality. **Gender** refers to the social definitions and expectations associated with being female or male and should be distinguished from **sex**, which refers to one's biological identity.

> In childhood, a woman must be subject to her father; in youth, to her husband; when her husband is dead, to her sons. A woman must never be free of subjugation.
>
> –Hindu Code

THE GLOBAL CONTEXT: THE STATUS OF WOMEN AND MEN

Although societies vary in the degree to which they regard men and women as equals, the 1997 United Nations *Human Development Report* concludes that "no society treats its women as well as its men" (p. 39). To assess the views on gender equality of a hundred university students from 14 countries (Canada, England, Finland, Germany, India, Italy, Japan, Malaysia, the Netherlands, Nigeria, Pakistan, Singapore, Venezuela, and the United States), Williams and Best (1990a) developed a series of 30 statements reflecting traditional and modern gender role positions. Agreement with such statements as "[T]he man's job is too important for him to get bogged down with household chores" reflected a traditional orientation. Agreement with such statements as "[M]arriage should not interfere with a woman's career any more than it does with a man's" reflected a modern orientation. Results indicate that the more highly developed the country, the more modern its gender role ideology.

> **National Data**
>
> When a random sample of American adults were asked if society generally favors one sex over the other, 62 percent reported "men over women," 24 percent reported "equally," and 10 percent reported "women over men;" 4 percent had no opinion.
>
> SOURCE: Newport 1993

Thus, in many underdeveloped countries, women do much of the physical labor, are forbidden to own land, can be divorced through the mere act of

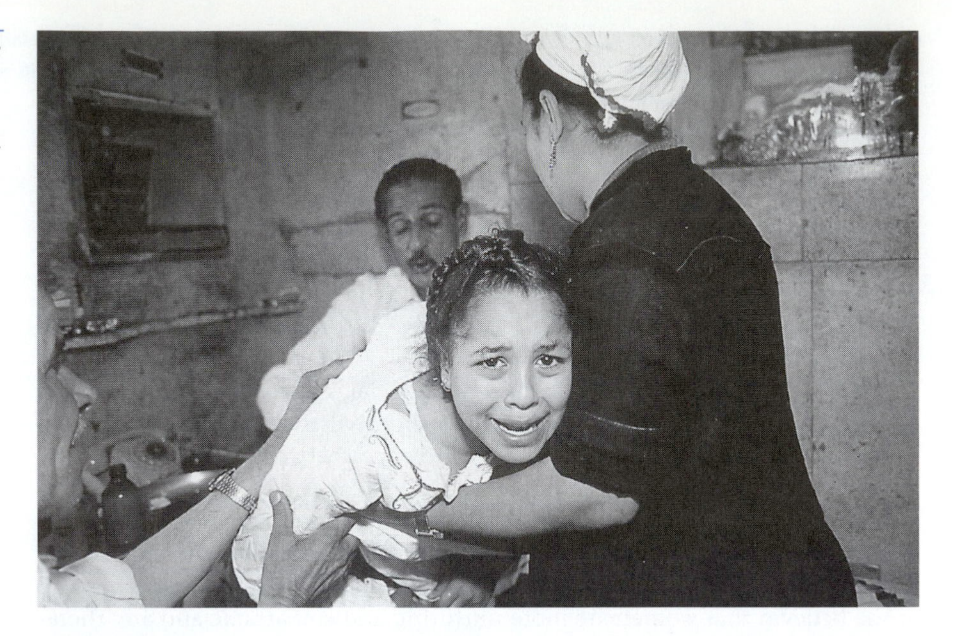

Millions of women and female children have been mutilated by genital operations throughout the world. The cultural beliefs behind female genital mutilation are economic, social, and religious.

repudiation, and earn as little as half of what a man earns. Even in countries where women have achieved some measure of success, gender inequality is evident. Although more than 69 percent of all physicians in the countries of the former Soviet Union are female, they constitute 90 percent of all pediatricians but only 6 percent of all surgeons (O'Kelly and Carney 1992). The subordinate status of women in many underdeveloped countries is further evidenced by the practice of female genital mutilation.

International Data

In one study in Bombay, India, of 8,000 abortions performed after amniocentesis, 7,900 were of female fetuses.

SOURCE: Anderson and Moore 1998

Clitoridectomy and infibulation are two forms of female genital mutilation. In a clitoridectomy, the entire glans and shaft of the clitoris and the labia minora are removed or excised. With infibulation the two sides of the vulva are stitched together shortly after birth, leaving only a small opening for the passage of urine and menstrual blood. After marriage, the sealed opening is reopened to permit intercourse and delivery. After childbirth, the woman is often reinfibulated. Worldwide, about 85 to 115 million women and girls have undergone genital mutilation (McCammon, Knox and Schacht 1998).

The societies that practice clitoridectomy and infibulation do so for a variety of economic, social, and religious reasons. A virgin bride can inherit from her father, thus making her an economic asset to her husband. A clitoridectomy increases a woman's worth because a woman whose clitoris is removed is thought to have less sexual desire and therefore to be less likely to be tempted to have sex before marriage. Older women in the community also generate income by performing the surgery so its perpetuation has an economic function (Kopelman 1994, 62). Various cultural beliefs also justify female genital mutilation. In Muslim cultures, for example, female circumcision is justified on both social and religious grounds. Muslim women are regarded as inferior to men: they cannot divorce their husbands, but their husbands can divorce them; they are restricted from buying and inheriting property; and they are not allowed to have custody of their children in the event of divorce. Female circumcision is merely an expression of the inequality and low social status women have in Muslim society.

The United States has a long history of gender inequality; women have had to fight for the right to vote, equal pay for comparable work, and other rights.

The history of mankind is a history of repeated injuries and usurpations on the part of man toward woman, having in direct object the establishment of a tyranny over her.

—Manifesto
Seneca Falls, New York
1848

Inequality in the United States

Although attitudes toward gender equality are becoming increasingly liberal (Loo and Thorpe 1999), the United States has a long history of gender inequality. Women have had to fight for equality: the right to vote, equal pay for comparable work, quality education, entrance into male-dominated occupations, and legal equality. Even today most U.S. citizens agree that American society does not treat women and men equally. As discussed later, many national statistics support the belief that men and women are not treated equally: women have lower incomes, hold fewer prestigious jobs, earn fewer academic degrees, and are more likely than men to live in poverty.

Americans' perceptions of the characteristics of men and women also reflect the inequality between the sexes in America. A Gallup poll asked whether various characteristics were "generally more true of men or more true of women" (DeStefano and Colasanto 1990). Respondents most frequently described men, in rank order, as aggressive, strong, proud, disorganized, courageous, confident, independent, ambitious, selfish, and logical. In contrast, women were most often described as emotional, talkative, sensitive, affectionate, patient, romantic, moody, cautious, creative, and thrifty. Notice that none of the top 10 characteristics were the same. About half of the respondents thought that biological factors were responsible for the differences, and half thought the differences were due to sociocultural factors.

International Data

Fifty percent of U.S. and European (Italy, England, Spain, France, and Germany) women believe they are better off than their mothers were at the same age; 55 percent of U.S. women consider themselves happier than their mothers compared with 47 percent of European women. Finally, 64 percent of American women compared with 50 percent of European women believe "that the next 20 years will herald a period of improvement for women."

SOURCE: Mogelonsky 1998

SOCIOLOGICAL THEORIES OF GENDER INEQUALITY

Both structural-functionalism and conflict theory concentrate on how the structure of society and, specifically, its institutions contribute to gender inequality. However, these two theoretical perspectives offer opposing views of the development and maintenance of gender inequality. Symbolic interactionism on the other hand, focuses on the culture of society and how gender roles are learned through the socialization process.

Structural-Functionalist Perspective

Structural-functionalists argue that preindustrial society required a division of labor based on gender. Women, out of biological necessity, remained in the home performing such functions as bearing, nursing, and caring for children. Men, who were physically stronger and could be away from home for long periods of time, were responsible for providing food, clothing, and shelter for their families. This division of labor was functional for society and, over time, became defined as both normal and natural.

Industrialization rendered the traditional division of labor less functional, although remnants of the supporting belief system still persist. Today, because of day care facilities, lower fertility rates, and the less physically demanding and dangerous nature of jobs, the traditional division of labor is no longer as functional. Thus, modern conceptions of the family have, to some extent, replaced traditional ones—families have evolved from extended to nuclear, authority is more egalitarian, more women work outside the home, and there is greater role variation in the division of labor. Functionalists argue, therefore, that as the needs of society change, the associated institutional arrangements also change.

Conflict Perspective

Many conflict theorists hold that male dominance and female subordination are shaped by the relationship men and women have to the production process. During the hunting and gathering stage of development, males and females were economic equals, each controlling their own labor and producing needed subsistence. As society evolved to agricultural and industrial modes of production, private property developed and men gained control of the modes of production, while women remained in the home to bear and care for children. Male domination was furthered by inheritance laws that ensured that ownership would remain in their hands. Laws that regarded women as property ensured that women would remain confined to the home.

As industrialization continued and the production of goods and services moved away from the home, the male-female gap continued to grow—women had less education, lower incomes, and fewer occupational skills and were rarely owners. World War II necessitated the entry of large numbers of women into the labor force, but in contrast to previous periods, many did not return home at the end of the war. They had established their own place in the workforce and, facilitated by the changing nature of work and technological advances, now competed directly with men for jobs and wages.

Conflict theorists also argue that the continued domination of males requires a belief system that supports gender inequality. Two such beliefs are that (1) women are inferior outside the home (e.g., they are less intelligent, less reliable, and less rational) and (2) women are more valuable in the home (e.g., they have maternal instincts and are naturally nurturing). Thus, unlike functionalists, conflict theorists hold that the subordinate position of women in society is a consequence of social inducement rather than biological differences that led to the traditional division of labor.

Symbolic Interactionist Perspective

While some scientists argue that gender differences are innate, symbolic interactionists emphasize that through the socialization process both females and males are taught the meanings associated with being feminine and masculine. Gender assignment begins at birth as a child is classified as either female or male. However, the learning of gender roles is a lifelong process whereby individuals acquire society's definitions of appropriate and inappropriate gender behavior.

Gender roles are taught in the family, the school, and in peer groups, and by media presentations of girls and boys, and women and men (see The Social Construction of Gender Roles on p. 179). Most importantly, however, gender roles are learned through symbolic interation as the messages others send us reaffirm or challenge our gender performances. As Lorber (1998, 213) notes:

> **Gender is so pervasive that in our society we assume it is bred into our genes. Most people find it hard to believe that gender is constantly created and recreated out of human interaction, out of social life, and is the texture and order of social life. Yet gender, like culture, is a human production that depends on everyone constantly "doing gender" . . .**

Conceptions of gender are, thus, socially constructed as societal expectations dictate what it *means* to be female or what it *means* to be male. Although race and class variations exist, in general, women are socialized into expressive or nurturing and emotionally supportive roles, and males are more often socialized into instrumental or task-oriented roles. These roles are then acted out in countless daily interactions as boss and secretary, doctor and nurse, football player and cheerleader "do gender."

GENDER STRATIFICATION: STRUCTURAL SEXISM

As structural-functionalists and conflict theorists argue, the social structure underlies and perpetuates much of the sexism in society. **Structural sexism**, also known as "institutional sexism," refers to the ways in which the organization of society, and specifically its institutions, subordinate individuals and groups based on their sex classification. Structural sexism has resulted in significant differences between the education and income levels, occupational and political involvement, and civil rights of women and men.

Education and Structural Sexism

Literacy rates worldwide indicate that women are less likely to be able to read and write than males. In developing countries the literacy rate for females is 61 percent compared to 79 percent for males (United Nations 1997). In the United States, more than half of all bachelor's and master's degrees are awarded to women (*Statistical Abstract of the United States: 1998*, Tables 325, 326). However, women still earn fewer doctorates, and fewer degrees in medicine, dentistry, and law than men (see Table 7.1).

One explanation for why women earn fewer advanced degrees than men is that women are socialized to choose marriage and motherhood over long-term career preparation (Olson, Frieze, and Detlefsen 1990). From an early age, women are exposed to images and models of femininity that stress the importance of domestic family life. When 821 undergraduate women were asked to identify their lifestyle preference, less than 1 percent selected being unmarried

National Data

Boys and girls have the same mathematics and science proficiency at age 9; by age 13, males outperform females in science, although not math. By age 17, males outperform females in both math and science. Male students are twice as likely to aspire to be a scientist or an engineer as female students.

SOURCE: Bae and Smith 1998

Table 7.1 **Percentages of Advanced Degrees Granted to Women, 1995**

Type of Degree	Percentage of Degrees Earned by Women
Doctorate (Ph.D.)	39
Medical degree (M.D.)	39
Dentistry degree (D.D.S. or D.M.D.)	36
Law degree (LL.B or J.D.)	43

SOURCE: *Statistical Abstract of the United States: 1998,* 118th ed. U.S. Bureau of the Census. Washington, D.C.: U.S. Government Printing Office, Tables 326 and 327.

and working full-time. In contrast, 53 percent selected "graduation, full-time work, marriage, children, stop working at least until youngest child is in school, then pursue a full-time job" as their preferred lifestyle sequence (Schroeder, Blood, and Maluso 1993, 243). Only 6 percent of 535 undergraduate men selected this same pattern. This lack of career priority on the part of women is reflected in the lack of priority they give to education as preparation for a career.

Structural limitations also discourage women from advancing in the educational profession itself. For example, women seeking academic careers may find that promotion in higher education is more difficult than for men. Long, Allison, and McGinnis (1993) examined the promotions of 556 men and 450 women with Ph.D.s in biochemistry. They found that women were less likely to be promoted to associate or full professor, were held to a higher standard than men, and were particularly disadvantaged in more prestigious departments. Even in public schools, where women comprise 75 percent of all classroom teachers, they are 45 percent of principals and assistant principals (*Statistical Abstract of the United States: 1998,* Table 277).

National Data

Projections indicate that the number of women entering college will continue to exceed the number of men into the next century, with the enrollment of men expected to increase at half the rate of women. Between 1995 and 2007 it is predicted that 4.2 million men (13 percent increase) and 5.4 million women (30 percent increase) will enroll in college full time.

SOURCE: Dortch 1997

Income and Structural Sexism

In general, the higher one's education, the higher one's income. Yet even when men and women have identical levels of educational achievement and both work full-time, women still tend to earn, on average, about 60 percent of what men earn (see Table 7.2). As in the United States, there is a general trend worldwide for income differences between females and males to decrease as educational differences between the two decrease (Wootton 1997; Educational Indicators 1998).

Tomaskovic-Devey (1993) examined the income differences between males and females and found that the percentage of females in an occupation was the best predictor of an income gender gap—the higher the percentage of females, the lower the pay. Supporting this observation, a team of researchers (Kilbourne et al. 1994) analyzed data from the National Longitudinal Survey that included more than 5,000 women and 5,000 men. They concluded that occupational pay is gendered and that "occupations lose pay if they have a higher percentage of female workers or require nurturant skills" (p. 708).

Two hypotheses are frequently cited in the literature as to why the income gender gap continues to exist. One is called the **devaluation hypothesis.** It argues that women are paid less because the work they perform is socially defined as less valuable than the work performed by men. The other hypothesis, the **human capital hypothesis**, argues that female-male pay dif-

Table 7.2 **Effect of Education and Sex on Income, 1997**

Level of Educational Attainment	Average Earnings Women	Men	Women's Earning as Percentage of Men's Percentage
Some high school, no diploma	$10,421	$17,826	58
High school diploma	16,161	27,642	59
Some college, no degree	17,475	30,057	58
Associate degree	22,480	35,484	63
Bachelor's degree or more	28,701	46,702	62
Master's degree	36,483	62,145	59

SOURCE: *Statistical Abstract of the United States: 1998*, 118th ed. U.S. Bureau of the Census. Washington, D.C.: U.S. Government Printing Office, Table 263.

ferences are a function of differences in women's and men's levels of education, skills, training, and work experience.

Tam (1997), in testing these hypotheses, concludes that human capital differences are more important in explaining the income gender gap than the devaluation hypothesis. Marini and Fan (1997) also found support for the human capital hypothesis, although their research supports a third category of variables as well. They found that organizational variables (characteristics of the business, corporation, or industry) explain, in part, the gender income gap. For example, women and men upon career entry are channeled by employers into sex-specific jobs that carry different wage rates.

Work and Structural Sexism

Work is highly gendered. As a group, women tend to work in jobs where there is little prestige and low or no pay, where no product is produced, and where women are the facilitators for others. Women are also more likely to hold positions of little or no authority within the work environment. Investigating the gender gap in organizational authority in seven countries (Australia, Canada, Japan, Norway, Sweden, United Kingdom, and United States), Wright, Baxter, and Birkelund (1995, 419) conclude that in every country, "[W]omen are less likely than men to be in the formal authority hierarchy, to have sanctioning power over subordinates, or to participate in organizational policy decisions." Women of color may be even less likely to hold positions of power. In an investigation of female and male African-American and white firefighters, black women were the most subordinated group as black males and white females relied on their superordinate gender and race statuses, respectively (Yoder and Aniakudo 1997).

No matter what the job, if a woman does it, it is likely to be valued less than if a man does it. For example, in the early 1800s, 90 percent of all clerks were men, and being a clerk was a very prestigious profession. As the job became more routine, in part because of the advent of the typewriter, the pay and prestige of the job declined and the number of female clerks increased. Today, female clerks predominate, and the position is one of relatively low pay and prestige.

The concentration of women in certain occupations and men in other occupations is referred to as **occupational sex segregation** (see Table 7.3). For

International Data

Between 1950 and 1995, the percentage of women in the labor force increased dramatically worldwide; 22 percent in North America, 15 percent in Latin America, 10 percent in Europe, and 5 percent in Asia. The percentage of women in the labor force between 1950 and 1995 in Africa remained unchanged.

SOURCE: United Nations 1997

Table 7.3 Highly Sex-Segregated Occupations, 1997

Female-Dominated Occupations	Percentage of Female Workers
Child care workers	97
Cleaners and servants	95
Dental hygienists	98
Dietitians	89
Elementary school teachers	84
Librarians	77
Prekindergarten and kindergarten teachers	98
Receptionists	97
Registered nurses	94
Secretaries	98
Speech therapists	95
Teacher's aides	93
Typists	94
Male-Dominated Occupations	**Percentage of Male Workers**
Airplane pilots and navigators	99
Architects	82
Automobile mechanics	97
Clergy	86
Construction	97
Dentists	83
Engineers	90
Firefighters	97
Mechanics and repairers	96
Physicians	74
Police and detectives	84

SOURCE: *Statistical Abstract of the United States: 1998,* 118th ed. U.S. Bureau of the Census. Washington, D.C.: U.S. Government Printing Office, Table 672.

". . . everyone assumed I was someone's wife . . . It took years before my mail was addressed to Ms. instead of Mr."

–Lynn Kimmel
Owner of Indianapolis-based multimillion dollar business

example, women are overrepresented in semiskilled and unskilled occupations, and men are disproportionately concentrated in professional, administrative, and managerial positions (Steiger and Wardell 1995; United Nations 1997).

In some occupations, sex segregation has decreased in recent years. For example, the percentage of female physicians increased from 16 percent to 26 percent between 1983 and 1997, female dentists increased from 7 to 17 percent, female engineers increased from 6 to 10 percent, and female clergy increased from 6 to 14 percent (*Statistical Abstract of the United States: 1998,* Table 672).

Nevertheless, despite these and other changes, women are still heavily represented in low-prestige, low-wage **"pink-collar" jobs** that offer few benefits. Even those women in higher-paying jobs are often victimized by a **glass ceiling**—an invisible barrier that prevents women and other minorities from moving into top corporate positions. A recent study of Fortune 500 companies found that less than 11 percent of all seats on the Fortune 500 company boards are held by women (Klein 1998b). Interestingly, Cianni and Romberger's investigation (1997) of Asian, African-American, and Hispanic women and men in Fortune 500 companies indicates that gender has more of a role in "organizational treatment" than race.

International Data

In the Ukraine, the second largest country in Europe, women comprise 69 percent of the white collar labor force but hold only 5 percent of the managerial, director, or department head positions.

SOURCE: Tserkonivnitska 1997

Sex segregation in occupations continues for several reasons (Martin 1992; Williams 1995). First, cultural beliefs about what is an "appropriate" job for a man or a woman still exist, as do socialization experiences in which males and females learn different skills and acquire different aspirations. Further, opportunity structures for men and women vary. Women have fewer opportunities in the more prestigious and higher-paying male dominated professions, resulting in women comprising more than 70 percent of all minimum wage earners. Women may also be excluded by male employers and employees who fear the prestige of their profession will be lessened with the entrance of women, or who simply believe that "the ideal worker is normatively masculine" (Martin 1992, 220).

Finally, since family responsibilities primarily remain with women, working mothers may feel pressure to choose professions that permit flexible hours and career paths, sometimes known as "mommy tracks." Thus, for example, women dominate the field of elementary education, which permits them to be home when their children are not in school. Nursing, also dominated by women, often offers flexible hours.

National Data

A recent survey indicates that 11 percent of mothers stayed out of the labor force 10 years after the birth of their first child; 13 percent stayed in the labor force the entire time; the remaining 76 percent alternated periods of working outside the home, remaining in the home, and/or going to school.

SOURCE: Mogelonsky, 1997

CONSIDERATION

Although women entering professions dominated by men are at a disadvantage, the reverse is not true. Williams (1995) interviewed 76 men and 23 women in four traditionally female professions—elementary school teacher, librarian, nurse, and social worker. She found that men who work in traditional female occupations often receive preferential treatment in hiring and promotion decisions. Williams concludes that men take their "superior" status with them, even into female-dominated professions.

Politics and Structural Sexism

Women received the right to vote in 1920 with the passage of the Nineteenth Amendment. Even though this amendment went into effect almost 80 years ago, women still play a rather minor role in the political arena. In general, the more important the political office, the lower the probability a woman will hold it (see Table 7.4). Although women comprise 52 percent of the population, the United States has never had a woman president or vice president and,

Women are barely tokens in the decision-making bodies of our nation, so the laws that govern us are made by men.

–National Organization for Women

Table 7.4 Percentage of Women Elected by Level and Type of Government Position, 1998

Level of Government/Positions	No. Seats	No. Women	Percent Held by Women
U.S. President	1	0	0.0
U.S. Vice President	1	0	0.0
U.S. Congress	535	62	11.6
House	435	53	12.2
Senate	100	9	9.0
Mayors of 100 Largest Cities	100	12	12.0
State Legislature	7,424	1,605	21.6
House	5,440	1,238	22.7
Senate	1,984	367	18.5
Statewide Elective Executive Offices	323	82	25.4

SOURCE: CAWP (Center for the American Women and Politics). 1998. "Women in Elective Office 1998." National Information Bank on Women in Public Office, Eagleton Institute of Politics. New Brunswick, N.J.: Rutgers University.

until 1993 when a second woman was appointed, had only one female U.S. Supreme Court justice. The highest ranking woman ever to serve in U.S. government is Madeleine Albright, who became the U.S. Secretary of State in 1997. Since the U.S. cabinet was established in 1789, of the 486 cabinet members only 21, or 4.3 percent, have been women (GenderGap 1998).

The relative absence of women in politics, as in higher education and in high-paying, high-prestige jobs in general, is a consequence of structural limitations. Running for office requires large sums of money, the political backing of powerful individuals and interest groups, and a willingness of the voting public to elect women. Thus, minority women have even greater structural barriers to election and, not surprisingly, represent an even smaller percentage of elected officials. For example, of the 82 women holding statewide elected executive offices, only 3, or 3.7 percent, are women of color (CAWP 1998).

International Data

Worldwide, women hold about 10 percent of all seats in legislative bodies—12 percent in industrialized countries, 10 percent in developing countries, and 8 percent in transitional countries.

SOURCE: United Nations 1997

Civil Rights and Structural Sexism

The 1963 Equal Pay Act and Title VII of the 1964 Civil Rights Act made it illegal for employers to discriminate in wages or employment on the basis of sex. Nevertheless, such discrimination still occurs as evidenced by the thousands of grievances filed each year with the Equal Employment Opportunity Commission (EEOC). One technique used to justify differences in pay is the use of different job titles for the same work. The courts have ruled, however, that jobs that are "substantially equal," regardless of title, must result in equal pay.

Women are also discriminated against in employment. Discrimination, although illegal, takes place at both the institutional and the individual level. Institutional discrimination includes male-dominated recruiting networks, employment screening devices designed for men, hiring preferences for veterans, and the practice of promoting from within an organization, based on seniority. One of the most blatant forms of individual discrimination is sexual harassment, an issue that was brought to public attention by the televised confirmation hearings of Supreme Court nominee Clarence Thomas by the Senate Judiciary Committee and also by the resignation of Senator Bob Packwood.

Women are also discriminated against in a number of economic matters. Hindu and Muslim law in India, Bangladesh, and some African and Arab countries limit the inheritance rights of widows and daughters (United Nations 1997). In the United States, having lower incomes, shorter work histories, and less collateral, women often have difficulty obtaining home mortgages or rental property.

Discrimination takes place in other forms as well. Up until fairly recently, husbands who raped their wives were exempt from prosecution. Even today, some states require a legal separation agreement and/or separate residences for a raped wife to receive full protection under the law. And in 1998, several states criminalized alcohol consumption by pregnant women. For example, Wisconsin recently passed legislation that "allows pregnant women who drink or use drugs to be detained until they give birth" (Roth 1998). Women in the military have traditionally been restricted in the duties they can perform and, finally, since the U.S. Supreme Court's 1973 *Roe v. Wade* decision, the right of a woman to obtain an abortion has been limited. Today, the debate continues with several recent court decisions weakening the self-determination doctrine (see Chapter 14).

CONSIDERATION

The Convention to Eliminate All Forms of Discrimination against Women (CEDAW), also known as the International Women's Bill of Rights, was adopted by the United Nations in 1979. CEDAW establishes rights for women not previously recognized internationally in a variety of areas, including education, politics, work, law, and family life. The United States signed the document on July 17, 1980, although it has yet to be ratified by the required two-thirds vote of the U.S. Senate. Over 161 countries have ratified the treaty, including every country in the Western Hemisphere and every industrialized nation in the world with the exception of Switzerland and the United States. (CEDAW 1998; WOC Alert 1998).

THE SOCIAL CONSTRUCTION OF GENDER ROLES: CULTURAL SEXISM

As symbolic interactionists note, structural sexism is supported by a system of cultural sexism that perpetuates beliefs about the differences between women and men. **Cultural sexism** refers to the ways in which the culture of society—its norms, values, beliefs, and symbols—perpetuates the subordination of an individual or group because of the sex classification of that individual or group. Cultural sexism takes place in a variety of settings including the family, the school, and media as well as in everyday interactions.

Family Relations and Cultural Sexism

From birth, males and females are treated differently. For example, the toys male and female children receive convey different messages about appropriate gender roles. Research by Rheingold and Cook (1975) revealed how traditional gender role stereotypes are reflected in children's rooms in middle-class homes. Girls' rooms contained dolls, floral prints, and miniature home appliances such as stoves and refrigerators. Boys' rooms were more often decorated with a military or athletic motif and contained such items as building blocks, cars, trucks, planes, and boats. Overall, boys had more toys, more educational toys, and a greater variety in types of toys than girls. This study was replicated more than 10 years later with the same results (Stoneman, Brody, and MacKinnon 1986). The significance of these findings lies in the relationship between the toys and the activities they foster: active versus passive play. This chapter's *Focus on Technology* documents the negative consequences of seemingly harmless play differences.

> Men kinda have to choose between marriage and death. I guess they figure at least with marriage they get meals. Then they get married and find out we don't cook anymore.
>
> —Rita Rudner
> Comedian

Household Division of Labor Little girls and boys work within the home in approximately equal amounts until the age of 18, "when the 2-to-1 female to male ratio familiar to adults begins to emerge" (Robinson and Bianchi 1997, 4). Throughout adulthood, housework remains primarily a woman's responsibility (Merida and Vobejda 1998). The fact that women, even when working full time, contribute significantly more hours to home care than men is known as the "second shift" (Hochschild 1989). For example, in the Ukraine, women work an average of 73 hours per week—34 hours at the workplace and 39 hours at home. Men, on the other hand, work 60 hours a week—41 hours in the workplace and 19 hours at home (Tserkonivnitska 1997).

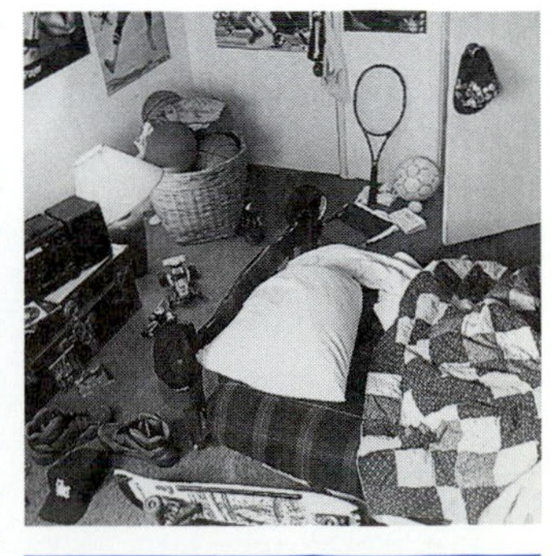

Is this a photo of a boy's bedroom or a girl's bedroom?

FOCUS ON TECHNOLOGY

WOMEN, MEN, AND COMPUTERS

Technology has changed the world in which we live. The technological revolution has brought the possibility of greater gender equality for, unlike tasks dominating industrialization, sex differences in size, weight, and strength are less relevant. Although feminists have long decried the gendering of technology, surely computers and other information technologies are gender neutral—or are they?

Before adolescence, boys and girls use computers equally. However, around the age of 13, cultural definitions of "computers are for boys" set in.

Because girls are less likely to play video games (to wit, *"Game Boy"*), software that appeals to girls is less likely to be manufactured—92 percent of video games have no female characters (O'Neal 1998)—leading to a self-fulfilling prophecy with very serious consequences.

In adolescence and beyond, girls are not as interested in computers as boys. When they are interested, unlike boys who define it as a toy to explore, something that's fun, girls define computers as a tool to accomplish a task, a kind of homework helper. Says Jane Margolis, a researcher at Carnegie Mellon (cited in Breidenbach 1997, 69):

> Girls want to do something constructive with computers, while boys get into hacking and using computers for their own sake . . . Computers are just one interest of many for girls, while they become an object of love and fascination for boys.

Knowing the difficulty in attracting girls to computers, researchers have developed several programs designed to involve them in the new technologies. For example, a $900,000 National Science Foundation grant facilitated training 200 teachers from all over the country who then went home and trained thousands more on how to encourage girls into computer classes. In one Oklahoma school where the program was initiated, the number of girls in computer classes increased from zero to 31 percent in 2 years (DeBare 1996).

If women do not pursue computer-based information technologies, they will have an "intellectual and workplace handicap that can only get worse as technology grows more prevalent" (Currid 1996, 114). For example, while clerical jobs in which women are heavily concentrated are projected to grow by 20 percent, computer engineer and systems analysts positions—two of the four fastest growing occupations between now and 2005—are projected to grow 92 and 90 percent, respectively (DeBare 1996).

The culture and structure of high-tech occupations, however, are often not conducive to female employment. Rapidly changing knowledge bases make taking time off for motherhood almost impossible; long hours make child care arrangements and time away from home difficult; and lingering stereotypes as "women can't handle stress" and women don't "understand technology as well as men" (*Informationweek* 1996) make advancement difficult. A Massachusetts Institute of Technology report on the gender gap in computing found that "women are often judged as less qualified than men even when their performance is identical" (Breidenbach 1997, 69).

There are some signs of optimism, however. Munk and Oliver (1996) report an increase in the number of venture capital grants submitted by women wanting to start high-tech businesses. The number of women getting computer science degrees—although in recent years declining—has increased over the last several decades (*Statistical Abstract of the United States: 1998,* Table 325). Some companies such as Sun Microsystems and Hewlett-Packard have begun aggressive recruitment and hiring programs for women and people of color (Wilde 1997), and the ratio of women to men on the Internet is projected to be 1:1 by the year 2000 (Grossman 1999). However, for significant changes to

FOCUS ON TECHNOLOGY

WOMEN, MEN, AND COMPUTERS *(continued)*

take place in reference to women, men, and computers we must recognize that computers specifically, and technology in general, are not gender neutral and, in fact, have emerged and flourished within the context of a male-dominated industry.

SOURCES: Susan Breidenbach. 1997. "Where Are All the Women?" *Network World* 14 (41):68–69; Cheryl Currid. 1996. "Bridging the Gender Gap: Women Will Lose out Unless They Catch up with Men in Technology Use." *Informationweek*, April 1, 114; Ilana DeBare. 1996. "Women and Computing." Parts I, III–VI. *The Sacramento Bee*. **http://www.sacbee.com/news/projects/women;** Lev Grossman. 1999. "The War over Women on the Web." *Time Daily*. January 29. **http://www.time.com;** *Informationweek*. 1996. "Women Gain in IS Ranks," September 16, 158; Nina Muck and Suzanne Oliver. 1996. "Women of the Valley." *Forbes*, December 30, 2–13; Glenn O'Neal. 1998. "Girls Often Dropped from Computer Equation." *USA Today*, March 10, D4; *Statistical Abstract of the United States: 1998*, 118th ed. U.S. Bureau of the Census. Washington, D.C.: U.S. Government Printing Office, Tables 325 and 326; Candee Wilde. 1997. "Women Cut through IT's Glass Ceiling." *Informationweek*, January 20, 83–86.

Calasanti and Bailey (1991) offer three explanations for the continued traditional division of labor in families. The first explanation is the "time-availability approach." Consistent with the structural-functionalist perspective, this position emphasizes that role performance is a function of who has the time to accomplish certain tasks. Because women are more likely to be at home, they are more likely to perform domestic chores.

A second explanation is the "relative resources approach." This explanation, consistent with a conflict perspective, suggests that the spouse with the least power is relegated the most unrewarding tasks. Since men have more education, higher incomes, and more prestigious occupations, they are less responsible for domestic labor. The husband who shouts at his wife, "I pay the bills, the least you can do is keep the house clean!" is expressing the relative resources ideology.

"Gender role ideology," the final explanation, is consistent with a symbolic interactionist perspective. It argues that the division of labor is a consequence of traditional socialization and the accompanying attitudes and beliefs. Females and males have been socialized to perform various roles and to expect their partners to perform other complementary roles. Women typically take care of the house, men the yard. This division of labor is learned in the socialization process through the media, schools, books, and toys. A husband's comment that washing the dishes is "women's work" reflects this perspective.

National Data

A survey of adult men and women indicates differences in the perception of gender inequality; the highest perception of inequality by both men and women is in reference to inequality in the household division of labor, followed by inequality in employment, parenting, and leisure activities.

SOURCE: Sanchez 1997

International Data

Worldwide, men's responsibilities for housework have increased in recent years. For example, in Korea between 1987 and 1990, men's attention to child care increased 20 percent and, in Japan, 5 percent. In Germany, between 1965 and 1992, men's share of household labor increased 12 percent in meal preparation, 14 percent in shopping, and 13 percent in child care.

SOURCE: United Nations 1997

The School Experience and Cultural Sexism

Sexism is also evident in the schools. It can be found in the books students read, the curricula and tests they are exposed to, and the different ways teachers interact with students.

Textbooks A study by the University of the Philippines concludes that textbooks and other instructional material perpetuate gender stereotypes (UP 1997). In the materials examined, men were portrayed in more interesting and active roles, women's roles were defined by their relationship to men (e.g., wife, daughter), and males were more often pictured in illustrations. Similarly, Purcell and Stewart (1990) analyzed 1,883 storybooks used in schools and found that they tended to depict males as clever, brave, adventurous, and income producing and females as passive and as victims. Females were more likely to be in need of rescue, and were also depicted in fewer occupational roles than males. In a study of 27 introductory psychology and 12 human development textbooks, Peterson and Kroner (1992) determined that the representation of the work, theory, and behavior of males significantly exceeded that of females. Even though 56 percent of psychologists and 78 percent of developmental psychologists receiving Ph.D.s are women, "the picture presented in textbooks is that the majority of persons working in the various domains of psychology are men" (p. 31). Even textbooks that were designed to promote equality are characterized by "subtle language bias, neglect of scholarship on women, omission of women as developers of history and initiators of events, and an absence of women from accounts of technological developments" (American Association of University Women 1992, 63).

Curricula and Testing Encouragement to participate in sports, academic programs, and extracurricular activities is gender biased despite Title IX of the 1972 Educational Amendments Act, which prohibits officials from "tracking" students by sex. Although women's and girls' participation in sports is now at an all-time high (Klein 1998a), there remain differences in the sports males and females play. Males are more likely to participate in competitive sports that emphasize traditional male characteristics such as winning, aggression, physical strength, courage, adventurousness, and dominance. Women are more likely to participate in sports that emphasize individual achievement (e.g., figure skating) or cooperation (e.g., synchronized swimming).

The differing expectations and/or encouragement that females and males receive also contribute to their varying abilities, as measured by standardized tests, in such disciplines as math and science. Women's lower scores on college entrance examinations can almost entirely be explained by the math portion of the test alone. Ironically, women "typically earn the same or higher grades as their male counterparts in math and other college courses despite having SAT-Math scores 30–50 points lower" (FairTest 1997). Not surprisingly, women comprise less than 30 percent of all mathematicians, natural scientists, engineers, and computer analysts (*Statistical Abstract of the United States: 1998*, Table 672). Research also indicates that standardized tests themselves are biased—almost exclusively being timed, multiple-choice tests—a format favoring males (FairTest 1997).

National Data

Compared to girls, boys are six times more likely to be labeled as having attention-deficit disorder and twice as likely to be diagnosed as learning disabled.

SOURCE: Goldberg 1999

Teacher-Student Interactions Sexism is also reflected in the way teachers treat their students. After interviewing 800 adolescents, parents, and teachers in three school districts in Kenya, Mensch and Lloyd (1997) report that teachers were more likely to describe girls as lazy and unintelligent. "And when the girls do badly," the researchers remark, "it undoubtedly reinforces teachers' prejudices, becoming a vicious cycle." Similarly, in the United States Sadker and Sadker (1990) observed that elementary and secondary school teachers pay more attention to boys than to girls. Teachers talk to boys more, ask them more questions, listen to them more, counsel them more, give them more extended directions, and criticize and reward them more frequently. Sadker and Sadker observed that this pattern continues at the postsecondary level.

Media Images and Cultural Sexism

One concern voiced by social scientists in reference to cultural sexism is the extent to which the media portrays females and males in a limited and stereo-

Women and girls alike are often portrayed in ornamental roles that emphasize their appearance rather than their abilities.

typical fashion, and the impact of such portrayals. A recent study by Signorielli (1998) analyzed gender images presented in six media: television, movies, magazines, music videos, TV commercials, and print media advertisements. The specific programs, movies, videos, etc. selected were those most often consumed by 12- to 17-year-old girls, for example, the 25 top watched television shows. The results indicate:

- Other than magazines directed at teenage girls, the medium with the highest representation of women is television (45 percent), followed by TV commercials (42 percent), movies (37 percent), and music videos (22 percent).
- In general, media content stresses the importance of appearance and relationships for girls/women and of careers and work for boys/men.
- Across the six media, 26 to 46 percent of women are portrayed as "thin or very thin" in contrast to 4 to 16 percent of men; 70 percent of girls wanted to look like, fix their hair like, or dress like a character on television, compared with 40 percent of boys.
- In a survey of boys and girls, both agreed that "worrying about weight, crying or whining, weakness, and flirting" are characteristics associated more with girls than with boys, and that "playing sports, being a leader, and wanting to kiss or have sex . . ." are more often characteristic of male characters.

National Data

Females are much more likely to suffer from anorexia or bulimia; 7 million women and girls suffer from eating disorders.

SOURCE: Wilmot 1999

International Data

In a survey of Russian youth, both male and female college students identified beauty and kindness as the ideal traits of a woman; strength and ability to provide for one's family were identified as the ideal traits of a man.

SOURCE: Attwood 1996

Language and Cultural Sexism

In subtle ways, both the words we use and the way we use them can reflect gender inequality. The term *nurse* carries the meaning of "a woman who . . ." and the term engineer suggests "a man who . . . ." The embedded meaning of words, as symbolic interactionists note, carry expectations of behavior.

Despite research that suggests females are most often the victim of negative stereotypes, at least one study concludes otherwise. Fiebert and Meyer (1997) had a sample of college students complete either the sentence "A woman is . . ." or "A man is . . . ." Responses were then coded as either positive, negative, or neutral. Males were significantly more likely to be stereotyped negatively.

CONSIDERATION

The terms "broad," "chick," "old maid," and "spinster" convey information about age as well as attaching a stigma to the recipient. There are no comparable terms for men. Further, women have traditionally been designated as either Mrs. or Miss, while Mr. conveys no social meaning about marital status. Language is so gender stereotyped that the placement of female or male before titles is sometimes necessary as in the case of "female police officer" or "male prostitute." There is a movement away from male-biased language. The American Sociological Association, for example, includes suggestions for using nonsexist language in its style manual for preparing articles to be submitted for publication.

Virginia Sapiro (1994) has shown how male-female differences in communication style reflect the structure of power and authority relations between men and women. For example, women are more likely to use disclaimers ("I could be wrong but . . . .") and self-qualifying tags ("That was a good movie, wasn't it?"), reflecting less certainty about their opinions. Communication differences between women and men also reflect different socialization experiences. Women are more often passive and polite in conversation; men are less polite, interrupt more often, and talk more (Tannen 1990).

SOCIAL PROBLEMS AND TRADITIONAL GENDER ROLE SOCIALIZATION

Cultural sexism, transmitted through the family, school, media, and language, perpetuates traditional gender role socialization. Although gender roles are changing, traditional gender roles are still dominant and have consequences for both women and men:

> **There is no denying that the gender system controls men too. Unquestionably, men are limited and restricted through narrow definitions of "masculinity." . . . They, too, face negative sanctions when they violate gender prescriptions. There is little value in debating which sex suffers or loses more through this kind of control; it is apparent that both do (Schur 1984, 12).**

These traditional socialization experiences lead to several social problems. These include the feminization of poverty, social-psychological costs, health costs, and relationship problems.

> The poor of America are women: the poor of the world are women.
>
> —Marilyn French
> Novelist/author

Poverty

Women have only about two-thirds of the income of men, which contributes to the "feminization of poverty" (see Chapter 10). Younger women with dependent children and older women who have outlived their spouses have the highest poverty rates. Of those 25 to 34 years of age, the poverty rate is 10 percent for men and 15 percent for women; of those 65 and over, the poverty rate is 14 percent for women and 6 percent for men (Bianchi and Spain 1996). As noted earlier, both individual and institutional discrimination contribute to the economic plight of women. Traditional gender role socialization also contributes to poverty among women. Women are often socialized to put family ahead of their education and careers. Women are expected to take primary responsibility for child care, which contributes to the alarming rate of single-parent poor families in the United States (see Figure 7.1). Hispanic and black female-headed households are the poorest of all families headed by a single woman. Further, a study of the relationship between martial status, gender, and poverty in the United States, Australia, Canada, and France indicates that never-married women compared to ever-married women in all four countries are more likely to live in poverty (Nichols-Casebolt and Krysik 1997).

Social-Psychological Costs

Many of the costs of traditional gender socialization are social-psychological in nature. Reid and Comas-Diaz (1990) noted that the cultural subordination of women results in women having low self-esteem and being dissatisfied with their roles as spouses, homemakers/workers, mothers, and friends. In a study of self-esteem among more than 1,160 students in grades 6 through 10, girls were significantly more likely to have "steadily decreasing self-esteem," while boys were more likely to fall into the "moderate and rising" self-esteem group (Zimmerman et al. 1997).

Not all researchers have found that women have a more negative self-concept than men. Summarizing their research on the self-concepts of women and men in the United States, Williams and Best (1990b) found "no evidence of an appreciable difference" (p. 153). They also found no consistency in the self-concepts of women and men in 14 countries: "[I]n some of the countries the men's perceived self was noticeably more favorable than the women's, whereas in others the reverse was found" (p. 152). More recent research also documents that women are becoming more assertive and desirous of controlling their own lives rather than merely responding to the wishes of others or the limitations of the social structure (Burger and Solano 1994).

Men also suffer from traditional gender socialization. Men experience enormous cultural pressure to be successful in their work and earn a high income. In a study of 601 adult men, those who reported feeling "more masculine" earned an average of $35,800; those who reported feeling less masculine earned an average of $28,300 (Rubenstein 1990). Further, as a consequence of traditional male socialization, males are more likely than females to value materialism and competition over compassion and self-actualization (Beutel and Marini 1995; McCammon, Knox and Schacht 1998). Traditional male socialization also

> I learned from my father how to work. I learned from him that work is life and life is work, and work is hard.
>
> —Philip Roth
> Author

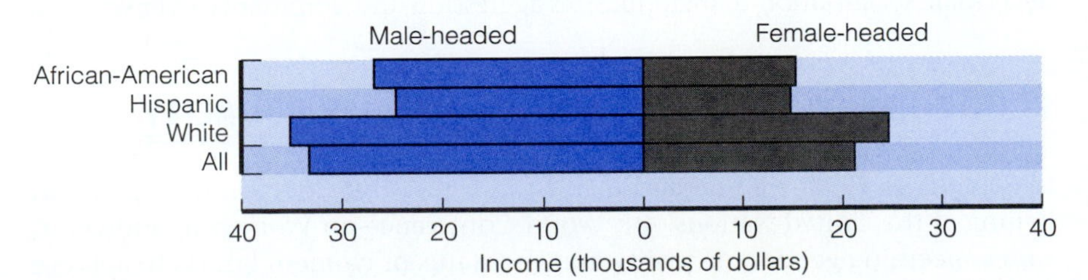

Figure 7.1 Family Households, Median Income by Race and Sex: 1996

SOURCE: *Statistical Abstract of the United States: 1998*, 118th ed. U.S. Bureau of the Census. Washington, D.C.: U.S. Government Printing Office, Table 743.

discourages males from expressing emotion. In a study of emotional expression among American women and men, Mirowsky and Ross (1995) conclude that "men keep emotions to themselves more than women, and that women express emotions more freely than men" (p. 449).

Health Costs

On the average, men in the United States die about six years earlier than women, although gender differences in life expectancy have been shrinking (Klein 1998c). Traditional male gender socialization is linked to males' higher rates of heart disease, cirrhosis of the liver, most cancers, AIDS, homicide, drug-induced deaths, suicide, and motor vehicle accidents (*Statistical Abstract of the United States: 1998,* Tables 140, 144, and 153).

National Data

Adolescent boys are four times more likely to commit suicide than teen girls and their numbers are increasing while the female rate is not.

SOURCE: Goldberg 1999

Are gender differences in morbidity and mortality a consequence of socialization differentials or physiological differences? Although both nature and nurture are likely to be involved, social rather than biological factors may be dominant. As part of the "masculine mystique," men tend to engage in self-destructive behaviors—heavy drinking and smoking, poor diets, a lack of exercise, stress-related activities, higher drug use, and a refusal to ask for help. Men are also more likely to work in hazardous occupations than women. For example, men are more likely to be miners than women—an occupation with one of the highest mortality rates in the United States.

Relationship Costs

I was more of a man with you as a woman than I have ever been with a woman as a man. I just need to learn to do it without the dress.

—Dustin Hoffman to Jessica Lange in the film *Tootsie*

Gender inequality also has an impact on relationships. For example, negotiating work and home life can be a source of relationship problems. While men in traditional versus dual-income relationships are more likely to report being satisfied with household task arrangements, women in dual-income families are the most likely to be dissatisfied with household task arrangements (Baker, Kriger, and Riley 1996). Further, the belief that your partner is not performing an equitable portion of the housework is associated with a reduction in the perception of spousal social support (Van Willigen and Drentea 1997).

There are also, of course, the practical difficulties of raising a family, having a career, and maintaining a happy and healthy relationship with a significant other. Successfully balancing work, marriage, and children may require a number of strategies, including: (1) a mutually satisfying distribution of household labor, (2) rejection of such stereotypical roles as "supermom" and "breadwinner dad," (3) seeking outside help from others (e.g., child care givers, domestic workers), and (4) a strong commitment to the family unit.

Finally, violence in relationships is gender specific (see Chapters 4 and 5). Although men are more likely to be victims of violent crime, women are more likely to be victims of rape and domestic violence. Violence against women reflects male socialization that emphasizes aggression and dominance over women.

STRATEGIES FOR ACTION: TOWARD GENDER EQUALITY

In September 1995, women and men from across the globe came together in Beijing at the United Nations' 4th World Conference on Women to address issues concerning gender inequality and the status of women. Efforts to achieve gender equality have been largely fueled by the feminist movement. Despite a conservative backlash, feminists, and to a lesser extent men's activists groups, have made some gains in reducing structural and cultural sexism in the workplace and in the political arena.

Be ready when the hour comes, to show that women are human and have the pride and dignity of human beings. Through such resistence our cause will triumph. But even if it does not, we fight not only for success, but in order that some inward feeling may have satisfaction. We fight that our pride, our self-respect, our dignity may not be sacrificed in the future as they have in the past.

—Christabel Pankhurst
American feminist, 1911

Grassroots Movements

Feminism and the Women's Movement The American feminist movement began in Seneca Falls, New York, in 1848 when a group of women wrote and adopted a women's rights manifesto modeled after the Declaration of Independence. Although many of the early feminists were primarily concerned with suffrage, feminism has its "political origins . . . in the abolitionist movement of the 1830s . . . ," where women learned to question the assumption of "natural superiority" (Andersen 1997, 305). Early feminists were also involved in the temperance movement, which advocated restricting the sale and consumption of alcohol, although their greatest success was the passing of the Nineteenth Amendment in 1920, which guaranteed women the right to vote.

The rebirth of feminism almost 50 years later was facilitated by a number of interacting forces: an increase in the number of women in the labor force, an escalating divorce rate, the socially and politically liberal climate of the 1960s, student activism, and the establishment of the Commission on the Status of Women by John F. Kennedy. The National Organization for Women (NOW) was established in 1966 and remains the largest feminist organization in the United States with more than 100,000 members. One of NOW's hardest-fought battles was the struggle to win ratification of the Equal Rights Amendment (ERA), which states that "[E]quality of rights under the law shall not be denied or abridged by the United States, or by any state, on account of sex." The proposed amendment passed both the House of Representatives and the Senate in 1972 but failed to be ratified by the required 38 states by the 1978 deadline, later extended to 1982. Thirty-five states have ratified the ERA, and it is presently in several state legislatures awaiting action.

Supporters of the ERA argue that its opponents used scare tactics—saying the ERA would lead to unisex bathrooms, mothers losing custody of their children, and mandatory military service for women—to create a conservative backlash. Susan Faludi in *Backlash: The Undeclared War against American Women* (1991) contends that the arguments against feminism today are the same as those levied against the movement a hundred years ago and that the negative consequences opponents of feminism predict (e.g., women unfulfilled and children suffering) have no empirical support.

Today, there is a new wave of feminism led by young women who grew up with the benefits won by their mothers but shocked by the realities of the Anita Hill hearing, the Tailhook scandal, and Paula Jones's accusations of sexual harassment against a sitting president. These young feminists are more inclusive than their predecessors, welcoming all who champion the cause of global equality. Not surprisingly, the new feminists are likely to attract a more diverse group of supporters than their predecessors. To measure your attitudes toward feminism look at this chapter's *Self & Society.*

International Data

The GDI Index (Gender-Related Index) developed by the United Nations measures the success of a country in moving toward gender equality. In 1997 the top-ranked countries were Canada, Norway, Sweden, Iceland, and the United States. The lowest-ranking countries were Sierra Leone, Niger, and Burkina Faso.

SOURCE: *Human Development Report* 1997

The Men's Movement As a consequence of the women's rights movement, men began to reevaluate their own gender status. In *Unlocking the Iron Cage: The Men's Movement, Gender Politics, and American Culture,* Michael Schwalbe (1996)

SELF & SOCIETY

Attitudes toward Feminism Scale

Following are statements on a variety of issues. Left of each statement is a place for indicating how much you agree or disagree. Please respond as you *personally* feel and use the following letter code for your answers:

A—Strongly Agree B—Agree C—Disagree D—Strongly Disagree

___ 1. It is naturally proper for parents to keep a daughter under closer control than a son.

___ 2. A man has the right to insist that his wife accept his view as to what can or cannot be afforded.

___ 3. There should be no distinction made between woman's work and man's work.

___ 4. Women should not be expected to subordinate their careers to home duties to any greater extent than men.

___ 5. There are no natural differences between men and women in sensitivity and emotionality.

___ 6. A wife should make every effort to minimize irritation and inconvenience to her husband.

___ 7. A woman should gracefully accept chivalrous attentions from men.

___ 8. A woman generally needs male protection and guidance.

___ 9. Married women should resist enslavement by domestic obligations.

___ 10. The unmarried mother is more immoral and irresponsible than the unmarried father.

___ 11. Married women should not work if their husbands are able to support them.

___ 12. A husband has the right to expect that his wife will want to bear children.

___ 13. Women should freely compete with men in every sphere of economic activity.

___ 14. There should be a single standard in matters relating to sexual behavior for both men and women.

___ 15. The father and mother should have equal authority and responsibility for discipline and guidance of the children.

___ 16. Regardless of sex, there should be equal pay for equal work.

___ 17. Only the very exceptional woman is qualified to enter politics.

___ 18. Women should be given equal opportunities with men for all vocational and professional training.

___ 19. The husband should be regarded as the legal representative of the family group in all matters of law.

___ 20. Husbands and wives should share in all household tasks if both are employed an equal number of hours outside the home.

___ 21. There is no particular reason why a girl standing in a crowded bus should expect a man to offer her his seat.

___ 22. Wifely submission is an outmoded virtue.

___ 23. The leadership of a community should be largely in the hands of men.

___ 24. Women who seek a career are ignoring a more enriching life of devotion to husband and children.

___ 25. It is ridiculous for a woman to run a locomotive and for a man to darn socks.

___ 26. Greater leniency should be adopted toward women convicted of crime than towards male offenders.

___ 27. Women should take a less active role in courtship than men.

___ 28. Contemporary social problems are crying out for increased participation in their solution by women.

___ 29. There is no good reason why women should take the name of their husbands upon marriage.

___ 30. Men are naturally more aggressive and achievement-oriented than women.

___ 31. The modern wife has no more obligation to keep her figure than her husband to keep down his waist line.

___ 32. It is humiliating for a woman to have to ask her husband for money.

___ 33. There are many words and phrases that are unfit for a woman's lips.

___ 34. Legal restrictions in industry should be the same for both sexes.

___ 35. Women are more likely than men to be devious in obtaining their needs.

___ 36. A woman should not expect to go to the same places or to have quite the same freedom of action as a man.

Attitudes toward Feminism Scale *(continued)*

___ 37. Women are generally too nervous and high-strung to make good surgeons.

___ 38. It is insulting to women to have the "obey" clause in the marriage vows.

___ 39. It is foolish to regard scrubbing floors as more proper for women than mowing the lawn.

___ 40. Women should not submit to sexual slavery in marriage.

___ 41. A woman earning as much as her male date should share equally in the cost of their common recreation.

___ 42. Women should recognize their intellectual limitations as compared with men.

List above reprinted by permission of Bernice Lott, Department of Psychology, University of Rhode Island.

SCORING

Score your answers as follows: A = +2, B = +1, C = −1, D = −2. Because half the items were phrased in a pro-feminist and half in an antifeminist direction, you will need to reverse the scores (+2 becomes −2, etc., for the following items: 1, 2, 6, 7, 8, 10, 11, 12, 17, 19, 21, 23, 25, 26, 27, 30, 33, 35, 36, 37, and 42. Now sum your scores for all the items. Scores may range from +84 to −84.

INTERPRETING YOUR SCORE

The higher your score, the higher your agreement with feminist (Lott used the term "women's liberation") statements. You may be interested in comparing your score, or that of your classmates, with those obtained by Lott (1973) from undergraduate students at the University of Rhode Island. The sample was composed of 109 men and 133 women in an introductory psychology class, and 47 additional older women who were participating in a special Continuing Education for Women (CEW) program. Based on information presented by Lott (1973), the following mean scores were calculated: Men = 13.07, Women = 24.30, and Continuing Education Women = 30.67.

More recently, Biaggio, Mohan, and Baldwin (1985) administered Lott's questionnaire to 76 students from a University of Idaho introductory psychology class and 63 community members randomly selected from the local phone directory. Although they did not present the scores of their respondents, they reported they did not find differences between men and women. Unlike Lott's students, in Biaggio et al.'s sample, women were not more pro-liberation than men. Biaggio, Mohan, and Baldwin (1985, p. 61) stated, "It seems that some of the tenets of feminism have taken hold and earned broader acceptance. These data also point to an intersex convergence of attitudes, with men's and women's attitudes toward liberation and child rearing being less disparate now than during the period of Lott's study." It would be interesting to determine if there are differences in scores between members of each sex in your class.

SOURCES: M. K. Biaggio, P. J. Mohan, and C. Baldwin. 1985. "Relationships among Attitudes toward Children, Women's Liberation, and Personality Characteristics." *Sex Roles* 12:47–62; B. E. Lott. 1973. "Who Wants the Children? Some Relationships among Attitudes toward Children, Parents, and the Liberation of Women." *American Psychologist* 28:573–82.

examines the men's movement as both participant and researcher. For three years, he attended meetings and interviewed active members. His research indicates that participants, in general, are white middle-class men who feel they have little emotional support, question relationships with their fathers and sons, and are overburdened by responsibilities, unsatisfactory careers, and what is perceived as an overly competitive society (Schwalbe 1996).

As with any grassroots movement, there are a variety of factions in the men's movement. Some men's organizations advocate gender equality; others developed to oppose "feminism" and what was perceived as male-bashing. For example, the Promise Keepers are part of a Christian men's movement that has often

> The world needs a man's heart.
>
> —Joseph Jastrab
> Author

been criticized as racially intolerant, patriarchal, and antifeminist. However, one woman researcher/author who attended meetings incognito, that is, as a man, reports: "I'm struck with how close it all sounds like feminism" (Leo 1997).

Today, issues of custody and fathers' rights, led by such groups as Dads against Discrimination, Texas Father's Alliance, and the National Coalition of Free Men (NCFM), headline the men's rights movement and have led to increased visibility (Goldberg 1997). Many members of such groups argue that society portrays men as "disposable," and that as fathers and husbands, workers and soldiers, they feel that they can simply be replaced by other men willing to do the "job." They also hold that there is nothing male-affirming in society and that the social reform of the last 30 years has "been the deliberate degradation and *dis-empowerment* of men economically, legally and socially" (NCFM 1998, 7). One reform advocated is in the area of reproductive rights (see this chapter's *The Human Side*).

National Data

A majority of dual worker couples are happy with their working partners; over 66 percent responded it gave them the freedom to change careers; 56 percent of men and 65 percent of women stated that a working spouse helped their own career, and 85 percent noted the advantages of two incomes. Lack of time was the biggest drawback cited. Respondents also wanted their companies to have flexible hours (85 percent), family leave time (75 percent), and company-supported child care (50 percent).

SOURCE: Kate 1998

Changes in the Workplace

Changes in the workplace are reflected in changes in the structure of the American family. In 1997, only 17 percent of households conformed to the traditional stereotype—breadwinner dad, stay-at-home mom, and children. Since the 1950s, the percentage of dual-career households has doubled, now representing 45 percent of the labor force (Kate 1998). Corporations have begun to accommodate changing gender roles and the increased emphasis on both work and family by offering a variety of new programs and benefits such as on-site child care, part-time employment, job sharing, flextime, telecommuting, and assistance with elderly parents.

More women have also begun to enter traditionally male occupations. Williams (1995) suggests that this is an important and essential step on the road to gender equality, because it gives women more economic opportunities and helps to break down limiting stereotypes about women's capabilities. However, Williams warns:

> **Well-meaning efforts directed at getting women to be more "like men" run the risk of reifying the male standard, making men the ultimate measure of success. If the aim is gender equality, then men should be encouraged to become more "like women" by developing, or feeling free to express, interests and skills in traditionally female jobs. (p. 179)**

International Data

In a study of workplace sexual harassment in Taiwan, 36 percent of the women and 13 percent of the men experienced sexual harassment, including, in order of frequency, unwanted sexual comments/jokes, unwanted deliberate touching, unwanted request/pressure for a date.

SOURCE: Luo 1996

Public Policy

A number of statutes have been passed to help reduce gender inequality. They include the 1963 Equal Pay Act, Title IX of the Educational Amendments Act, Title VII of the Civil Rights Act of 1964, the Displaced Homemakers Act, the 1978 Pregnancy Discrimination Act, and the Family Leave Act of 1993. The National Organization for Women (1995) encourages women to be politically active, to run for political office, and to participate in the decision-making processes of the nation. Recently, political activism has focused on the issues of sexual harassment and affirmative action.

Sexual Harassment During the 1980s and 1990s, the courts held that Title VII of the 1964 Civil Rights Act prohibited **sexual harassment** of males or females.

THE HUMAN SIDE

DECLARATION OF THE FATHER'S FUNDAMENTAL PRE-NATAL RIGHTS

WHEREAS, the function of men in parenting has been confined largely to the second-class role of material provider; and

WHEREAS, healthy relationships and healthy families require that men be equal participants in every facet of parenting, including responsible contraception and conception; and

WHEREAS, in a substantial number of abortions, the prospective fathers are excluded from all phases of the pregnancy termination process, including any prior knowledge of the initial decision to abort and even post-abortion notification; and

WHEREAS, purported "freedom of choice" is neither freedom nor choice as long as one of the two partners-in-conception can unilaterally impose a decision on the other without notification, discussion, consultation, or any other form of reasonable or humane discourse; and

WHEREAS, the question of the participation of the prospective father in any decision to abort is a separate and distinct matter from those positions normally identified as "pro-choice" and "pro-life"; and

WHEREAS, individuals may in good conscience fully embrace either position and still hold to the heartfelt principle that conception and pre-natal participation, as a vital part of a man's role in parenting, should be respected along with all the other facets of his parenting role; and

WHEREAS, the decision to abort or to carry to term is an intensely private matter, and, to the maximum extent reasonable, the decision to do either should be made solely by the partners-in-conception; and

WHEREAS, ensuring that all options are fully understood and supported by each partner-in-conception before irreversible decisions are implemented requires the complete and healthy involvement of both partners; and

WHEREAS, for the prospective father to participate fully in that process and to fully exercise his pre-natal role as nurturer and protector, notification of conception and establishment of paternity are necessary:

NOW THEREFORE, this organization, in recognition of the foregoing facts, hereby adopts the following:

I. The prospective father has the fundamental right to be informed by his partner-in-conception that conception has resulted from their union.

II. The prospective father has the fundamental right to participate with his partner-in-conception in any decision affecting the future of the fetus he helped create.

III. The prospective father has the fundamental right to consult with his partner-in-conception or the health care provider, and to be apprised of any relevant information concerning the pregnancy or the abortion process.

IV. A putative father has the fundamental right to a determination of paternity, during both pre-natal and post-natal periods, at the earliest practical time, and by the most conclusive methods reasonably available.

V. The prospective father has a fundamental right of custody

(continued)

THE HUMAN SIDE

DECLARATION OF THE FATHER'S FUNDAMENTAL PRE-NATAL RIGHTS *(continued)*

equal to that of his partner-in-conception and superior to that of any other.

VI. The prospective father has the right to personal guardianship of the fetus when required to protect the well-being of the fetus or to preserve the right to custody.

VII. The prospective father has the fundamental right, with the consent of his partner-in-conception, to be present at delivery.

VIII. The foregoing fundamental rights shall be neither abrogated nor abridged without due cause clearly and appropriately established.

IX. A prospective mother has the moral responsibility to respect and support the rights of her partner-in-conception.

X. Both public policy and medical ethics should seek to protect and advance the fundamental pre-natal rights of the father.

SOURCE: Resolution adopted by the National Coalition of Free Men on August 19, 1992. Reprinted by permission of the National Coalition of Free Men, **http://ncfm.org**

National Data

In a survey of women returning from active duty in the Gulf War, 7.3 percent reported sexual assault, 33.1 percent physical sexual harassment, and 66.2 percent reported verbal sexual harassment.

SOURCE: Wolfe et al. 1998

Reports of sexual harassment to the EEOC nearly doubled between 1991 and 1992 in response to the publicity surrounding the Anita Hill–Clarence Thomas controversy.

There are two types of sexual harassment: (1) *quid pro quo,* in which an employer requires sexual favors in exchange for a promotion, salary increase, or any other employee benefit, and (2) the existence of a hostile environment that unreasonably interferes with job performance, as in the case of sexually explicit comments or insults being made to an employee. According to a 1993 Supreme Court decision, a person no longer has to demonstrate "severe psychological damage" in order to win damages. Sexual harassment occurs at all occupational levels, and some research suggests that the incidents of sexual harassment are inversely proportional to the number of women in an occupational category (Fitzgerald and Shullman 1993). For example, female doctors (Schneider and Phillips 1997) and lawyers (Rosenberg, Perlstadt, and Phillips 1997) report high rates of sexual harassment, in the first case by male patients and in the second by male colleagues.

Accusations of obstruction of justice in the Paula Jones sexual harassment suit led to impeachment hearings against President Clinton.

Affirmative Action

The 1964 Civil Rights Act provided for **affirmative action** to end employment discrimination based on sex and race (see Chapter 8). Such programs require employers to make a "good faith effort" to provide equal opportunity to women and other minorities. However, in response to the growing sentiment that affirmative action programs constitute "reverse discrimination," recent court decisions have begun to dismantle affirmative action programs (see Chapter 8). The pending Civil Rights Act of 1997 would eliminate affirmative action programs in all federal programs and activities (LCCR 1998b). Although supporters of the bill argue that a majority of Americans oppose affirmative action, in fact, most polls show that the American people support affirmative action (LCCR 1998a). Many argue that as long as the culture and structure of society promote gender (as well as racial and class) inequality, the seemingly "neutral" positions advocated by reformers will result in continued gender inequities.

UNDERSTANDING GENDER INEQUALITY

> Give to every other human being every right that you claim for yourself.
>
> —Thomas Paine
> Political and social activist

Gender roles and the social inequality they create are ingrained in our social and cultural ideologies and institutions and are, therefore, difficult to alter. Nevertheless, as we have seen in this chapter, growing attention to gender issues in social life has spurred some change. For example, women who have traditionally been expected to give domestic life first priority are now finding it acceptable to be more ambitious in seeking a career outside the home. Men who have traditionally been expected to be aggressive and task oriented are now expected to be more caring and nurturing. Women seem to value gender equality more than men, however, perhaps because women have more to gain. For instance, 84 percent of 600 adult women said that the ideal man is caring and nurturing; only 52 percent of 601 adult men said that the ideal woman is ambitious (Rubenstein 1990, 160).

But men also have much to gain by gender equality. Eliminating gender stereotypes and redefining gender in terms of equality do not mean simply liberating women, but liberating men and our society as well. "What we have been talking about is allowing people to be more fully human and creating a society that will reflect that humanity. Surely that is a goal worth striving for" (Basow 1992, 359). Regardless of whether traditional gender roles emerged out of biological necessity as the functionalists argue or economic oppression as the conflict theorists hold, or both, it is clear today that gender inequality carries a high price: poverty, loss of human capital, feelings of worthlessness, violence, physical and mental illness, and death. Surely, the costs are too high to continue to pay.

CRITICAL THINKING

1. Some research suggests that "[Men] and women with more androgynous gender orientations—that is to say, those having a balance of masculine and feminine personality characteristics—show signs of greater mental health and more positive self-images." (Anderson 1997, 34). Do you agree or disagree? Why or why not?
2. The chapter indicates that there is a "gender gap" in the number of men and women entering college—women will be entering college at a rate of over twice that of men. While the number of females in the population is slightly higher, the difference does not explain the projected gap in enrollments. Why are women entering college at a higher rate than men?
3. What have been the interpersonal costs, if any, of sensitizing U.S. society to the "political correctness" of female-male interactions?
4. Why are women more likely to work in traditionally male occupations than men are to work in traditionally female occupations? Are the barriers that prevent men from doing "women's work" cultural, structural, or both? Explain.

KEY TERMS

affirmative action
cultural sexism
devaluation hypothesis
double or triple (multiple) jeopardy
gender
glass ceiling
human capital hypothesis
occupational sex segregation
pink-collar jobs
sex
sexism
sexual harassment
structural sexism

INTERNET

You can find more information on men's parental rights, feminism, occupational sex segregation, and the health and social-psychological costs of traditional gender roles at the *Understanding Social Problems* web site at **http://sociology.wadsworth.com.**

INFOTRAC COLLEGE EDITION

Either from the Wadsworth sociology resource at **http://sociology.wadsworth.com** or directly from your web browser, you may access InfoTrac College Edition, an online university library that includes over 700 popular and scholarly journals in which you can find articles related to the topics in this chapter. Suggested articles and questions relating to these articles are listed below.

Bottero, Wendy. 1998. "Clinging to the Wreckage? Gender and the Legacy of Class." *Sociology* 32(3):469–89.

1. What does the author see as the failings of Marxian class theory in relation to gender inequality?
2. What is the Neo-Weberian theory of class and gender in the economy?
3. What is the theory of gendered closure?

Cohen, Phillip N. 1998. "Black Concentration Effects of Black-White and Gender Inequality." *Social Forces* 77(1):207–30.

1. According to the author, what are some of the problems inherent in studying the interaction of gender and racial inequality?
2. How might employed white females benefit from racial inequalities in the working world?
3. What new finding concerning the interaction of gender and racial inequality resulted from this research?

Condravey, Jace, Ester Skirboli, and Rhoda Taylor. 1998. "Faculty Perceptions of Classroom Gender Dynamics." *Women and Language* 21(1):18–28.

1. In the many studies cited in this article, what is the general relationship between gender of instructor and gender of student?
2. In relation to how instructors view differences in gendered classroom behavior, what were the findings of this study?
3. According to the authors, what seems to be the leading factor in determining whether or not an instructor observes gender inequality in classroom discussions?

CHAPTER EIGHT

Race and Ethnic Relations

Is It True?

Is It True?

1. Many anthropologists and other scientists have concluded that "races" do not really exist.
2. In a 1997 Gallup poll, more than half of whites reported that they would approve of interracial marriage.
3. In 1997, the unemployment rate for foreign-born citizens was significantly higher than the unemployment rate for native-born Americans.
4. Affirmative action programs have resulted in higher unemployment rates for whites than for blacks.
5. In Sweden, it is against the law to give the Nazi salute.

Answers: 1 = T; 2 = T; 3 = F; 4 = F; 5 =

I am colored and America is colorblind. Do you see me?

James M. Jones
University of Delaware

In June 1998, James Byrd Jr., a 49-year-old father of three, was walking home from a niece's bridal shower in the small town of Jasper, Texas. According to police reports, three white men riding in a gray pickup truck saw Byrd, a black man, walking down the road and offered him a ride. The men reportedly drove down a dirt lane and, after beating Byrd, chained him to the back of the pickup truck and dragged him for 2 miles down a winding, narrow road. The next day, police found Byrd's mangled and dismembered body. The three men who were arrested had ties to white supremacist groups. James Byrd had been brutally murdered simply because he was black.

Although most Americans were shocked by the murder of James Byrd, many were not surprised. Sadie Hodge, a black woman who has lived in Jasper, Texas, for 64 years, said of Byrd's murder, "This is the most gruesome thing I've ever heard of, but I'm not surprised... There's a lot that goes on but it's pushed under the rug" (Jones 1998, 2A). Indeed, when asked about the future of black-white relations in the United States, a majority of both races in a national Gallup poll (54 percent of whites and 58 percent of blacks) said that relations between blacks and whites will "always be a problem" (Gallup Organization 1997).

In this chapter, we discuss the nature and origins of prejudice and look "under the rug" to uncover the extent of discrimination and its consequences for both racial and ethnic minorities. We also discuss strategies designed to reduce prejudice and discrimination. We begin by examining racial and ethnic diversity worldwide and in the United States, emphasizing first that the concept of race is based on social rather than biological definitions.

National Data

In a national survey of first-year college students, one in five agreed that "racial discrimination is no longer a major problem in America."

SOURCE: "This Year's Freshmen: A Statistical Profile" 1999

THE GLOBAL CONTEXT: DIVERSITY WORLDWIDE

A first-grade teacher asked the class, "What is the color of apples?" Most of the children answered red. A few said green. One boy raised his hand and said "white." The teacher tried to explain that apples could be red, green or sometimes golden, but never white. The boy insisted his answer was right and finally said, "Look inside" (Goldstein 1999). Like apples, human beings may be similar on the "inside," but are often classified into categories according to external appearance. After examining the social construction of racial categories, we review patterns of interaction among racial and ethnic groups and overview racial and ethnic diversity in the United States.

Mary Nell Verrett remembers her brother, James Byrd Jr., who was brutally murdered in a racial hate crime in Jasper, Texas.

The Social Construction of Race

The concept **race** refers to a category of people who are believed to share distinct physical characteristics that are deemed socially significant. Racial groups are sometimes distinguished on the basis of such physical characteristics as skin color, hair texture, facial features, and body shape and size. Some physical variations among people are the result of living for thousands of years in different geographical regions (Molnar 1983). For example, humans living in regions with hotter climates developed darker skin from the natural skin pigment, melanin, which protects the skin from the sun's rays. In regions with moderate or colder climates, people had no need for protection from the sun and thus developed lighter skin.

Cultural definitions of race have taught us to view race as a scientific categorization of people based on biological differences between groups of individuals. Yet, racial categories are based more on social definitions than on biological differences. Anthropologist Mark Cohen (1998) explains that distinctions among human populations are graded, not abrupt. Skin color is not black or white, but rather ranges from dark to light with many gradations of shades. Noses are not either broad or narrow, but come in a range of shapes. Physical traits such as these, as well as hair color and other both visible and invisible characteristics, come in an infinite number of combinations. For example, a person with dark skin can have any blood type and can have a broad nose (a common combination in West Africa), a narrow nose (a common trait in East Africa), or even blond hair (a combination found in Australia and New Guinea) (Cohen 1998).

The science of genetics also challenges the notion of race. Geneticists have discovered that "the genes of black and white Americans probably are 99.9 percent alike" (Cohen 1998, B4). Furthermore, genetic studies indicate that genetic variation is greater *within* racially classified populations than *between* racial groups (Keita and Kittles 1997). Classifying people into different races fails to recognize that over the course of human history, migration and intermarriage have resulted in the blending of genetically transmitted traits. Thus, there are no "pure" races; people in virtually all societies have genetically mixed backgrounds.

The American Anthropological Association has passed a resolution stating that "differentiating species into biologically defined 'race' has proven meaningless and unscientific" (Etzioni 1997, 39). Scientists who reject the race concept now speak of **populations** when referring to groups that most people would call races (Zack 1998).

The problem is . . . we don't believe we are as much alike as we are. Whites and blacks, Catholics and Protestants, men and women. If we saw each other as more alike, we might be very eager to join in one big human family in this world, and to care about that family the way we care about our own.

—Morrie Schwartz
Sociologist
(from *Tuesdays with Morrie*)

CONSIDERATION

The concept of race is so embedded in our culture that even those who advocate abandoning racial classification continue to use racial terminology. For example, in his book *The New Colored People: The Mixed-Race Movement in America,* Professor Jon Spencer (1997) argues for abandoning racial classification, yet he uses racial terminology throughout the book (even in the last chapter entitled "Thou Shalt Not Racially Classify"). Keita and Kittles (1997) suggest that the concepts of biological race and racial categories continue to exist and be utilized "in part, due simply to old habits. Categorical thinking is entrenched . . ." (p. 591).

Patterns of Racial and Ethnic Group Interaction

When two or more racial or ethnic groups come into contact, one of several patterns of interaction may occur, including genocide, expulsion or population transfer, slavery, colonialism, segregation, acculturation, assimilation, pluralism, and amalgamation. These patterns of interaction may occur when two or more groups exist in the same society or when different groups from different societies come into contact.

- **Genocide** refers to the deliberate, systematic annihilation of an entire nation or people. The European invasion of the Americas, beginning in the sixteenth century, resulted in the decimation of most of the original

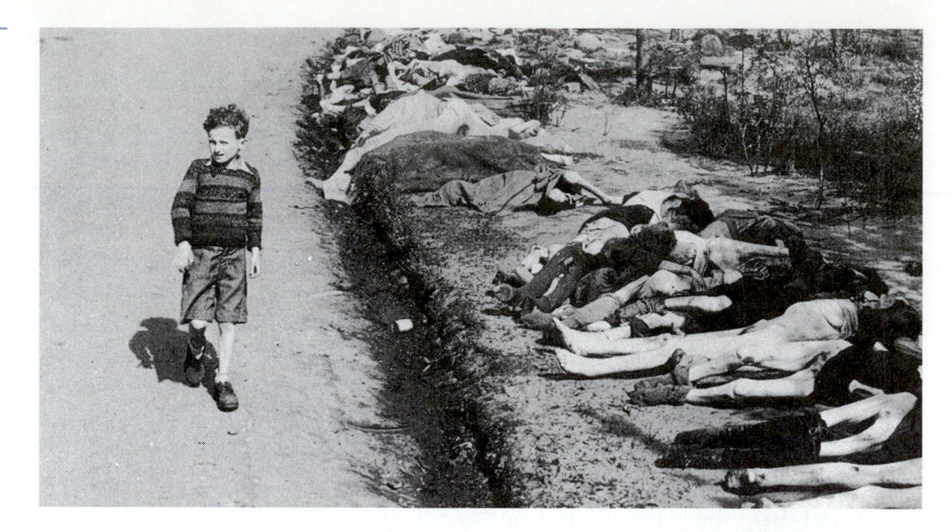

The most cited twentieth-century example of genocide is the "extermination" of over 12 million Jews and others considered by Hitler not to be members of his "superior race."

inhabitants of North and South America. Some native groups were intentionally killed, whereas others fell victim to diseases brought by the Europeans. In the twentieth century, Hitler led the Nazi extermination of more than 12 million people, including over 6 million Jews, in what has become known as the Holocaust. More recently, ethnic Serbs have attempted to eliminate Muslims from parts of Bosnia—a process they call "ethnic cleansing."

- **Expulsion or population transfer** occurs when a dominant group forces a subordinate group to leave the country or to live only in designated areas of the country. The 1830 Indian Removal Act called for the relocation of eastern tribes to land west of the Mississippi River. The movement, lasting more than a decade, has been called the Trail of Tears because tribes were forced to leave their ancestral lands and endure harsh conditions of inadequate supplies and epidemics that caused illness and death. After Japan's attack on Pearl Harbor in 1941, President Franklin Roosevelt authorized the removal of any people considered threats to national security. All people on the West Coast of at least one-eighth Japanese ancestry were transferred to evacuation camps surrounded by barbed wire, where 120,000 Japanese Americans experienced economic and psychological devastation. In 1979, Vietnam expelled nearly 1 million Chinese from the country as a result of long-standing hostilities between China and Vietnam.
- **Slavery** exists when one group treats another group as property to exploit for financial gain. The dominant group forces the enslaved group to live a life of servitude, without the basic rights and privileges enjoyed by the dominant group. In early American history, slavery was tolerated and legal for three centuries. Enslavement of Africans also occurred in Canada from 1689 to 1833 (Schaefer 1998).
- **Colonialism** occurs when a racial or ethnic group from one society takes over and dominates the racial or ethnic group(s) of another society. The European invasion of North America, the British occupation of India, and the Dutch presence in South Africa prior to the end of apartheid are examples of outsiders taking over a country and controlling the native population. As a territory of the United States, Puerto

Rico is essentially a colony whose residents are U.S. citizens but cannot vote in presidential elections unless they move to the mainland.

- **Segregation** refers to the physical separation of two groups in residence, workplace, and social functions. Segregation may be **de jure** (Latin meaning "by law") or **de facto** ("in fact"). Between 1890 and 1910, a series of U.S. laws that came to be known as **Jim Crow laws** were enacted that separated blacks from whites by prohibiting blacks from using "white" buses, hotels, restaurants, and drinking fountains. In 1896, the U.S. Supreme Court (in *Plessy v. Ferguson*) supported de jure segregation of blacks and whites by declaring that "separate but equal" facilities were constitutional. Blacks were forced to live in separate neighborhoods and attend separate schools. Beginning in the 1950s, various rulings overturned these Jim Crow laws, making it illegal to enforce racial segregation. Although de jure segregation is illegal in the United States, de facto segregation still exists in the tendency for racial and ethnic groups to live and go to school in segregated neighborhoods.

  Segregation is not unique to the United States. In Germany, for example, Turkish immigrants are concentrated in ghettos, and in Sweden, Greek, Chilean, and Turkish immigrants are largely isolated from the rest of the population.
- **Acculturation** refers to learning the culture of a group different from the one in which a person was originally raised. Acculturation may involve learning the dominant language, adopting new values and behaviors, and changing the spelling of the family name. In some instances, acculturation may be forced, as in the California decision to discontinue bilingual education and force students to learn English in school.
- **Assimilation** is the process by which formerly distinct and separate groups merge and become integrated as one. One form of assimilation is referred to as the **melting pot**, whereby different groups come together and contribute equally to a new, common culture. Although the United States has been referred to as a "melting pot," in reality, many minorities have been excluded or limited in their cultural contributions to the predominant white Anglo-Saxon Protestant tradition.

  There are two types of assimilation: secondary and primary. **Secondary assimilation** occurs when different groups become integrated in public areas and in social institutions, such as neighborhoods, schools, the workplace, and in government. **Primary assimilation** occurs when members of different groups are integrated in personal, intimate associations such as friends, family, and spouses.

  The degree of acculturation and assimilation that occurs between majority and minority groups depends in part on (1) whether minority group members have voluntary or involuntary contact with the majority group and (2) whether majority group members accept or reject newcomers or minority group members. Groups that *voluntarily immigrate* and who are *accepted* by "host" society members will experience an easier time acculturating and assimilating than those who are forced (through slavery, frontier expansion, or military conquest) into contact with and are rejected by the majority group.
- **Pluralism** refers to a state in which racial and ethnic groups maintain their distinctness, but respect each other and have equal access to social resources. In Switzerland, for example, four ethnic groups—French,

National Data

Less than one percent of U.S. married couples are interracial.

SOURCE: *Statistical Abstract of the United States 1998* Table 67

National Data

In a 1997 Gallup poll, 61 percent of whites reported that they would approve of interracial marriage (compared to only 4 percent in 1958).

SOURCE: Gallup Organization 1997

Italians, Germans, and Swiss Germans—maintain their distinct cultural heritage and group identity in an atmosphere of mutual respect and social equality. In the United States, the political and educational recognition of multiculturalism reflects efforts to promote pluralism. Given the level of prejudice and discrimination toward racial and ethnic minorities (discussed later in this chapter), however, the United States is far from pluralistic.

- **Amalgamation**, also known as **marital assimilation**, occurs when different ethnic or racial groups become married or pair-bonded and produce children. But in most societies, the norm of **endogamy** influences individuals to marry within their social group. For many years, some states had **antimiscegenation laws** that prohibited interracial marriages, but in 1967, the Supreme Court (in *Loving v. Virginia*) declared these laws unconstitutional and required all states to recognize interracial marriage (Reid 1995). In the United States, marriages between individuals with different ethnic backgrounds are not unusual, but interracial marriages are relatively rare.

Racial Diversity in the United States

The first census in 1790 divided the U.S. population into four groups: free white males, free white females, slaves, and other persons (including free blacks and Indians). In order to increase the size of the slave population, the **"one drop of blood rule"** appeared, which specified that even one drop of Negroid blood defined a person as black and, therefore, eligible for slavery. In 1960, the census recognized only two categories: white and nonwhite. In 1970, the census categories consisted of white, black, and "other" (Hodgkinson 1995). In 1990, the U.S. Bureau of the Census recognized four racial classifications: (1) white, (2) black, (3) American Indian, Aleut, or Eskimo, and (4) Asian or Pacific Islander. The 1990 census also included the category of "other."

In 1997, the U.S. Office of Management and Budget revised federal guidelines for collecting data on race and ethnicity. According to these guidelines, respondents to federal surveys and the census have the option of officially identifying themselves as being more than one race, rather than checking only one racial category.

Table 8.1 presents U.S. census data on the racial composition of the United States. The data presented are based on the prior system of selecting

Table 8.1 U.S. Racial Composition and Projected Composition for Selected Years (Percentage of Total Population)

	1990	1998	2005	2050
White	83.9	82.6	81.3	72.8
Black	12.3	12.7	13.2	15.7
American Indian, Eskimo, and Aleut	0.8	0.9	0.9	1.1
Asian and Pacific Islander	3.0	3.8	4.6	10.3

SOURCES: *Statistical Abstract of the United States: 1995.* U.S. Bureau of the Census. Washington, D.C.: U.S. Government Printing Office. U.S. Bureau of the Census. 1996 (March). "Resident Population of the United States: Middle Series Projections, 2001–2005, by Sex, Race, and Hispanic Origin, with Median Age." **http://www.census.gov/population/projections/nation/nsrh/nprh0105.txt** (May 1, 1998). U.S. Bureau of the Census. 1998 (May 29). "Resident Population of the United States: Estimates by Sex, Race, and Hispanic Origin, with Median Age." **http://www.census.gov/population/estimates/nation/intfile3-1.txt** (June 3, 1998).

one racial category (data based on multiracial categories were not available at the time of this writing).

> Anthropologists tell us that we are nearly all multiracial. If we were to be perfectly accurate, we would all check the box marked "Other."
>
> —Glenda Valentine
> Research associate,
> Teaching Tolerance

CONSIDERATION

A mixed-race option for self-identification avoids putting children of mixed-race parents in the difficult position of choosing the race of one parent over the other when filling out race data on school and other forms. It also avoids impairment of children's self-esteem and social functioning that comes from choosing the racial category of "Other." Such a category implies that the society does not recognize and respect mixed-race individuals, and thus "children growing up within mixed families may feel ashamed of their 'irregular' racial makeup and may experience rejection and alienation in the wider social community" (Zack 1998, 23).

However, will the wide-scale recognition of mixed-race identity decrease the numbers within minority groups and disrupt the solidarity and loyalty based on racial identification? What will happen, for example, to organizations and movements devoted to equal rights for blacks if much of the "black" population acquires a new mixed-racial identity?

Finally, the recognition of mixed-race is criticized as perpetuating scientifically unfounded classifications by race. This criticism suggests that mixed-race categorization is just as unscientific as the concept of race itself. However, mixed-race classification may encourage the realization that racial categorization is meaningless:

> Recognized mixed-race identity would undo the assumption that everyone is racially pure, and may, in the process, undo the assumption that everyone belongs to a race or that race is a meaningful way to type people. (Zack 1998, 27)

Golf pro Tiger Woods has referred to himself as "Cabrinasian"—reflecting his mixed heritage that includes Caucasian, Black, Indian, and Asian.

If you look at me, I'm black. But if you're asking my ethnicity, it's Haitian. The term "black" just doesn't cut it.

–Denise Bernard
President of
Queens College
student government

Ethnic Diversity in the United States

Ethnicity refers to a shared cultural heritage or nationality. Ethnic groups may be distinguished on the basis of language, forms of family structures and roles of family members, religious beliefs and practices, dietary customs, forms of artistic expression such as music and dance, and national origin.

Two individuals with the same racial identity may have different ethnicities. For example, a black American and a black Jamaican have different cultural, or ethnic, backgrounds. Conversely, two individuals with the same ethnic background may identify with different races. Hispanics, for example, may be white or black. The current Census Bureau classification system does not allow people of mixed Hispanic/Latino ethnicity to identify themselves as such. Individuals with one Hispanic and one non-Hispanic parent still must say they are either Hispanic or not Hispanic. And Hispanics must select one country of origin, even if their parents are from different countries.

U.S. citizens come from a variety of ethnic backgrounds. More than 20 percent of the U.S. population come from German ancestry, more than 10 percent are of Irish ancestry, and more than 10 percent have English ancestry. In 1998, about 11 percent of the U.S. population were Hispanic (of any race) (U.S. Bureau of the Census 1998).

National Data

Hispanics are the fastest-growing segment of the U.S. population. The majority of U.S. Hispanics are Mexican in origin (63 percent), with the balance being Puerto Rican, Cuban, and Central and South American.

SOURCE: Galper 1998

CONSIDERATION

The use of racial and ethnic labels is often misleading. The ethnic classification of "Hispanic/Latino," for example, lumps together such disparate groups as Puerto Ricans, Mexicans, Cubans, Venezuelans, Colombians, and others from Latin American countries. The racial term "American Indian" includes more than 300 separate tribal groups that differ enormously in language, tradition, and social structure. The racial label "Asian American" includes individuals with ancestors from China, Japan, Korea, India, the Philippines, or one of the countries of Southeast Asia.

Race and ethnicity are in the eye of the beholder.

–Harold L. Hodgkinson
Director of the Center
for Demographic Policy

U.S. Immigration

The growing racial and ethnic diversity of the United States is largely due to immigration (as well as the higher average birthrates among many minority groups). For the first hundred years of U.S. history, all immigrants were allowed to enter and become permanent residents. Initially, most U.S. immigrants were from England, Ireland, and Germany, but by the early and mid-1900s, they were predominantly from southern and eastern Europe and Mexico. In 1996, leading countries of immigrant origin included Mexico, Philippines, India, Vietnam, and China (Immigration & Naturalization Service 1997). Although there is currently a limit for legal immigration, it is not a ceiling because "immediate relatives" of U.S. citizens—the largest category of legal immigrants—are not limited (see Figure 8.1).

A major concern regarding immigration is the extent of people entering the United States illegally. In 1986, Congress approved the Immigration Reform and Control Act, which made hiring illegal aliens an illegal act punishable by fines and even prison sentences. This act also prohibits employers from discriminating against legal aliens who are not U.S. citizens. Under the 1996 Illegal Immigration Reform and Immigrant Responsibility Act, illegal immigrants may not benefit from programs such as Social Security and welfare.

National Data

In 1997, nearly 1 in 10 residents of the United States was foreign-born.

SOURCE: U.S. Bureau of the Census 1998

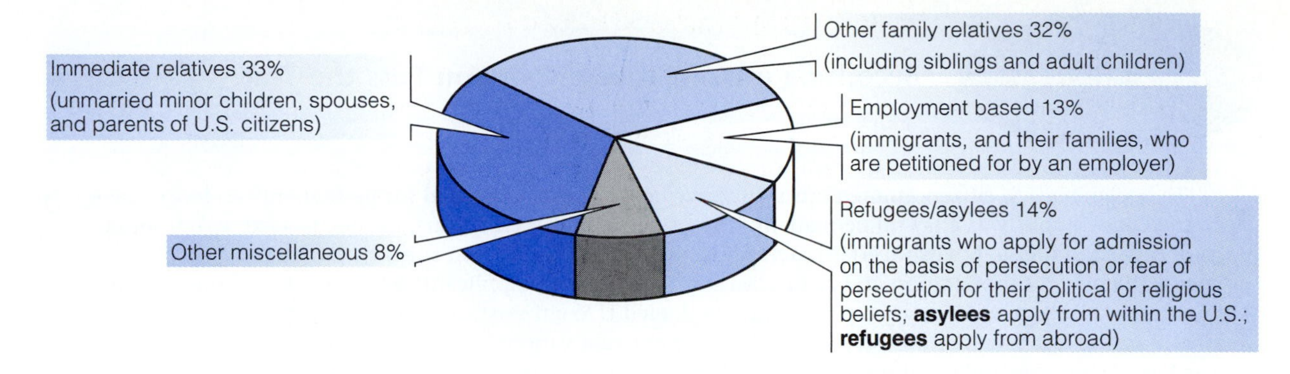

Figure 8.1 **Major Categories of Legal Immigrants: 1996**

SOURCE: Immigration and Naturalization Service Statistics Division. 1997 (May). "Immigration Overview." **http://www.fairus.org/04121604.htm** April 6, 1998

CONSIDERATION

According to a study by the National Academy of Sciences, immigration produces economic benefits for the United States as a whole ("Study Finds Benefits from Immigration" 1997). Economist James Smith explained: "It's true that some Americans are now paying more taxes because of immigration, and native-born Americans without a high school education have seen their wages fall slightly because of the competition sparked by lower-skilled, newly arrived immigrants. But the vast majority of Americans are enjoying a healthier economy as a result of the increased supply of labor and lower prices that result from immigration" (p. 4A).

National Data

In 1996, of the 1,216,000 immigrants who entered the United States, an estimated 25% were illegal and 75% were legal. In 1996, the total illegal U.S. population reached more than 5,000,000.

SOURCE: Immigration & Naturalization Service Statistics Division 1997

Many immigrants struggle to adjust to life in the United States. But despite the prejudice, discrimination, and lack of social support they experience, many foreign-born U.S. residents work hard to succeed educationally and occupationally. Foreign born residents of the United States may or may not apply for, and be granted, U.S. citizenship. In 1997, of all foreign born U.S. residents, 35% were **naturalized citizens** (immigrants who applied and met the requirements for U.S. citizenship); 65% were not U.S. citizens (Current Population Survey, U.S. Bureau of the Census 1997). Requirements for becoming a U.S. citizen are discussed in this chapter's *Self & Society* feature.

National Data

In 1997, about 1 in 4 of the U.S. foreign-born population age 25 and over had completed four or more years of college—the same rate as the native born. (However, 35 percent of the foreign-born population had not completed high school, compared to 16 percent of native born.) In the same year, the unemployment rate for foreign-born citizens was 4.3 percent, lower than the 5.4 percent rate for native born citizens.

SOURCE: U.S. Bureau of the Census 1998

SOCIOLOGICAL THEORIES OF RACE AND ETHNIC RELATIONS

Some theories of race and ethnic relations suggest that individuals with certain personality types are more likely to be prejudiced or to direct hostility toward minority group members. Sociologists, however, concentrate on the impact of the structure and culture of society on race and ethnic relations. Three major sociological theories lend insight into the continued subordination of minorities.

Structural-Functionalist Perspective

Functionalists emphasize that each component of society contributes to the stability of the whole. In the past, inequality between majority and minority groups was functional for some groups in society. For example, the belief in

Becoming a U.S. Citizen: Can You Pass the Test?

To become a U.S. citizen, immigrants must have been lawfully admitted for permanent residence; have resided continuously as a lawful permanent U.S. resident for at least 5 years; be able to read, write, speak, and understand basic English (certain exemptions apply); and they must show that they have "good moral character" (Immigration and Naturalization Service 1998). Applicants who have been convicted of murder or an aggravated felony are permanently denied U.S. citizenship. In addition, applicants are denied if in the last 5 years they have engaged in any one of a variety of offenses, including prostitution, illegal gambling, controlled substance law violation (except for a single offense of possession of 30 grams or less of marijuana), habitual drunkardness, willful failure or refusal to support dependents, and criminal behavior involving "moral turpitude."

To become a U.S. citizen, one must take the oath of allegiance and swear to support the Constitution and obey U.S. laws, renounce any foreign allegiance, and bear arms for the U.S. military or perform services for the U.S. government when required.

Finally, applicants for U.S. citizenship must pass an examination administered by the U.S. Immigration and Naturalization Service. The following questions are typical of those on the examination given to immigrants seeking U.S. citizenship. Applicants may choose between an oral and a written test. On the oral test, they must answer all the questions correctly. On the written test, they must correctly answer 12 of 20 questions. Based on your answers to these questions, would you be granted U.S. citizenship if you were an immigrant (that is, could you correctly answer 6 out of the following 10 questions)? After selecting an answer for each of the following questions, check your answer using the answer key provided.

SAMPLE QUESTIONS FOR BECOMING A U.S. CITIZEN

1. Who becomes the president of the United States if the president and vice president should die?
   a. The speaker of the House of Representatives
   b. The Senate majority leader
   c. The chairman of the Joint Chiefs of Staff
   d. The chief justice of the Supreme Court
2. Who said, "Give me liberty or give me death"?
   a. George Washington
   b. Benjamin Franklin
   c. Patrick Henry
   d. Thomas Jefferson
3. How many branches are there in our government?
   a. 2
   b. 3
   c. 4
   d. 6
4. Which countries were our enemies during World War II?
   a. Iraq, Libya, and Turkey
   b. Germany, Japan, and the Soviet Union
   c. Japan, Italy, and Germany
   d. Italy, Germany, and France
5. What are the duties of Congress?
   a. To execute laws
   b. To naturalize citizens
   c. To sign bills into law
   d. To make laws

Becoming a U.S. Citizen: Can You Pass the Test? *(continued)*

6. Which list contains three rights or freedoms guaranteed by the Bill of Rights?
   a. Right to life, right to liberty, right to the pursuit of happiness
   b. Freedom of speech, freedom of press, freedom of religion
   c. Right to protest, right to protection under the law, freedom of religion
   d. Freedom of religion, right to elect representatives, human rights
7. How many times may a senator be reelected?
   a. There is no limit
   b. Once
   c. Twice
   d. 4 times
8. Who signs bills into law?
   a. The Supreme Court
   b. The president
   c. Congress
   d. The Senate
9. How many changes or amendments are there to the Constitution?
   a. 5
   b. 9
   c. 13
   d. 27
10. Who has the power to declare war?
    a. Congress
    b. The president
    c. Chief justice of the Supreme Court
    d. Chairman of the Joint Chiefs of Staff

ANSWER KEY

1 = a; 2 = c; 3 = b; 4 = c; 5 = d; 6 = b; 7 = a; 8 = b; 9 = d; 10 = a

SOURCES: Immigration and Naturalization Service, U.S. Department of Justice. 1989. *By the People . . . U.S. Government Structure.* Washington, D.C.: U.S. Government Printing Office; Immigration and Naturalization Service, 1998 (February 17). "General Naturalization Requirements." **http://www.ins.usdoj.gov/natz/general.html** (April 6, 1998).

the superiority of one group over another provided moral justification for slavery, supplying the South with the means to develop an agricultural economy based on cotton. Further, members of the majority perpetuated the belief that minority members would not benefit from changing structural conditions (Nash 1962). Thus, southern whites felt that emancipation would be detrimental for blacks, who were highly dependent upon their "white masters" for survival.

Functionalists recognize, however, that racial and ethnic inequality is also dysfunctional for society (Schaefer 1998; Williams and Morris 1993). A society that practices discrimination fails to develop and utilize the resources of minority members. Prejudice and discrimination aggravate social problems such as crime and violence, war, poverty, health problems, urban decay, and drug use—problems that cause human suffering as well as financial burdens on individuals and society.

In the 21st century—and that's not far off—racial and ethnic groups in the U.S. will outnumber whites for the first time. The "browning of America" will alter everything in society, from politics and education to industry, values and culture.

—William Henry

Conflict Perspective

Conflict theorists emphasize the role of economic competition in creating and maintaining racial and ethnic group tensions. Majority group subordination of racial and ethnic minorities reflects perceived or actual economic threats by the minority. For example, between 1840 and 1870, large numbers of Chinese immigrants came to the United States to work in mining (California Gold Rush of 1848), railroads (transcontinental railroad completed in 1860), and construction. As Chinese workers displaced whites, anti-Chinese sentiment rose, resulting in increased prejudice and discrimination and the eventual passage of the Chinese Exclusion Act of 1882, which restricted Chinese immigration until 1924.

Further, conflict theorists suggest that capitalists profit by maintaining a surplus labor force, that is, having more workers than are needed. A surplus labor force assures that wages remain low, for someone is always available to take a disgruntled worker's place. Minorities who are disproportionately unemployed serve the interests of the business owners by providing surplus labor, keeping wages low, and, consequently, enabling them to maximize profits.

Conflict theorists also argue that the wealthy and powerful elites foster negative attitudes toward minorities in order to maintain racial and ethnic tensions among workers. As long as workers are divided along racial and ethnic lines, they are less likely to join forces to advance their own interests at the expense of the capitalists. In addition, the "haves" perpetuate racial and ethnic tensions among the "have-nots" to deflect attention away from their own greed and exploitation of workers.

Symbolic Interactionist Perspective

The symbolic interactionist perspective focuses on how meanings and definitions contribute to the subordinate position of certain racial and ethnic groups. The different connotations of the colors white and black are a case in point. The white knight is good, and the black knight is evil; angel food cake is white, devil's food cake is black. Other negative terms associated with black include black sheep, black plague, black magic, black mass, blackballed, and blacklisted. The continued use of such derogatory terms as Jap, Gook, Spic, Frogs, Kraut, Coon, Chink, Wop, and Mick also confirms the power of language in perpetuating negative attitudes toward minority group members.

The labeling perspective directs us to consider the role that negative stereotypes play in race and ethnicity. **Stereotypes** are exaggerations or generalizations about the characteristics and behavior of a particular group. Negative stereotyping of minorities leads to a self-fulfilling prophecy. As Schaefer (1998, 17) explains:

> **Self-fulfilling prophecies can be devastating for minority groups. Such groups often find that they are allowed to hold only low-paying jobs with little prestige or opportunity for advancement. The rationale of the dominant society is that these minority individuals lack the ability to perform in more important and lucrative positions. Training to become scientists, executives, or physicians is denied to many subordinate group individuals, who are then locked into society's inferior jobs. As a result, the false definition becomes real. The**

subordinate group has become inferior because it was defined at the start as inferior and was therefore prevented from achieving the levels attained by the majority.

PREJUDICE AND RACISM

Prejudice refers to an attitude or judgment, usually negative, about an entire category of people. Prejudice may be directed toward individuals of a particular religion, sexual orientation, political affiliation, age, social class, sex, race, or ethnicity.

At the extreme, racial prejudice takes the form of **racism:** the belief that certain groups or races are innately superior to other groups. The perception that certain groups have inferior traits serves to justify subordination and mistreatment of those groups.

> It is never too late to give up your prejudices.
>
> –Henry David Thoreau
> Writer/Civil activist

Measuring Prejudice and Racism

The study and assessment of prejudice and racism usually involves two aspects: (1) the content of stereotypes and (2) people's reported willingness to interact with various racial and ethnic groups in specified social situations. For example, Emory Bogardus (1968) developed a social distance scale (frequently referred to as the Bogardus scale) that presents situations representing different degrees of social contact or social distance. The items, with their corresponding distance scores, follow. People are asked if they would be willing to admit various racial and ethnic groups:

> Racism . . . has been the most persistent and devisive element in this society and one that has limited our growth and happiness as a nation.
>
> –James E. Jones Jr.
> Director, Center for the Study of Affirmative Action

- To close kinship by marriage (1.00)
- To my club as personal chums (2.00)
- To my street as neighbors (3.00)
- To employment in my occupation (4.00)
- To citizenship in my country (5.00)
- As only visitors to my country (6.00)

A score of 1.00 for any group would indicate no prejudice toward that group. Schaefer (1998) summarized the results of studies using the social distance scale in the United States in 1926, 1966, and 1991. In general, average scores for all three time periods indicate little to no prejudice against white Americans and northern Europeans, moderate prejudice toward eastern and southern Europeans, and the highest degree of prejudice toward racial minorities. Comparing social distance scores across three time periods suggests that prejudice has decreased over the last several decades. Dovidio and Gaertner (1991) suggest, however, that "what may have changed across time . . . is what people regard as socially desirable rather than racial attitudes per se" (p. 125). In other words, survey data on racial attitudes may reflect what respondents view as "politically correct" rather than what people really believe or feel.

> One of the great divides in this country is between those Americans who see only blatant racism and those who see the subtle forms as well.
>
> –David K. Shipler
> Author, researcher

Modern Racism

A number of scholars believe that prejudice in the United States has not declined, but rather, has taken on a more subtle and complex form known as **modern racism.** Shipler (1998) explains:

Today, when explicit discrimination is prohibited and blatant racism is no longer fashionable in most circles, much prejudice has gone underground. It may have diminished in some quarters, but it is far from extinct. Like a virus searching for a congenial host, it mutates until it finds expression in a belief, a statement, or a form of behavior that seems acceptable. (p. 1)

People who have modern racist views have negative feelings toward minority groups but do not necessarily endorse the idea of genetic or biological inferiority or adhere to traditional stereotypes. The modern racist believes that serious discrimination in America no longer exists, that any continuing racial inequality is the fault of minority group members, and that demands for preferential treatment or affirmative action for minorities are unfair and unjustified. "Modern racism tends to 'blame the victim' and place the responsibility for change and improvements on the minority groups, not on the larger society" (Healey 1997, 55).

> We are not born with hatred; we learn to hate.
>
> —A Holocaust survivor

Sources of Prejudice

Sources of prejudice include cultural transmission, stereotypes, and the media.

Cultural Transmission Prejudice is taught and learned through the socialization process. One of the first sources of prejudice is the family. Research indicates, for example, that parents who are prejudiced are more likely to have children who are also prejudiced. Other institutions also foster prejudice. Some religious doctrines teach intolerance of racial and ethnic groups. Further, school curricula have traditionally been Eurocentric—that is, biased toward white Europeans—often perpetuating the belief that minority group members are inferior.

> Why are blacks stereotyped as lazy? Wasn't it whites who enslaved blacks to do the hard work on the plantations? If you ask me, whites were too lazy to do the work themselves.
>
> —Student in a sociology class

Stereotypes Prejudicial attitudes toward racial and ethnic groups are often the result of stereotypes—exaggerations or generalizations about a category of individuals. These generalizations, which become deeply embedded in the culture, are either untrue or are gross distortions of reality.

When Americans in a 1990 National Opinion Research Center poll were asked to evaluate various racial and ethnic groups, blacks were rated least favorably (Shipler 1998). Most of the respondents labeled blacks as less intelligent than whites (53 percent), lazier than whites (62 percent), and more likely than whites to prefer being on welfare than being self-supporting (78 percent).

CONSIDERATION

Shipler (1998) suggests that negative stereotyping of minorities enhances the self-esteem of majority group members. "If blacks are less intelligent, in whites' belief, then it follows that whites are more intelligent. If blacks are lazier, whites are harder working. If blacks would prefer to live on welfare, then whites would prefer to be self-supporting" (p. 3).

> The great enemy of truth is very often not a lie—deliberate, continued, and dishonest—but the myth, persistent, persuasive, and unrealistic.
>
> —John F. Kennedy
> Former U.S. President

Media The media contribute to prejudice toward racial and ethnic minorities by portraying minorities in negative and stereotypical ways, or by not portraying them at all. On prime-time television, for example, Hispanics are disproportionately portrayed as criminals (Lichter et al. 1987). When blacks are portrayed in magazine advertisements, they are disproportionately pictured as musicians, athletes, and objects of charity (Langone 1993, 30). An analysis of more than 2,000 children's storybooks published between 1937 and 1993 revealed that only 15 percent had at least one black character in the story (Pescosolido, Grauerholz, and Milkie 1997). The researchers found that the depiction of intimate, egalitarian interracial interaction and the portrayal of black adults as main characters were rare.

Some music also fuels racial hatred, especially among the industry's primary consumer group—youth. Consider the following music lyrics:

> **. . . Niggers just hit this side of town, watch my property values go down. Bang, gang, watch them die, watch those niggers drop like flies. . . . —Berserkr.**

The Internet has provided another avenue for the proliferation of ideologies based on racial hatred. Web sites that represent racial hate groups are designed to recruit new members and give racists an empowering sense of community.

National Data

There are more than 25 racist bands in the United States, and over 100 worldwide.

SOURCE: *Intelligence Report* 1998

International Data

Resistance Records, a Detroit area company that sells music with white power lyrics, sells 50,000 compact discs a year in Europe, South Africa, South America, and the United States.

SOURCE: *Intelligence Report* 1998

DISCRIMINATION AGAINST MINORITIES

While prejudice refers to attitudes, **discrimination** refers to behavior that involves treating categories of individuals unequally. A person may be prejudiced and not discriminate; conversely, a person may discriminate but not be prejudiced.

Individual versus Institutional Discrimination

Individual discrimination occurs when individuals treat persons unfairly or unequally due to their group membership. There are two types of individual discrimination—overt and adaptive. In **overt discrimination** the individual discriminates because of his or her own prejudicial attitudes. For example, a white landlord may refuse to rent to a Mexican-American family because of her own prejudice against Mexican Americans. Or, a Taiwanese-American college student who shares a dorm room with an African-American student may request a roommate reassignment from the student housing office because he is prejudiced against blacks.

National Data

In March 1995 the first neo-Nazi site was posted on the World Wide Web. By 1998, there were at least 163 U.S.-based web sites preaching racial hatred.

SOURCE: *Intelligence Report* 1998

Suppose a Cuban-American family wants to rent an apartment in a predominantly white neighborhood. If the landlord is prejudiced against Cubans and does not allow the family to rent the apartment, that landlord has engaged in overt discrimination. But what if the landlord is not prejudiced against Cubans but still refuses to rent to a Cuban family? Perhaps that landlord is engaging in **adaptive discrimination,** or discrimination that is based on the prejudice of others. In this example, the landlord may fear that if he rents to a Cuban-American family, other renters who are prejudiced against Cubans may move out of the building or neighborhood and leave the landlord with unrented apartments. Overt and adaptive individual discrimination may coexist. For example, a landlord may not rent an apartment to a Cuban family because of her own prejudices *and* the fear that other tenants may move out.

Institutional discrimination occurs when normal operations and procedures of social institutions result in unequal treatment of minorities. Institutional discrimination is covert and insidious and maintains the subordinate position of minorities in society. When businesses move out of inner-city areas, they are removing employment opportunities for America's highly urbanized minority groups. When schools use standard intelligence tests to decide which children will be placed in college preparatory tracks, they are limiting the educational advancement of minorities whose intelligence is not fairly measured by culturally biased tests developed from white middle-class experiences. Institutional discrimination is also found in the criminal justice system, which more heavily penalizes crimes that are more likely to be committed by minorities. For example, the penalties for crack cocaine, more often used by minorities, have traditionally been higher than those for other forms of cocaine use even though the same prohibited chemical substance is involved. As conflict theorists emphasize, majority group members make rules that favor their own group.

Racial and ethnic minorities experience discrimination and its effects in almost every sphere of social life. Next, we look at discrimination in education, employment and income, housing, and politics. Finally, we expose the extent and brutality of hate crimes against minorities.

> Recent studies indicate that a majority of the nation's black children attend schools that are more than 90 percent black which has caused many to wonder which side actually won in *Brown vs. Board of Education.*
>
> —Robert Pratt

Educational Discrimination

With the exception of Asian Americans, minorities tend to achieve lower levels of education than nonminority whites (see Chapter 12). As we explain in Chapter 12, minorities are often segregated in inner-city schools that are poorly funded. The institutional discrimination in education results in inner-city students (who are largely minorities) receiving a much poorer quality education than students who live in middle- and upper-middle-class America (Kozol 1991).

Minorities also experience individual discrimination in the schools, due to continuing prejudice among teachers. One college student completing a teaching practicum reported that "some of the teachers in her school often spoke openly about children of color . . . as *wild kids who slam doors in your face* . . . She reported that she heard teachers refer to children of color as *them* and complain that they *seem to be getting more and more of 'them'* at the school" (Lawrence 1997, 111). It is likely that teachers who are prejudiced against minorities discriminate against them, giving them less teaching attention and less encouragement. One report found that minority students in the Cincinnati School System were more than twice as likely to be punished in school as members of the majority group (Hull 1994).

Employment and Income Discrimination

In comparison to U.S. whites, racial and ethnic minorities are more likely to be unemployed. As Figure 8.2 indicates, differences in educational level do not completely explain differences in unemployment rates.

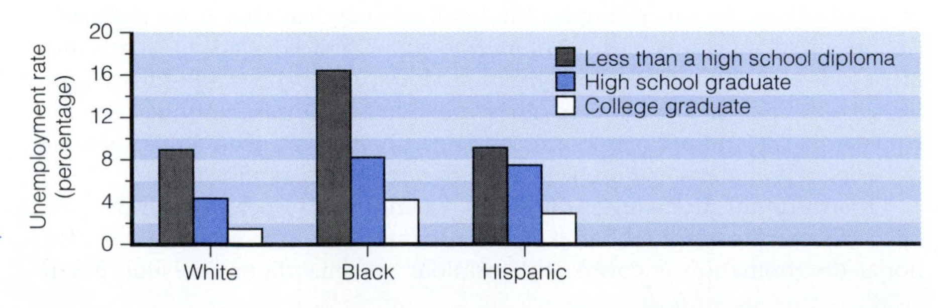

Figure 8.2 Unemployment Rates among Whites, Blacks, and Hispanics by Educational Level, 1997

SOURCE: *Statistical Abstract of the United States: 1998,* 118th ed. U.S. Bureau of the Census. Washington, D.C.: U.S. Government Printing Office, Table 681.

Minority unemployment is, in part, a function of institutional discrimination. Neckerman and Kirschenman (1991) studied the hiring practices of 185 Chicago area businesses and found that recruitment efforts largely targeted white neighborhoods.

Minorities are also more likely to be fired, work in low-paying and low-status jobs, and earn less income than the majority. A study of job dismissals at a large northeastern city postal office found that blacks were twice as likely to be fired as whites (Zwerling and Silver 1993).

Even when workplaces seem integrated, there is often segregation within as whites—particularly white males—tend to occupy the positions of power and authority. Social segregation contributes to minority disadvantage in the workplace, as illustrated in the following account (Shipler 1998):

> **A black man worked for IBM for three years before learning that every evening a happy hour was taking place in a nearby bar. Only white men from the office were involved—no women, no minorities. Had it been strictly social it would have been merely offensive, but it was also professionally damaging, for business was being done over drinks, plans were being designed, connections made. Excluded from that network, the black man was excluded from opportunity for advancement, and he left the job. (p. 2)**

National Data

In 1997, the U.S. unemployment rate was 10 percent for Blacks, 7.7 percent for Hispanics, and 4.2 percent for whites.

SOURCE: Mishel, Bernstein, and Schmitt 1999

How much do we really value the lives of children? . . . On the basis of what I've seen in the South Bronx, it's hard to believe we love these children. The physical degradation of their neighborhood, the chaotic and dysfunctional medical care we give them, the apartheid school system we provide them, the poisons we pump into their neighborhood by putting every kind of toxic-waste incinerator in their neighborhood, all that together does not give me the impression we value their lives.

–Jonathon Kozol
Author and children's advocate

Housing Segregation and Discrimination

U.S. minorities, who are disproportionately represented among the poor, tend to be segregated in concentrated areas of low-income housing, often in inner-city areas of concentrated poverty (Massey and Denton 1993). As explained in Chapter 13, the more affluent suburbs restrict development of affordable housing in order to keep out "undesirables" and maintain their high property values. Suburban zoning regulations that require large lot sizes, minimum room sizes, and single family dwellings serve as barriers to low-income development in suburban areas.

Segregation also results from discriminatory practices such as redlining, racial steering, and restrictive home covenants. **Redlining** occurs when mortgage companies deny loans for the purchase of houses in minority neighborhoods, arguing that the financial risk is too great. Racial steering occurs when realtors discourage minorities from moving into certain areas by showing them homes only in "their" kind of neighborhood. Restrictive home covenants, illegal since the 1950s, involve a pact between residents that they will not sell their homes to minority group members.

Although housing discrimination is illegal, it is not uncommon. In a study in Jacksonville, Florida, white and African-American volunteer applicants, or "testers," with similar backgrounds tried to rent the same apartments. More than half of the apartment owners tested in the study broke laws against racial discrimination (Halton 1998). Black testers were quoted higher rents and security deposits, told that units were not available or were not granted appointments or applications.

National Data

In 1995, 8 percent of members of Congress were black; 1 percent were Asian or Pacific Islander, and 3 percent were Hispanic.

SOURCE: *Statistical Abstract of the United States 1998,* Table 469

Political Discrimination

In general, members of the majority group deprive minorities from participating fully in the political-legal process through discriminatory practices in voter registration, candidate qualifications, and jury selection. Exclusionary techniques

National Data

In the 1996 presidential election, 56 percent of whites reported that they voted, compared to 51 percent of blacks and 27 percent of Hispanics.

SOURCE: *Statistical Abstract of the United States 1998,* Table 483

include changing elected offices into appointed positions, redrawing the boundaries of voting districts, changing polling places, requiring voters to take literacy tests, and using automobile registration for jury selection. All of these practices make it difficult for minorities to exercise their right to be politically involved.

Blacks have been discouraged from political involvement by segregated primaries, poll taxes, literacy tests, and threats of violence. However, tremendous strides have been made since the passage of the 1965 Voting Rights Act, which prohibited literacy tests and provided for poll observers. Blacks have won mayoral elections in Atlanta, Cleveland, Washington, D.C., and New York City and the governorship in Virginia. However, racial minorities and Hispanics continue to be underrepresented in political positions and voting participation.

Hate Crime Victimization

Those who hate you don't win unless you hate them back; and then you destroy yourself.

—Richard M. Nixon
Former U.S. president

Perhaps the most brutal form of discrimination takes the form of **hate crimes**—acts of violence motivated by prejudice against racial, ethnic, religious, and sexual orientation groups. The brutal murder of James Byrd Jr. described in the opening of this chapter exemplifies the horrific nature of some hate crimes. Hate crimes are often less severe than brutal murder, yet are nonetheless harmful, inhumane, and against the law. For example, six black students at Kent University were assaulted by several men carrying baseball bats; a white high school student with black friends received threats and found pictures of burning crosses, Klan symbols, and swastikas in her locker; a black doll was found hanging from a tree at Duke University; and a girl of Mexican descent was slashed with a pocketknife by a man who said he hated Mexicans (*Intelligence Report* 1998).

Jackie Burden and Michael James died because three soldiers stationed at Fort Bragg . . . decided it was a good night to commit a hate crime.

—Howard Chua-eoan
Journalist

According to an FBI report, the majority (62 percent) of reported hate crimes in 1996 were motivated by racial bias, and 11 percent were motivated by ethnicity/national origin bias (U.S. Department of Justice 1998).

Levin and McDevitt found that the motivations for hate crimes were of three distinct types: thrill, defensive, and mission. Thrill hate crimes are committed by offenders who are looking for excitement and attack victims for the "fun of it." Defensive hate crimes involve offenders who view their attacks as necessary to protect their community, workplace, or college campus from "outsiders." Perpetrators of defensive hate crimes are trying to send a message that their victims do not belong in a particular community, workplace, or campus and that anyone in the victim's group who dares "intrude" could be the next victim. Mission hate crimes are perpetrated by offenders who have dedicated their lives to bigotry. In Levin and McDevitt's (1995) study of hate crimes in Boston, the most common type of hate crime was thrill hate crime (58 percent) followed by defensive hate crime (41 percent). The least common, but most violent, type of hate crime is mission hate crime. Levin and McDevitt (1995) describe the 1995 Oklahoma City bombing as "probably the quintessential mission hate crime, perpetrated by individuals whose lives had become consumed with hatred not only for the federal government but possibly for blacks and Jews as well" (p. 9).

I'm not against Blacks. I'm against all non-whites.

—Ku Klux Klan member

Mission hate crimes are often committed by members of white supremacist organizations that endorse racist beliefs and violence against minority group members. The Ku Klux Klan, the first major racist, white supremacist group in the United States, began in Tennessee shortly after the Civil War. Klansmen have threatened, beat, mutilated, and lynched blacks as well as whites who dared to oppose them. White Aryan Resistance (WAR), another white supremacist group, fosters hatred that breeds violence. The following

Mission hate crimes are typically committed by members of white supremacist groups. This scene from the movie "Mississippi Burning" depicts a Ku Klux Klan member brutally attacking a black man.

message is typical of one received by calling a White Aryan Resistance phone number in one of many states (Kleg 1993, 205):

> **This is WAR hotline. How long, White men, are you going to sit around while these non-White mud races breed you out of existence? They have your jobs, your homes, and your country. Have you stepped outside lately and looked around while these niggers and Mexicans hep and jive to this Africanized rap music? While these Gooks and Flips are buying up the businesses around you? . . . This racial melting pot is more like a garbage pail. Just look at your liquor stores. Most of them are owned by Sand niggers from Iraq, Egypt, or Iran. Most of the apartments are owned by the scum from India, or some other kind of raghead . . . [Jews] are like maggots eating off a dead carcass. When you see what these Jews and their white lackeys have done, the gas chambers don't sound like such a bad idea after all. For more information write us at . . . .**

White supremacy has to die for humanity to live.

—Louis Farrakhan
Nation of Islam leader

Other racist groups known to engage in hate crimes are the Identity Church Movement, neo-Nazis, and the skinheads. While Klan members have traditionally concealed their identities under white hoods and robes, racist skinheads are usually identifiable through their shaved heads, steel-toed boots, jeans, and suspenders. They often have tattoos of Nazi or satanic emblems.

Hate crimes have become a growing threat to the well-being of our society—on college campuses, in the workplace, and in our neighborhoods.

—Jack Levin, Sociologist
Jack McDevitt, Criminologist

CONSIDERATION

Not all skinheads are racist. Many youth have adopted the skinhead "look" and lifestyle but do not endorse racism or violence. One nonracist skinhead said: "Being a skinhead does not mean being a Nazi. I happen to have no hair, a black leather jacket, and army boots, and I get stopped all the time by people trying to preach nonviolence to me. I am a pacifist." (Quoted in Landau 1993, 43.)

National Data

The Intelligence Project documents 474 hate groups and chapters in the United States, up about 20 percent from 1996.

SOURCE: *Intelligence Report* 1998

In this chapter's *The Human Side*, a former member of a racist hate group describes how he became involved in the skinhead movement, how he

THE HUMAN SIDE

AN INTERVIEW WITH A FORMER RACIST SKINHEAD

After spending 15 years in the skinhead movement, Thomas (T.J.) Leyden renounced racism and went to work for the Simon Wiesenthal Center in Los Angeles. Since joining the human rights organization in June 1996, Leyden has given speeches at more than 100 high schools, the Pentagon, FBI headquarters, and police agencies. In the following interview excerpted from the Southern Poverty Law Center's *Intelligence Report* (Winter 1998), Leyden talks about his life in a racist hate group, his views on why youth join hate groups, and why he left the racist skinhead movement.

Intelligence Report: What brought you into the Skinhead movement?

T.J. Leyden: I was hanging out in the punk rock scene in the late '70s and early '80s . . . In 1980, my parents got a divorce, and I started to hang out in the street. I was venting a lot of my frustration and anger over the divorce. I went around attacking kids, punching them and beating them up. A group of older kids who were known as Skinheads saw this, and I got in with them . . . In 1981, four big-time racist bands came into the Skinhead movement . . . We started to listen to their music, and that broke the Skinhead movement into two factions, SHARPs [Skinheads Against Racial Prejudice] and the neo-Nazi Skinheads. Since I lived in a very upper-middle class, white neighborhood, we decided to establish one of the first neo-Nazi Skinhead gangs in Southern California.

If we caught somebody black, Hispanic or Asian, we'd attack them, beat them for sure . . .

Intelligence Report: Did your racism come partly from your parents?

T.J. Leyden: My mom was nonracist and my dad was a stereotypical man. I mean, if somebody cut him off on the freeway, if they were black, he'd use the word 'nigger' . . . But the racism I really learned came from my grandfather, a staunch Irish Catholic. He would say, "You don't bring darkies home" and "Jews killed Christ."

Intelligence Report: What are the circumstances that lead teenagers to join neo-Nazi gangs?

T.J. Leyden: We were middle-class to rich, bored white kids. We had a lot of time on our hands so we decided to become gang members. When a kid doesn't have something else constructive to do, he's going to find something, whether it's football, baseball or hanging with neo-Nazi Skinheads . . .

Intelligence Report: When did you start to really learn the ideology of racism?

T.J. Leyden: After I joined the Marine Corps in 1988. They teach a philosophy that if you do something, you do it all the way, not half-assed. So since I was a racist, I started reading everything I could about Nazism . . . I was recruiting, organizing Marines to join the racist movement . . . Eventually, I was kicked out for alcohol-related incidents—not for being a racist. If you look at my military packet you're not going to find anything about me being a racist. And I had two-inch high Nazi SS bolts tattooed on my neck! Once I got cut, I decided to be a [Skinhead] recruiter. I was going to get younger kids to be street soldiers.

THE HUMAN SIDE

AN INTERVIEW WITH A FORMER RACIST SKINHEAD *(continued)*

Intelligence Report: How did recruitment work?

T.J. Leyden: We incited violence on high school campuses. We'd put out literature that got black kids to think the white kids were racist. Then the black kids would attack the white kids and the white kids would say, "I'm not going to get beat up by these black guys anymore." They'd start fighting back, and we'd go and fight with them. They'd say, "God, these guys are really cool" . . . That put my foot in the door. Then I could start talking to them, giving them comic books with racist overtones or CDs of racist music. And I would just keep talking to them, giving them literature, indoctrinating them over a period of time. . . .

Intelligence Report: What finally brought you to leave the racist movement?

T.J. Leyden: It was an incident with my son that woke me up more than anything. We were watching a Caribbean-style show. My 3-year-old walked over to the TV, turned it off and said, "Daddy, we don't watch shows with niggers." My first impression was, "Wow, this kid's pretty cool." Then I started seeing something different. I started seeing my son acting like someone 10 times tougher than I was, 10 times more loyal, and I thought he'd end up actually doing something and going to prison. Or he was going to get hurt or killed.

I started looking at the hypocrisy. A white guy, even if he does crystal meth and sells crack to kids, if he's a Nazi he's okay. And yet this black gentleman here, who's got a Ph.D. and is helping out white kids, he's still a "scummy nigger."

In 1996, when I was at the Aryan Nations Congress [in Hayden Lake, Idaho], I started listening to everybody and I felt like, "God, this is pathetic." I asked the guy sitting next to me, "If we wake up tomorrow and the race war is over and we've won, what are we going to do next?" And he said, "Oh, come on, T.J., you know we're going to start with hair color next, dude."

I laughed at it, but when I drove home, 800 miles, that question and answer kept popping into my head. I thought that kid was so right. Next it'll be you have black hair so you can't be white, or you have brown eyes so somebody in your past must have been black, or you wear glasses so you have a genetic defect.

A little over two years after my son said the thing about the "niggers" on TV, I left the racist movement . . .

Intelligence Report: What has been the personal cost of your involvement in the movement?

T.J. Leyden: A little bit of my dignity. I look at myself as two people, who I am now and who I was then. I see the destruction I did to people by bringing them into the movement, the families I hurt. I ruined a lot of lives. That's the biggest thing I have to pay back. I don't forgive myself. Only my victims can forgive me.

SOURCE: Excerpted from "A Skinhead's Story: An Interview with a Former Racist." From 1998 *Intelligence Report.* Winter, Issue no. 89, pp. 21–23. Montgomery, Ala.: Southern Poverty Law Center. Used by permission of the Southern Poverty Law Center.

recruited others into the movement, and what led him to leave the movement and denounce racial hatred.

STRATEGIES FOR ACTION: RESPONDING TO PREJUDICE, RACISM, AND DISCRIMINATION

In June 1997, President Clinton announced "One America in the 21st Century: The President's Initiative on Race." Clinton convened an advisory board to recommend creative ways to resolve racial tensions with the help and input of community leaders across the country. Clinton's vision is to have a diverse, democratic society in which members respect and celebrate their differences. Next, we look at various strategies that address problems of prejudice, racism, and discrimination. These include multicultural education, political strategies, affirmative action programs, and diversity training in the workplace.

Multicultural Education

> I have a dream that my four little children will one day live in a nation where they will not be judged by the color of their skin, but by the content of their character.
>
> –Martin Luther King Jr.
> Civil rights leader

Much educational material is biased toward white Europeans. For example, Zinn (1993) observes, "[T]o emphasize the heroism of Columbus and his successors as navigators and discoverers, and to de-emphasize their genocide, is not a technical necessity but an ideological choice. It serves, unwittingly—to justify what was done" (p. 355). **Multicultural education** focuses on the need to represent all racial and ethnic groups in the school curriculum (see also Chapter 12). With multicultural education, the school curriculum reflects the diversity of American society and fosters an awareness and appreciation of the contributions of different racial and ethnic groups to American culture. Multicultural education also works to dispel myths, stereotypes, and ignorance about minorities. One minority student commented: "What bothers me a lot is ignorance—people not realizing the difference between someone who's Japanese, someone who's Chinese, someone who's Korean, and just tending to classify them as one big word—Asian" ("In Our Own Words" 1996, 54).

> Teach tolerance. Because open minds open doors for *all* our children.
>
> –Radio Public Service Announcement

Many colleges and universitites have made efforts to promote awareness and appreciation of diversity by offering courses and degree programs in racial and ethnic studies, and multicultural events and student organizations. Research suggests that college attendance tends to result in increased tolerance and support for diversity (Pascarella et al. 1996). This chapter's *Focus on Social Problems Research* feature presents a study that identifies various factors that influence students' openness to diversity in the first year of college.

Political Strategies

> Perhaps the racial tolerance that is prominent in America's consciousness today may tomorrow become part of America's unconscious.
>
> –John F. Dovidio and
> Samuel L. Gaertner
> Psychologists

Various political strategies have been implemented or suggested to reduce prejudice and discrimination. Although some individuals argue that laws cannot change attitudes, there is evidence that changing behaviors can change attitudes. When the military, schools, and housing were integrated, the attitudes of whites toward non-whites became more positive. Hence, there is a need for laws that prohibit discrimination, for those laws to be strictly enforced, and for sanctions to be imposed when the laws are violated.

One political strategy to decrease discrimination involves increasing minority representation in government. This requires increasing political involvement,

FOCUS ON SOCIAL PROBLEMS RESEARCH

What Influences College Students' Openness to Diversity?

A study by Pascarella et al. (1996) sought to determine how college students' openness to diversity is influenced by four different sets of factors: student background or precollege characteristics, environmental emphases of the institution attended, measures of the students' academic experience, and measures of students' social involvement.

Sample and Methods

The researchers collected data from 2,290 first-year students at 18 colleges and universities. The dependent variable was a scale designed to measure openness to diversity (see scale). This scale not only assesses an individual's openness to cultural, racial, and value diversity, it also measures the extent to which an individual enjoys being challenged by different ideas, values, and perspectives.

OPENNESS TO DIVERSITY/ CHALLENGE SCALE

(Scored on a Likert-type scale: 5 = strongly agree to 1 = strongly disagree)

1. I enjoy having discussions with people whose ideas and values are different from my own.
2. The real value of a college education lies in being introduced to different values.
3. I enjoy talking with people who have values different from mine because it helps me understand myself and my values better.
4. Learning about people from different cultures is a very important part of my college education.
5. I enjoy taking courses that challenge my beliefs and values.
6. The courses I enjoy the most are those that make me think about things from a different perspective.
7. Contact with individuals whose background (e.g., race, national origin, sexual orientation) is different from my own is an essential part of my college education.
8. I enjoy courses that are intellectually challenging.

Four sets of independent variables were developed, each of which consisted of numerous measures. These included (1) precollege variables (including a measure of precollege openness to diversity and precollege academic ability); (2) environmental emphasis of the college/university (including a measure of the degree of racial discrimination at the institution); (3) student academic experiences (including number of social science courses taken and self-reported number of hours spent studying per week); and (4) student social/nonacademic experiences (including involvement in clubs and organizations and assessment of students' peer interactions and topics of conversation).

Findings

The precollege measure of openness to diversity/challenge had the strongest effect on openness to diversity/challenge after the first year of college. Women and nonwhite students had higher levels of openness to diversity/challenge than men and white students.

The extent to which students perceived their institution as having a nondiscriminatory racial environment had a positive impact on openness to diversity. Hours spent studying had a small positive effect, while the number of mathematics courses taken during the first year of college had a small negative impact.

Living on campus, participating in a racial or cultural awareness workshop, and hours worked per week had positive effects on openness to diversity/challenge, while joining a fraternity or sorority had a negative effect. In addition, "the more students interact with diverse peers and the greater the extent to which such interactions focus on controversial or value-laden issues that may engender a change in perspective or opinion, the greater one's development of openness to diversity and challenge" (p. 188).

The findings of this study suggest that colleges and universities that offer racial or cultural awareness workshops can foster students' appreciation and acceptance of cultural, racial, and value diversity. Encouraging openness to diversity may also be achieved by the institution establishing policies and programs that sensitize students and personnel to racial discrimination and demonstrate that such discrimination is not acceptable. Colleges and universities may also consider interventions to counteract the negative influence of membership in fraternities or sororities on openness to diversity.

Finally, the fact that precollege openness to diversity had the largest effect on openness to diversity among first-year college students points to the need to foster openness to diversity in the elementary and secondary grades. This may be achieved through multicultural programs, educational approaches, and school policies that discourage and sanction prejudice and discrimination.

SOURCE: Based on "Influences on Students' Openness to Diversity and Challenge in the First Year of College" by Ernest T. Pascarella, Marcia Edison, Amaury Nora, Linda Serra Hagedorn, and Patrick T. Terenzini, *The Journal of Higher Education* 67(2):174–93.

action, and participation. For example, the Asian American Voters Coalition has successfully mobilized voters in states with high concentrations of Asian Americans—California, Hawaii, New York, and Texas (Feagin and Feagin 1993). However, the 1995 Supreme Court ruling (in *Miller v. Johnson*) that the creation of congressional districts for the sole purpose of establishing "minority majorities" is unconstitutional will make future minority representation more difficult.

> If we are to move the society toward freedom, away from racism, we must choose leaders deeply committed to that objective.
>
> —Eddie William and Milton D. Morris
> Joint Center for Political and Economic Studies

Affirmative Action

Affirmative action refers to programs that provide or seek to provide opportunities or other benefits to persons on the basis of their membership in a specified group. Most affirmative action programs have involved one of the following: (1) programs for the enrollment of minorities or women in higher education and professional schools, (2) the setting aside of a percentage of contracts for minority or female subcontractors, or (3) the imposition of quota goals for minority or female employment.

Opponents of affirmative action argue that these programs constitute **reverse discrimination**. Some opponents of affirmative action are white males who feel they have been treated unfairly as a result of efforts to provide minorities with educational and employment opportunities. Bobo and Kluegel (1993) argue that much of the white opposition to race-targeted policies such as affirmative action is due to self-interest and prejudice. But some African Americans are also critical of affirmative action, arguing that it perpetuates feelings of inferiority among minorities and fails to help the most impoverished of minorities (Shipler 1998; Wilson 1987).

National Data

In a 1997 Gallup poll, 53 percent of blacks and 22 percent of whites agreed that government should increase affirmative action programs; 12 percent of blacks and 37 percent of whites agreed that government should decrease affirmative action.

SOURCE: Gallup Organization 1997

CONSIDERATION

Shipler (1998) notes that "while most whites think that under affirmative action less qualified blacks are hired and promoted over more qualified whites, most blacks think that *without* affirmative action, less qualified *whites* are hired and promoted over more qualified *blacks*" (p. 3).

Some advocates of affirmative action point to the moral and compensatory justification for such programs. In response to the charge that affirmative action is as discriminatory to whites as prior discrimination was to blacks, Herman Schwartz (1992) argues:

> **No one can honestly equate a remedial preference for a disadvantaged (and qualified) minority member with the brutality inflicted on blacks and other minorities by Jim Crow laws and practices. The preference may take away some benefits from some white men, but none of them is being beaten, lynched, denied the right to use a bathroom, a place to sleep or eat, being forced to take the dirtiest jobs or denied any work at all, forced to attend dilapidated and mind-killing schools, subjected to brutally unequal justice, or stigmatized as an inferior being. (pp. 193–94)**

> You do not take a person who for years has been hobbled by chains, and liberate him, bring him up to the starting line, and then say, "You are free to compete with all others."
>
> —Lyndon Baines Johnson
> Former U.S. president

Other affirmative action advocates suggest that affirmative action provides minority role models. "Nonwhites in educational and professional positions where they were previously not present function as models for other, especially younger, members of their racial group who can identify with them and form realistic goals to occupy the same roles themselves" (Zack 1998, 51).

Numerous legal battles have been fought over affirmative action. In 1990, a Hispanic student sued the University of Maryland for denying him a scholarship that was limited to black students. A lower court ruled the black scholarship program was unconstitutional and constituted "reverse discrimination." In 1995, the U.S. Supreme Court refused to hear the appeal, which means that any race-based scholarship in the United States can now be challenged. In Texas, racial quotas have been banned in college admissions procedures. Proposition 209, which outlaws race- and gender-based preferences in state hiring, was passed by California voters in 1996. If the trend toward dismantling affirmative action programs continues, some fear that the social disadvantages of minorities will worsen.

> While we react to those wearing white sheets, it is those who wear black robes who take away our protection.
>
> –Jesse Jackson
> Civil rights leader

Diversity Training

Increasingly, corporations have begun to implement efforts to reduce prejudice and discrimination in the workplace through diversity training. Broadly defined, **diversity training** involves "raising personal awareness about individual 'differences' in the workplace and how those differences inhibit or enhance the way people work together and get work done" (Wheeler 1994, 10). Diversity training may address such issues as stereotyping and cross-cultural insensitivity, as well as provide workers with specific information on cultural norms of different groups and how these norms affect work behavior and social interactions.

In a survey of 45 organizations that provide diversity training, Wheeler (1994) found that for 85 percent of the respondents, the primary motive for offering diversity training was to enhance productivity and profits. In the words of one survey respondent, "The company's philosophy is that a diverse work force that recognizes and respects differing opinions and ideas adds to the creativity, productivity, and profitability of the company" (p. 12). Only 4 percent of respondents said they offered diversity training out of a sense of social responsibility.

> If there is such a thing as human capital, our society cannot afford to continue to underutilize what is likely to become an ever-increasing share of its human resources.
>
> –James E. Jones Jr.
> Director, Center for the Study of Affirmative Action

Religion's Role in Promoting Racial Harmony

Beginning in the early 1990s, a racial reconciliation movement emerged among white and black evangelical Christians. Religious organizations that have launched initiatives advocating racial harmony include the Southern Baptist Convention, the National Association of Evangelicals, and the Promise Keepers (Glynn 1998). Efforts to achieve racial harmony among religious organizations include the following (Glynn 1998):

- In 1995, the Southern Baptist Convention passed a resolution apologizing for past support of slavery and racism.
- In the same year, the president of the National Association of Evangelicals publicly confessed and repented past racism by white evangelicals.
- In 1996, the Promise Keepers organization sponsored a gathering of more than 39,000 male pastors of diverse racial, ethnic, and denominational backgrounds under the theme "Breaking Down the Walls." One of the seven promises that members of Promise Keepers make is a promise to overcome racial and denominational differences. "Promise Keepers materials encourage members to go out of their way to engage with those of different races and ethnic groups for purposes of advancing reconciliation" (Glynn 1998, 840–41).

> Politics strives to transform people by altering the structure of society; religion strives to change society by transforming individuals.
>
> –Patrick Glynn
> Institute for Communitarian Policy Studies
> George Washington University

tol•er•ance *n.:* the capacity for or the practice of recognizing and respecting the beliefs or practices of others.

–*The American Heritage Dictionary*

Unlike political strategies to promote racial harmony, religious-based racial reconciliation efforts provide a spiritual imperative. The religious reconciliation movement promotes belief in God's ability to provide healing and forgiveness for past racism. Glynn (1998) suggests that "the mere belief in the possibility of divine forgiveness and in divine aid at arriving at reconciliation could provide a strong psychological impetus for positive group interaction" (p. 840).

Given the recency of the religious racial reconciliation movement, the extent and duration of the effects of this movement are not known. Glynn (1998) predicts that social change will come increasingly from grassroots community and religious activists, "who strive to change the nature of society one community, and one soul, at a time" (p. 841).

UNDERSTANDING **RACE AND ETHNIC RELATIONS**

Ethnic diversity is an opportunity rather than a problem.

–Andrew Greeley
Sociologist/Author

After considering the material presented in this chapter, what understanding about racial and ethnic relations are we left with? First, we have seen that racial categories are socially constructed with no scientific validity. Racial, and ethnic, categories are largely arbitrary, imprecise, and misleading. While some scholars suggest we abandon racial and ethnic labels, others advocate adding new categories—multiracial and multiethnic—to reflect the identities of a growing segment of the U.S. population.

Conflict theorists and functionalists agree that prejudice, discrimination, and racism have benefitted certain groups in society. It is also true, however, that racial and ethnic disharmony has created tensions that disrupt social equilibrium. Further, as symbolic interactionists emphasize, lowered expectations of minority group members, negative labeling, and the use of pejoratives to describe racial and ethnic minority members contribute to their subordinate position.

Prejudice, racism, and discrimination are debilitating forces in the lives of minorities. In spite of these negative forces, many minority group members succeed in living productive, meaningful, and prosperous lives. But many others cannot overcome the social disadvantages associated with their minority status and become victims of a cycle of poverty (see Chapter 10). Minorities are disproportionately poor, receive inferior education, and, with continued discrimination and prejudice in housing and in the workplace, have difficulty improving their standard of living.

Social acceptance of interracial relationships has increased in the past few decades. Nevertheless, interracial couples are still targets for prejudice and discrimination.

Thus, alterations in the structure of society that increase opportunities for minorities—in education, employment and income, and political participation—are crucial to achieving racial and ethnic equality. While government policies or regulations may provide structural opportunities for minorities, policies and regulations often do not change attitudes and beliefs about minority groups embedded in the culture. Changing cultural attitudes begins with the socialization process in the home, school, and place of religious worship. The same socialization process through which children learn prejudice and discrimination may be used to teach children acceptance, compassion, and appreciation for people with varied ethnic and racial backgrounds.

Harmonious racial and ethnic relations may also be achieved through increased contact with different racial and ethnic group members, and through multicultural education and diversity training. Civil rights activist Lani Guinier (1998) suggests that "the real challenge is to . . . use race as a window on issues of class, issues of gender, and issues of fundamental fairness, not just to talk about race as if it's a question of individual bigotry or individual prejudice. The issue is more than about making friends—it's about making change." But, as Shipler (1998) alludes to, making change requires that members of society recognize that change is necessary, that there is a problem that needs rectifying.

> **One has to perceive the problem to embrace the solutions. If you think that racism isn't harmful unless it wears sheets or burns crosses or bars blacks from motels and restaurants, you will support only the crudest anti-discrimination laws and not the more refined methods of affirmative action and diversity training. (p. 2)**

That both black and white in our country can today say we are to one another brother and sister, a united rainbow nation that derives its strength from the bonding of its many races and colours, constitutes a celebration of the oneness of the human race.

—Nelson Mandela
President, Republic of South Africa

CRITICAL THINKING

1. A national survey of first-year college students found that men were more likely than women to think that "racial discrimination is no longer a major problem in America" (24 percent of men, versus 16 percent of women). Why do you think an individual's perception of racial discrimination may be affected by gender?
2. Lieberman (1997) asked university faculty members in five disciplines (biology, biological anthropology, cultural anthropology, psychological anthropology, and developmental psychology) to indicate agreement or disagreement with the statement "There are biological races in the species Homo sapiens." In each of the disciplines, women were more likely than men to reject race as a biological reality. Why do you think women in Lieberman's study were more likely than men to reject the concept of race?
3. Should race be a factor in adoption placements? Should people be discouraged from adopting a child that is of a different race than the adoptive parents? Why or why not?
4. Under Swedish law, giving Nazi salutes is a crime (Lofthus 1998). Do you think that the social benefits of outlawing racist expressions outweighs the impingement of free speech? Do you think such a law should be proposed in the United States? Why or why not?
5. Do you think that there will ever be a time when a racial classification system will no longer be used? Why or why not? What arguments can be made for discontinuing racial classification? What arguments can be made for continuing it?

KEY TERMS

acculturation
adaptive discrimination
affirmative action
amalgamation
antimiscegenation laws
assimilation (primary and secondary)
asylees
colonialism
de facto segregation
de jure segregation
discrimination
diversity training
endogamy
ethnicity
expulsion
genocide
hate crime
individual discrimination
institutional discrimination
Jim Crow laws
marital assimilation
melting pot
modern racism
multicultural education
naturalized citizen
one drop of blood rule
overt discrimination

pluralism
population transfer
populations
prejudice
race
racism
redlining
refugees
reverse discrimination
segregation
slavery
stereotype

INTERNET

You can find more information on interracial relationships, hate crimes, affirmative action, and the National Association for the Advancement of Colored People (NAACP) at the *Understanding Social Problems* website at **http://sociology.wadsworth.com**.

INFOTRAC COLLEGE EDITION

Either from the Wadsworth sociology resource center at **http://sociology.wadsworth.com** or directly from your web browser, you may access InfoTrac College Edition, an online university library that includes over 700 popular and scholarly journals in which you can find articles related to the topics in this chapter. Suggested articles and questions relating to these articles are listed below.

Cose, Ellis. 1997. "Census and the Complex Issue of Race." *Society* 34(6):9–14.

1. What percentage of "black blood" makes a person black in Louisiana?
2. What changes in racial definitions have interested groups advocated to the Census Bureau?
3. What are some of the arguments made against adding a multi-racial category?

Drummond-Hammond, Linda. 1998 (Spring). "Unequal Opportunity: Race and Education." *Brookings Review* 16(2):28–33.

1. What differences did Kozol discover between urban and suburban schools and whom they serve?
2. What characteristics differentiate predominantly minority schools from predominantly white schools?
3. What changes, at the federal and state level, have been implemented to address the inequality seen in the education of minority children?

Platt, Anthony M. 1997 (Spring). "The Land that Never Has Been Yet: U.S. Race Relations at the Crossroads." *Social Justice* 24(1):7–22.

1. What measures have California and Texas taken to reduce or eliminate race-based college admission?
2. Why has debate over race relations been relegated to discussions in academia rather than the political arena?
3. What does the author see as the four major challenges to racial equality?

CHAPTER NINE

Sexual Orientation

Is It True?

Is It True?

1. The majority of Americans say that gays should have equal rights in the workplace.
2. People who believe that gay individuals are "born that way" tend to be more tolerant of gays than people who believe that gay individuals choose their sexual orientation.
3. Most countries throughout the world have laws that protect gay individuals from discrimination on the basis of sexual orientation.
4. The American Psychiatric Association currently classifies homosexuality as a mental disorder.
5. The majority of Americans support same-sex marriage.

Answers: 1 = T; 2 = T; 3 = F; 4 = F; 5 = F

I have friends. Some of them are straight . . . Year after year I continue to realize that the facts of my life are irrelevant to them and that I am only half listened to, that I am an appendage to the doings of a greater world, a world of power and privilege, . . . a world of exclusion. "That's not true," argue my straight friends. There is only one certainty in the politics of power; those left out beg for inclusion, while the insiders claim that they already are. Men do it to women, whites do it to blacks, and everyone does it to queers.

Gay Pride Parade Flier

On October 6, 1998, Matthew Shepard, a 21-year-old student at the University of Wyoming, was abducted and brutally beaten. He was found tied to a wooden ranch fence by two motorcyclists who had initially thought that he was a scarecrow. His skull had been smashed, and his head and face had been slashed. The only apparent reason for the attack: Matthew Shepard was gay. On October 12, Matthew died of his injuries. Media coverage of his brutal attack and subsequent death focused nationwide attention on sexual orientation minorities—gay men, lesbians, and bisexuals—and their treatment in society.

It is beyond the scope of this chapter to explore how sexual diversity and its cultural meanings vary throughout the world. Rather, this chapter focuses on Western conceptions of diversity in sexual orientation. The term "**sexual orientation**" refers to the classification of individuals as heterosexual, bisexual, or homosexual, based on their emotional and sexual attractions, relationships, self-identity, and lifestyle. **Heterosexuality** refers to the predominance of emotional and sexual attraction to persons of the other sex. **Homosexuality** refers to the predominance of emotional and sexual attraction to persons of the same sex and **bisexuality** involves emotional and sexual attraction to members of both sexes.

In this chapter we examine prejudice and discrimination toward homosexual (or gay) women (also known as **lesbians**), homosexual (or gay) men, and bisexual individuals. We begin by summarizing the legal status of lesbians and gay men around the world. Then, we discuss the prevalence of homosexuality, heterosexuality, and bisexuality in the United States, review biological and environmental explanations for sexual orientation diversity, and apply sociological theories to better understand societal reactions to sexual diversity. The chapter ends with a discussion of strategies to reduce antigay prejudice and discrimination.

In 1998, the brutal hate crime murder of Matthew Shepard brought national attention to hate crimes against sexual orientation minorities.

THE GLOBAL CONTEXT: A WORLD VIEW OF LAWS PERTAINING TO HOMOSEXUALITY

The International Lesbian and Gay Association (ILGA) sponsored a survey of 210 countries throughout the world (excluding the United States) to investigate laws and social attitudes regarding homosexuality (*ILGA Annual Report* 1996/1997). Data were not available for 19 countries. This survey found that female homosexuality was legal in 49 countries and illegal in 44 countries. Male homosexuality was legal in 58 countries and illegal in 84 countries. Thus, most laws prohibiting homosexual behavior apply to male rather than female homosexuality. Female homosexuality was not mentioned in the laws of 98 countries, whereas male homosexuality was not mentioned in the laws of 49 countries. In countries where homosexual behavior is legal, ages of consent vary. Also, in many countries where homosexuality is legal, there are no laws that protect lesbians and gay men from discrimination.

In general, oppression of gays is worse in Africa than in other regions of the world. However, in 1996 South Africa became the first country in the world to include in its constitution a clause banning discrimination based on sexual orientation. Fiji, Canada, and Ecuador also have constitutions that ban discrimination based on sexual orientation ("Constitutional Protection" 1999).

Social acceptance of homosexuality is generally higher in Europe than in the rest of the world. At least 12 European countries, as well as Canada, Brazil, and Australia, have national laws banning discrimination on the basis of sexual orientation. In 1996, Buenos Aires, Argentina, also drafted a statute that included a clause recognizing "the right to be different" and condemning all forms of discrimination, including those based on sexual orientation (Sarda 1998).

In Hungary, same-sex couples may have a common-law marriage. And in other European countries (Denmark, Iceland, Norway, Sweden, and the Netherlands) "registered partnership" laws convey legal status to same-sex couples (Comparative Survey 1998). The Netherlands is moving toward becoming the first nation in the world to let same-sex couples marry under regular marriage laws (*ILGA Annual Report* 1996/1997).

A global perspective on laws and social attitudes regarding homosexuality reveals that countries vary tremendously in their treatment of homosexuals—from intolerance and criminalization to acceptance and legal protection. Later in this chapter, we examine social attitudes and laws regarding homosexuality in the United States.

Prevalence of Homosexuality, Heterosexuality, and Bisexuality in the United States

The prevalence of homosexuality, heterosexuality, and bisexuality is difficult to determine. Due to embarrassment, a desire for privacy, or fear of social disapproval, many individuals are not willing to answer questions about their sexuality honestly. In addition, estimates of the prevalence of sexual orientations vary due to differences in the way researchers define and measure them. Classifying individuals as heterosexual, homosexual, or bisexual is not as clear-cut as some people believe. A person's self-identity may not correspond with her or his behavior. For example, substantial numbers of individuals who view themselves as heterosexual have had same-sex attractions and relations. Definitional problems also arise from the fact that sexual attractions, behavior, and self-identity may change over time.

Nevertheless, research data have yielded rough estimates of prevalence rates of homosexuality and bisexuality. In a national study of U.S. adults aged 18 to 59, researchers focused on three aspects of homosexuality: sexual attraction to persons of the same sex, sexual behavior with people of the same sex, and homosexual self-identification (Michael et al. 1994). The researchers found that not all people who are sexually attracted to or have had sexual relations with individuals of the same sex view themselves as homosexual or bisexual.

Other data reflect the prevalence of same-sex sexual behavior among a large sample of high school students in Massachusetts (Faulkner and Cranston 1998). Over 3,000 students in grades 9 through 12 completed a self-administered anonymous questionnaire that included the question: "The person(s) with whom you have had sexual contact is (are) (1) female(s), (2) males(s), (3) females(s) and male(s), (4) I have not had sexual contact with anyone." Of the sexually experienced students (which did not necessarily imply having had intercourse), 6.4% reported same-sex contact, either exclusively or bisexually. This survey found equal numbers of male and

Two, four, six, eight, how do you know your grandma's straight?

–Gay slogan

National Data

Based on a national survey of U.S. adults, less than 3 percent of U.S. men and less than 2 percent of U.S. women identify themselves as homosexual or bisexual. Yet, 4 percent of women and 6 percent of men said they are sexually attracted to individuals of the same sex. And, 4 percent of women and 5 percent of men reported that they had sexual relations with a same-sex partner after age 18.

SOURCE: Michael et al. 1994

female individuals with same sex experience, and equal numbers of students reporting exclusively same-sex activity and bisexual experience.

> We are maintaining that all the ingredients, biological, individual, and social, codetermine the effects that each has on the other and on the shaping of human sexual preference.
>
> —John P. De CeCecco and John P. Elia
> Sex researchers

Origins of Sexual Orientation Diversity: Nature or Nurture?

One of the prevailing questions raised regarding homosexuality and bisexuality centers on its origin or "cause." Despite the growing research on this topic, a concrete cause of sexual orientation diversity has yet to be discovered. Many researchers believe that an interaction of biological and environmental forces is involved in the development of one's sexual orientation (De Cecco and Parker 1995). After presenting research findings on the biological bases of sexual orientation, we look briefly at environmental explanations.

Biological Origins of Sexual Orientation Gays and gay rights advocates tend to support the position that homosexuality is an inherited, inborn trait. In a national study of homosexual men, 90 percent believe that they were born with their homosexual orientation; only 4 percent believe that environmental factors are the sole cause (Lever 1994). Gallup poll findings reveal an increasing trend in the belief that homosexuality is inborn, from 13 percent in 1977 to 31 percent in 1996 (Saad 1996).

> I knew in my bones that my own sexuality was not a decision but a natural part of who I am.
>
> —Jonathon Tolins

Biological explanations of sexual orientation diversity usually focus on genetic or hormonal differences between heterosexuals and homosexuals. In an overview of genetics research on homosexual and heterosexual orientations, Pillard and Bailey (1998) conclude that genes account for at least half of the variance in sexual orientation. Their review of family, twin, and adoptee studies indicate that homosexuality (and thus heterosexuality) runs in families.

Environmental Explanations of Sexual Orientation According to Doell (1995), ". . . we all probably develop, from infancy, the capacity to have heterosexual, homosexual, or bisexual relationships" (p. 352). Environmental theories propose that such factors as availability of sexual partners, early sexual experiences, and sexual reinforcement influence subsequent sexual orientation. The degree to which early sexual experiences have been negative or positive has been hypothesized as influencing sexual orientation. Having pleasurable same-sex experiences would be likely to increase the probability of a homosexual orientation. By the same token, early traumatic sexual experiences have been suggested as causing fear of heterosexual activity. However, a study that compared sexual histories of lesbian and heterosexual women found no differences in the incidence of traumatic experiences with men (Brannock and Chapman 1990).

CONSIDERATION

Beliefs about the causes of homosexuality are related to people's attitudes towards homosexuals. A national poll of Americans found that "those who believe homosexuals choose their sexual orientation are far less tolerant of gays and lesbians and more likely to conclude homosexuality should be illegal than those who think sexual orientation is not a matter of personal choice" (Rosin and Morin 1999, 8).

SOCIOLOGICAL THEORIES OF SEXUAL ORIENTATION

Although sociological theories do not explain the origin or "cause" of sexual orientation diversity, they help explain societal reactions to homosexuality and bisexuality. In addition, the symbolic interaction perspective sheds light on the process of adopting a gay, lesbian, or bisexual identity.

Structural-Functionalist Perspective

Structural-functionalists, consistent with their emphasis on institutions and the functions they fulfill, emphasize the importance of monogamous heterosexual relationships for the reproduction, nurturance, and socialization of children. From a functionalist perspective, homosexual relations, as well as heterosexual nonmarital relations, are defined as "deviant" because they do not fulfill the family institution's main function of producing and rearing children. Clearly, however, this argument is less salient in a society in which other institutions, most notably schools, have supplemented the traditional functions of the family and in which reducing (rather than increasing) population is a societal goal.

Some functionalists argue that antagonisms between heterosexuals and homosexuals may disrupt the natural state, or equilibrium, of society. Durkheim, however, recognized that deviation from society's norms may also be functional. As Durkheim observed, deviation ". . . may be useful as a prelude to reforms which daily become more necessary" (Durkheim [1938] 1993, 66). Specifically, the gay rights movement has motivated many people to reexamine their treatment of sexual orientation minorities and has produced a sense of cohesion and solidarity among members of the gay population (although bisexuals have often been excluded from gay and lesbian communities and organizations). Gay activism has been instrumental in advocating for more research on HIV and AIDS, more and better health services for HIV and AIDS patients, protection of the rights of HIV-infected individuals, and HIV/AIDS public education. Such HIV/AIDS prevention strategies and health services benefit the society as a whole.

Finally, the structural-functionalist perspective is concerned with how changes in one part of society affect other aspects. With this focus on the interconnectedness of society, we note that urbanization has contributed to the formation of strong social networks of gays and bisexuals. Cities "acted as magnets, drawing in gay migrants who felt isolated and threatened in smaller towns and rural areas" (Button, Rienzo, and Wald 1997, 15). Given the formation of gay communities in large cities, it is not surprising that the gay rights movement first emerged in large urban centers.

Conflict Perspective

Conflict theorists, particularly those who do not emphasize a purely economic perspective, note that the antagonisms between heterosexuals and nonheterosexuals represent a basic division in society between those with power and those without power. When one group has control of society's institutions and resources, as in the case of heterosexuals, they have the authority to dominate other groups. The recent battle over gay rights is just one example of the political struggle between those with power and those without it.

A classic example of the power struggle between gays and straights took place in 1973 when the American Psychiatric Association (APA) met to revise its classification scheme of mental disorders. Homosexual activists had been appealing to the APA for years to remove homosexuality from its list of mental illnesses but with little success. The view of homosexuals as mentally sick contributed to their low social prestige in the eyes of the heterosexual majority. In 1973, the APA's board of directors voted to remove homosexuality from its official list of mental disorders. The board's move encountered a great deal of resistance from conservative APA members and was put to a referendum, which reaffirmed the board's decision (Bayer 1987).

More currently, gays and lesbians are waging a political battle to win civil rights protections in the form of laws prohibiting discrimination on the basis of sexual orientation (discussed later in this chapter). Conflict theory helps to explain why many business owners and corporate leaders oppose civil rights protection for gays and lesbians. Employers fear that such protection would result in costly lawsuits if they refused to hire homosexuals, regardless of the reason for their decision. Business owners also fear that granting civil rights protections to homosexual employees would undermine the economic health of a community by discouraging the development of new businesses and even driving out some established firms (Button, Rienzo, and Wald 1997).

However, some companies are recognizing that implementing antidiscrimination policies that include sexual orientation is good for the "bottom line." A 1993 National Gay and Lesbian Task Force survey of Fortune 1000 companies found that 72 percent had a nondiscrimination policy that included sexual orientation (reported in Button, Rienzo, and Wald 1997). These companies realized that "many of their most talented employees might leave if they did not provide a more tolerant setting" (Button, Rienzo, and Wald 1997, 127).

In summary, conflict theory frames the gay rights movement and the opposition to it as a struggle over power, prestige, and economic resources. Recent trends toward increased social acceptance of homosexuality may, in part, reflect the corporate world's competition over the gay and lesbian consumer dollar—the "lavender dollar."

Symbolic Interactionist Perspective

Symbolic interactionism focuses on the meanings of heterosexuality, homosexuality, and bisexuality and how these meanings are socially constructed. The meanings we associate with same-sex relations are learned from society—from family, peers, religion, and the media. Freedman and D'Emilio (1990) observed that ". . . sexual meanings are subject to the forces of culture. Human beings learn how to express themselves sexually, and the content and outcome of that learning vary widely across cultures and across time" (p. 485). Historical and cross-cultural research on homosexuality reveals the socially constructed nature of homosexuality and its meaning. Although many Americans assume that same-sex romantic relationships have always been taboo in our society, during the nineteenth century, "romantic friendships" between women were encouraged and regarded as preparation for a successful marriage. The nature of these friendships bordered on lesbianism. President Grover Cleveland's sister Rose wrote to her friend Evangeline Whipple in 1890: ". . . It makes me heavy with emotion . . . all my whole being leans out to you . . . I dare not think of your arms" (Goode and Wagner 1993, 49).

The symbolic interactionist perspective also points to the effects of labeling on individuals. Once individuals become identified or labeled as lesbian, gay, or bisexual, that label tends to become their **master status.** In other words, the dominant heterosexual community tends to view "gay," "lesbian," and "bisexual" as the most socially significant statuses of individuals who are identified as such. Esterberg (1997) notes that "unlike heterosexuals, who are defined by their family structures, communities, occupations, or other aspects of their lives, lesbians, gay men, and bisexuals are often defined primarily by what they do in bed. Many lesbians, gay men, and bisexuals, however, view their identity as social and political as well as sexual" (p. 377).

CONSIDERATION

Because of the stigma of being labeled as lesbian, gay or bisexual, many gays and bisexuals struggle through the process of developing a gay or bisexual identity, particularly those who are also members of a racial or ethnic minority group. However, others feel that they have always known and accepted their sexual orientation. One woman, for example, claimed: "I have never not felt that I was bisexual" (quoted in Esterberg 1997, p. 38). In addition, some bisexuals resist defining themselves with a sexual orientation label. In interviews with bisexual women, Esterberg (1997) noted: "for a number of women, what was important was not taking on a label and an identity, but having the freedom to define themselves and live their lives as they please" (p. 157). This resistance to sexual orientation labels may be analogous to the resistance among some individuals to using racial labels to describe themselves and who, instead, leave questions about their race blank or check "other."

HETEROSEXISM, HOMOPHOBIA, AND BIPHOBIA

The United States, along with many other countries throughout the world, is predominantly heterosexist. **Heterosexism** refers to "an ideological system that denies, denigrates, and stigmatizes any nonheterosexual form of behavior, identity, relationship, or community" (Herek 1990, 316). Heterosexism is based on the belief that heterosexuality is superior to homosexuality and results in prejudice and discrimination against homosexuals and bisexuals. Prejudice refers to negative attitudes, whereas discrimination refers to behavior that denies individuals or groups equality of treatment. Before reading further, you may wish to complete this chapter's *Self & Society* feature: the Homophobia Scale.

I am one behind a closet door.
Scared, embarrassed and different for sure.
I know I'm not alone, but it often feels like I am.
I want my family to accept me instead of not giving a damn.
I want to step out of this dark closet now,
I'm running out of time, will someone show me how?

—Excerpts from a poem written by an anonymous student

Homophobia

SIECUS, the Sex Information and Education Council of the United States, affirms the sexual orientation of all persons. "Individuals have a right to accept, acknowledge, and live in accordance with their sexual orientation, be they bisexual, heterosexual, gay, or lesbian" (SIECUS 1996, 22). Nevertheless, negative attitudes toward homosexuality are reflected in the high percentage of the U.S. population that disapproves of homosexuality.

The term **homophobia** is commonly used to refer to negative attitudes toward homosexuality. Other terms that refer to negative attitudes toward homosexuality include "homonegativity" and "anti-gay bias."

The Homophobia Scale

Directions: Indicate the extent to which you agree or disagree with each statement by placing a check mark on the appropriate line.

	Strongly Agree	Agree	Undecided	Disagree	Strongly Disagree
1. Homosexuals contribute positively to society.					
2. Homosexuality is disgusting.					
3. Homosexuals are just as moral as heterosexuals.					
4. Homosexuals should have equal civil rights.					
5. Homosexuals corrupt young people.					
6. Homosexuality is a sin.					
7. Homosexuality should be against the law.					

SCORING

Assign scores of 0, 1, 2, 3, and 4 to the five choices respectively ("strongly agree" through "strongly disagree"). Reverse-score items 2, 5, 6, and 7 (0 = 4; 1 = 3; 2 = 2; 3 = 1; 4 = 0). All items are summed for the total score. The possible range is 0 to 28; high scores indicate greater homophobia.

COMPARISON DATA

The Homophobia Scale was administered to 524 students enrolled in introductory psychology courses at the University of Texas. The mean score for men was 15.8; for women, it was 13.8. The difference was statistically significant.

From Richard A. Bouton, P. E. Gallagher, P. A. Garlinghouse, T. Leal, et al., 1987. "Scales for Measuring Fear of AIDS and Homophobia." *Journal of Personality Assessment*, 67(1):609.

National Data

More than half (57 percent) of Americans in a national poll said that homosexuality is unacceptable.

SOURCE: Rosin and Morin 1999

In general, certain categories of people are more likely to have negative attitudes toward homosexuals. Persons who are older, less educated, living in the South or Midwest, living in lightly populated rural areas, and Protestant are the most likely to have negative attitudes. In contrast, people who are younger, more educated, never married, living in the West, living in heavily populated urban areas, and Jewish are least likely to have antigay attitudes (Klassen, Williams, and Levitt 1989). Also, positive contact with homosexuals or having homosexuals as friends is associated with less homophobia (Simon 1995).

Public opinion surveys also indicate that men are more likely than women to have negative attitudes toward gays (Moore 1993). But many studies on attitudes toward homosexuality do not distinguish between attitudes toward gay men and attitudes toward lesbians (Kite and Whitley 1996). Research that has assessed attitudes toward male versus female homosexuality has found that heterosexual women and men hold similar attitudes toward lesbians, but men are more negative toward gay men (Louderback and Whitley 1997; Price and Dalecki 1998).

CONSIDERATION

The homophobic and heterosexist social climate of our society is often viewed in terms of how it victimizes the gay population. However, heterosexuals are also victimized by homophobia and heterosexism. Due to the antigay climate, heterosexuals, especially males, are hindered in their own self-expression and intimacy in same-sex relationships. "The threat of victimization (i.e., antigay violence) probably also causes many heterosexuals to conform to gender roles and to restrict their expressions of (nonsexual) physical affection for members of their own sex" (Garnets, Herek, and Levy 1990, 380).

Some cases of rape and sexual assault are related to homophobia and compulsory heterosexuality. For example, college men who participate in gang rape, also known as "pulling train," entice each other into the act "by implying that those who do not participate are unmanly or homosexual" (Sanday 1995:399). Homonegativity also encourages early sexual activity among adolescent men. Adolescent male virgins are often teased by their male peers, who say things like "You mean you don't do it with girls yet? What are you, a fag or something?" Not wanting to be labeled and stigmatized as a "fag," some adolescent boys "prove" their heterosexuality by having sex with girls.

National Data

In a national survey of first-year U.S. college students, 43 percent of the men, but only 25 percent of the women, agreed with the statement, "It is important to have laws prohibiting homosexual relations."

SOURCE: "This Year's Freshmen: A Statistical Profile" 1999

Cultural Origins of Homophobia

Why is homosexuality viewed so negatively in the United States? Antigay bias has its roots in various aspects of U.S. culture.

1. *Religion.* Most Americans who view homosexuality as unacceptable say they object on religious grounds (Rosin and Morin 1999). Although some religious groups (such as the Quakers) accept homosexuality, many religions teach that homosexuality is sinful and prohibited by God. The Roman Catholic church rejects all homosexual expression and resists any attempt to validate or sanction the homosexual orientation. Some fundamentalist churches have endorsed the death penalty for homosexual people and teach the view that AIDS is God's punishment for engaging in homosexual sex (Nugent and Gramick 1989). Some religions are divided on the issue of homosexuality. For example, while the official position of the United Methodist Church is one that condemns homosexuality, some Methodist ministers advocate acceptance of and equal rights for lesbians and gay men. In January 1999, 95 United Methodist Church ministers violated the Church's ban on "ceremonies that celebrate homosexual unions" as they officiated at a same-sex union church service in Sacramento, California. The United Methodist Church has considered splitting into two organizations: one in favor of equal rights for gays and lesbians and the other opposed (*The United Methodist Church and Homosexuality* 1999).

God made Adam and Eve, not Adam and Steve.

—Poster at a conservative Christian rally

2. *Marital and procreative bias.* Many societies have traditionally condoned sex only when it occurs in a marital context that provides for the possibility of producing and rearing children. Although laws prohibiting same-sex marriages have been challenged in the courts (which we discuss later in this chapter), as of this writing, same-sex marriages are not legally recognized in the United States. Further, even though assisted reproductive technologies make it possible for gay individuals and couples to have children, many people believe that these advances should only be used by heterosexual married couples only (Franklin 1993).

3. *Concern about HIV and AIDS.* Although most cases of HIV and AIDS worldwide are attributed to heterosexual transmission, the rates of HIV and AIDS in the United States are much higher among gay and bisexual men than among other groups. Because of this, many people associate HIV and AIDS with homosexuality and bisexuality. This association between AIDS and homosexuality has fueled antigay sentiments. During a 1985 mayoral campaign in Houston, the challenging contender was asked what he would do about the AIDS problem. His answer: he would "shoot the queers" (Button, Rienzo, and Wald

1997, 70). Lesbians, incidentally, have a very low risk for sexually transmitted HIV—a lower risk than heterosexual women.

4. *Threat to the power of the majority.* Like other minority groups, the gay minority threatens the power of the majority. Fearing loss of power, the majority group stigmatizes homosexuals as a way of limiting their power.

5. *Rigid gender roles.* Antigay sentiments also stem from rigid gender roles. When Cooper Thompson (1995) was asked to give a guest presentation on male roles at a suburban high school, male students told him that the most humiliating put-down was being called a "fag." The boys in this school gave Thompson the impression they were expected to conform to rigid, narrow standards of masculinity in order to avoid being labeled in this way.

From a conflict perspective, heterosexual men's subordination and devaluation of gay men reinforces gender inequality. "By devaluing gay men . . . heterosexual men devalue the feminine and anything associated with it" (Price and Dalecki 1998, 155–56). Negative views toward lesbians also reinforce the patriarchal system of male dominance. Social disapproval of lesbians is a form of punishment for women who relinquish traditional female sexual and economic dependence on men. Not surprisingly, research findings suggest that individuals with traditional gender role attitudes tend to hold more negative views toward homosexuality (Louderback and Whitley 1997).

6. *Psychiatric labeling.* As noted earlier, prior to 1973 the APA defined homosexuality as a mental disorder. Treatments for the "illness" of homosexuality included lobotomies, aversive conditioning, and, in some cases, castration. The social label of mental illness by such a powerful labeling group as the APA contributed to heterosexuals' negative reactions to gays. Further, it created feelings of guilt, low self-esteem, anger, and depression for many homosexuals. Thus, the psychiatric care system is now busily treating the very conditions it, in part, created.

7. *Myths and negative stereotypes.* Prejudice toward homosexuals may also stem from some of the unsupported beliefs and negative stereotypes regarding homosexuality. One negative myth about homosexuals is that they are sexually promiscuous and lack "family values" such as monogamy and commitment to relationships. While some homosexuals do engage in casual sex, as do some heterosexuals, many homosexual couples develop and maintain long-term committed relationships.

Another myth that is not supported by data is that homosexuals, as a group, are child molesters. The ratio of heterosexual to homosexual child molesters is approximately eleven to one (Moser 1992). Most often the abuser is a father, stepfather, or heterosexual relative of the family. When a father rapes his daughter, the media do not report that something is wrong with heterosexuality or with traditional families. But when a homosexual child molestation is reported, it is viewed as confirmation of "the way homosexuals are" (Mohr 1995, 404).

Biphobia

Just as the term "homophobia" is used to refer to negative attitudes toward gay men and lesbians, **biphobia** refers to "the parallel set of negative beliefs about and stigmatization of bisexuality and those identified as bisexual" (Paul 1996, 449). Although both homosexual- and bisexual-identified individuals are often rejected by heterosexuals, bisexual-identified women and men also face rejection from many homosexual individuals. Thus, bisexuals experience "double discrimination."

Lesbians seem to exhibit greater levels of biphobia than do gay men. This is because many lesbian women associate their identity with a political stance against sexism and patriarchy. Some lesbians view heterosexual and bisexual women who "sleep with the enemy" as traitors to the feminist movement.

One negative stereotype that encourages biphobia is the belief that bisexuals are, by definition, nonmonogamous. However, many bisexual women and men prefer and have long-term committed monogamous relationships.

DISCRIMINATION AGAINST SEXUAL ORIENTATION MINORITIES

Like other minority groups in American society, homosexuals and bisexuals experience various forms of discrimination. Next, we look at sexual orientation discrimination in the workplace, in family matters, and in violent expressions of hate. But first we examine discrimination in the application and enforcement of sodomy laws.

Sodomy Laws

Sodomy laws, sometimes referred to as "crimes against nature," prohibit what are considered to be "unnatural acts" such as oral and anal sex. In 1986, the U.S. Supreme Court ruled in *Bowers v. Hardwick* that the Constitution allows states to criminalize sodomy. Sodomy laws were once on the books in all 50 U.S. states, but they have been repealed or struck down by courts in more than 30 states. Eleven states still ban oral and anal sex between consenting adults. Penalties for violating sodomy laws range from a $200 fine to 20 years imprisonment.

In 5 states, sodomy laws target only same-sex acts ("Status of U.S. Sodomy Laws" 1999). These states include Arkansas, Kansas, Missouri, Oklahoma, and Texas. In states that criminalize both same- and opposite-sex sodomy, sodomy laws are usually not used against heterosexuals but are primarily used against gay men and lesbians. For example, some courts have taken children away from lesbian and gay parents on the basis that these parents violate sodomy laws and are, therefore, lawbreakers.

CONSIDERATION

When asked in a national poll if homosexual relations between consenting adults should be legal, 55 percent said "yes" and 34 percent answered "no." However, when reminded that people could be prosecuted for engaging in these activities in their own homes, more than half of those who first said homosexuality should be illegal changed their minds (Rosin and Morin 1999).

National Data

More than four out of five Americans (87 percent) now say gays should have equal rights in terms of job opportunities, up from 74 percent in 1992, 59 percent in 1982, and 56 percent in 1977.

SOURCE: Rosin and Morin 1999; Saad 1996

Discrimination in the Workplace

In recent years, the percentage of Americans who express approval of equal employment rights for homosexuals has increased. As we discuss later in the section "Strategies for Action," most states still do not offer protection against employment discrimination based on sexual orientation.

> I know now that I had more than one teacher who happened to be gay. I also know I wouldn't be a member of Congress today had it not been for those great teachers.
>
> –Representative Christopher Shays (R-Conn.)

> I've made the choice that my personal liberty and the emancipation that I felt about coming out was definitely more important than my career at that point.
>
> –kd Lang Music artist

> The Constitution does not allow government to subordinate a class of persons simply because others do not like them.
>
> –Chief Judge Abner Mikava

When asked if homosexuals should be hired for specific occupations, Americans make distinctions (see Table 9.1).

Many people who are gay fear being fired, denied salary increases, or not being promoted. The Cracker Barrel restaurant chain fired at least 11 openly gay and lesbian employees in 1991. Cracker Barrel stated that it would refuse employment to people "whose sexual preferences fail to demonstrate normal heterosexual values which have been the foundation of families in our society" (cited in Button, Rienzo, and Wald 1997, 126). In 1996, a manager at an Illinois Red Lobster fired Dale Hall after repeatedly ridiculing him for being gay. The Cook County Commission on Human Rights ruled that Red Lobster violated Chicago's antidiscrimination law and ordered Red Lobster to rehire Dale Hall and pay him $95,000 in back pay and damages (*Lawbriefs* 1999). In 1998, the company that owns Red Lobster filed a suit to have Chicago's antidiscrimination law overturned ("1998 in Review" 1999). Another publicized example of employment discrimination based on sexual orientation involves the case of a woman who was offered a job as a lawyer in the Georgia attorney general's office but lost the job offer after the state attorney general learned that she was planning a commitment ceremony with her lesbian partner. In January 1998, the Supreme Court decided not to hear her appeal (Greenhouse 1998). In 1998, Oklahoma passed a measure to ban homosexuals from working in public schools ("1998 in Review" 1999).

Occupational discrimination also occurs in the military. In 1992, a gay Navy serviceman was fired after revealing his homosexuality but was reinstated by orders of a federal judge. In 1993, President Clinton instituted a "Don't ask, don't tell" policy in which recruiting officers are not allowed to ask about sexual orientation, and homosexuals are encouraged not to volunteer such information. Although the "Don't ask, don't tell" policy was intended to protect homosexuals from discrimination in the military, military discharges due to homosexuality have increased in recent years, from 757 in 1995, 858 in 1996, 997 in 1997, to 1,145 in 1998 ("Discharges of Gay Troops Up," 1998; Human Rights Campaign News 1999).

Discrimination in Family Relationships

In addition to discrimination in the workplace, sexual orientation minorities experience discrimination in policies concerning marriage, child custody and visitation, and adoption.

Table 9.1 Attitudes toward Hiring Homosexuals for Various Jobs

A 1996 Gallup poll asked a national sample of U.S. adults, "Do you think homosexuals should or should not be hired for each of the following occupations?" The percentages listed indicate the percentage who said homosexuals *should* be hired.

	1977	1987	1996
Salespersons	68%	72%	90%
Doctors	44%	49%	69%
Armed forces	51%	55%	65%
Elementary school teachers	27%	33%	55%
Clergy	36%	42%	53%

SOURCE: Saad, Lydia. 1996. "Americans Growing More Tolerant of Gays." The Gallup Organization. Princeton, N.J. **http://198.175.140.8/POLL_ARCHIVES/961214.htm**

Same-Sex Marriage In late 1996, Hawaii circuit court judge Kevin Chang made history when he ruled (in *Baehr v. Miike*) that Hawaii's refusal to grant marriage licenses to same-sex couples violated the state's constitution. In November 1998, Hawaii voters passed a constitutional amendment that gives the state legislature the power to reserve marriage for opposite-sex couples. As of this writing, the issue of whether same-sex couples will be allowed to marry in Hawaii is pending before the Hawaii State Supreme Court. In 1996, Congress passed and President Clinton signed **The Defense of Marriage Act**, which states that marriage is a "legal union between one man and one woman" and denies federal recognition of same-sex marriage. In effect, this law allows states to either recognize or not recognize same-sex marriages performed in other states. As of May 1998, 28 states had passed anti-gay marriage laws, declaring that they will not recognize same-sex marriages ("Anti-Marriage Bills Fact Sheet" 1998).

National Data

In a national poll of U.S. adults, only 23 percent supported same-sex marriage.

SOURCE: Rosin and Morin 1999

Advocates of same-sex marriage argue that banning same sex marriages, or refusing to recognize same-sex marriages granted in other states, denies same-sex couples the many legal and financial benefits that are granted to heterosexual married couples. For example, married couples have the right to inherit from a spouse who dies without a will; to avoid inheritance taxes between spouses; to make crucial medical decisions for a partner and take family leave in order to care for a partner in the event of the partner's critical injury or illness; to receive Social Security survivor benefits; and to include a partner in his or her health insurance coverage. Other rights bestowed to married (or once married) partners include assumption of spouse's pension, bereavement leave, burial determination, domestic violence protection, reduced rate memberships, divorce protections (such as equitable division of assets and visitation of partner's children), automatic housing lease transfer, and immunity from testifying against a spouse.

> We don't object to domestic partnerships. We don't agree that it is a substitute for same-sex marriage. . . . Domestic partnership can be a way of helping to eliminate forms of discrimination, but it has nothing to do with marriage. We don't want separate but equal, and domestic partnership is not equal. It doesn't offer the same benefits and recognition.
>
> –Dan Foley
> Honolulu attorney

While advocates of same-sex marriage argue that as long as same-sex couples cannot be legally married, they will not be regarded as legitimate families by the larger society, opponents do not want to legitimize homosexuality as an acceptable, legitimate lifestyle. Opponents of same-sex marriage who view homosexuality as unnatural, sick, and/or immoral do not want their children to learn that homosexuality is an accepted "normal" lifestyle. The most common argument against same-sex marriage is that it subverts the stability and

These gay men have been in a committed relationship for over 20 years and consider themselves a "married" couple. They exchanged vows during a "celebration of commitment" ceremony, which was attended by family and friends.

Hate is not a family value.

—Bumper sticker

integrity of the heterosexual family. However, Sullivan (1997) suggests that homosexuals are already part of heterosexual families:

> **[Homosexuals] are sons and daughters, brothers and sisters, even mothers and fathers, of heterosexuals. The distinction between "families" and "homosexuals" is, to begin with, empirically false; and the stability of existing families is closely linked to how homosexuals are treated within them. (p. 147)**

National Data

In a national survey of first-year U.S. college students, about half (49.4 percent) agreed that same-sex couples should have the right to legal marital status. Women were more likely than men to support same-sex marriage (56 percent versus 41 percent).

SOURCE: "This Year's Freshmen: A Statistical Profile" 1999

Some states, counties and cities allow unmarried couples, including gay couples, to register as domestic partners. The benefits of **domestic partnerships** vary from place to place but may include coverage under a partner's health and pension plan, rights of inheritance and community property, and tax benefits. In 1991 the Lotus Development Corporation became the first major American firm to extend domestic partnership benefits to gay and lesbian employees. By 1999, more than 500 U.S. firms had extended medical and often other benefits to gay and lesbian domestic partners. However, these benefits represent a fraction of those granted to married couples.

Child Custody, Visitation, and Adoption Rights Homosexual and bisexual individuals become involved in heterosexual relationships for a variety of reasons, including genuine love for a spouse, wanting to have children, family pressure to marry, and the desire to live a socially approved heterosexual lifestyle. Some individuals do not realize that they are homosexual or bisexual until after they are married. In cases of divorce, openly gay, lesbian, and bisexual individuals experience discrimination in child custody and visitation rights. Judges are less apt to give a gay father sole or joint custody than a heterosexual father.

> **The courts are homophobic, prejudiced, and biased against gay fathers. Two major cultural fears work against gay fathers who want full or joint custody of their children or a liberal visitation schedule. The first of these fears is that gay fathers will molest their children, and the second is that gay fathers will "recruit" their children into a gay lifestyle. (Knox 1998, 180)**

National Data

The U.S. General Accounting Office identified 1,049 federal laws in which benefits, rights, and privileges are contingent on marital status.

SOURCE: Lambda Legal Defense and Education Fund 1997

Lesbian or bisexual women may also be denied custody of their children on the sole basis of their sexual orientation, despite the fact that research shows that children of lesbian mothers are just as likely to be well-adjusted as those of heterosexual mothers (Patterson 1997; Tasker and Golombok 1997). In a highly publicized case, Sharon Bottoms lost custody of her son Tyler after a Virginia court awarded custody to Sharon's mother. The court based its custody decision on the fact that Sharon Bottoms is a lesbian.

Most cases of child adoption by gays and lesbians involve one partner adopting the biological child of the other. At least 21 states have granted second-parent adoptions to lesbian and gay couples to ensure that the children can enjoy the benefits of having two legal parents, especially if one parent dies or becomes incapacitated (ACLU Fact Sheet 1997). In addition, at least 22 states allow lesbians and gay men to adopt children through state-run or private adoption agencies (Beauvais-Godwin and Godwin 1997). Four states—New Hampshire, Florida, Colorado and Connecticut—specifically ban gay and lesbian adoptions ("Idaho Court to Take Up First Case of Adoption by Same-Sex

National Data

An estimated 6 to 14 million children nationwide are living with at least one gay parent.

SOURCE: ACLU Fact Sheet 1997

Some lesbian couples artificially inseminate one partner with the sperm of a male relative of the other so the child will be genetically related to both partners.

Couple" 1999). In early 1998, New Jersey became the first state to set a policy allowing same-sex (and unmarried heterosexual) couples to jointly adopt children under the same qualification standards as married couples.

International Data

In a Newsweek poll of U.S. adults, 65 percent said that same-sex couples should not be allowed to adopt children.

SOURCE: Ingrassia and Rossi 1994

Hate Crimes against Sexual Orientation Minorities

In eighteenth century America, where laws against homosexuality often included the death penalty, violence against gays and lesbians was widespread and included beatings, burnings, various kinds of torture, and execution (Button, Rienzo, and Wald 1997). Although such treatment of sexual orientation minorities is no longer legally condoned, gays, lesbians, and bisexuals continue to be victimized by hate crimes. Surveys indicate that as many as one-fourth of lesbians and gay men report having been victims of physical attacks because of their sexual orientation (Herek 1989). A survey of more than 3,000 Massachusetts high school students found that students who reported having engaged in same-sex relations were more than three times as likely to report not going to school because they felt unsafe and more than twice as likely to report having been threatened or injured with a weapon at school (Faulkner and Cranston 1998). These students were also significantly more likely to report that their property was deliberately damaged or stolen at school.

As long as the anti-gay campaigners continue to spread their message of ignorance and hate, our nation will remain a hostile, dangerous and sometimes deadly place for us, our friends and millions of America's gay and lesbian citizens.

—Eric Marcus
Gay activist

Hate-motivated violence toward sexual orientation minorities (also known as "gay-bashing") is often brutal and horrific. In April 1996, 34-year-old Ron Daugherty and a friend went to a bar in Myrtle Beach, South Carolina, to hear a rock 'n' roll band. Daugherty's friend accidentally bumped into someone's bar stool, leading to an argument in which Daugherty and his friend were called "faggots." Daugherty describes what happened later:

> **After we left the bar, I realized we weren't alone. I saw a guy I had seen in the bar coming toward us. He was wearing brass knuckles. . . . Someone grabbed me from behind and started hitting me. Someone was screaming, "I'm gonna kill you faggot." Then someone**

National Data

The National Coalition of Anti-Violence Programs (1998) documented 2,445 incidents of violence against sexual orientation minorities in 1997.

National Data

As of February 1999, 21 states and the District of Columbia have hate crimes laws that include sexual orientation. Nineteen states have hate crimes laws that do not include sexual orientation and ten states have no hate crimes laws at all.

SOURCE: Cipriaso 1999

hit me and knocked me out. When I came to, I was lying in a trash shed. The friend who was with me was lying on top of me, his throat slashed, with blood all over. (reported in Mills 1997, 9)

CONSIDERATION

From 1996 to 1997, the number of *heterosexuals* who reported that they were victims of anti-gay hate crimes rose 36 percent. "This underscores the fact that hate crimes are crimes of perception" (National Coalition of Anti-Violence Programs 1998, 4). Victims are chosen not necessarily because they are lesbian, gay or bisexual, but because the perpetrator perceives them to be. Therefore, "no one is safe from hate crimes and . . . it is in everyone's interest to stop this epidemic of hate" (p. 4).

The Hate Crimes Sentencing Enhancement Act (part of the Violent Crime Control and Law Enforcement Act of 1994) calls for tougher sentencing when prosecutors can prove that the crime committed was a hate crime. However, because federal law enforcement agencies do not have jurisdiction over anti-gay hate crimes, this law only applies to hate crime based on sexual orientation when the offense occurs on federal property, such as a national park ("Fighting Anti-Gay Hate Crimes" 1998).

STRATEGIES FOR ACTION: REDUCING ANTIGAY PREJUDICE AND DISCRIMINATION

Many of the efforts to change policies and attitudes regarding sexual orientation minorities have been spearheaded by organizations such as the Human Rights Campaign, the National Gay and Lesbian Task Force (NGLTF), Gay and Lesbian Alliance against Defamation (GLAAD), AIDS Coalition to Unleash Power

Actress Ellen DeGeneres spoke during the Human Rights Campaign national dinner in 1997, at which she received a civil rights award. President Clinton also spoke at this event, becoming the first U.S. president to speak before the nation's largest lesbian and gay rights organization.

(ACT-UP), Fund for Human Dignity, Lambda Legal Defense and Education Fund, National Lesbian and Gay Health Foundation, National Black Gay and Lesbian Leadership Forum, Lesbian Avengers, and Astraea Lesbian Action Foundation. These organizations are politically active in their efforts to achieve equal rights for gays and lesbians. The gay rights movement is also active in promoting HIV/AIDS research, adequate health care for AIDS victims, and the rights of HIV-infected individuals. Other issues on the "gay agenda" include providing programs and services for gay and lesbian students.

Next we highlight some of the strategies for reducing and responding to prejudice and discrimination toward sexual orientation minorities. These strategies are divided into those aimed at influencing local, state, and federal legislation and those aimed at implementing policies and programs within the educational setting.

National Data

In 1981, Wisconsin became the first state to ban discrimination on the basis of sexual orientation. By 1998, 19 states had enacted laws or executive orders banning such discrimination. However, Maine repealed its executive order in early 1998, leaving 18 states banning discrimination on the basis of sexual orientation.

SOURCE: Lambda 1998

Legislative Action and the Gay Rights Movement

As of this writing, the federal government has not passed the Employment NonDiscrimination Act prohibiting discrimination against sexual orientation minorities. In May 1998, President Clinton did sign an executive order banning sexual orientation discrimination among federal employees ("1998 in Review" 1999). In addition, gay rights advocates have influenced various states to enact legislation, executive orders, or ordinances prohibiting such discrimination. The scope of these ordinances or policies varies from protection against discrimination in public employment only to comprehensive protection against discrimination in public and private employment, education, housing, public accommodations, credit, and union practices (see Table 9.2).

Table 9.2 States Prohibiting Discrimination Based on Sexual Orientation

State	Public Employment	Public Accommodations	Private Employment	Education	Housing	Credit	Union Practices
California	X	X	X	X			
Colorado*	X						
Connecticut	X	X	X	X	X	X	X
Hawaii	X		X				
Louisiana*	X						
Maryland*	X						
Massachusetts	X	X	X	X	X	X	X
Minnesota	X	X	X	X	X	X	
New Hampshire	X	X	X		X		
New Jersey	X	X	X	X	X		
New Mexico*	X						
New York*	X						
Ohio*	X						
Pennsylvania*	X						
Rhode Island	X	X	X		X	X	
Vermont	X	X	X	X	X	X	X
Washington*	X						
Wisconsin	X	X	X	X	X	X	X

*These states have executive orders prohibiting sexual orientation discrimination in public employment. These orders do not carry the same weight or have the same permanence as laws.

SOURCE: Lambda Legal Defense and Education Fund. 1998. From "States Prohibiting Discrimination Based on Sexual Orientation." **www.lambdalegal.org** Reprinted by permission of Lambda Legal Defense and Education Fund, Inc., 1998.

In addition to state legislation granting legal protection to lesbians and gay men, antigay discrimination legislation has also been enacted at the local level. As of May 1998, over 160 U.S. cities and counties had enacted some form of ban on discrimination based on sexual orientation.

A main priority of the gay rights movement is to promote passage of the federal **Employment NonDiscrimination Act** (ENDA), which would prohibit discrimination, preferential treatment, and quotas in the workplace on the basis of sexual orientation (Stachelberg 1997). The ENDA would not apply to religious organizations, businesses with fewer than 15 employees, or the military.

Another item on the gay rights legislative agenda is passage of the Hate Crimes Prevention Act of 1998, which would make hate crimes against homosexual and bisexuals, as well as gender-based hate crimes and hate crimes against the disabled, a federal offense. This chapter's *Focus on Social Problems Research* feature looks at a landmark study of the various social effects of sexual orientation antidiscrimination legislation.

Barney Frank (1997), an openly gay U.S. representative, emphasizes the importance of political participation in influencing social outcomes. He notes that demonstrative and cultural expressions of gay activism, such as "gay pride" celebrations, marches, demonstrations, or other cultural activities promoting gay rights are important in organizing gay activists. However, he notes:

> **Too many people have seen the cultural activity as a substitute for democratic political particpation. In too many cases over the past decades we have left the political arena to our most dedicated opponents [of gay rights], whose letter writing, phone calling, and lobbying have easily triumphed over our marching, demonstrating, and dancing. The most important lesson . . . for people who want to make America a fairer place is that politics—conventional, boring, but essential politics—will ultimately have a major impact on the extent to which we can rid our lives of prejudice. (Frank 1997, xi).**

Defying prejudice in colorful and expressive ways brings a good deal more immediate satisfaction than registering voters, lobbying public officials, educating people who may genuinely be ignorant about the realities of the prejudice facing gay men and lesbians, and campaigning for candidates . . .

—Barney Frank
U.S. representative

In a study by Button, Rienzo, and Wald (1997), opposition groups attempted to overturn local antigay discrimination measures in one-fourth of the U.S. cities and counties that had such measures. The Christian right, religious and church groups, conservative family groups, and other conservative organizations and their political allies will continue to crusade against gay rights. But opposition groups are up against a powerful pro–gay rights movement. In 1997, the pro–gay rights Human Rights Campaign political action committee (PAC) was the twelfth largest political action committee in the nation (Salkind and Layton 1997). A total of 83 percent of the candidates it supported in the last election cycle won the election.

Educational Strategies: Policies and Programs in the Schools

National Data

Antigay violence at schools and colleges rose 34 percent from 1996 to 1997.

SOURCE: National Coalition of Anti-Violence Programs 1998

In 1989, the U.S. Department of Health and Human Services released a report indicating that gay and lesbian youth are two to three times more likely than heterosexual youth to attempt suicide. A more recent survey of more than 3,000 Massachusettes public high school students also found that students who reported having had same-sex relations were twice as likely to have attempted suicide at least once in the past year (Faulkner and Cranston 1998). Students reporting same-sex behavior were also more likely to report heavy

FOCUS ON SOCIAL PROBLEMS RESEARCH

Effects of Sexual Orientation Antidiscrimination Legislation

In *Private Lives, Public Conflicts: Battles over Gay Rights in American Communities,* researchers James Button, Barbara Rienzo, and Kenneth Wald (1997) present survey data of all U.S. cities and counties with laws or policies prohibiting discrimination on the basis of sexual orientation. When the survey began in mid-1993, there were 126 such cities and counties.

Sample and Methods

To examine the effects of gay rights legislation in these communities, the researchers mailed questionnaires to either the director of the enforcement agency that was responsible for the antidiscrimination policy or ordinance or the local official who was perceived to be most knowledgeable about the legislation and its effect. Surveys were also sent to public school officials. For comparison purposes, the researchers also surveyed a random sample of 125 U.S. communities without antigay discrimination legislation. In addition, the researchers interviewed a variety of individuals who were knowledgeable about the passage and impact of gay rights legislation, including elected city officials and administrators, gay and lesbian activists, religious and other leaders of opposition groups, members of the business community, minority group leaders, and school personnel. Among other topics, this research investigated the various effects of antidiscrimination legislation.

Findings

The most frequently cited (by 27 per cent of respondents) positive effect of antidiscrimination legislation was that such legislation reduced discrimination against lesbians and gays in public employment and other institutions covered by the law.

Antidiscrimination legislation also made lesbians and gay men feel safer, more secure, comfortable, and generally more accepted in the community. As a result, lesbians and gay men were more likely to "come out of the closet"–to reveal their sexual orientation to others (or to at least stop hiding it). Legal measures banning discrimination based on sexual orientation also gave gays and lesbians a sense of legitimacy that helps them overcome **"internalized homophobia"**–a sense of personal failure and self-hatred among lesbians and gay men due to social rejection and stigmatization.

The second most frequently cited (by 22 percent of respondents) positive effect of antigay discrimination legislation was the recognition that discrimination based on sexual orientation is legally wrong and not permissible. One survey respondent noted, "The law sends a message that no kind or type of harassment or hiring discrimination against gays will be tolerated" (p. 118). Respondents in this study indicated that through public hearings and publicity surrounding proposed antibias legislation, gay and lesbian groups were able to describe how they have been discriminated against by employers, landlords, the police, and others. "With many Americans expressing ignorance and confusion about the degree to which gays are stigmatized and faced with discrimination, these public messages provided an invaluable means by which to educate the public" (p. 130).

Based on their research, Button, Rienzo and Wald concluded that antidiscrimination legislation has had only modest effects on police prejudice and discrimination against lesbians and gay men. "Indications of progess in law enforcement . . . seem fragile and incomplete . . . Police raids on gay gatherings still occur, often with violence and abuse" (p. 124). According to a Cincinnati city official, "There's still real prejudice here against gays, and intolerance among police on the street. . . . Gay jokes are still common. The police don't treat gays well generally" (p. 124).

The majority (51 percent) of survey respondents in this study reported no negative effects of antidiscrimination legislation. However, "a modest number of surveyed public officials mentioned that the debate over and passage of legislation protecting gays resulted in divisions in the community, greater controversy or tension, or the increased mobilization of those opposed to gay rights" (p. 120).

SOURCE: James W. Button, Barbara A. Rienzo, and Kenneth D. Wald. 1997. *Private Lives, Public Conflicts: Battles over Gay Rights in American Communities.* Washington, D.C.: CQ Press.

drinking and regular use of marijuana, cocaine, and other illegal drugs. Students reporting same-sex behavior were also more likely to be victimized by hate crimes and were more than three times as likely to report not going to school because they felt unsafe. These findings suggest that if schools are to promote the health and well-being of all students, they must address the needs of gay, lesbian, and bisexual youth and promote acceptance of sexual orientation diversity within the school setting. Otherwise, students will continue to experience the harassment and rejection—which lead to violence, substance abuse, and suicide—reflected in the examples in this chapter's *The Human Side* feature.

THE HUMAN SIDE

VOICES OF GAY AND LESBIAN YOUTH IN SCHOOL

The following excerpts describe the experiences of gay, lesbian, and bisexual youth in the school setting (cited in Button, Rienzo, and Wald 1997, 140–41, and Mathison 1998, 3–4).

> I was repeatedly tripped—everyday—in my classrooms. When students tripped me, they would call me "faggot," "pussy," etc. Nearly everyday, someone would push up against me. Out of 8 teachers, I had only one teacher who would intervene in these anti-gay incidents. (16-year-old male)
>
> I had a history teacher tell me that I couldn't come to class because I was gay. . . . She wanted me out of there. Teachers would often gossip and make anti-gay hand motions . . . I expected teachers to help me out and to protect me—not to be the main ones to ridicule me. (17-year-old male)
>
> One health teacher tells her students that homosexuality is wrong and evil. The other health teachers avoid the subject when teaching sex education. Almost all of my teachers have at one point in the year made some kind of joke about homosexuals or have made a derogatory comment. Some do it on a regular basis. (female high school student)
>
> In my school, out of 2000 students, I only knew five other gay students. I felt very isolated, very alone, and I had no one I could turn to for support. (18-year-old female)
>
> I felt like I didn't belong anymore. I felt like I was a small child in a sea of hate and that people didn't know me; they only knew of my sexual preference for women. (18-year-old female)
>
> There were these junior bikers that just talked about how they were going to beat up fags whenever they found them . . . and I would just be listening to it and every now and then they would try to include me in their conversation. I would go "ah, well, you know, I just live and let live." And then they say, "I hear a wimp," and I hated being there because they would try to drag me down a road I didn't wanna go. If only they knew. (Juan)

One strategy for promoting tolerance for diversity among students involves establishing and enforcing a school policy prohibiting antigay behavior. In 1994, the Massachusetts state legislators passed such a policy prohibiting discrimination against gay and lesbian students. This law permits students who have suffered antigay discrimination, and who were not protected by the school administration, to bring lawsuits against their schools.

Another strategy for addressing the needs of homosexual and bisexual youth is having school-based support groups. Such groups can help students increase self-esteem, overcome their sense of isolation, provide information and resources, and provide a resource for parents.

School counselors may be trained to work with gay, lesbian, and bisexual youth and their parents. Education about sexual orientation can be implemented in sex education or health education classes or in conflict resolution or diversity curricula. In-service training for teachers and other staff is important and may include examining the effects of antigay bias, dispeling myths about homosexuality, and brainstorming ways to create a more inclusive environment (Mathison 1998). However, most public schools offer little to no support and education regarding sexual orientation diversity. Most schools have no support groups or special counseling services for gay and lesbian youth, and the majority of schools do not have any policies prohibiting antigay harassment (Button, Rienzo, and Wald 1997).

Campus Policies Regarding Homosexuality

Student groups have been active in the gay liberation movement since the 1960s. More than 400 gay student groups are organized in community colleges, and public gay student groups have amassed a body of judicial opinion protecting their right to assembly under the First Amendment (D'Emilio 1990).

D'Emilio (1990) suggests that colleges and universities have the ability and the responsibility to promote gay rights and social acceptance of homosexual people:

> **For reasons that I cannot quite fathom, I still expect the academy to embrace higher standards of civility, decency, and justice than the society around it. Having been granted the extraordinary privilege of thinking critically as a way of life, we should be astute enough to recognize when a group of people is being systematically mistreated. We have the intelligence to devise solutions to problems that appear in our community. I expect us also to have the courage to lead rather than follow. (p. 18)**

Among the policies colleges and universities might adopt to meet this challenge are the following:

1. Develop outreach programs to provide support for victims of harassment and violence
2. Provide training of local law enforcement personnel in lesbian and gay issues
3. Create institutional policies that clearly affirm the unacceptability of discrimination on the basis of sexual orientation

UNDERSTANDING **SEXUAL ORIENTATION**

As both functionalists and conflict theorists note, alternatives to a heterosexual lifestyle are threatening, for they challenge traditional definitions of family, childrearing, and gender roles. The result is economic, social, and legal discrimination by the majority. Gay, lesbian, and bisexual individuals are also victimized by hate crimes. In some countries, homosexuality is formally sanctioned with penalties ranging from fines to imprisonment and even death.

In the past, homosexuality was thought to be a consequence of some psychological disturbance. More recently, some evidence suggests that homosexuality, like handedness, may have a biological component. The debate between biological and social explanations is commonly referred to as the "nature versus nurture" debate. Research indicates that both forces affect sexual orientation, although debate over which is dominant continues. Sociologists are interested in society's response to sexual orientation diversity and how that response impacts the quality of life of society's members. Because individuals' views toward homosexuality are related to their beliefs about what "causes" it, the question of the origins of sexual orientation diversity has sociological significance.

> We must never let ourselves believe that our diversity is a weakness, for it is our greatest strength.
>
> —Bill Clinton
> In a 1998 letter to the National Black Lesbian and Gay Leadership Forum

Prejudice and discrimination toward sexual orientation minorities are rooted in various aspects of culture, such as religious views, rigid gender roles, and negative myths and stereotypes. Hate crimes against homosexuals and bisexuals and lack of protection against job discrimination are examples of discrimination that even those who disapprove of homosexuality do not endorse.

The gay rights movement has made significant gains in the last few decades and has suffered losses and defeat due to gay rights opposition groups and politicians. Politicians, religious leaders, courts, and educators will continue to make decisions that either promote the well-being of sexual orientation minorities or hinder it. Ultimately, however, each individual must decide to embrace either an inclusive or an exclusive ideology; collectively, those individual decisions will determine the future treatment of sexual minorities in the United States.

CRITICAL THINKING

1. How do you think the legalization of same-sex marriages in the United States would affect public attitudes toward homosexuals?
2. Through the use of the Internet, sexual orientation minorities today can readily gain access to support organizations and networks. How do you think this use of the Internet has affected the gay rights movement?
3. How is the homosexual population similar to and different from other minority groups?
4. Do you think that social acceptance of homosexuality leads to the creation of laws that protect lesbians and gays? Or does the enactment of laws that protect lesbians and gays help to create more social acceptance of gays?
5. In the Virginia court ruling that denied Sharon Bottoms's custody of her son, the court awarded custody to Sharon's mother. Some people support this court decision on the grounds that they believe that lesbian mothers are more likely to influence their children to be homosexual. Do you see any contradictions in the court's ruling in the Sharon Bottoms case?

KEY TERMS

biphobia
bisexuality
Defense of Marriage Act
domestic partnerships
Employment Non-Discrimination Act
heterosexism
heterosexuality
homophobia
homosexuality
internalized homophobia
lesbians
master status
sexual orientation
sodomy laws

You can find more information on religion and homosexuality, the Human Rights Campaign, the National Gay and Lesbian Task Force, the Bisexual Resource Center, and the National Federation of Parents and Friends of Lesbians and Gays at the *Understanding Social Problems* website at **http://sociology.wadsworth.com**.

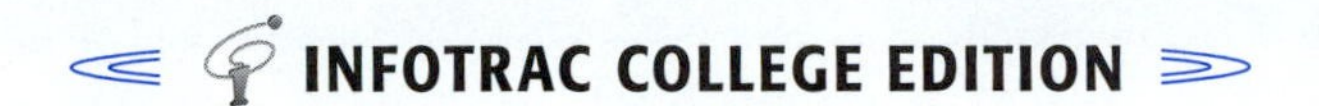

Either from the Wadsworth sociology resource center at **http://sociology.wadsworth.com** or directly from your web browser, you may access InfoTrac College Edition, an online university library that includes over 700 popular and scholarly journals in which you can find articles related to the topics in this chapter. Suggested articles and questions relating to these articles are listed below.

Pillard, Richard C. and J. Michael Bailey. 1998. "Human Sexual Orientation Has a Heritable Component." *Human Biology* 70(2):347–66.

1. Where did the oft-cited statistic "one out of every 10 men are exclusively homosexual" come from?
2. What is "gender atypicality" and how is it evidenced in childhood and adulthood?
3. What methods have researchers used in their search for the gay gene?

Postrel, Virginia. 1998. "The Claims of Nature: The 'Can Gays Change' Debate." *Reason* 30(October)(5):4–7.

1. What problem does the author see with claims of homosexuals being converted to heterosexuality?
2. What was the homosexual community's reaction to such claims of conversion?
3. What parallel does the author see between homosexuals and some American Jews?

Weissberg, Robert. 1998 (November–December). "The Abduction of Tolerance." *Society* 36(1):8–15.

1. What is the classical versus contemporary meaning of tolerance?
2. Why does the author believe that we, as a democracy, need to be tolerant?
3. What measures do educators feel need to be taken in order to educate future generations about tolerance?

Section 3

Problems of Inequality and Power

CHAPTER 10
The Haves and the Have-Nots

CHAPTER 11
Work and Unemployment

CHAPTER 12
Problems in Education

CHAPTER 13
Cities in Crisis

The story of Harrison Bergeron is set in a futuristic society where absolute equality is rigidly enforced (Vonnegut 1968). If people can run faster, they must wear weights in their shoes; if they are brighter, disruptive transistors are implanted in their brains; if they have better vision, blinders are placed over their eyes. The point of the story is that equality, although a cultural ideal, in reality is not always the optimal way to live. The quality of life becomes so unbearable for Harrison Bergeron that he decides that it is better to commit one courageous act of grace and beauty and be killed than to live in a society where everyone is equal.

Differences between people in and of themselves are not what is meant by inequality as a social problem, for few would want to live in a society where everyone was the same. Rather, problems of inequality concern inequities in the quality of life—between the haves and the have-nots (Chapter 10). Social inequality affects the opportunity to work and prosper (Chapter 11), to attend quality schools (Chapter 12), and to live in a healthy

and safe environment (Chapter 13).

To a large extent, differences between the rich and the poor are a consequence of changes in the nature of work and the resulting increase in unemployment for some people, in some jobs, in some parts of the world. A shift to a global economy and the accompanying deconcentration and deindustrialization of urban areas has resulted in high inner city unemployment, displaced workers, deteriorating neighborhoods, and eroding tax bases. Those who can get out do so, as evidenced by patterns of suburbanization; those who must remain are involuntarily segregated.

As a consequence of these urban processes, city schools, which are heavily dependent on municipal funds, attract fewer and fewer good teachers, facilities erode, the quality of education declines, and the number of poor minority students increases. Debates begin over the relevance of traditional curricula and the need for multiculturalism. Students graduate without knowing how to read, work simple math problems, or resolve interpersonal conflicts.

The chapters in Section 3—"The Haves and the Have-Nots," "Work and Unemployment," "Problems in Education," and "Cities in Crisis"—are highly interrelated and speak to the need for examining both the cultural and the structural underpinnings of the social problems described. Further, all three of the sociological perspectives, structural-functionalism, conflict theory, and symbolic interactionism, are used to understand various aspects of the four problem areas discussed.

CHAPTER TEN

The Haves and the Have-Nots

- The Global Context: Poverty and Economic Inequality around the World
- Sociological Theories of Poverty and Economic Inequality
- Wealth, Economic Inequality, and Poverty in the United States
- Consequences of Poverty and Economic Inequality
- *The Human Side: Life on the Streets: New York's Homeless*
- Strategies for Action: Antipoverty Programs, Policies, and Proposals
- *Focus on Social Problems Research: Making Ends Meet: Survival Strategies among Low-Income and Welfare Single Mothers*
- *Self & Society: Reasons for Contributing or Not Contributing to Charity*
- Understanding the Haves and the Have-Nots

Is It True?

Is It True?

1. The United States has the lowest poverty rate of all industrialized nations.
2. About a quarter of the world's population live in dire poverty.
3. In 1998, a U. S. worker with a full-time minimum-wage job could keep a family of three above the poverty line.
4. In the mid-1990s, CEOs in the United States earned 135 times what the average factory worker earned.
5. The age group with the highest poverty rate in the United States includes individuals aged 65 and older.

Answers: 1 = F; 2 = T; 3 = F; 4 = T; 5 = F

> **If all of the afflictions of the world were assembled on one side of the scale and poverty on the other, poverty would outweigh them all.**
>
> Rabba, Mishpatim 31:14

The blockbuster success of the 1997 film *Titanic* depicted how the luxury liner divided passengers into the "haves" and the "have-nots." Scenes of lower-deck passengers standing in water behind locked gates emphasized the fatal consequences of not having the resources to afford first-class passage and lifeboat access. Of the 1,500 who died in the icy waters of the Atlantic, most were from the lower decks.

Despite the fact that the United States is a nation of wealth, U.S. society is characterized by persistent economic inequalities that divide the population into haves and have-nots. This chapter examines the extent of poverty globally and in the United States, focusing on the consequences of poverty for individuals, families, and societies. Theories of poverty and economic inequality are presented and strategies for rectifying economic inequality and poverty are considered.

THE GLOBAL CONTEXT: POVERTY AND ECONOMIC INEQUALITY AROUND THE WORLD

Who are the poor? Are rates of world poverty increasing, decreasing, or remaining stable? The answers depend on how we define and measure poverty.

Defining and Measuring Poverty

Poverty can be loosely defined as lacking resources for an "adequate" standard of living. Some sociologists make a distinction between absolute poverty and relative poverty. **Absolute poverty** is characterized by the lack of basic necessities of life, such as food, clean water, and housing. **Relative poverty** refers to having a deficiency in material and economic resources *compared with* some other population. However, even absolute poverty is, to some degree, relative, because standards for adequacy vary among and within societies and change across time. Thus, there is no universally accepted objective definition of poverty. Although starvation may be globally accepted as an indicator of poverty, other living conditions are not widely agreed upon as indicators of poverty. For example, while living in a hut without indoor plumbing and electricity may be an indicator of poverty in the United States, such living conditions are considered normal and adequate among certain populations in less developed countries. In less developed countries where large segments of the population die of starvation and infectious diseases, poverty is equated with living on the brink of survival. Yet, the 13 percent of Americans who were classified as living in poverty in 1997 did not face starvation, though many had inadequate diets.

Various measures of poverty are used by governments, researchers, and organizations. Next, we discuss ways in which poverty is measured throughout the world and in the United States.

International Measures of Poverty The World Bank sets a "poverty threshold" at $1 a day to compare poverty in most of the developing world, labeling population groups with a per capita income above $1 a day as "nonpoor."

A poverty line of $2 a day is often used for Latin America and the Caribbean; $4 a day in Eastern Europe and the Commonwealth of Independent States (CIS); and $14.40 a day in industrial countries (which corresponds to the

> Whether there is proved to be life on Mars, and whether you may conduct your affairs electronically without leaving your armchair, the new century is not going to be a new century at all in terms of progress of humanity if we take along with us acceptance of the shameful shackles of the past. The shackles of poverty are not just a metaphor.
>
> —Nadine Gordimer
> Speaking at the United Nations on the occasion of International Day for the Eradication of Poverty

> Human poverty is more than income poverty—it is the denial of choices and opportunities for living a tolerable life.
>
> —*Human Development Report 1997*

> I didn't agree to take a vow of poverty.
>
> —William Bennett
> Declining to take a job as national chairman of the Republican party at a salary of $125,000 per year

The accumulation of material goods is at an all-time high, but so is the number of people who feel an emptiness in their lives.

—Al Gore
U.S. vice president

income poverty line in the United States). Another poverty measure used by the World Health Organization (WHO) is based on a household's ability to meet the minimum calorie requirements of its members. According to this poverty measure, a household is considered poor if it cannot meet 80 percent of the minimum calorie requirements (established by WHO), even when using 80 percent of its income to buy food.

In industrial countries, national poverty lines are sometimes based on the median household income of a country's population. According to this relative poverty measure, members of a household are considered poor if their household income is less than 50 percent of the median household income in that country.

In the *Human Development Report 1997*, the United Nations Development Programme proposes a new composite measure of poverty: the **Human Poverty Index (HPI)**. Rather than measure poverty by income, three measures of deprivation are combined to yield the Human Poverty Index: (1) deprivation of life, which is measured by the percentage of people expected to die before the age of 40; (2) deprivation of knowledge, which is measured by the percentage of adults who are illiterate; and (3) deprivation in living standards, which is measured as a composite of three variables: the percentage of people without access to health services, the percentage of people without access to safe water, and the percentage of malnourished children under 5. The Human Poverty Index is a useful complement to income measures of poverty and "will serve as a strong reminder that eradicating poverty will always require more than increasing the income of the poorest" (*Human Development Report* 1997, 19).

International Data

An estimated 1.3 billion people live on $1 a day.

SOURCE: *Human Development Report* 1997

U.S. Measures of Poverty In 1964 the Social Security Administration devised a poverty index based on a 1955 Agriculture Department survey that estimated the cost of an economy food plan for a family of four. Because families with three or more members spent one-third of their income on food at the time, the poverty line was set at three times the minimum cost of an adequate diet. Poverty thresholds differ by the number of adults and children in a family and, for some family types, by the age of the family head of household. Poverty thresholds are adjusted each year for inflation. Anyone living in a household with income (before tax) below the official poverty line is considered "poor." Individuals living in households that are above the poverty line, but not very much above it, are classified as "near poor."

National Data

The 1998 U.S. poverty threshold for one person (under age 65) was $8,480. For two people (no children) the poverty line was $10,915 and for three persons (two adults, one child) the poverty line was $13,120. The poverty line for a couple with two children was $16,530.

SOURCE: U.S. Census Bureau 1999

The U.S. poverty line has been criticized on several grounds. First, the poverty line is based on the assumption that low-income families spend one-third of their household income on food. That was true in 1955, but because other living costs (such as housing, medical care, and child care) have risen more rapidly than food costs, low-income families today spend closer to one-fifth (rather than one-third) of their income on food (Pressman 1998). So, current poverty lines should be based on multiplying food costs by five rather than three. This would raise the official poverty line by two-thirds, making the poverty level consistent with public opinion regarding what a family needs to escape poverty.

Another shortcoming of the official poverty line is that it is based solely on money income and does not take into consideration noncash benefits received by many low-income persons, such as food stamps, Medicaid, and public housing. Family assets, such as savings and property, are also excluded in official poverty calculations. The poverty index also fails to account for tax burdens that affect the amount of disposable income available to meet basic needs. The U.S. poverty line also disregards regional differences in the cost of living. Lastly, because poverty rates are based on surveys of households, the homeless—the most destitute of the poor—are not counted among the poor.

The Extent of Global Poverty and Economic Inequality

Economic and social progress in recent decades has eliminated mass poverty in economically advanced countries and significantly reduced poverty in many developing countries. The 1997 *Report on the World Social Situation* notes that "infant mortality has fallen almost steadily in all regions and life expectancy has risen all over the globe. Educational attainment is rising, health care and living conditions are improving in most countries and the quantity, quality and range of goods and services available to a large majority of the world's population is increasing" (United Nations 1997, 80). But the report goes on to say that "not everyone has shared in this prosperity. Economic growth has been slow or non-existent in many of the world's poorest countries . . . The plight of the poor stands in stark contrast to the rising standards of living enjoyed by those favoured by growing abundance" (p. 80).

For those who live in a wealthy nation such as the United States, it may be difficult to imagine the depth and scope of poverty in many less developed countries throughout the world. According to the *Human Development Report 1997*, about one-third of the population living in developing countries—1.3 billion people—live on incomes of less than $1 a day.

South Asia has the most people affected by poverty; sub-Saharan Africa has the highest proportion of people in poverty. In 2000, an estimated half of people in sub-Saharan Africa will be living in poverty. Eastern Europe and the countries of the Commonwealth of Independent States have experienced the greatest increase in poverty in the last decade. In that time period, the proportion of people living in income poverty grew from a small part of the population to about a third. Between 1992 and 1998, the average Russian household lost more than one-half of its income, and the male life expectancy dropped from 65.5 years to 57 years (Weisbrot 1998).

Poverty also affects industrial countries, where 100 million people live below the income poverty line (set at half the individual median income). More than 37 million are jobless, and more than 5 million people in industrialized countries are homeless (*Human Development Report* 1997).

Globally, income inequality has increased, widening the gap between the haves and the have-nots. In 1960, the ratio of the income of the richest 20 percent of the world to that of the poorest 20 percent was 30 to 1. In 1994, this ratio had increased to 78 to 1 (*Human Development Report* 1997).

International Data

The World Bank estimates that about a quarter of the world's population live in dire poverty.

SOURCE: *Human Development Report* 1997

International Data

In 1997, the poorest 20 percent of the world's people accounted for a meager 1.1 percent of global income, down from 1.4 percent in 1991 and 2.3 percent in 1960.

SOURCE: *Human Development Report* 1997

SOCIOLOGICAL THEORIES OF POVERTY AND ECONOMIC INEQUALITY

The three main theoretical perspectives in sociology—structural-functionalism, conflict theory, and symbolic interactionism—offer insights into the nature, causes, and consequences of poverty and economic inequality.

> In a nation as smart, inventive, and rich as America, the continuation of widespread poverty is a choice, not a necessity.
>
> —Michael Katz
> University of Pennsylvania

Structural-Functionalist Perspective

According to the structural-functionalist perspective, poverty and economic inequality serve a number of positive functions for society. Decades ago, Davis and

> We talk about the American Dream, and want to tell the world about the American Dream, but what is that dream, in most cases, but the dream of material things? I sometimes think that the United States for this reason, is the greatest failure the world has ever seen.
>
> —Eugene O'Neill
> Playwright

Moore (1945) argued that because the various occupational roles in society require different levels of ability, expertise, and knowledge, an unequal economic reward system helps to assure that the person who performs a particular role is the most qualified. As people acquire certain levels of expertise (e.g., B.A., M.A., Ph.D., M.D.), they are progressively rewarded. Such a system, argued Davis and Moore, motivates people to achieve by offering higher rewards for higher achievements. If physicians were not offered high salaries, for example, who would want to endure the arduous years of medical training and long, stressful hours at a hospital?

The structural-functionalist view of poverty suggests that a certain amount of poverty has positive functions for society. Although poor people are often viewed as a burden to society, having a pool of low-paid, impoverished workers ensures that there will be people willing to do dirty, dangerous, and difficult work that others refuse to do. Poverty also provides employment for those who work in the "poverty industry" (such as welfare workers) and supplies a market for inferior goods such as older, dilapidated homes and automobiles (Gans 1972).

The structural-functionalist view of poverty and economic inequality has received a great deal of criticism from contemporary sociologists, who point out that many important occupational roles (such as child care workers) are poorly paid, whereas many individuals in nonessential roles (such as professional sports stars and entertainers) earn astronomical sums of money. Functionalism also accepts poverty as a necessary evil and ignores the role of inheritance in the distribution of rewards.

Conflict Perspective

> Wealth . . . means power, it means leisure, it means liberty.
>
> —James Russell Lowell
> American poet

Conflict theorists regard economic inequality as resulting from the domination of the **bourgeoisie** (owners of the means of production) over the **proletariat** (workers). The bourgeoisie accumulate wealth as they profit from the labor of the proletariat, who earn wages far below the earnings of the bourgoisie. The U.S. educational institution furthers the ideals of capitalism by perpetuating the belief in equal opportunity, the "American Dream," and the value of the work ethic. The proletariat, dependent on the capitalistic system, continue to be exploited by the wealthy and accept the belief that poverty is a consequence of personal failure rather than a flawed economic structure.

Conflict theorists pay attention to how laws and policies benefit the wealthy and contribute to the gap between the haves and the have-nots. Laws and policies that favor the rich—sometimes referred to as **wealthfare** or **corporate welfare**—include low-interest government loans to failing businesses, special subsidies and tax breaks to corporations, and other laws and policies that benefit corporations and the wealthy. For example, wealthy home owners can deduct up to $1 million in mortgage interest. Lowering this ceiling to $250,000 would affect the wealthiest 5 percent of Americans, but would save taxpayers $10 billion a year (reported in Albelda and Tilly 1997). A 1998 *Time* magazine series of special reports on corporate welfare gave national visibility to the issue. In one report, *Time* revealed that between 1990 and 1997, Seaboard Corporation, an agribusiness corporate giant, received at least $150 million in economic incentives from federal, state, and local governments to build and staff poultry- and hog-processing plants in the United States, support its operations in foreign countries, and sell its products (Barlett and Steele 1998). Taxpayers picked up the tab not just for the corporate welfare, but also for the costs of new classrooms and teachers (for schooling the children of

Seaboard's employees, many of whom are immigrants), homelessness (due to the inability of Seaboard's low-paid employees to afford housing), and dwindling property values resulting from smells of hog waste and rotting hog carcasses in areas surrounding Seaboard's hog plants. Meanwhile, wealthy investors in Seaboard have earned millions in increased stock values.

Corporations and wealthy individuals buy political influence with huge financial contributions to politicians and political candidates. For example, in the 1990s the timber industry gave more than $8 million in political contributions. Since 1991, the Forest Service purchaser road credit program has given the timber industry a $458-million subsidy to build logging roads in national forests. In 1997, the Clinton budget proposed to eliminate this subsidy, but instead Congress not only saved, but expanded the timber industry's logging subsidy at the expense of the U.S. taxpayer (McBride 1998). Another example of corporate welfare involves the pharmaceutical industry, which gave more than $18 million in political contributions in the 1990s. Congress voted to let drugmakers hold on to their drug patents longer, costing consumers as much as $550 million a year from loss of access to generic drugs (Common Cause 1998). Securities and investment interests—a top contributor to both political parties in the 1995–96 election cycle—received significant victories in the 1997 budget and tax deal, including a long-sought cut in the capital gains tax rate (Common Cause 1997). These examples illustrate how wealthy interests use money to influence politicians to enact corporate welfare policies that benefit the wealthy and increase economic inequality.

Conflict theorists also note that throughout the world, "free-market" economic reform policies have been hailed as a solution to poverty. Yet, while such economic reform has benefited many wealthy corporations and investors, it has also resulted in increasing levels of global poverty. As companies relocate to countries with abundant supplies of cheap labor, wages decline. Lower wages lead to decreased consumer spending, which leads to more industries closing plants, going bankrupt, and/or laying off workers (downsizing). This results in higher unemployment rates and a surplus of workers, enabling employers to lower wages even more. Chossudovsky (1998) suggests that "this new international economic order feeds on human poverty and cheap labor" (p. 299). Yet, the increasing levels of global poverty are masked by World Bank poverty statistics that, according to Chossudovsky (1998), "blatantly misrepresent . . . the seriousness of global poverty" in order to portray a positive view of free-market reform policies (p. 298).

Symbolic Interactionist Perspective

Symbolic interactionism focuses on how meanings, labels, and definitions affect and are affected by social life. This view calls attention to ways in which wealth and poverty are defined and the consequences of being labeled as "poor." Individuals who are viewed as poor—especially those receiving public assistance (i.e., welfare)—are often stigmatized as lazy; irresponsible; and lacking in abilities, motivation, and moral values. Wealthy individuals, on the other hand, tend to be viewed as capable, motivated, hard working, and deserving of their wealth.

As noted earlier, definitions of wealth and poverty vary across societies and across time. For example, the Dinka are the largest ethnic group in the sub-Saharan African country of Sudan. By global standards, the Dinka are among the poorest of the poor, being among the least modernized peoples of the

> In today's economy a woman is considered lazy when she's at home taking care of her children. And to me that's not laziness . . . Let some of these men that work in the government, let some of them stay home and do that. They'll find that a woman is not lazy when she's taking care of her family.
>
> —Denise Turner
> Welfare recipient

world. In the Dinka culture, wealth is measured in large part according to how many cattle a person owns. But, to the Dinka, cattle have a social, moral, and spiritual value as well as an economic value. In Dinka culture, a man pays an average "bridewealth" of 50 cows to the family of his bride. Thus, men use cattle to obtain a wife to beget children, especially sons, to ensure continuity of their ancestral lineage and, according to Dinka religious beliefs, their linkage with God. Although modernized populations might label the Dinka as poor, the Dinka view themselves as wealthy. As one Dinka elder explained, "It is for cattle that we are admired, we, the Dinka . . . All over the world, people look to us because of cattle . . . because of our great wealth; and our wealth is cattle" (Deng 1998, 107). Deng (1998) notes that many African peoples who are poor by U.S. standards resist being labeled as poor.

The symbolic interactionist perspective emphasizes that norms, values, and beliefs are learned through social interaction. Social interaction also influences the development of one's self-concept. Lewis (1966) argued that, over time, the poor develop norms, values, and beliefs and self-concepts that contribute to their own plight. According to Lewis, the **culture of poverty** is characterized by female-centered households, an emphasis on gratification in the present rather than in the future, and a relative lack of participation in society's major institutions. "The people in the culture of poverty have a strong feeling of marginality, of helplessness, of dependency, of not belonging . . . Along with this feeling of powerlessness is a widespread feeling of inferiority, of personal unworthiness" (Lewis 1998, 7). Early sexual activity, early marriage, and unmarried parenthood are considered normal and acceptable among individuals living in a culture of poverty. Certain groups, according to this view, remain poor over time as the culture of poverty is transmitted from one generation to the next. Critics of the culture of poverty approach argue that it blames the victim rather than the structure of society for poverty, justifies the status quo, and perpetuates inequality (Ryan 1992).

> **In an environment such as the American Black ghetto, where families with a steady, employed breadwinner have become the exception rather than the rule, the chance to contact families and institutions that represent conventional role models is small. Hence, people are not introduced into jobs, and they do not learn from experience about the behavior and norms that belong to steady work and stable family life. Being surrounded by people who have to rely on other, often illegal strategies to survive, these strategies come to be seen as a way of life. (Van Kempen 1997, 434)**

WEALTH, ECONOMIC INEQUALITY, AND POVERTY IN THE UNITED STATES

The United States is a nation of tremendous economic variation ranging from the very rich to the very poor. Signs of this disparity are visible everywhere, from opulent mansions perched high above the ocean in California to shantytowns in the rural South where people live with no running water or electricity.

National Data

Nearly 20 percent of U.S. households possess no net wealth (debts offset the value of assets), or have negative wealth (their debts exceed their assets).
SOURCE: Weinstein 1999

Wealth in the United States

Wealth refers to the total assets of an individual or household, minus liabilities (mortgages, loans, and debts). Wealth includes the value of a home, investment real estate, the value of cars, unincorporated business, life insurance (cash value), stocks/bonds/mutual funds/trusts, checking and savings accounts, individual retirement accounts (IRAs), and valuable collectibles.

According to a 1998 Forbes magazine report, Bill Gates, chief of Microsoft Corporation, was the wealthiest American for the fifth year in a row ("Bill Gates Tops Forbes Billionaires for 5th Year" 1998). Even after the 500-point stock market drop in August of 1998, Gates's net worth was $58.4 billion. Investor Warren Buffet was second at $29.4 billion. In September 1998, there were 189 billionaires in the United States, up from 170 in 1997. Forbes reported that of the richest 400 U.S. individuals in 1998, only 58 were women.

Bill Gates can lose more in an hour in the stock market than most people earn in a lifetime.

Economic Inequality in the United States

The 1990s was a decade of U.S. economic growth; interest rates were down, unemployment low, and stock market averages reached record levels. At the close of the twentieth century, the United States had experienced the longest period of peacetime economic expansion in history. But contrary to the adage that "a rising tide lifts all boats," economic prosperity has not been equally distributed in the United States. Economic inequality—the gap between the haves and the have-nots—has increased considerably.

From 1950 to 1978, all U.S. social classes enjoyed increases in economic prosperity. Family income for the bottom fifth of the U.S. population increased substantially more than for the top fifth of the population (a 138 percent increase for the former versus a 99 percent increase for the latter) (Briggs 1998). However, between 1979 and the end of the 1990s, inflation-adjusted income of the top 20 percent of the population grew by 26 percent while for the poorest it *decreased* by 9 percent.

In 1997, U.S. Secretary of Labor Robert Reich observed that, "over 15 years ago, inequality of income, wealth, and opportunity began to widen and the gap today is wider than at any time in living memory" (Reich 1997, E-13). Labor economist Richard Freeman noted that "inequality has jumped to levels that raise doubts about the health of the U.S. economy and its ability to deliver to all the American dream of rising living standards" (Freeman 1997, 1).

National Data

In 1998, the wealthiest 20 percent of U.S. households earned nearly 50 percent of aggregate income. In contrast, the poorest 20 percent of U.S. households earned less than 4 percent.

SOURCE: Lord 1998

Patterns of Poverty in the United States

Although poverty is not as widespread or severe in the United States as it is in many less developed countries, it nevertheless represents a significant social problem. Poverty is not equally distributed; certain demographic populations are more likely to experience poverty than others. Poverty rates vary according to age, education, gender, family structure, race/ethnicity, and labor force participation.

National Data

In 1970 the poverty rate among U.S. elderly (age 65+) was 24.6 percent; this rate fell to 15.7 percent in 1980, 12.0 percent in 1988, and 10.5 percent in 1997.

SOURCE: Levitan, Mangum, and Mangum 1998; U.S. Census Bureau 1998c

Age and Poverty

The age group most affected by poverty are children. The poverty rate for young children is at least one-third higher and usually two to three times as high in the United States as in any other Western industrialized nation (Levitan, Mangum, and Mangum 1998). In previous decades, the elderly have suffered high rates of poverty. Since 1970, poverty among the elderly has experienced a downward trend, largely due to more Social Security benefits and the growth of private pensions (see also Chapter 6).

Table 10.1 Poverty Rates of U.S. Adults Aged 25 and Older, by Educational Attainment, 1997

Level of Education	Percent Living in Poverty
No High School Diploma	24.5%
High School Diploma (no college)	9.9%
Some College (no bachelors degree)	6.5%
Bachelors Degree or More	3.1%

SOURCE: U.S. Census Bureau. 1998b. "Annual Demographic Survey, March Supplement. Table 7." **http://ferret.bls.census.gov/macro/031998/pov/new7_000.htm** (November 20, 1998)

National Data

In 1997, 13.3 percent of the U.S. population (35.6 million people) lived below the official poverty line. Poverty rates vary considerably among the 50 states, from 6.9 percent in New Hampshire to 24.0 percent in New Mexico.

SOURCE: U.S. Census Bureau 1998c

Education and Poverty

Education is one of the best insurance policies to protect against an individual living in poverty. In general, the higher a person's level of educational attainment, the less likely that person is to be poor (see also Chapter 12). Among adults over age 25, those without a high school diploma are the most vulnerable to poverty (see Table 10.1).

Gender and Poverty

Women are more likely than men to live below the poverty line—a phenomenon referred to as the **feminization of poverty** (see Figure 10.1). As discussed in Chapter 7, women are less likely than men to pursue advanced educational degrees and tend to be concentrated in low-paying jobs, such as service and clerical jobs. However, even with the same level of education and the same occupational role, women still earn significantly less than men. Women who are minorities and/or who are single mothers are at increased risk of being poor.

National Data

In 1997, 1 in 5 U.S. children under 18 lived below the poverty line.

SOURCE: U.S. Census Bureau 1998c

Family Structure and Poverty

Poverty is much more prevalent among female-headed single-parent households than among other types of family structures.

The relationship between family structure and poverty helps to explain why women and children have higher poverty rates than men. The high rate of divorce and out-of-wedlock births (see also Chapter 5) means that 60 percent of all children (and 90 percent of black children) will spend some time in a single-parent household (Levitan, Mangum, and Mangum 1998). The 1997 poverty rate of children under age 6 who lived in female-headed families with no spouse present was 59 percent (Center on Budget and Policy Priorities 1998b).

In other countries, poverty rates of female-headed families are lower than those in the United States. For example, poverty rates of female-headed households are less than 10 percent in Belgium, France, Great Britain, Ireland, Luxembourg, Netherlands, Norway, and Poland (Pressman 1998). Unlike the United States, these countries offer a variety of supports for single mothers,

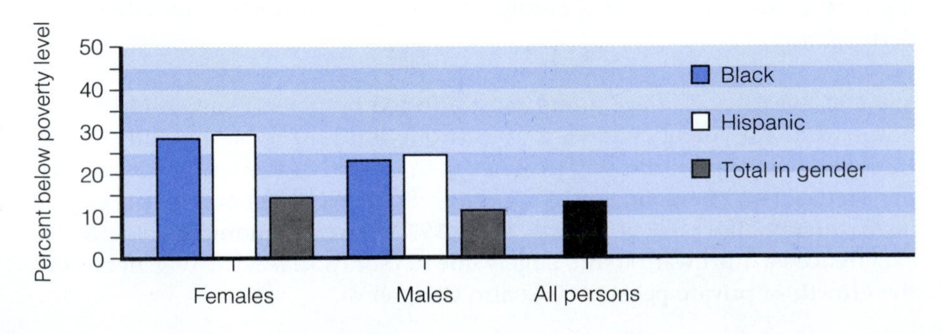

Figure 10.1 Poverty Status by Gender, Race, and Ethnicity, 1997

SOURCE: U.S. Census Bureau. 1998a. "Annual Demographic Survey, March Supplement." Table 1. **http://ferret.bls.census.gov/macro/031998/pov/new1_001.htm** (November 20, 1998)

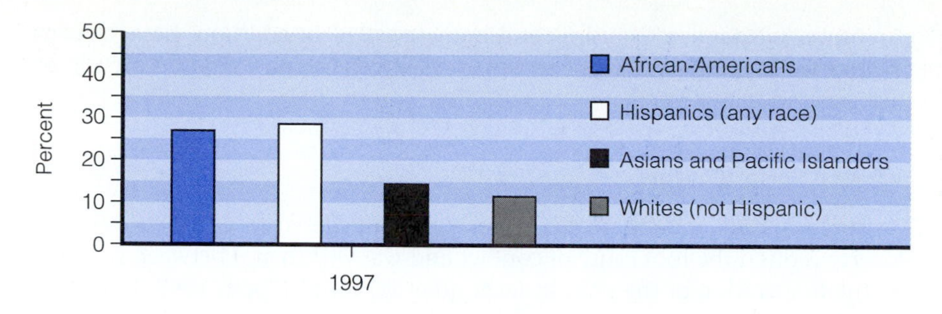

Figure 10.2 U.S. Poverty Rates by Race and Hispanic Ethnicity, 1997

SOURCE: U.S. Census Bureau. 1998c. *Poverty in the United States: 1997.* Washington, D.C.: U.S. Government Printing Office.

such as income supplements, tax breaks, universal child care, national health care, and higher wages for female-dominated occupations.

CONSIDERATION

Levitan, Mangum, and Mangum (1998) note that "it is still an open question whether family structure is the cause or the victim of poverty. How many males disappear because they cannot support the children they have fathered? Are single-parent families poor because they are headed by a woman, or because the wage structure allows few mothers to earn their way out of poverty?" (p. 22). Although poverty is often seen as resulting from the rise of single female–headed families, Hernandez (1997) argues that low earnings for fathers continue to be major determinants of childhood poverty, both because of their direct effect on family income and because of their indirect contribution to mother-only families.

National Data

The poverty rate of U.S. married-couple families was 5.2 percent in 1997, compared with 31.6 percent for families headed by a woman with no husband present.

SOURCE: U.S. Census Bureau 1998c

Race/Ethnicity and Poverty Nearly half (46 percent) of the U.S. poor are non-Hispanic whites (U.S. Census Bureau 1998c). However, as displayed in Figure 10.2, poverty rates are higher among blacks and Hispanics than among non-Hispanic whites.

Past and present discrimination has contributed to the persistence of poverty among minorities (see also Chapter 8). Other contributing factors include the loss of manufacturing jobs from the inner city, the movement of whites and middle-class blacks out of the inner city, and the resulting concentration of poverty in predominantly minority inner-city neighborhoods (Massey 1991; Wilson 1987). Finally, blacks and Hispanics are more likely to live in female-headed households with no spouse present—a family structure that is associated with high rates of poverty.

National Data

In 1996, one in five U.S. poor (7.4 million persons) were classified as "working poor."

SOURCE: Bureau of Labor Statistics 1997

Labor Force Participation and Poverty A common image of the poor is that they are jobless and unable or unwilling to work. Although the U.S. poor are primarily children and adults who are not in the labor force, many U.S. poor are classified as "working poor." The **working poor** are individuals who spend at least 27 weeks a year in the labor force (working or looking for work), but whose income falls below the official poverty level.

CONSEQUENCES OF POVERTY AND ECONOMIC INEQUALITY

Poverty and economic inequality have enormous negative consequences for individuals, families, communities, and societies. These include war and social conflict, health problems, problems in education, problems in families and

> Although I have made a fortune in the financial markets, I now fear that the untrammeled intensification of laissez-faire capitalism and the spread of market values into all areas of life is endangering our open and democratic society. The main enemy . . . is no longer the communist threat but the capitalist threat.
>
> —George Soros
> Billionaire financier

parenting, substandard housing, and homelessness. In addition, poverty contributes to the perpetuation of poverty across generations, feeding a cycle of intergenerational poverty.

War and Social Conflict

Modern wars are fought mainly in poorer countries (see also Chapter 16). Poverty is often the root cause of conflict and war within and between nations, as "the desperation of the poor is never quiet for long" (Speth 1998, 281). Not only does poverty breed conflict and war, but war contributes to poverty. For example, war contributes to homelessness as individuals and families are forced to flee from their homes. Military and weapons spending associated with war also diverts resources away from economic development and social spending on health and education.

Briggs (1998) suggests that the widening inequalities between the haves and the have-nots presents a threat to social order. He asks how long the United States can maintain social order "when increasing numbers of persons are left out of the banquet while a few are allowed to gorge?" (p. 474). Although Karl Marx predicted that the have-nots would revolt against the haves, Briggs does not foresee a revival of Marxism; "the means of surveillance and the methods of suppression by the governments of industrialized states are far too great to offer any prospect of success for such endeavors" (p. 476). Instead, Briggs predicts that American capitalism and its resulting economic inequalities will lead to social anarchy—a state of political disorder and weakening of political authority.

National Data

In 1960, CEOs earned 12 times the average factory wage. By 1974, the multiplier had risen to 35 and in the mid-1990s, CEOs earned 135 times what the average factory worker earns.

SOURCE: Piven 1996

Health Problems and Poverty

In Chapter 2, we noted that poverty has been identified as the world's leading health problem. Persistant poverty is associated with higher rates of infant mortality and childhood deaths and lower life expectancies among adults. Poverty often causes chronic malnutrition, which can result in permanent brain damage, learning disabilities, and mental retardation in infants and children (Hill 1998). Poor children and adults also receive inadequate and inferior health care, which exacerbates their health problems (see also Chapter 2). For example, poor children are less likely to receive vaccinations against childhood diseases than are nonpoor children.

Economic inequality also affects psychological and physical health. Streeten (1998) cited research that suggests that "perceptions of inequality translate into psychological feelings of lack of security, lower self-esteem, envy and unhappiness, which, either directly or through their effects on life-styles, cause illness" (p. 5). People who live in states with the greatest gap between the rich and the poor are much more likely to rate their own health as poor or fair than people who live in states where income is more equitably distributed (Kennedy et al. 1998). This finding was true not only among poor people, but among middle-income adults as well.

> No society can surely be flourishing and happy, of which the far greater part of the members are poor and miserable.
>
> —Upton Sinclair
> Author

Educational Problems and Poverty

The various adverse physical and mental outcomes of infant and childhood poverty combine to cause learning failure and to limit academic performance. Children from poor families score lower on tests of cognitive skill than children from affluent families (Mayer 1997a). Because school districts are determined by residential patterns, the poor often attend schools that are characterized by lower-quality facilities, overcrowded classrooms, and a higher teacher turnover rate (see also Chapter 12). Because poor parents have less schooling, on average, than do nonpoor parents, they may be less able to encourage and help their children succeed in school. However, research suggests that family income is a stronger predictor of ability and achievement outcomes than are measures of parental schooling or family structure (Duncan and Brooks-Gunn 1997). Poor parents have fewer resources to provide educational experiences (such as travel), private tutoring, books, and computers for their children. Not surprisingly, poor students are less likely to graduate from high school or go to college, creating a vicious cycle whereby subsequent generations remain poor over time.

Poverty also presents obstacles to educational advancement among poor adults. Women and men who want to further their education in order to escape poverty may have to work while attending school or may be unable to attend school due to unaffordable child care, transportation, and/or tuition/fees/books.

National Data

For every year that 14.5 million U.S. children continue to experience poverty, their lifetime contribution to the economy will decline by an estimated $130 billion because poor children grow up to be less educated, less productive workers.

SOURCE: Children's Defense Fund 1998c

Family and Parenting Problems Associated with Poverty

In some cases, family problems contribute to poverty. For example, domestic violence causes some women to flee from their homes and live in poverty without the economic support of their husbands. Poverty also contributes to family problems. The stresses associated with poverty contribute to substance abuse, domestic violence, child abuse, divorce, and questionable parenting practices. For example, poor parents unable to afford child care expenses are more likely to leave children home without adult supervision. Poor parents are more likely than other parents to use harsh disciplinary techniques, such as physical punishment, more likely to value obedience, and less likely to be supportive of their children (Mayer 1997a).

Another family problem associated with poverty is teenage pregnancy. Poor adolescent girls are more likely to have babies as teenagers or become young single mothers. Early childbearing is associated with numerous problems, such as increased risk of premature or low-birthweight babies, dropping out of school, and lower future earning potential due to lack of academic achievement. Luker (1996) notes that "the high rate of early childbearing is a measure of how bleak life is for young people who are living in poor communities and who have no obvious arenas for success" (p. 189). For poor teenage women who have been excluded from the American dream and disillusioned with education, "childbearing . . . is one of the few ways . . . such women feel they can make a change in their lives . . ." (p. 182).

> **Having a baby is a lottery ticket for many teenagers: it brings with it at least the dream of something better, and if the dream fails, not much is lost. . . . In a few cases it leads to marriage or a stable relationship; in many others it motivates a woman to push herself for her baby's sake; and in still other cases it enhances the woman's self-esteem, since it enables her to do something productive, something nurturing and socially responsible . . . To the extent that babies can be ill or impaired, mothers can be unhelpful or unavailable, and boyfriends can be unreliable or punitive, childbearing can be just another risk gone wrong in a life that is filled with failures and losses. (Luker 1996, 182)**

Some cities and communities have laws or ordinances that prohibit the homeless from sleeping on public benches and soliciting money.

Substandard Housing and Homelessness

Housing conditions for the U.S. poor are largely inadequate. Many poor families and individuals live in housing units that lack central heating and air conditioning, sewer or septic systems, electric outlets in one or more rooms, and that have no telephone. Housing units of the poor are also more likely to have holes in the floor, a leaky roof, and open cracks in the wall or ceiling. In addition, poor individuals are more likely than the nonpoor to live in high-crime neighborhoods (Mayer 1997b).

Even substandard housing would be a blessing to many people who live without conventional housing—the homeless. Homelessness has become a growing problem in the United States. The homeless population is diverse (see Table 10.2). Some homeless individuals have been forced out of their houses or apartments by rising rents or the inability to pay the mortgage. The homeless population also includes individuals who have been released from mental hospitals as a result of the movement to deinstitutionalize individuals with psychiatric disorders. This chapter's *The Human Side* feature presents glimpses of what it is like to be homeless.

National Data

Over 2 million men, women, and children in the United States are homeless.

SOURCE: "Homelessness and Poverty in America" 1998

Table 10.2 Characteristics of the Homeless in the United States

25–40% work
37% are families with children
25% are children
25–30% are mentally disabled
30% are veterans
40% are drug or alcohol dependent

SOURCE: "Homelessness and Poverty in America." 1998. National Law Center on Homelessness and Poverty in the America. **http://www.nlchp.org/h&pusa.htm** (November 18, 1998)

THE HUMAN SIDE Human Side

LIFE ON THE STREETS: NEW YORK'S HOMELESS

While a sociology graduate student at Columbia University, Gwendolyn Dordick undertook a study of homeless people living in New York City. She spent 15 months with four groups of homeless people: inhabitants of a large bus terminal, residents of a shantytown, occupants of a large public shelter, and clients of a small, church-run private shelter. In this *Human Side* feature, we present some of Dordick's observations, as well as excerpts from her conversations with homeless individuals she encountered at the Station and the Shanty.

The Station

The Station, located in Manhattan's West Side, encompasses bus terminals and depots, ticketing windows, shops, and fast-food restaurants. Scattered among the commuters and visitors are homeless women and men, and young adolescent boys and girls who have either run away or were kicked out by their guardians. One homeless man described living on the edge:

> Living on the streets makes you do a lot of things that you wouldn't normally do . . . Comin' into this environment I've done a lot of things I said I wouldn't do . . . There was some people that came along in a van and just threw sandwiches on the street and I picked them up and ate them . . . The guilt almost killed me . . . but my stomach said, "Hey, listen, you better eat this food." (pp. 5–6)

Homeless individuals often rely on one another, offering each other companionship, friendship, and protection. Although they have little to share, they often share what they have with their fellow homeless friends.

> These people, when I got down here, these people reached out to me because they knew, they already knew what it was like. They're not afraid to help their fellow man. As soon as I got down here I met Ron and the fellows and they didn't push me away. I mean I didn't know where to go, I didn't know where to eat, I didn't know where to sleep. They just invited me right in. And ever since then at least I've been healthy, and I've been clean since I've met them. (p. 13)

Homeless individuals are often treated harshly by police. One homeless man lamented:

> You may have an invalid laying down here. He's got problems and the [Station] cops will come up and kick him. Like he's an animal with no rights. (p. 11)

Pregnancy is common among homeless women who do not have access to or cannot afford contraception. Pregnancy can have disastrous consequences for homeless women, who fear having a baby and having to care for a baby. According to one man in the Station:

> It's one thing being homeless, but pregnant and homeless? Some women have their babies right out here; others get rid of them . . . Some of them abort theirself by sticking hangers up their vaginas. I've seen that myself. This young girl didn't want a baby and she stuck a hanger up her vagina. She had to go to the hospital . . . (p. 25)

(continued)

THE HUMAN SIDE Human Side

LIFE ON THE STREETS: NEW YORK'S HOMELESS *(continued)*

The Shanty

A barricaded makeshift community of 20 to 25 residents sits on a formerly vacant lot visible from the nearby streets and a bridge that crosses the East River. The Shanty consists of 15 makeshift dwellings, or "huts" as they are called by the residents, which are made of a variety of discarded materials such as pieces of wood and boards, cardboard, mattresses, fabric, and plastic tarps. The materials are fastened with nails, twine, or fabric. One of the residents has tapped into a source of electricity by running a wire from a lamppost into several huts, providing electricity for light and heat. The researcher notes that as in the case of the residents of the Station, welfare plays a minimal role in the lives of residents of the Shanty. "So difficult is negotiating the system that most forgo their entitlements" (p. 58). One resident explains:

> I don't get welfare. I just can't . . . do it. I hate those people in there. They make you . . . sit and ask you questions that don't make any sense . . . You're homeless but you have to have an address. What kind of shit is that? Give me a break. They want you to get so . . . upset that you do get up and walk out. They test you. And if you do get up and walk out, that means you really don't want it. (p. 58)

Although drug use is common among residents of the Shanty, using drugs in public is a violation of norms of "etiquette." One resident explained that using drugs in the presence of a nonuser is disrespectful:

> For me to just take my works out and shoot, I would feel uncomfortable in front of you. It's not right. If I was just to turn around and put a needle in my arm, it's disrespectful. . . . very disrespectful. God forbid, I could be an influence. I could cause you to do it. You never know. (pp. 71–72)

Residents of the Shanty look out for one another and barter and trade with each other. "Residents often help build and fix others' huts in return for food, drugs, or cash" (pp. 77–78). Although survival among the homeless requires hustling, buying, trading, and selling, not everything is for sale. Some belongings have sentimental value that outweighs their economic value. One resident of the Shanty treasures a small gold key:

> There's a golden rule about gifts. You treasure them. You don't give them away, you don't sell them. I have right here a little key, a skeleton key. A little kid handed it to me four years ago. And every time somebody see that and say, "what is that"? I say it's a key to the world. I wouldn't give it to anyone if it was given to me. I treasure it. (p. 78)

Dordick notes that friendships and love relationships suffer from the stresses of drug addictions and impoverished conditions. Nevertheless, Dordick explains that the homeless "survive through their personal relationships" (p. 193). Relationships are critical to securing the material resources needed to survive and to creating—to the extent that it is possible—a safe and secure environment. For some homeless individuals, being involved in a committed love relationship is an important source of security and satisfaction. One resident of the

THE HUMAN SIDE

LIFE ON THE STREETS: NEW YORK'S HOMELESS *(continued)*

Shanty, Richie, conveys the importance of a love relationship:

> . . . Regardless of what people might think and say, most of us that might have a woman it's all we want. We don't look for anyone else. We really don't . . . I'm happy just to take care of my woman. I'm satisfied with her. I happen to love my girl . . . Really, I know it sounds corny, but that's the truth. I was married for ten years, but I finally found the right woman. As a matter of fact, we're gonna be married soon.

A Final Note

Virtually all the homeless individuals Dordick encountered expressed the desire to escape homelessness. In most cases, the homeless want to be self-reliant—and they want to be understood. In the words of one homeless man at the Station:

> You never see me sleep in the street. I worked 32 years of my life. Went to prison in '85. . . . I was brought up with a certain degree of independence through my parents. . . . And now all I need is two dollars to go and sit in a movie all night long. My pride is too good to beg. I don't want you to help me, Miss. I want you to understand me. (p. 5)

SOURCE: Gwendolyn Dordick. 1997. Excerpted and reprinted from *Something Left to Lose: Personal Relations and Survival among New York's Homeless* by Gwendolyn Dordick by permission of Temple University Press.

Intergenerational Poverty

As we have seen, problems associated with poverty, such as health and educational problems, create a cycle of poverty from one generation to the next. Poverty that is transmitted from one generation to the next is called **intergenerational poverty**. In a study of intergenerational poverty using a national longitudinal survey of families, researchers found considerable mobility out of childhood poverty: three-quarters of white poor children and over half of black poor children escaped poverty in early adulthood (Corcoran and Adams 1997). However, both white and black children in poor families were still much more likely to be poor in early adulthood than were children raised in nonpoor families.

Intergenerational poverty creates a persistently poor and socially disadvantaged population sometimes referred to as the **underclass**. The term *underclass* usually refers to impoverished individuals who have low educational attainment, have criminal records, are unmarried welfare mothers, and/or are ghetto residents. Although the underclass is stereotyped as being composed of minorities living in inner city or ghetto communities, the underclass is a heterogeneous population that includes poor whites living in urban and nonurban communities (Alex-Assensoh 1995).

Mead (1992) argues that intergenerational poverty may be caused by welfare dependency. According to Mead, when poor adults rely on welfare, the stigma of welfare fades, and welfare recipients develop poor work ethics that are passed on to their children. William Julius Wilson attributes intergenerational

> All too many of those who live in affluent America ignore those who exist in poor America; in doing so, the affluent American will eventually have to face themselves with the question . . . : How responsible am I for the well-being of my fellows? To ignore evil is to become an accomplice to it.
>
> —Martin Luther King Jr.
> Civil rights activist

poverty and the underclass to a variety of social factors, including the decline in well-paid jobs and their movement out of urban areas, the resultant decline in the availability of marriageable males able to support a family, declining marriage rates and an increase in out-of-wedlock births, the migration of the middle-class to the suburbs, and the impact of deteriorating neighborhoods on children and youth (Wilson 1987, 1996).

STRATEGIES FOR ACTION: ANTIPOVERTY PROGRAMS, POLICIES, AND PROPOSALS

In the United States, federal, state, and local governments have devoted considerable attention and resources to antipoverty programs for the last 50 years. Here we describe some of these programs and proposals, discuss international responses to poverty, and note the role of charity and the nonprofit sector in poverty alleviation.

Government Public Assistance and Welfare Programs in the United States

Many public assistance programs stipulate that households are not eligible for benefits unless their income and/or assets fall below a specified guideline. Programs that have eligibility requirements based on income are called **means-tested programs**. In 1996, 57.5 million people (21.6 percent of Americans) would have been poor if government benefits were not counted as part of their income. When government benefits are counted as income, these figures drop by nearly one-half, to 30.5 million people, or 11.5 percent of Americans. These figures indicate that government benefits lifted 27 million people out of poverty in 1996 (Center on Budget and Policy Priorities 1998a). Government public assistance programs designed to help the poor include cash support, food programs, housing assistance, medical care, educational assistance, job training programs, child care, child support enforcement, and the earned income tax credit (EITC).

In 1996, President Clinton signed into law the **Personal Responsibility and Work Opportunity Reconciliation Act (PRWOR)** with the promise of "ending welfare as we know it." As you read further, you will note that the 1996 welfare reform legislation has affected numerous public assistance programs, primarily in the form of cutbacks and eligibility restrictions.

National Data

In 1960, 1.7 percent of the U.S. population received either AFDC or TANF. The percentage rose to 4.1 percent in 1970, 4.7 percent in 1980, and 5.5 percent in 1994. After the passage of 1996 welfare reform laws, the percentage of the U.S. population receiving welfare dropped to 2.9 percent in 1998.

SOURCE: U.S. Department of Health and Human Services 1999b

Cash Support Two publicly funded cash support programs include Supplemental Security Income (SSI) and Temporary Aid to Needy Families (TANF).

Supplemental Security Income Federal SSI, administered by the Social Security Administration, provides a minimum income to poor people who are aged 65 or older, blind, or disabled. Under the 1996 welfare reforms, the definition of disability has been sharply restricted and the eligibility standards tightened. As a result, an estimated 850,000 families will lose SSI benefits (reported in Handler and Hasenfeld, 1997).

Temporary Aid to Needy Families Before 1996, a cash assistance program called **Aid to Families with Dependent Children** (AFDC) provided single parents (primarily women) and their children with a minimum monthly income. The 1996 welfare reform legislation marked the end of the AFDC program and replaced it with a program called **Temporary Aid To Needy Families (TANF)**. Under the new welfare legislation, after 2 consecutive years of receiving cash aid, welfare recipients are required to work at least 20 hours per week or to participate in a state-approved work program (exceptions may be made). A lifetime limit of 5 years is set for families receiving benefits, and able-bodied recipients aged 18 to 50 and without dependents have a 2-year lifetime limit. States are permitted to exempt 20 percent of their caseloads, however. To qualify for TANF benefits, unwed mothers under the age of 18 are required to live in an adult-supervised environment (such as with their parents) and to receive education and job training. Legal immigrants (with few exceptions) are not eligible to receive benefits. Some states have implemented "behavior modification" measures, linking welfare receipt to a set of expected behaviors. Such state programs include (Albelda and Tilly 1997):

- Learnfare, which suspends aid if a child misses a certain number of days of school or gets a failing grade
- The Family Cap, which freezes benefits at their current level when a poor mother receiving cash aid has another child (rather than raising benefits as was standard in the AFDC program)
- Incentives for implanting the contraceptive Norplant
- Bridefare (or Wedfare), which gives monetary benefits for marrying the father of a child
- Shotfare, in which a family loses benefits if immunization records are incomplete for any child.

National Data

Of adult welfare recipients in 1997, 8 percent were exempt from work participation because they were single custodial parents with a child under 12 months. Another 20 percent were exempt because of a disability or health problem.

SOURCE: U.S. Department of Health and Human Services 1999c

Food Benefits Food benefits include food stamps, school lunch and breakfast programs, the Special Supplemental Food Program for Women, Infants, and Children (WIC), and nutrition programs for the elderly. The largest food assistance program is the food stamp program, which issues monthly benefits through coupons or Electronic Benefits Transfer (EBT), using a plastic card similar to a credit card and a personal identification number (PIN). The WIC program provides low-income pregnant and postpartum women, infants, and children up to 5 years of age with vouchers to purchase food items such as milk, cheese, fruit juice, and cereal, which are deemed important for pregnant and nursing mothers and their young children.

Welfare reform legislation of 1996 has resulted in cutbacks in federal food assistance programs, especially the food stamp program. Families with children will lose an average of 13 percent of their food stamp benefits by 2002, or about $45 each month. In addition, over 1 million people—largely unemployed adults and legal immigrants—will lose their eligibility to receive food stamps (Smallwood 1998).

People on welfare are just like you and me. They have the same basic hopes and fears. They want a job that brings self-worth and validation. They want to support their families and contribute to their communities. They want pride and dignity, just like those of us who have been lucky enough never to need public assistance.

—Alexis M. Herman
Secretary, U.S. Department of Labor

Housing Assistance Housing costs represent a major burden for the poor. The median cost of housing for poor households is 60 percent of household income; for nonpoor households the median cost of housing is only 20 percent of household income (Levitan, Mangum, and Mangum 1998). Various federal programs provide housing assistance to low-income individuals. These include public housing and subsidized housing. Housing assistance programs typically require tenants to pay 30 percent of their adjusted income in rent; the federal government pays the difference between the tenant's payment and the actual rent.

The **public housing** program, initiated in 1937, provides federal subsidies for low-income housing units built, owned, and operated by local public housing authorities (PHAs). Public housing has been plagued by problems. To save costs and avoid public opposition, high-rise public housing units were built in

National Data

In a two-year period after the enactment of the 1996 welfare reform law, the number of U.S. families on welfare decreased 34 percent.

SOURCE: U.S. Department of Health and Human Services 1999a

inner-city projects. The concentration of poor families in deteriorating neighborhoods led to increases in crime, drugs, and vandalism. Rising costs of construction and operating costs and public opposition has led to decreased funding for new public housing construction. The waiting list for public housing is long, and applicants often wait years before a housing unit is available.

Subsidized housing involves leasing of private housing units by local housing authorities or direct payments of rent supplements to low-income families. Rather than build new housing units for low-income families, subsidized housing relies on existing housing. Unlike public housing that confines low-income families to high-poverty neighborhoods, subsidized housing *attempts* to disperse low-income families. However, due to opposition by residents in middle-class neighborhoods, most subsidized housing units remain in low-income areas. In many communities, low-income families may remain on a waiting list for subsidized housing for as long as 5 years.

National Data

Over half (60 percent) of poor U.S. renters spend more than half their income on rent and utilities. The average poor renter not receiving a housing subsidy spends more than 77 percent of income on shelter.

SOURCE: DeParle 1999

> Unfortunately, there are a lot more $6-an-hour jobs than $6-an-hour apartments.
>
> —Andrew Cuomo
> Secretary of Housing and Urban Development

Medical Care

Medical care assistance programs include Indian Health Services, maternal and child health services, and Medicaid, which provides medical services and hospital care for the poor through reimbursing physicians and hospitals. However, as discussed in Chapter 2, many low-income individuals and families do not qualify for Medicaid and either cannot afford health insurance or cannot pay the deductible and co-payments under their insurance plan. In the earlier AFDC welfare program, all recipients were automatically entitled to Medicaid. Under the TANF program, states decide who is eligible for Medicaid; eligibility for cash assistance does not automatically convey eligibility for Medicaid. A provision of the 1996 welfare reform legislation guarantees at least 1 year of transitional Medicaid when leaving welfare for work.

National Data

Only about one-fourth of all people eligible to receive housing assistance actually receive it.

SOURCE: Kingsley 1998

Educational Assistance

Educational assistance includes Head Start and Early Head Start programs and college assistance programs (see also Chapter 12). Head Start and Early Head Start programs provide educational services for disadvantaged infants, toddlers and preschool-age children and their parents.

Evaluations of Head Start programs indicate that they improve school performance and employability in adulthood; "compared with children in similar circumstances who do not receive early education, Head Start enrollees are more likely to graduate from high school and to find work" (Levitan, Mangum, and Mangum 1998, 163). However, due to inadequate funding of Head Start programs, only one in five poor children are enrolled (Levitan, Mangum, and Mangum 1998).

To alleviate economic barriers for low-income persons wanting to attend college, the federal government offers grants, loans, and work opportunities. The Pell grant program aids students from low-income families. The guaranteed student loan program enables college students and their families to obtain low-interest loans with deferred interest payments. The federal college-work-study program provides jobs for students with "demonstrated need." The federal government pays 70 to 80 percent of student wages.

> In our society, it is murder, psychologically, to deprive a man of a job or an income. You are in effect saying to that man that he has no right to exist.
>
> —Martin Luther King Jr.
> Civil rights activist

Job Training Programs

Various employment and job training programs are available to help individuals out of poverty (see also Chapter 11). These include summer youth employment programs, Job Corps, and training for disadvantaged adults and youth. These programs fall under the Job Training and Partnership Act (JTPA), a federally funded program passed in 1982 and amended in

1992. A primary shortcoming of job training programs has been that "they spread too little money among too many trainees, with the result that few are in training long enough for it to make a sufficient impact on their posttraining wages" (Levitan, Mangum, and Mangum 1998, 29).

Child Care Assistance Most Western industrialized countries provide free, high-quality preschools for 3-to-5-year olds, and many also subsidize childcare for infants and toddlers (Albelda and Tilly 1997). In France, for example, public nursery schools enroll almost all children between ages 3 and 6 and public child care centers care for infants and toddlers. French law also assists women in caring for their own children in infancy; a woman giving birth to her first or second child is entitled to 16 weeks of *paid* maternity leave, and when she has a third child, she gets 26 weeks of paid leave (Bergmann 1996). By comparison, the U.S. Family and Medical Leave Act requires employers to offer *unpaid* leave to new parents of up to 12 weeks. And the United States does not offer universal child care for children under 5—the typical age at which children begin kindergarten.

> If we are serious about rewarding work and helping people stay off welfare and keep their jobs, then we must make quality child care affordable now.
>
> —Marian Wright Edelman
> President of the Children's Defense Fund

Lack of affordable quality child care is a major obstacle to employment for single parents and a tremendous burden on dual-income families and employed single parents. Poor families with a preschool child and an employed mother who pays for child care spend almost one-quarter of their income on child care (Albelda and Tilley 1997).

National Data

Full-day child care costs $4,000 to $10,000 per year.

SOURCE: Children's Defense Fund 1998a

Some public and private sector programs and policies provide limited assistance with child care. The Dependent Care Assistance Plan provisions of the 1981 Economic Recovery Tax Act permits individuals to exclude the value of employer-provided child care services from their gross income. However, few employers provide on-site child care or subsidies for child care. At the same time, Congress increased the amount of the child care tax credit and modified the federal tax code to allow taxpayers to shelter pretax dollars for child care in "flexible spending plans." The Family Support Act of 1988 offered additional funding for child care services for the poor (in conjunction with mandatory work requirements) and the Child Care and Development Block Grant, which became law in 1990, targeted child care funds to low-income groups. The Personal Responsibility and Work Opportunity Reconciliation Act of 1996 appropriated $16 billion over 5 years for child care, yet this amount is insufficient (Michel 1998). In 1998, no state had sufficient funds to provide child care to all families eligible under federal guidelines. According to Sonya Michel (1998), "the reluctance to make adequate provision for childcare is . . . symptomatic of a deeper aversion on the part of many legislators and public officials to helping poor and low-income women become truly economically independent, a status which is, in turn, essential to their ability to form autonomous households" (pp. 47–48).

Many qualifying families do not receive child care assistance—they are put on long waiting lists, are turned away due to inadequate funds, or are not informed of available assistance. And, because the amount of child-care assistance many families do receive is so small, families often cannot find child-care providers who will serve their children. The states pay for child care at such low rates that parents are either limited in the providers who are willing to serve their children or are forced to pay providers additional fees on top of their existing co-payments. Finally, many hardworking, low-income families are not eligible for child-care assistance due to low state eligibility cutoffs (Children's Defense Fund 1998b).

National Data

Only one in ten eligible children needing child care are receiving assistance for child care services.

SOURCE: Children's Defense Fund 1998a

Child Support Enforcement The Personal Responsibility and Work Opportunity Act of 1996 requires the states to set up child support enforcement programs. The new law establishes a Federal Case Registry and National Directory of New Hires to track delinquent parents across state lines, expands and streamlines procedures for direct withholding of child support from wages, and streamlines procedures for establishing paternity. Individuals who fail to cooperate with paternity establishment will have their monthly cash assistance reduced by at least 25 percent. The law allows for tough penalties for failure to pay child support, enabling states to seize assets and to revoke drivers and professional licenses for parents who fail to pay child support. However, due to the low wages of many absent fathers, collecting child support offers minimal economic relief to many single mothers. Sorensen (1995) found that with a perfect child support system, access to absent fathers' income would have reduced the number of children officially classified as poor in 1989 by less than 10 percent. The official poverty rate for all U.S. children in 1989 would have been reduced by 1 to 2 percentage points, from 19.6 percent to about 18 percent.

National Data

As of September 1998, 10 states offered earned income tax credits.

SOURCE: Johnson and Lazere 1998

Earned Income Tax Credit The federal **earned income tax credit (EITC)**, created in 1975, is a refundable tax credit based upon a working family's income and number of children. The EITC is designed to offset adverse effects of Social Security and Medicare payroll taxes on the working poor families and to strengthen work incentives. Increases in the EITC were enacted in 1986, 1990, and 1993. In 1997, maximum benefits were $2,210 for families with one child and $3,656 for families with two or more children (Johnson and Lazere 1998).

National Data

In 1996, 38.9 percent of Hispanic children classified as poor lived in families where someone worked full-time, compared with 31.9 percent of poor white children and 17.7 percent of poor black children.

SOURCE: Johnson and Lazere 1998

The federal EITC lifts more children out of poverty than any other program (Johnson and Lazere 1998). The EITC is most effective in lifting Hispanic children out of poverty because poor Hispanic children are more likely than other poor children to live in families where a family member works full time. This is significant because poor families with a full-time worker receive the largest EITC benefits.

Some states also offer an earned income tax credit to offset state personal income taxes on poor families. In 1997, income taxes were levied on below-poverty families in 21 of the 41 states with a personal income tax. The average income tax burden in these states was $243 for a family of four below the poverty line. Other state and local tax codes impose an even greater burden on poor families. Most states rely heavily on revenue from sales taxes. These taxes are **regressive**, meaning that they absorb a much higher proportion of the incomes of lower-income households than of higher-income households. In 1995, average state and local taxes on the poorest fifth of married, nonelderly families was 12.5 percent of income. By contrast, the wealthiest 1 percent of such families spent an average of 7.9 percent of income for state and local taxes (Johnson and Lazere 1998). State EITCs help to offset state and local tax burdens on poor families.

Welfare in the United States: Myths and Realities

Public attitudes toward welfare assistance and welfare recipients are generally negative. Rather than view poverty as the problem, many Americans view welfare as the problem. What are some of the common myths about welfare that perpetuate negative images of the welfare and welfare recipients?

MYTH 1 People receiving welfare are lazy and have no work ethic.

Reality First of all, single parents on welfare already *do* work—they do the work of parenting. Albelda and Tilly (1997) emphasize that "raising children is work. It requires time, skills, and commitment. While we as a society don't place a monetary value on it, it is work that is invaluable—and indeed, essential to the survival of our society" (p. 111). Second, most adults receiving public assistance are either employed or in the labor force looking for work. In 1997, nearly one in five adult welfare recipients were employed (earning an average of $592 per month) and another 40 percent were seeking employment (U.S. Department of Health and Human Services 1999c). But, there are not enough jobs—especially for low-skill workers.

National Data

For every job opening, there are three job seekers.

SOURCE: Weisbrot, 1997

MYTH 2 Welfare benefits are granted to many people who are not really eligible to receive them.

Reality Although some people obtain welfare benefits through fraudulent means, it is much more common for people who are eligible to receive welfare to not receive benefits. For example, among those eligible for benefits, 25 percent failed to receive AFDC and 50 percent failed to receive food stamps (Kim and Mergoupis 1997). A main reason for not receiving benefits is lack of information; people don't know they are eligible. Many people who are eligible for public assistance do not apply for it because they do not want to be stigmatized as lazy people who just want a "free ride" at the taxpayers' expense—their sense of personal pride prevents them from receiving public assistance. Others want to avoid the administrative and transportation hassles involved in obtaining it.

MYTH 3 Most welfare parents are teenagers.

Reality In 1997, the average age of adult welfare recipients was 30 years. Only 8 percent of adult recipients were teenagers; 18 percent were 40 years and older (U.S. Department of Health and Human Services 1999c). In addition, nearly one-fourth (23 percent) of TANF families had *no* adult recipients—only children were eligible to receive assistance.

MYTH 4 Most welfare mothers have large families with many children.

Reality Mothers receiving welfare have no more children, on average, than mothers in the general population. In 1996, 44 percent of welfare families had one child; 30 percent had two children, 15 percent had three children, and 9 percent had four children (U.S. Department of Labor 1998).

MYTH 5 Unmarried women have children so they can receive benefits. And if single mothers already receive benefits, they have additional children in order to receive increased benefits.

Reality Research consistently shows that receiving welfare does not significantly increase out-of-wedlock births (Albelda and Tilley 1997). In states that had the lowest AFDC cash benefits to single mothers, the teenage birthrates were among the highest (reported in Albelda and Tilley 1997).

The economic, social, and psychological situation in which women on welfare find themselves is simply not conducive to desiring more children. Such women would appear to be motivated by cost-benefit considerations, but it is the costs that outweigh the benefits, not the reverse. Becoming pregnant and having a child are perceived as making the situation worse, not better.

—Mark Rank
Sociologist

MYTH 6 Most welfare families become dependent on public assistance and stay on welfare for a long time.

Reality One-third (34 percent) of welfare families have stayed on welfare for 1 year or less; 27 percent stayed on welfare for 1 to 3 years, 15 percent for 3 to 5 years, and 22 percent for 5 or more years (U.S. Deptartment of Labor 1998).

MYTH 7 Public assistance to the poor creates an enormous burden on taxpayers.

National Data

In 1997, most TANF families received monthly average assistance of $359.00.

SOURCE: U.S. Department of Health and Human Services 1999c

Reality Public assistance to the poor is a small part of government spending. In the 1999 federal budget, 6 percent was allocated to Medicaid and another 6 percent was allocated to other assistance to the poor, including food stamps, child nutrition, cash aid, and the earned income tax credit (Office of Management and Budget 1998). Meanwhile 15 percent of the 1999 federal budget was allocated for national defense and 23 percent was earmarked for providing Social Security benefits. Although Social Security benefits are paid to disabled workers and their children, they are also paid to retired workers and dependents and survivors, even those who are affluent. In 1994, 29 percent of families with an income of $150,000 or more received government benefits averaging more than $16,000 through programs such as Social Security. Families with incomes of less than $10,000, in contrast, received public aid averaging less than $8,000 (reported in Albelda, Folbre, and the Center for Popular Economics 1996). In addition, the estimated cost of tax breaks to corporations and wealthy individuals in 1996 was $440 billion, more than 17 times combined state and federal spending on AFDC (Collins 1995). Despite the public perception that welfare benefits are too generous, cash and other forms of assistance to the poor do not meet the basic needs for many individuals and families who receive such benefits. This chapter's *Focus on Social Problems Research* feature describes a research study that documents the hardships experienced by poor single mothers and the survival strategies they use to "make ends meet."

Minimum Wage Increase and "Living Wage" Laws

As noted earlier, many families that leave welfare for work are still living in poverty due to involuntary part-time work and low wages. One strategy for improving the standard of living for low-income individuals and families is to increase the minimum wage. In 1968, minimum wage workers earned 111 percent of the federal poverty level; in 1998 minimum wage earnings measured up to only 78 percent of the poverty line. The 1998 federal minimum wage—$5.15 per hour—had 20 percent less purchasing power than the minimum wage of $1.60 did in 1968. To keep pace with inflation, the minimum wage in 1998 would need to be $7.21 to afford the same purchasing power as in 1968 (Pitcoff 1998). In September 1998 the Senate voted against minimum wage legislation that would have increased the minimum wage to $6.15 an hour by the year 2000.

> In the scholarly debate about welfare reform, the voices and lived experiences of single mothers are often drowned out by reams of statistics, usually aggregate numbers that, while useful, can distance us from the daily struggles poor single women face as they try to both parent and provide for their children.
>
> –Kathryn Edin and Laura Lein
> Researchers

Those opposed to increasing the minimum wage argue that such an increase would result in higher unemployment and fewer benefits for low-wage workers; businesses would reduce wage costs by hiring fewer employees and

Making Ends Meet: Survival Strategies among Low-Income and Welfare Single Mothers

Previous research has shown that nearly one-half of families with incomes below the poverty line reported that their expenditures on food, housing, and medical care exceeded their incomes. This lead researchers Kathryn Edin and Laura Lein (1997) to ask how poor families—particularly poor single-mother families—make ends meet.

Sample and Methods

The sample consisted of 379 African-American, white, and Mexican-American single mothers from four cities: Chicago, San Antonio, Boston, and Charleston, South Carolina. Two basic populations of single mothers were sampled: mothers who received AFDC (N = 214) and nonrecipients who held low-wage jobs earning $5 to $7 an hour between 1988 and 1992 (N = 165). Assuming that welfare recipients are as likely to hide their unreported income from survey researchers as from welfare officials, Edin and Lein recruited welfare mothers with the assistance of trusted community residents, including members of churches, local charities, and neighborhood and community organizations. Personal introductions to welfare mothers were crucial in gaining the mothers' trust. Edin and Lein used a "snowball sampling" technique in which each mother who was interviewed was asked to refer researchers to one or two friends who might also participate in interviews. Nearly 90 percent of the mothers contacted agreed to be interviewed.

Researchers collected data through conducting multiple semistructured in-depth interviews with women in the sample. Interview topics included the mothers' income and job experience, types and amount of welfare benefits they received, spending behavior, housing situation, use of medical care and child care, and hardships the women and their children experienced due to lack of financial resources.

Interviewing the mothers more than once was an important research strategy in gathering accurate information. Mothers who were unclear about their expenditures in the first interview could keep careful track of what they spent between interviews and give a more precise accounting of their spending in a later interview. Also, the more conversations the researchers had with a mother, the more likely she was to reveal sensitive information. Edin and Lein explain that, "the whole story was hard to get not only because mothers were sometimes hesitant to tell it but also because we did not always know how to interpret their words when they did" (pp. 12–13). For example, when researchers asked mothers if they engaged in prostitution to make ends meet, nearly every woman denied doing so. However, some women reported that they had several "boyfriends" who would "help them out" with a bill in exchange for "going out." Also, some mothers who insisted they received no child support later revealed that the child's father "helped out" every week by providing cash. "Most mothers only termed absent fathers' cash contributions as 'child support' if it was collected by the state" (p. 13).

Findings

As indicated in the table low-wage earning single mothers had a higher monthly reported income than welfare-reliant mothers. However, the expenses of wage-earning mothers were also higher. This is because employed mothers usually have

Average Monthly Reported Income and Expenses of 214 Welfare-Reliant and 165 Wage-Reliant Mothers

	Welfare-Reliant Mothers	Wage-Reliant Mothers
Income		
Main job	NA	$777
AFDC	$307	NA
Food stamps	222	57
SSI	36	3
Total Welfare Benefits	**$565**	**$60**
Earned Income Tax Credit	3	25
Total Reported Income	**$568**	**$862**
Monthly Expenses		
Housing	$213	$341
Food	262	249
Medical	18	56
Clothing	69	95
Transportation	62	129
Child Care	7	66
Other Necessities	179	223
Nonessentials*	64	84
Total Expenses	**$876**	**$1,243**
Reported Income Minus Total Expenses	**−$306**	**−$381**

*Nonessentials include entertainment (such as video rentals), cable TV, cigarettes and alcohol, eating out, and lottery tickets.

SOURCE: Based on Kathryn Edin and Laura Lein. 1997. *Making Ends Meet* by Kathryn Edin and Laura Lein, © 1997 Russell Sage Foundation, New York, New York. Used with permission.

(continued)

FOCUS ON SOCIAL PROBLEMS RESEARCH *(continued)*

to pay for child care, transportation to work, and additional clothing to wear to work. If newly employed mothers have a federal housing subsidy, every extra $100 in cash income raises their rent by $30 (Jencks 1997). And employed mothers are usually not eligible for Medicaid, which means that they have more out-of-pocket medical expenses and often go uninsured.

The monthly expenses of both groups of women exceeded their reported monthly income, forcing women to use various strategies to make ends meet. Cash welfare and food stamps covered only three-fifths of welfare-reliant mothers' expenses. The main job of low-wage earning mothers covered only 63 percent of their expenses. Edin and Lein found that women relied on three basic strategies to make ends meet: work in the formal, informal, or underground economy; cash assistance from absent fathers, boyfriends, relatives, and friends; and cash assistance or help from agencies, community groups, or charities in paying overdue bills. Welfare recipients had to keep their income-generating activities hidden from their welfare caseworkers and other government officials. Otherwise, their welfare checks would be reduced by nearly the same amount as their earnings. Many of the wage-earning mothers also concealed income generated "on the side" in order to maintain food stamps, housing subsidies, or other benefits that would have been reduced or eliminated if they had reported this additional income.

Most of the single mothers in the study described experiencing serious material hardship during the prior 12 months. Material hardships included not having enough food and clothes, not receiving needed medical care, not having health insurance, having the utilities or phone cut off, not having a phone, and being evicted and/or homeless. An important finding was that wage-reliant mothers experienced more hardship than welfare-reliant mothers. In addition to the increased financial pressures of child care costs, transportation, health care, and work clothing, employed mothers worried about not providing adequate supervision of their children and struggled with balancing work and parenting responsibilities, especially when their children were sick. Nevertheless, almost all of the mothers said they would rather work than rely on welfare. They believed that work provided important psychological benefits and increased self-esteem, avoided the stigma of welfare, and enabled them to be good role models for their children.

Harvard University scholar Christopher Jencks (1997) comments on the implications of Edin and Lein's (1997) research:

> As the new time limits on welfare receipt begin to take effect, more and more single mothers will have to take jobs. Most of these newly employed mothers will have more income than they had on welfare, so their official poverty rate will fall. But they will also have more expenses than they had on welfare, and they will get fewer non-cash benefits. Edin and Lein's findings dramatize the likely result. Between 1988 and 1992, mothers who held low-wage jobs reported substantially more income than those who collected welfare, but they also reported more hardship. If this pattern persists in the years ahead, time limits will probably bring both a decline in the official poverty rate and an increase in material hardship. p. (x)

National Data

More than 20 municipalities and counties have living wage laws on the books, and at least 35 locations are working toward such measures.

SOURCE: Pitcoff 1998

providing fewer benefits. However, in a review of research on minimum wage increase, Card and Krueger (1995) found no evidence that employers reduced benefits to compensate for wage increases and employment actually increased slightly, rather than decreased.

In some cities and states, "living wage laws" are under consideration. "**Living wage laws**" require state or municipal contractors, recipients of public subsidies or tax breaks, or, in some cases, all businesses to pay employees wages significantly above the federal minimum, enabling families to live above the poverty line.

In 1994, Baltimore became one of the first cities to pass a living wage law. The ordinance mandated a minimum hourly wage of $6.10 for employees working on a city service contract effective July 1, 1995, increasing to $6.60 after July 1, 1996, and $7.70 in 1999. Researchers found that of companies interviewed before and after the law went into effect, none reported reducing staffing levels in response to the higher wage requirement (Weisbrot and Sforza-Roderick 1996). The cost to taxpayers was minimal—

about 17¢ per person annually. Finally, researchers found no evidence that businesses reacted negatively to the ordinance. In fact, the value of business investment in Baltimore increased substantially in the year after the law passed.

Charity, Nonprofit Organizations, and Nongovernmental Organizations

Various types of aid to the poor are provided through individual and corporate donations to charities and nonprofit organizations. Before reading further, you may want to complete this chapter's *Self & Society* feature: "Reasons for Contributing or Not Contributing to Charity."

In 1995, 69 percent of U.S. households contributed to charity; the average contribution per household was $1,017 (Hodgkinson and Weitzman 1996). In 1998, Bill Gates donated $100 million to provide vaccinations for poor children in developing countries. Religious organizations receive over half of household charitable contributions.

Corporations also donate to charitable causes. In 1993, U.S. corporations contributed $5.2 billion to charities (Hodgkinson and Weitzman 1996).

Charity involves giving not only money, but time and effort in the form of volunteering. The National Student Campaign against Hunger and Homelessness (NSCAHH) is the largest student network fighting hunger and homelessness in the United States, with more than 600 participating campuses in 45 states.

Nongovernmental organizations (NGOs) play an important role in helping to reduce poverty and often collaborate with governmental poverty reduction efforts. In Thailand, for example, NGOs work with the government in such efforts as HIV/AIDS prevention and housing rights for slum dwellers. In Mumbai, Bombay, the nongovernmental organization Yuva—Youth for Unity and Voluntary Action—organizes youth and women for social action in housing, education, and health (*Human Development Report* 1997).

International Responses to Poverty

Alleviating worldwide poverty continues to be a major concern of both developing and developed countries. Approaches to poverty reduction include promoting economic growth and investing in "human capital." Conflict resolution and the promotion of peace is also important for reducing poverty worldwide.

> Poverty is no longer inevitable. The world has the material and natural resources, the know-how and the people to make a poverty-free world a reality in less than a generation.
>
> —*Human Development Report 1997*

Promoting Economic Growth Economic growth, over the long term, generally reduces poverty (United Nations 1997). An expanding economy creates new employment opportunities and increased goods and services. As employment prospects improve, individuals are able to buy more goods and services. The increased demand for goods and services, in turn, stimulates economic growth. As emphasized in Chapter 15, economic development requires controlling population growth and protecting the environment and natural resources, which are often destroyed and depleted in the process of economic growth.

However, economic growth does not always reduce poverty; in some cases it increases it. For example, growth resulting from technological progress may

Reasons for Contributing or Not Contributing to Charity

Directions: After reading each of the following reasons for contributing or not contributing to charity, indicate how important you think each reason is according to the following scale:

1 = very or somewhat important   2 = not too important or not at all important

PART ONE: REASONS FOR CONTRIBUTING TO CHARITY

___ 1. Being asked to give by someone you know well.
___ 2. Because you volunteered at the organization.
___ 3. Being asked by clergy to give.
___ 4. Reading or hearing a news story.
___ 5. Being asked at work to give.
___ 6. Someone coming to the door asking you to give.
___ 7. Being asked to give in a telethon/radiothon.
___ 8. Receiving a letter asking you to give.
___ 9. Receiving a phone call asking you to give.
___ 10. Reading a newspaper or magazine advertisement.
___ 11. Seeing a television commercial.

PART TWO: REASONS FOR NOT CONTRIBUTING TO CHARITY OR NOT CONTRIBUTING MORE

___ 12. I could not afford to give money.
___ 13. Because I already give as much as I can.
___ 14. Because I'm making less money this year than last.
___ 15. Because I'm unsure about having a job next year.
___ 16. Because I lost my job.
___ 17. I would rather spend my money in other ways.
___ 18. Because no one I know personally asked me to give.
___ 19. Because no charitable organization contacted me asking for a contribution.
___ 20. Because I didn't get around to it.

Comparison Data: You may compare your responses with those of a nationally representative sample of 2,700 U.S. adults. (Percentages are rounded and do not total 100 percent due to some respondents not responding to the item).

	Percentage who said reason was somewhat or very important	*Percentage who said reason was not too important or not at all important*
1.	72	26
2.	61	36
3.	59	39
4.	43	55
5.	38	54
6.	36	61
7.	30	69
8.	29	70
9.	23	75
10.	17	81

Reasons for Contributing or Not Contributing to Charity *(continued)*

	Percentage who said reason was somewhat or very important	*Percentage who said reason was not too important or not at all important*
11.	17	81
12.	56	43
13.	45	20
14.	40	40
15.	29	33
16.	29	37
17.	26	30
18.	17	26
19.	16	27
20.	15	29

SOURCE: Virginia A. Hodgkinson and Murray S. Weitzman. 1996. *Giving and Volunteering in the United States, 1996.* **http://www.indepsec.org/p...public_attitude.html**. Reprinted by permission of Independent Sector.

reduce demand for unskilled workers. Growth does not help poverty reduction when public spending is diverted away from meeting the needs of the poor and instead is used to pay international debt or finance military operations. Thus, "economic growth, though essential for poverty reduction, is not enough. Growth must be pro-poor, expanding the opportunities and life choices of poor people" (*Human Development Report* 1997, pp. 72–73). Because three-fourths of poor people in most developing countries depend on agriculture for their livelihoods, economic growth to reduce poverty must include raising the productivity of small-scale agriculture. Not only does improving the productivity of small-scale agriculture create employment, it also reduces food prices. The poor benefit the most because about 70 percent of their income is spent on food (*Human Development Report* 1997).

Investing in Human Capital Promoting economic development in a society requires having a productive workforce. Yet, in many poor countries, large segments of the population are illiterate and without job skills, and/or are malnourished and in poor health. Thus, a key feature of poverty reduction strategies involves investing in human capital. The term **human capital** refers to the skills, knowledge, and capabilities of the individual. Investments in human capital involve programs and policies that enhance the individual's health, skills, knowledge, and capabilities. Such programs and policies include those that provide adequate nutrition, sanitation, housing, health care (including reproductive health care and family planning), and educational and job training. Nobel Laureate Gary Becker has concluded that "the case is overwhelming that investments in human capital are one of the most effective ways to raise the poor to decent levels of income and health" (reported in Hill 1998, p. 279).

Poor health is both a consequence and a cause of poverty; improving the health status of a population is a significant step toward breaking the cycle of poverty. Increasing the educational levels of a population better prepares

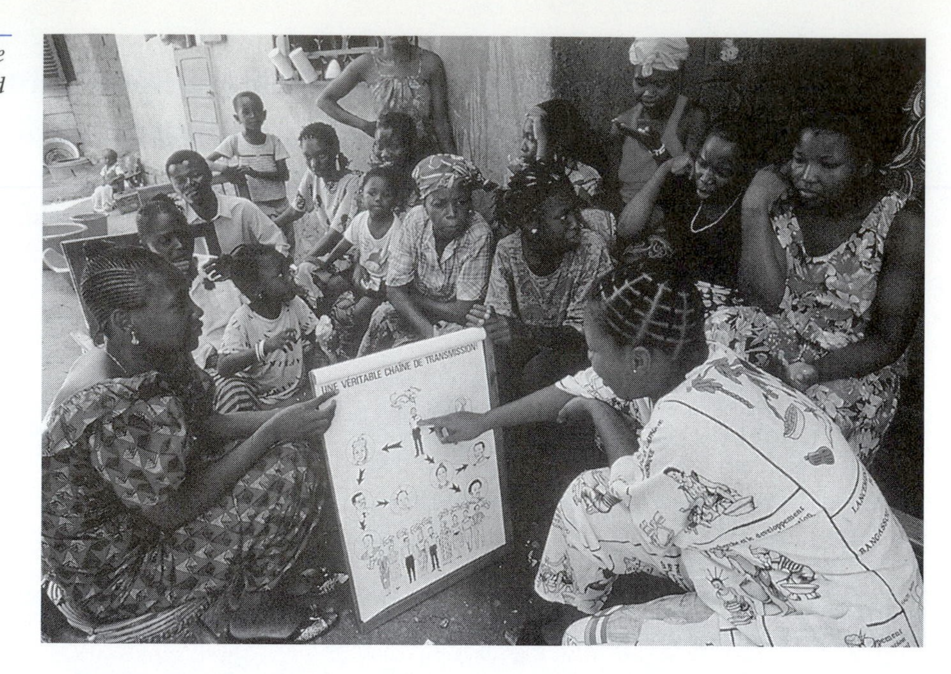

Because poverty offers a fertile breeding ground for the spread of HIV, the incidence of HIV in poorer countries is more than 10 times that in industrial countries. Widespread HIV infection leads to further economic and social disintegration. Thus, efforts to prevent the spread of HIV is an important strategy for poverty reduction in poor countries.

individuals for paid employment and for participation in political affairs that affect poverty and other economic and political issues. Improving the educational level and overall status of women in developing countries is also associated with lower birth rates, which in turn fosters economic development (see also Chapter 15).

One way to help poor countries invest in human capital and reduce poverty is to provide debt relief. If African countries were relieved of their national debts, they would have funds that would save the lives of millions of children and provide basic education to millions of girls and women. Providing debt relief to the 20 worst-affected countries would cost between $5.5 billion and $7.7 billion—less than the cost of one Stealth bomber and roughly the cost of building the Euro-Disney theme park in France (*Human Development Report* 1997).

UNDERSTANDING **THE HAVES AND THE HAVE-NOTS**

> We must face the fact that families with extremely low wages do not earn enough to raise their children out of poverty. Without help like child care, transportation, training, and wage supplements, families are one crisis away from joblessness or hunger.
>
> –Deborah Weinstein
> Children's Defense Fund, Family Income Division Director

As we have seen in this chapter, economic prosperity has not been evenly distributed; the rich have become richer, while the poor have become poorer. Meanwhile, the United States has implemented welfare reform measures that essentially weaken the safety net for the impoverished segment of the population—largely children. The decrease in assistance to the poor is a major reason why poverty rates have remained high, despite the economic prosperity of the country. Welfare reform legislation of 1996 has achieved its goal of reducing welfare rolls across the country. Although the long-term effects of the 1996 welfare reform are not yet known, early reports contain disturbing findings. A joint report released by the Children's Defense Fund and the National Coalition for the Homeless (1998) announced that up to half of the families leaving welfare rolls don't have jobs. Among those who do find jobs, 71 percent earn less than $250 per week, less than the poverty level for a family of three. Many families leaving welfare report struggling to get food, shelter, medical care, child care, and transportation. The effects of poverty on children perpetuate the cycle of poverty. Larin (1998) notes that "while it always has been possible for individuals to move up the economic ladder, the odds against it are high for individuals who suffer from the ill effects of poverty during childhood" (p. 26). Given the association between poverty and poor health and low educational attainment, poor children often grow up to be poor adults who are unable to escape poverty.

A common belief among U.S. adults is that the rich are deserving and the poor are failures. Blaming poverty on individual rather than structural and cultural factors implies not only that poor individuals are responsible for their plight, but also that they are responsible for improving their condition. If we hold individuals accountable for their poverty, we fail to make society accountable for making investments in human capital that are necessary to alleviate poverty. Such human capital investments include providing health care, adequate food and housing, education, child care, and job training. Economist Lewis Hill (1998) believes that "the fundamental cause of perpetual poverty is the failure of the American people to invest adequately in the human capital represented by impoverished children" (p. 299). Blaming the poor for their plight also fails to recognize that there are not enough jobs for those who want to work and that many jobs fail to pay wages that enable families to escape poverty. And lastly, blaming the poor for their condition diverts attention away from the recognition that the wealthy—individuals and corporations—receive far more benefits in the form of wealthfare or corporate welfare, without the stigma of welfare.

Ending or reducing poverty begins with the recognition that doing so is a worthy ideal and an attainable goal. Imagine a world where everyone had comfortable shelter, plentiful food, adequate medical care, and education. If this imaginary world were achieved, and absolute poverty were effectively eliminated, what would the effects be on such social problems as crime, drug abuse, family problems (such as domestic violence, child abuse, divorce, and unwed parenthood), health problems, prejudice and racism, and international conflict? But it would be too costly to eliminate poverty—or would it? According to one source, the cost of eradicating poverty worldwide would be only about 1 percent of global income—and no more than 2 to 3 percent of national income in all but the poorest countries (*Human Development Report* 1997). Certainly the costs of allowing poverty to continue are much greater than that.

> A decent provision for the poor is the true test of civilization.
>
> –Samuel Johnson
> English essayist and poet

CRITICAL THINKING

1. Does a decline in the official U.S. poverty rate necessarily mean that fewer people are experiencing economic hardship? Or is it possible for a decline in the poverty rate to be accompanied by an *increase* in the numbers of individuals experiencing economic hardship?
2. Should someone receiving welfare benefits be entitled to spend some of his or her money on "nonessentials" such as cigarettes, eating out, lottery tickets, and cable TV? Why or why not?
3. Parenti (1998) points out that reports of income inequality based on U.S. census data are misleading because they do not take into account the super rich— the top 1 percent of income earners. For years, the Census Bureau

never interviewed anyone who had an income higher than $300,000. The reportable upper limit of $300,000 was the top figure allowed by the bureau's computer program. In 1994, the bureau raised the upper limit to $1 million. But this figure still excludes the richest 1%—the hundreds of billionaires and thousands of multimillionaires who make many times more than $1 million a year. "The super rich simply have been computerized out of the Census Bureau's picture" (Parenti 1998, 36). How does the exclusion of the superrich from census data distort reports of economic inequality? Who benefits from this distortion?

4. The U.S. poor have low rates of voting and thus have minimal influence on elected government officials and the policies they advocate. What strategies might be effective in increasing voter participation among the poor?
5. Oscar Lewis (1998) noted that "some see the poor as virtuous, upright, serene, independent, honest, secure, kind, simple and happy, while others see them as evil, mean, violent, sordid and criminal" (p. 9). Which view of the poor do you tend to hold? How have various social influences, such as parents, peers, media, social class, and education, shaped your views toward the poor?

KEY TERMS

absolute poverty
Aid to Families with Dependent Children (AFDC)
bourgeoisie
corporate welfare
culture of poverty
earned income tax credit (EITC)
feminization of poverty
human capital
Human Poverty Index (HPI)
intergenerational poverty
living wage laws
means-tested programs
Personal Responsibility and Work Opportunity Reconciliation Act (PRWOR)
poverty
proletariat
public housing
regressive taxes
relative poverty
subsidized housing
Temporary Aid to Needy Families (TANF)
underclass
wealth
wealthfare
working poor

INTERNET

You can find more information on corporate welfare, Temporary Aid to Needy Families (TANF), child support enforcement programs, and housing assistance at the *Understanding Social Problems* web site at **http://sociology.wadsworth.com.**

INFOTRAC COLLEGE EDITION

Either from the Wadsworth sociology resource center at **http://sociology.wadsworth.com** or directly from your web browser, you may access InfoTrac College Edition, an online university library that includes over 700 popular and scholarly journals in which you can find articles related to the topics in this chapter. Suggested articles and questions relating to these articles are listed below.

Briggs Jr., Vernon M. 1998. "American-style Capitalism and Income Disparity." *Journal of Economic Issues* 32(June):473–81.

1. What recent business trends have exacerbated the income inequality we see in modern America?
2. What institution bolstered capitalism in its fledgling days, and how has this institution changed due to capitalism?
3. Why does the author feel that social anarchy will result from income inequality rather than a Marxian revolution?

Dugger, William M. 1998. "Against Inequality." *Journal of Economic Issues* 32(2):286–304.

1. What problem does the author see with governmental redistribution of wealth?
2. According to the author, what three methods of social control are used to keep the downtrodden in their place?
3. What is "folly" and how is it different from inequality?

Niggle, Christopher J. 1998. "Equality, Democracy, Institutions and Growth." *Journal of Economic Issues* 32(2):523–31.

1. What did classical economists see as the key to economic growth?
2. In the new theory of the political economy of growth, what effects does inequality have on business?
3. What is the weakness the author identifies in the political economy of growth model?

CHAPTER ELEVEN

Work and Unemployment

Is It True?

Is It True?

1. Child labor and sweatshops are found in poor countries but are virtually nonexistent in the United States.
2. Federal regulations mandate that employers allow employees to use the bathroom when they need to.
3. The world's top 200 corporations control over one-fourth of all sales in the global economy but employ less than 1 percent of the world's workforce.
4. Since 1983, U.S. labor union membership has increased steadily.
5. Since 1960, the rate of U.S. workers killed on the job has increased dramatically.

Answers: 1 = F; 2 = T; 3 = T; 4 = F; 5 = F

All that serves labor serves the nation. All that harms is treason . . . If a man tells you he loves America, yet hates labor, he is a liar . . . There is no America without labor, and to fleece one is to rob the other.

Abraham Lincoln

On December 10, 1948, the General Assembly of the United Nations adopted and proclaimed the Universal Declaration of Human Rights. Among the articles of that declaration are the following:

> *Article 23.* Everyone has the right to work, to free choice of employment, to just and favourable conditions of work and to protection against unemployment.
> Everyone, without any discrimination, has the right to equal pay for equal work.
> Everyone who works has the right to just and favourable remuneration ensuring for himself and his family an existence worthy of human dignity, and supplemented, if necessary, by other means of social protection.
> Everyone has the right to form and to join trade unions for the protection of his interests.
> *Article 24.* Everyone has the right to rest and leisure, including reasonable limitation of working hours and periodic holidays with pay.

More than half a century later, workers around the world are still fighting for these basic rights as proclaimed in the Universal Declaration of Human Rights. In this chapter, we examine problems of work and unemployment, including child labor, health and safety hazards in the workplace, job dissatisfaction and alienation, work/family concerns, and declining labor unions. We begin by looking at the global economy. *Note*: Before reading further, you may want to complete the "Attitudes toward Corporations" survey in the *Self & Society* feature of this chapter. It might be interesting to retake this survey after reading this chapter and see how your attitudes may have changed.

THE GLOBAL CONTEXT: THE ECONOMY IN THE TWENTY-FIRST CENTURY

In January 1999, 11 of the 15 European Union nations began making the transition from their national currency to a new common currency—the euro—which will lock them together financially. The euro and national currencies will operate in parallel until 2002, after which German marks, French francs, and Italian lire will be replaced by single currency notes and coins. Some economists predict the euro could rival the U.S. dollar as the dominant international currency (Reed 1999). The adoption of the euro reflects the increasing globalization of economic institutions. The **economic institution** refers to the structure and means by which a society produces, distributes, and consumes goods and services.

In recent decades, innovations in communication and information technology have spawned the emergence of a **global economy**—an interconnected network of economic activity that transcends national borders and spans across the world. The globalization of economic activity means that increasingly our jobs, the products and services we buy, and our nation's political policies and agendas influence and are influenced by economic activities occurring around the world. After summarizing the two main economic systems in the world—capitalism and socialism—we look at the emergence of corporate multinationalism. Then we describe how industrialization and postindustrialization have changed the nature of work.

Capitalism and Socialism

> Capitalism is the extraordinary belief that the nastiest of men, for the nastiest of reasons, will somehow work for the benefit of us all.
>
> —John Maynard Keynes
> Economist

The principal economic systems in the world are capitalism and socialism. Under **capitalism** private individuals or groups invest capital (money, technology, machines) to produce goods and services to sell for a profit in a competitive market. Capitalism is characterized by economic motivation through profit, the determination of prices and wages primarily through supply and demand, and the absence of governmental intervention in the economy. Critics of capitalism argue that it creates too many social evils, including alienated workers, poor working conditions, near-poverty wages, unemployment, a polluted and depleted environment, and world conflict over resources.

Socialism emphasizes public rather than private ownership. Theoretically, goods and services are equitably distributed according to the needs of the citizens. Whereas capitalism emphasizes individual freedom, socialism emphasizes social equality.

Both capitalism and socialism claim that they result in economic well-being for society and its members. In reality, capitalist and socialist countries have been unable to fulfill their promises. Although the overall standard of living is higher in capitalistic countries, so is economic inequality. Some theorists have suggested that capitalist countries will adopt elements of socialism, and socialist countries will adopt elements of capitalism. This idea, known as the **convergence hypothesis**, is reflected in the economies of Germany, France, and Sweden, which are sometimes called "integrated economies" because they have elements of both capitalism and socialism.

CONSIDERATION

The Great Depression caused the U.S. federal government to reassess its hands-off attitude toward the economy and led to the view that some governmental involvement in the economy is necessary. Today, the U.S. government regulates the stock market, the minimum wage, communications networks, and transportation (air, rail, truck). The government also controls economic programs such as Social Security, unemployment compensation, Medicaid, and Medicare. The government also monitors the safety and quality of products, services, and the work environment.

Corporate Multinationalism

Corporate multinationalism is the practice of corporations to have their home base in one country and branches, or affiliates, in other countries. Corporate multinationalism allows businesses to avoid import tariffs and costs associated with transporting goods to another country. Access to raw materials, cheap foreign labor, and the avoidance of government regulations also drive corporate multinationalism. "By moving production plants abroad, business managers may be able to work foreign employees for long hours under dan-

Multinational corporations market their goods and services in countries throughout the world.

gerous conditions at low pay, pollute the environment with impunity, and pretty much have their way with local communities. Then the business may be able to ship its goods back to its home country at lower costs and bigger profits" (Caston 1998, 274–75).

Although multinationalization provides jobs for U.S. managers, secures profits for U.S. investors, and helps the United States compete in the global economy, it also has "far-reaching and detrimental consequences" (Epstein, Graham, and Nembhard 1993, 206) First, it contributes to the trade deficit in that more goods are produced and exported from outside the United States than from within. Thus, the United States imports more than it exports. Second, multinationalism contributes to the budget deficit. The United States does not get tax income from U.S. corporations abroad, yet multinationals pressure the government to protect their foreign interests; as a result, military spending increases. Third, multinationalism contributes to U.S. unemployment by letting workers in other countries perform labor that could be performed by U.S. employees. Finally, corporate multinationalization must take its share of the blame for an array of other social problems such as poverty resulting from fewer jobs, urban decline resulting from factories moving away, and racial and ethnic tensions resulting from competition for jobs.

Industrialization, Postindustrialization, and the Changing Nature of Work

The nature of work has been shaped by the Industrial Revolution, the period between the mid–eighteenth century and the early nineteenth century when the factory system was introduced in England. **Industrialization** dramatically altered the nature of work: machines replaced hand tools; steam, gasoline, and electric power replaced human or animal power. Industrialization also led to the development of the assembly line and an increased division of labor as goods began to be mass-produced. The development of factories contributed to the emergence of large cities where the earlier informal social interactions dominated by primary relationships were replaced by formal interactions centered

Attitudes toward Corporations

PART ONE

How good a job do you think corporations are doing these days? Using letter grades like those in school, give corporations an A, B, C, D, or F in:

	Letter Grade
1. Paying their employees good wages	_____
2. Being loyal to employees	_____
3. Making profits	_____
4. Keeping jobs in America	_____

PART TWO

Here are some things some large corporations are doing that some people think are serious problems, while others think they are not serious problems. For each of the following practices, indicate whether you think this is a serious problem or not.

	Serious Problem	Not a Serious Problem	Don't Know
5. Not providing health care and pensions to employees			
6. Not paying employees enough so that they and their families can keep up with the cost of living			
7. Paying CEOs 200 times what their employees make			
8. Laying off large numbers of workers even when they are profitable.			

PART THREE

Which of the following statements comes closer to your view? (check one)

9A. A major problem with the economy today is government waste and inefficiency. Excessive government spending and high taxes burden middle class families and slow economic growth. Our government debt drives up interest rates, making it much harder for businesses to invest and create jobs.

OR

9B. A major problem with the economy today is politicians catering to the interests of powerful corporations and wealthy campaign contributors at the expense of working families. That is why politicians are not doing anything to stop large corporations from laying off large numbers of employees, denying health benefits, moving jobs overseas, and raiding pension funds.

9A. _____ 9B. _____

10A. Wasteful and inefficient government is preventing the middle class from getting ahead and doing better. Excessive government spending and high taxes burden working families and slow economic growth. The budget deficit drives up interest rates and taxes, hurts consumers and business, and reduces job-creating investments. Red tape and excessive regulation are hurting business.

OR

10B. Corporate greed is preventing the middle class from getting ahead and doing better. In the past when people did their jobs well they could earn a decent wage and provide a better life for their children. Now, corporate America is squeezing their employees—cutting wages, downsizing jobs, and eliminating pensions and health benefits. Companies say they can't afford to treat employees better, but many have growing profits, record stock prices, and huge salaries for their executives.

10A. _____ 10B. _____

Attitudes toward Corporations *(continued)*

11A. Large corporations are laying people off, cutting benefits, and moving jobs overseas mainly because they have gotten greedy and are squeezing employees to maximize profits.

OR

11B. Large corporations are laying people off, cutting benefits, and moving jobs overseas mainly because they have to in order to stay in business and provide jobs.

11A. ____ 11B. ____

RESULTS OF A NATIONAL SAMPLE

You may want to compare your answers to this survey with responses from a national sample of U.S. adults.

PART ONE (Percentages do not total 100 due to individuals who responded "Don't know.")

1.	2.	3.	4.
A, 11%	A, 10%	A, 52%	A, 12%
B, 26%	B, 16%	B, 26%	B, 16%
C, 36%	C, 28%	C, 9%	C, 31%
D, 12%	D, 23%	D, 3%	D, 20%
F, 7%	F, 19%	F, 2%	F, 18%

PART TWO

	Serious	Not Serious	Don't Know
5.	82%	15%	3%
6.	76%	19%	5%
7.	79%	14%	7%
8.	81%	14%	5%

PART THREE

9A. 33% 9B. 40% (21% answered "both," and 6% answered "Don't know")

10A. 28% 10B. 46% (22% answered "both," and 4% answered "Don't know")

11A. 70% 11B. 22% (7% answered "Don't know")

SOURCE: Adapted from "Corporate Irresponsibility: There Ought to be Laws." 1996. EDK Poll, Washington, D.C.: Preamble Center for Public Policy. **http://www.preamble.org/polledk.html** (December 12, 1998). Used by permission.

around secondary groups. Instead of the family-centered economy characteristic of an agricultural society, people began to work outside the home for wages.

Postindustrialization refers to the shift from an industrial economy dominated by manufacturing jobs to an economy dominated by service-oriented, information-intensive occupations. Postindustrialization is characterized by a highly educated workforce, automated and computerized production methods, increased government involvement in economic issues, and a higher standard of living (Bell 1973). Like industrialization before it, postindustrialization has transformed the nature of work.

There are three fundamental work sectors: primary, secondary, and tertiary. These work sectors reflect the major economic transformations in society—the

National Data

Between 1996 and 2006, virtually all U.S. job growth will be in service-producing industries. The 10 fastest growing occupations include 6 health-related and 4 computer-related occupations.

SOURCE: Bureau of Labor Statistics 1997

Industrial Revolution and the Postindustrial Revolution. The **primary work sector**, involves the production of raw materials and food goods. In developing countries with little industrialization, about 60 percent of the labor force works in agricultural activities; in the United States less than 3 percent of the workforce is in farming (Bracey 1995; *Report on the World Social Situation* 1997). The **secondary work sector** involves the production of manufactured goods from raw materials (e.g., paper from wood). The third sector is the **tertiary work sector**, which includes professional, managerial, technical-support, and service jobs. In a postindustrial society, the highest proportion of jobs is in the tertiary sector.

In a postindustrial society, highly skilled and technological personnel are needed, but many U.S. workers, particularly women and minorities, are not educated and skilled enough for many of these positions (Koch 1998a). U.S. employers claim that a shortage of skilled high-tech workers is hurting business, and the employers have lobbied Congress to allow them to admit more foreign workers. But critics believe that employers want more foreign workers because they are cheaper (Koch 1998a). In developing countries, many individuals with the highest level of skill and education leave the country in search of work abroad, leading to the phenomenon known as the **brain drain.** Although U.S. employers benefit as they pay lower wages to foreign workers, U.S. workers are displaced and developing countries lose valuable labor.

SOCIOLOGICAL THEORIES OF WORK AND THE ECONOMY

Numerous theories in economics, political science, and history address the nature of work and the economy. In sociology, structural-functionalism, conflict theory, and symbolic interactionism serve as theoretical lenses through which we may better understand work and economic issues and activities.

Structural-Functionalist Perspective

According to the structural-functionalist perspective, the economic institution is one of the most important of all social institutions. It provides the basic necessities common to all human societies, including food, clothing, and shelter. By providing for the basic survival needs of members of society, the economic institution contributes to social stability. After the basic survival needs of a society are met, surplus materials and wealth may be allocated to other social uses, such as maintaining military protection from enemies, supporting political and religious leaders, providing formal education, supporting an expanding population, and providing entertainment and recreational activities. Societal development is dependent on an economic surplus in a society (Lenski and Lenski 1987).

Although the economic institution is functional for society, elements of it may be dysfunctional. For example, prior to industrialization, agrarian societies had a low division of labor in which few work roles were available to members of society. Limited work roles meant that society's members shared similar roles and thus developed similar norms and values (Durkheim [1893] 1966). In contrast, industrial societies are characterized by many work roles, or a high division of labor, and cohesion is based not on the similarity of people and their roles but on their interdependence. People in industrial societies need the skills and services that others provide. The lack of common norms and values in industrialized societies may result in *anomie*—a state of normlessness—which is linked to a variety of social problems including crime, drug addiction, and violence (see Chapters 3 and 4).

> Our principal motivations in world affairs have been largely economic.
>
> –Richard J. Caston
> Sociologist

Conflict Perspective

According to Karl Marx, capitalism is responsible for the inequality and conflict within and between societies. The ruling class controls the economic system for its own benefit and exploits and oppresses the working masses. While structural-functionalism views the economic institution as benefitting society as a whole, conflict theory holds that capitalism benefits an elite class that controls not only the economy but other aspects of society as well—the media, politics and law, education, and religion.

As an indication of the ties between business and government, conflict theorists point to the growing level of corporate influence in U.S. politics. As noted in Chapter 10, huge corporate financial contributions to politicians and political parties continue to buy political influence that favors corporate interests.

> The country is governed for the richest, for the corporations, the bankers, the land speculators, and for the exploiters of labor.
>
> –Helen Keller, 1911
> Social activist

Corporate wealth and power have grown at the expense of workers. Between 1980 and 1992, the 500 largest U.S. corporations increased their assets 227 percent, but during that same time they cut jobs by 28 percent (Danaher 1998).

Corporate power is also reflected in the policies of the International Monetary Fund (IMF) and the World Bank, which pressure developing countries to open their economies to foreign corporations, promoting export production at the expense of local consumption, encouraging the exploitation of labor as a means of attracting foreign investment, and hastening the degradation of natural resources as countries sell their forests and minerals to earn money to pay back loans. Ambrose (1998) asserts that "for some time now, the IMF has been the chief architect of the global economy, using debt leverage to force governments around the world to give big corporations and billionaires everything they want—low taxes, cheap labor, loose regulations—so they will locate in their countries" (p. 5). Treaties such as the North American Free Trade Agreement (NAFTA), the General Agreement on Tariffs and Trade (GATT), and the Multilateral Agreement on Investments (MAI) also benefit corporations at the expense of workers by providing U.S. corporations greater access to foreign markets. "These laws increasingly allow corporations to go anywhere and do anything they like, and prohibit workers and the governments that supposedly represent them from doing much about it" (Danaher 1998, 1).

International Data

The world's 200 top corporations control 28 percent of all sales in the global economy, yet they employ less than 1 percent of the world's workforce.

SOURCE: Danaher 1998

According to the conflict perspective, work trends that benefit employees, such as work site health promotion programs and work-family policies (discussed later in this chapter) are not the result of altruistic or humanitarian concern for workers' well-being. Rather, corporate leaders recognize that these programs and policies result in higher job productivity and lower health care costs and are thus good for the "bottom line."

Symbolic Interactionist Perspective

According to symbolic interactionism, the work role is a central part of a person's identity. When making a new social acquaintance, one of the first questions we usually ask is, "What do you do?" The answer largely defines for us who that person is. For example, identifying a person as a truck driver provides a different social meaning than identifying someone as a physician. In addition, the title of one's work status—maintenance supervisor or president of the

> When a man tells you that he got rich through hard work, ask him whose.
>
> –Don Marquis
> Journalist

> No race can prosper til it learns there is as much dignity in tilling a field as in writing a poem.
>
> —Booker T. Washington
> Address to the
> Atlanta Exposition,
> September 18, 1895

United States—also gives meaning and self-worth to the individual. An individual's job is one of his or her most important statuses; for many, it comprises a "master status," that is, the most significant status in a person's social identity.

As symbolic interactionists note, definitions and meanings influence behavior. Meanings and definitions of child labor (discussed later) contribute to its perpetuation. In some countries, children learn to regard working as a necessary and important responsibility and rite of passage, rather than an abuse of human rights. Some children look forward to becoming bonded to a master "in the same way that American children look forward to a first communion or getting a driver's license" (Silvers 1996, 83).

Symbolic interactionism emphasizes that attitudes and behavior are influenced by interaction with others. The applications of symbolic interactionism in the workplace are numerous—employers and managers are concerned with using interpersonal interaction techniques that achieve the attitudes and behaviors they want from their employees; union organizers are concerned with using interpersonal interaction techniques that persuade workers to unionize; and job training programs are concerned with using interpersonal interaction techniques that are effective in motiving participants.

PROBLEMS OF WORK AND UNEMPLOYMENT

Next, we examine unemployment and other problems associated with work. The problem of discrimination in the workplace based on gender, race and ethnicity, and sexual orientation is addressed in other chapters. Minimum wage issues are discussed in Chapter 10. Here we discuss problems concerning child labor, health and safety hazards in the workplace, job dissatisfaction and alienation, work/family concerns, declining labor unions, and unemployment and underemployment.

Child Labor: A Global Problem

Youth employment is common is the United States; 80 percent of U.S. youth are employed at some point before they leave high school (Davis 1997). While economic necessity compels many U.S. youth to work, the majority work for spending money (see also Chapter 6). Employment can be a valuable experience for youth, as it encourages psychosocial development and the acquisition of work skills. But for most working children around the world, work is an oppressive, dehumanized way of life.

Child labor involves children performing work that is hazardous, that interferes with a child's education, or that harms a child's health or physical, mental, spiritual, or moral development (U.S. Department of Labor 1995). Even though virtually every country in the world has laws that limit or prohibit the extent to which children can be employed, child labor persists throughout the world.

Child labor is most prevalent in Africa, Asia, and Central and South America. India has the largest child labor force in the world, with between 20 and 80 million working children (Parker 1998). This chapter's *The Human Side* feature depicts child labor in Pakistan.

Child laborers work in factories, workshops, construction sites, mines, quarries, and fields, on deep-sea fishing boats, at home, and on the street. They make

International Data

According to the International Labour Organization (ILO), nearly 250 million children around the world work. In less developed countries, nearly one-quarter of children aged 5 to 14 work.

SOURCE: Parker 1998

THE HUMAN SIDE

CHILD LABOR IN PAKISTAN

Like most other countries, Pakistan has laws prohibiting child labor and indentured servitude. However, these laws are largely ignored, and about 11 million children aged 4 to 14 work under brutal and squalid conditions. Children make up about a quarter of the unskilled work force in Pakistan and can be found in virtually every factory, field, and workshop. They earn on average a third of the adult wage. Reading the following excerpts from an *Atlantic Monthly* report on child Labor in Pakistan (Silvers 1996) will no doubt leave you outraged, sad, and shocked that such conditions persist in our modern world.

The median age of children now entering the Pakistani work force is seven. Two years ago it was eight. Two years from now it may be six. In the lowest castes, children become laborers almost as soon as they can walk. Much of the nation's farmland is worked by toddlers, yoked teams of three-, four-, and five-year-olds who plough, seed, and glean fields from dawn to dusk. . . .

In rural areas children are raised without health care, sanitation, or education; many are as starved for affection as for food. As soon as they're old enough to have an elementary understanding of their circumstances, their parents teach them that they are expected to pay their way, to make sacrifices, and, if necessary, to travel far from home and live with strangers. "When my children were three, I told them they must be prepared to work for the good of the family," says Asthma, a Sheikhupura villager who bonded her five children to masters in distant villages. . . .

Soon after I arrived in Pakistan, I arranged a trip to a town whose major factories were rumored to enslave very young children. I found myself hoping during the journey there that the children I saw working in fields, on the roads, at the marketplaces, would prepare me for the worst. They did not. No amount of preparation could have lessened the shock and revulsion I felt on entering a sporting-goods factory in the town of Sialkot, seventy miles from Lahore, where scores of children, most of them aged five to ten, produce soccer balls by hand for forty rupees, or about $1.20, a day. The children work eighty hours a week in near-total darkness and total silence. According to the foreman, the darkness is both an economy and a precautionary measure; child-rights activists have difficulty taking photographs and gathering evidence of wrongdoing if the lighting is poor. The silence is to ensure product quality: "If the children speak, they are not giving their complete attention to the product and are liable to make errors." The children are permitted one thirty-minute meal break each day; they are punished if they take longer. They are also punished if they fall asleep, if their workbenches are sloppy, if they waste material or miscut a pattern, if they complain of mistreatment to their parents or speak to strangers outside the factory. . . . Punishments are doled out in a storage closet at the rear of the factory. . . . Children are hung upside down by their knees, starved, caned, or lashed. . . . The punishment room is a standard feature of a Pakistani factory, as common as a lunchroom at a Detroit assembly plant.

(continued)

THE HUMAN SIDE

CHILD LABOR IN PAKISTAN *(continued)*

The town's other factories are no better, and many are worse. Here are brick kilns where five-year-olds work hip-deep in slurry pits, where adolescent girls stoke furnaces in 160 degree heat. Here are tanneries where nursing mothers mix vats of chemical dye, textile mills where eight-year-olds tend looms and breathe air thick with cotton dust. . . .

A carpet workshop . . . was . . . about the size of a subway car, and about as appealing. The long, narrow room contained a dozen upright looms. On each rough-hewn workbench between the looms squatted a carpet weaver. The room was dark and airless. Such light as there was came from a single ceiling fixture, two of its four bulbs burned out. A thermometer read 105 degrees, and the mud walls were hot to the touch. . . .

Of the twelve weavers, five were eleven to fourteen, and four were under ten. The two youngest were brothers named Akbar and Ashraf, aged eight and nine. They had been bonded to the carpet master at age five, and now worked six days a week at the shop. Their workday started at 6:00 A.M. and ended at 8:00 P.M., except, they said, when the master was behind on his quotas and forced them to work around the clock. They were small, thin, malnourished, their spines curved from lack of exercise and from squatting before the loom. Their hands were covered with calluses and scars, their fingers gnarled from repetitive work. Their breathing was labored, suggestive of tuberculosis. Collectively these ailments, which pathologists call captive-child syndrome, kill half of Pakistan's working children by age twelve. . . .

A hand-knotted carpet is made by tying short lengths of fine colored thread to a lattice of heavier white threads. The process is labor-intensive and tedious: a single four-by-six-foot carpet contains well over a million knots and takes an experienced weaver four to six months to complete. . . . Each carpet Akbar completed would retail in the United States for about $2,000—more than the boy would earn in ten years. Abkar revealed that, "the master screams at us all the time, and sometimes he beats us. . . . We're slapped often. Once or twice he lashed us with a cane. I was beaten ten days ago after I made many errors of color in a carpet. He struck me with his fist quite hard on the face. . . . I was fined one thousand rupees and made to correct the errors by working two days straight." The fine was added to Akbar's debt, and would extend his "apprenticeship" by several months. . . Akbar declared that "staying here longer fills me with dread. I know I must learn a trade. But my parents are so far away, and all my friends are in school. My brother and I would like to be with our family. We'd like to play with our friends. This is not the way children should live."

SOURCE: Jonathan Silvers, 1996. From "Child Labor in Pakistan." © 1996 by Jonathan Silvers as first published in *The Atlantic Monthly*, February 1996. Reprinted by permission.

bricks, shoes, soccer balls, fireworks and matches, furniture, toys, rugs, and clothing. They pick crops and tend livestock. They work in manufacturing of brass, leather goods, and glass. Kids rummage through garbage searching for rags to sell.

Illegal and oppressive employment of children also occurs in the United States in agriculture, in restaurants, grocery stores, meatpacking industries, and in sweatshops in urban garment districts. **Sweatshops** are work environments that are characterized by less than minimum wage pay, excessively long hours of work (often without overtime pay), unsafe or inhumane working conditions, abusive treatment of workers by employers, and/or the lack of worker organizations aimed at negotiating better work conditions. (Adults—particularly immigrants and women—also work in U.S. sweatshops.)

Extreme forms of child labor include bonded labor and commercial sexual exploitation. **Bonded labor** refers to the repayment of a debt through labor. Typically, an employer loans money to parents, who then give the employer their children as laborers to repay the debt. Sometimes the child is taken far away from the family to work; other times the child works in the same village and continues to live at home. The children are unable to work off the debt due to high interest rates, low wages, and wage deductions for meals, lodging, and mistakes made at work (U.S. Department of Labor 1995). Bonded labor is like slavery; a bonded worker is not free to leave the workplace. About 10 to 20 million children in the world are forced to work as bonded laborers (Parker 1998). Bonded labor is most common in India, Nepal, Bangladesh, and Pakistan.

Child prostitution occurs throughout the world and is particularly prevalent in Asia and Central and South America. In poor societies, the sexual services of children are often sold by their families in an attempt to get money. Some children are kidnapped or lured by traffickers with promises of employment, only to end up in a brothel. An estimated one-quarter of all visitors using child prostitutes in Asia are American businessmen or military personnel (Kennedy 1996).

National Data

In the United States, an estimated 230,000 children are working in agriculture, and 13,000 children are working in sweatshops.

SOURCE: Global March against Child Labor 1998

National Data

In the United States, over 100,000 children are involved in child prostitution.

SOURCE: Kennedy 1996

Causes of Child Labor

Poverty, economic exploitation, social values, and lack of access to education are factors contributing to the persistence of child labor. One mother in Bangladesh whose 12-year-old daughter works up to 14 hours a day in a garment sweatshop explained, "Children shouldn't have to work . . . But if she didn't, we'd go hungry" (Parker 1998, 47). The economic advantages to industries that profit from child labor also perpetuate the practice. However, some employers and government officials in Eastern countries claim that child labor is not a violation of social values. In the words of one employer in Pakistan who uses child labor, "Child labor is a tradition the West cannot understand and must not attempt to change" (Silvers 1996, 86). Finally, child labor results from failure to provide education to all children. The education system in Pakistan, for example, can only accommodate about one-third of the country's school-age children, leaving the remainder to join the child labor pool (Silvers 1996).

Consequences of Child Labor

Child laborers are at risk for a wide variety of health problems such as injuries, stunted growth, and many diseases. Child carpet weavers develop gnarled fingers from the repetitive work, and their spines are curved from sitting at looms all day. Young brickworkers breathe in dust from the dry bricks and sand, causing scarring of the lungs and early death. Child farmworkers are exposed to harmful pesticides. Child prostitutes are at high risk for acquiring HIV and suffer the emotional scars of their exploitation. Child laborers are fed inadequate diets and must endure harsh punishment from their employers. One girl who was forced into prostitution in Bankok said, "One time I refused to sleep with a man and they slapped me, hit me with a cane and bashed my head against the wall. One of my friends tried to run away but unfortunately she was caught and very badly beaten" (Parker 1998, 42).

Iqbal Masih, who worked in a carpet factory in Pakistan when he was just 4 years old, escaped the factory when he was 10. With the help of the Bonded Labor Liberation Front (BLLF), Iqbal entered school and worked with the BLLF to enforce labor laws, free bonded children, and educate the public about child labor. In 1995, at age 12, Iqbal was shot and killed in a rural village where he was visiting relatives. Although the killer was never found, some suspect that Iqbal's murder was arranged by the carpet manufacturers, who were losing child workers as a result of Iqbal's campaigning and activism.

Child labor also has detrimental consequences for the larger society. Although poverty drives parents to send their children to work, child labor also increases poverty by depressing already low wages. Parker (1998) explains,

> **For every child who works, there may be an adult who cannot find a job. Children are usually paid less than adult workers—sometimes only one-third of what adults earn. As a result, adult workers' wages stay low or go down. When parents cannot find jobs, they are more likely to send their children to work. They have more children in the hope of increasing their income. Each generation of poor, uneducated child workers becomes the next genre of poor parents who must send their kids to work. Then the cycle of poverty and illiteracy continues. (Parker 1998, 48)**

Although cheap child labor may fuel economic growth for some countries, it also hinders industrial development, especially in the use of advanced technologies. Rather than invest in labor-saving technology, manufacturers use a cheaper alternative: child labor.

National Data

In 1997, cumulative trauma disorders made up 64 percent of the 430,000 total workplace illness cases. Women represent two-thirds of reported cases.

SOURCES: Bureau of Labor Statstics 1998e; Center for Workplace Health Information 1998

Health and Safety Hazards in the U.S. Workplace

Accidents at work and hazardous working conditions contribute to illnesses, injuries, and deaths. Of the 6.1 million nonfatal workplace injuries and illnesses in 1997, nearly 5.7 million resulted in either lost work time, medical treatment, loss of consciousness, restriction of work or motion, or transfer to another job (Bureau of Labor Statistics 1998e). Manufacturing industries have the highest rate of occupational illnesses and injuries (National Safety Council 1997).

The most common type of workplace illness is **cumulative trauma disorders** (see Table 11.1), which are muscle, tendon, vascular, and nerve injuries

Table 11.1 Occupational Illness Rates in the United States, 1997

Occupational Illness	Rate per 10,000 Full-Time Workers
ALL ILLNESSES	49.8
Cumulative trauma disorders	32.0
Skin diseases, disorders	6.7
Respiratory conditions due to toxic agents	2.4
Disorders due to physical agents	1.9
Poisoning	0.6
Dust diseases of the lungs	0.3
All other occupational diseases	5.8

SOURCE: Bureau of Labor Statistics. 1998 (December). "Workplace Injuries and Illnesses in 1997." **http://www.osha.gov/oshstats/bls/osnr0007.txt** (February 23, 1999).

> As long as the existence of hazardous jobs is tolerated, the most economically and socially disadvantaged workers will continue to be at the greatest risk.
>
> —Dana Loomis and David Richardson
> University of North Carolina at Chapel Hill

that result from repeated or sustained actions or exertions of different body parts. Jobs that are associated with high rates of upper body cumulative trauma disorders include computer programming, manufacturing, meatpacking, poultry processing, and clerical/office work (National Safety Council 1997). Common medical conditions that have been related to cumulative trauma disorder include noise-induced hearing loss, tendinitis (inflammation of the tendons), epicondylitis ("tennis elbow"), and carpal tunnel syndrome (a wrist disorder that can cause numbness, tingling, and severe pain). Cumulative trauma disorders are classified as illness, not as injury, because they are not sudden, instantaneous traumatic events.

> It's difficult to think of an area of my life that has not been affected. . . . I have changed everything from going from a pump hair spray to an aerosol, to a new bra I could hook, to a new car I didn't have to shift.
>
> —Alli Robertson, who developed a cumulative trauma disorder while programming software at a Utah company

Workplace Fatalities Unintentional work-injury death rates have declined significantly over the last several decades (see Figure 11.1). Although many workplaces are safer today than in generations past, fatal occupational injuries and illnesses still occur in troubling numbers.

The most common cause of fatal occupational injury is highway traffic accidents, followed by homicides. Together, these two events totaled over a third of work-injury deaths in 1996 (Occupational Safety and Health Administration 1997). Most workplace homicides—80 percent—are committed in the course of robberies (National Safety Council 1997). Other causes of workplace fatalities include falls, being struck with objects and equipment, contact with electric current, and exposure to harmful substances or environments.

Risk of fatal workplace injuries is much higher among men than among women, as men are more likely to work in jobs that have high fatal injury rates.

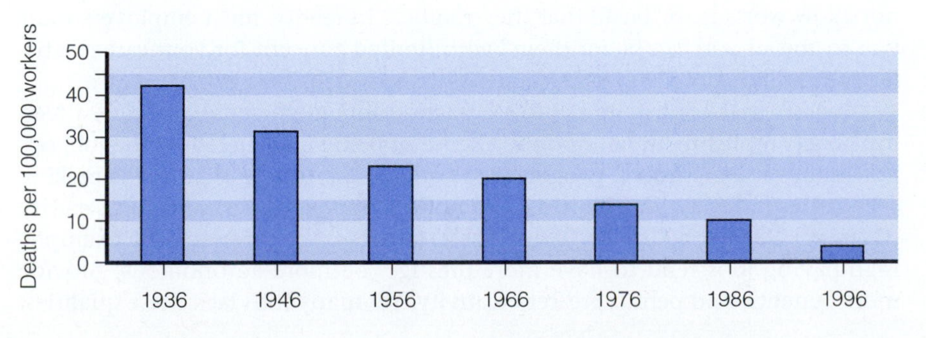

Figure 11.1 Unintentional U.S. Work-Injury Death Rates* in Selected Years

SOURCE: National Safety Council. 1997. *Accident Facts 1997 Edition.* Itasca, Ill.: National Safety Council.

National Data

In 1996, there were 6,112 fatal work injuries in the United States. Of these fatal workplace injuries, 4,800 were unintentional.

SOURCES: Occupational Safety and Health Administration 1997; National Safety Council 1997

Minorities also are at higher risk of fatal workplace injuries. One study found that African-American workers' rate of death from injuries on the job was 30 percent to 50 percent higher than the rate among white workers (Loomis and Richardson 1998). This is primarily because black men are concentrated in more hazardous jobs.

Many occupational health and safety hazards are attributed to company negligence or willful disregard of government and safety laws. For example, the dangers of asbestos were known as early as 1918, when American insurance companies stopped selling life insurance policies to asbestos workers. Nevertheless, the asbestos industry took little action until the 1960s. Of the half-million workers exposed to "significant doses of asbestos," 100,000 will die from lung cancer, 35,000 from mesothelioma, and 35,000 from asbestosis (Coleman 1994, 79).

Another work-related health hazard is job stress and chronic fatigue. Prolonged job stress, also known as **job burnout**, can cause physical problems, such as high blood pressure, ulcers, and headaches, as well as psychological problems. In a survey of 1,298 employees (Bond, Galinsky, and Swanberg 1997), nearly one-quarter of employees felt nervous or stressed often or very often in the 3 months prior to the survey; 13 percent had difficulty coping with the demands of of everyday life often or very often. The survey also found that substantial numbers of employees felt burned-out by their jobs. In the 3 months prior to the survey, 26 percent felt emotionally drained by their work often or very often, and 36 percent felt used up at the end of the workday often or very often. In Japan, as many as 10,000 workers each year perish from *karoshi*—literally translated as "death from overwork" (Bettelheim 1998). The Japanese government now recognizes *karoshi* as an industrial disease involving a blend of fatigue, high blood pressure, and hardening of the arteries.

National and International Data

Vacation time from one's job helps individuals unwind and refuel. While the average vacation time for full-time U.S. workers is about 16 days a year, several European countries have legally mandated paid vacation times of 21 to 32 days annually.

SOURCE: Mishel, Bernstein, and Schmitt 1999

Job stress, burnout, and chronic fatigue are related to the increasing demands of the workplace. A survey of 1,298 employees found that employees are working longer hours today than employees 20 years ago worked. Among employees working 20 hours or more per week, hours worked at all jobs increased from 43.6 hours in 1977 to 47.1 hours in 1997 (Bond, Galinsky, and Swanberg 1997). The survey also found that nearly one in five employees are required to work paid or unpaid overtime hours at least once a week with little or no notice.

Dissatisfaction and Alienation

An advertisement for Army recruits claims that joining the Army enables one to "be all that you can be." Indeed, if you read the classified ad section of any newspaper, you are likely to find job advertisements that entice applicants with claims such as "discover a rewarding and challenging career . . ." and "we offer opportunities for advancement and travel. . . ." Unfortunately, most jobs do not allow workers to "be all that they can be." In reality, most employers want you to "be all you can be for them" with limited concern for your career satisfaction. Millions of U.S. workers are dissatisfied with their work.

Without work, all life goes rotten, but when work is soulless, life stifles and dies.

—Albert Camus
Philosopher

Factors that contribute to job satisfaction include income, prestige, a feeling of accomplishment, autonomy, a sense of being challenged by the job, opportunities to be creative, congenial coworkers, the feeling that one is making a contribution, pay and benefits, promotion opportunities, and job security (Gordon 1996; Robie et al. 1998). These factors often overlap—for example, high-paying jobs tend to have more prestige, be more autonomous, provide more benefits, and permit greater creativity. Yet many jobs lack these qualities, leaving workers dissatisfied. About 3 out of 10 employees think it is somewhat or very likely they will lose their jobs in the next couple of years (Bond, Galinsky, and Swanberg 1997). Thirty percent feel that their company does not have a strong sense of loyalty to them. Workers are also dissatisfied with declining wages (see Chapter 10). The decline in wages in the 1980s and 1990s occurred in spite of the increased economic productivity achieved in the same time period. Economist Lester Thurow (1996) commented that "never before have a majority of American workers suffered real wage reductions while the real per capita GDP was advancing" (p. 24).

> Clearly the most unfortunate people are those who must do the same thing over and over again, every minute, or perhaps twenty to the minute. They deserve the shortest hours and the highest pay.
>
> –John Kenneth Galbraith
> American economist

One form of job dissatisfaction is a feeling of **alienation.** Work in industrialized societies is characterized by a high division of labor and specialization of work roles. As a result, workers' tasks are repetitive and monotonous and often involve little or no creativity. Limited to specific tasks by their work roles, workers are unable to express and utilize their full potential—intellectual, emotional, and physical. According to Marx, when workers are merely cogs in a machine, they become estranged from their work, the product they create, other human beings, and themselves. Marx called this estrangement "alienation."

Alienation usually has four components: powerlessness, meaninglessness, normlessness, and self-estrangement. Powerlessness results from working in an environment in which one has little or no control over the decisions that affect one's work. Meaninglessness results when workers do not find fulfillment in their work. Workers may experience normlessness if workplace norms are unclear or conflicting. For example, many companies that have family leave policies informally discourage workers from using them. Or workplaces that officially promote nondiscrimination in reality practice discrimination. Alienation also involves a feeling of self-estrangement, which stems from the workers' inability to realize their full human potential in their work roles and lack of connections to others. In general, traditional women's work is more alienating than men's work (Ross and Wright 1998). "Homemaking exposes women to routine, unfulfilling, isolated work; and part-time employment exposes them to routine, unfulfilling work, with little decision-making autonomy" (p. 343). This chapter's *Focus on Social Problems Research* feature examines job satisfaction among male and female lawyers.

National Data

In 1997, 65 percent of mothers with children under 6 were in the labor force; 58 percent of mothers with children under a year old were employed or looking for employment.

SOURCE: Bureau of Labor Statistics 1998f

Work/Family Concerns

More than three out of four married employees have partners who are also employed—an increase from 66 to 78 percent over the past 20 years. And, in three-quarters of these couples both partners work full-time (Bond, Galinsky, and Swanberg 1997). Understandably, for many employed parents, balancing work and family demands is a daily challenge. When Hochschild (1997) asked a sample of employed parents, "Overall, how well do you feel you can balance the demands of your work and family?" only 9 percent said "very well" (pp. 199–200). (Work/family concerns are also discussed in Chapters 5, 7, and 10.)

National Data

The annual hours worked by all family members in married couple families with children grew from 3,236 hours per year in 1979 to 3,851 hours per year in 1996—an increase of more than 13 weeks of full-time work.

SOURCE: Mishel, Bernstein, and Schmitt 1999

CONSIDERATION

The demands of the workplace may also affect the amount of time and energy adults have to get involved in volunteer work and social activism. One employed mother said,

I used to be an activist, leading the recycling movement in Spokane, Washington. I helped get out the vote for local environmental candidates. And I always thought

FOCUS ON SOCIAL PROBLEMS RESEARCH

What Determines Job Satisfaction among Male and Female Lawyers?

Researcher Charlotte Chiu (1998) wanted to investigate the effects of gender on job satisfaction. Despite women's often subordinate occupational statuses, most research suggests there is little difference in job satisfaction between men and women. The most commonly cited reasons for this finding include the following: (1) women have lower job expectations, thereby lower dissatisfaction levels, (2) women, due to traditional socialization, are less likely to voice discontent and thus appear to have higher job satisfaction rates, or (3) men and women value the variables that determine job satisfaction (for example, salary, job security) differently. The present study investigates the validity of the third explanation by examining gender and its relationship to job satisfaction in an oft-neglected sample of respondents—private practice attorneys.

Sample and Methods

The sample of respondents Chiu (1998) used was from the 1990 National Survey of Lawyers' Career Satisfaction sponsored by the American Bar Association. The survey was a self-administered questionnaire. In order to maximize homogeneity of the sample, respondents were restricted to full-time private practice attorneys who had graduated from law school within the 10 years prior to the survey date. The final sample included 326 lawyers; 27 percent of the sample were women and 12 percent other minorities. Statistics indicate that male and female respondents were similar to one another in terms of age, years since law school graduation, average number of hours worked per week, and law firm size. Men, however, were more likely to be married, to have children, and to be white.

Findings

Contrary to much of the literature, female attorneys reported overall lower rates of job satisfaction than their male counterparts. Specifically, women were significantly more likely to report dissatisfaction with (1) opportunities to succeed and influence decisions, (2) financial rewards, and (3) a competitive environment.

However, when respondents were asked to rank determinants of job satisfaction, there were no statistically significant gender differences. Both men and women ranked opportunity to succeed and influence decisions as the most important determinant, followed by a noncompetitive environment. For both men and women, financial rewards were relatively low on the list of what makes a job satisfying.

The research results thus lead to several important conclusions. First, despite similarities in background characteristics (i.e., age, hours worked, etc.), female lawyers reported lower levels of job satisfaction than male lawyers. Second, women's lower job satisfaction rate was predominantly a function of a stated lack of opportunity to succeed and influence decisions, and inadequate financial rewards. Third, and most importantly, women's lower job satisfaction rates were *not* a function of valuing job satisfaction determinants differently. There was no statistically significant gender differences in what makes a job satisfying.

What then explains the lowered job satisfaction rates of female versus male attorneys? Since both males and females ranked opportunity to succeed and influence decisions as the most important determinant of job satisfaction, and women were significantly more likely to express dissatisfaction on this variable, the author concludes that the "results support the discrimination thesis, and show that inequality in the workplace explains much more of women's dissatisfaction" than the traditional argument that men and women value job satisfaction determinants differently (1998, 538). Chiu (1998) also suggests that future research examine whether the resulting relationship between gender and job satisfaction rates among lawyers is also true among other professional and nonprofessional occupational groups.

SOURCE: Charlotte Chiu. 1998. "Do Professional Women Have Lower Job Satisfaction than Professional Men? Lawyers as a Case Study." *Sex Roles: A Journal of Research* 38:521–38.

when I had children, I'd work on a community garden with them planting, working in nature, and I'd throw out our TV. I wouldn't want them to be exposed to all that junk, ads, and violence. But now that I'm working these hours, Diane watches TV after school . . . and we're not doing a thing about recycling or gardening (Hochschild 1997, p. 220)

International Data

Worldwide unemployment affects 1 billion people, or nearly one-third of the global workforce.

SOURCE: Chossudovsky 1998

Unemployment and Underemployment

Measures of unemployment consider an individual to be unemployed if he or she is currently without employment, is actively seeking employment, and is available for employment. Unemployment figures do not include discouraged workers, who have given up on finding a job and are no longer looking for employment. Unemployment is a worldwide problem.

Compared to other industrialized countries, the United States has a low rate of unemployment (see Table 11.2). (However, other industrialized countries have higher wages than the United States and provide more social supports, such as universal health care, for their citizens.)

Unemployment rates are based on the percentage of unemployed persons at any given time who are actively seeking work. Another measure of unemployment is the **work-experience unemployment rate**, or the percentage of persons who participated in the labor force in a given year and experienced some unemployment during that year.

National Data

U.S. unemployment decreased from 5.4 percent in 1996 to 4.9 percent in 1997. Unemployment in mid-1998 was 4.3 percent—the lowest level since the 1960s.

SOURCES: Bureau of Labor Statistics 1998b; Mishel, Bernstein, and Schmitt 1999

Types and Causes of Unemployment

There are two types of unemployment: discriminatory and structural. **Discriminatory unemployment** involves high rates of unemployment among particular social groups, such as racial and ethnic minorities and women (see Chapters 7 and 8). **Structural unemployment** exists when there are not enough jobs available for those who want them. Structural unemployment is the result of social factors rather than personal inadequacies of the unemployed.

For example, unemployment results from **corporate downsizing**—the corporate practice of discharging large numbers of employees. Simply put, the term *downsizing* is a euphemism for mass firing of employees (Caston 1998).

According to Brecher (1996), declining markets or sales are not the primary cause of downsizing. Brecher cites an annual survey conducted by the American Management Association that asked major U.S. firms to list one or more reasons for their downsizing. In 1995, only 6 percent of downsizing companies reported that an actual or anticipated business downswing was the sole cause of job cuts, although 38 percent listed business downturn as a contributing factor. Further, only 20 percent of all surveyed companies reporting layoffs had a net operating loss in the year of the cuts (Brecher 1996). Over half of companies that downsized in the survey created new positions at the same time. But these new positions were largely for low-wage part-time and temporary employees.

National Data

The 1997 unemployment rates for U.S. white, black, and Hispanic males were 3.9 percent, 10.3 percent, and 7.1 percent, respectively. Unemployment rates for white, black, and Hispanic females were 3.8 percent, 10.0 percent, and 9.1 percent, respectively.

SOURCE: Mishel, Bernstein, and Schmitt 1999

Another cause of U.S. unemployment is **job exportation**, the relocation of U.S. jobs to other countries where products can be produced more cheaply. In 1990, for example, Levi-Strauss Co. moved its largest Texas plant to Costa Rica and laid off 300 workers immediately and hundreds more within months (Cockburn 1992).

Automation, or the replacement of human labor with machinery and equipment, is another feature of the work landscape that contributes to unemployment (see also Chapter 14). For example, recorded phone trees, automated

Table 11.2 Unemployment Rates in Nine Countries, 1997

U.S.	Canada	Australia	Japan	France	Germany	Italy	Sweden	United Kingdom
4.9	9.2	8.6	3.4	12.4	7.8	12.3	9.8	7.0

SOURCE: Bureau of Labor Statistics. 1998. (December 4). "Unemployment Rates in 9 Countries: 1997." **http://www.bls.gov/flsdata.htm** (December 15, 1998).

National Data

In the 1990s, over a half million workers lost their jobs due to downsizings.

SOURCE: Caston 1998

Many blue-collar jobs have been replaced by automation—machines and robots that perform work previously done by workers.

> No business which depends for existence on paying less than living wages to its workers has any right to continue in this country. By living wages I mean more than a bare subsistence level—I mean the wages of decent living.
>
> –F.D. Roosevelt

teller machines (ATMs), and automatic car washes do jobs that otherwise would be performed by workers. By 1990, 32 percent of the painters and 17 percent of the welders in the auto industry had been replaced by robots (Hodson and Sullivan 1990).

In addition to the problem of unemployment, many individuals experience **underemployment,** which refers to employment in a job that is underpaid; is not commensurate with one's skills, experience, and/or education; and/or involves working fewer hours than desired. Many of those underemployed are **contingent workers** (also called "disposable workers")—involuntary part-time workers, temporary employees, and workers who do not perceive themselves as having an explicit or implicit contract for ongoing employment.

Effects of Unemployment and Underemployment The personal consequences of unemployment (and in some cases, underemployment) include lowered self-esteem and confidence, anxiety, depression, and alcohol abuse (Feather 1990; Liem and Liem 1990). But unemployment has consequences for families and communities as well. Unemployment has also been linked to increased family violence (Straus 1980). Unemployment may also mean losing health care benefits for one's self and one's family, and underemployed individuals rarely get health care benefits from their employer. Unemployment and underemployment result in a decline in one's standard of living. Even when displaced workers find other employment, they often do not earn what they did in their previous job. In the early 1970s, most people who lost their jobs were able to find employment paying as well or better than the jobs they had lost; by the mid-1990s, only about 35 percent could do so (Caston 1998). Unemployment and underemployment also compel many to seek public assistance, which can be a frustrating and humiliating experience (see Chapter 10). Welfare-to-work legislation of 1996 requires individuals to work after 2 consecutive years of public aid (with a lifetime limit of 5 years). Yet, many unemployed persons looking for work do not find employment.

National Data

In 1997, 13.6 percent of persons who looked for work did not find employment during the year.

SOURCE: Bureau of Labor Statistics 1998c

Declining Labor Union Representation

Labor unions originally developed to protect workers and represent them at negotiations between management and labor. But the strength and membership of unions in the United States, as well as in other advanced capitalist countries, has declined in recent decades (Western 1995). **Union density**—the percentage of workers who belong to unions—grew in the 1930s and peaked in the 1940s and 1950s, when 35 percent of U.S. workers were unionized. In the 1960s and 1970s, U.S. corporations mounted an offensive attack on labor unions, "aiming to tame them or maim them" (Gordon 1996, 207). Corporations hired "management consultants" to help them develop and implement antiunion campaigns. They threatened unions with decertification, fired union leaders and organizers, and threatened to relocate their plants unless the unions and their members "behaved."

> **One management consultant firm . . . was unusually blunt in broadcasting its methods. A late-1970s blurb promoting its manual promised: "We will show you how to screw your employees (before they screw you)—how to keep them smiling on low pay—how to maneuver them into low-pay jobs they are afraid to walk away from. (Gordon 1996, 208).**

The Reagan and Bush administrations supported policies that strengthened corporations against the demands of their employees. In 1981, Reagan halted the air traffic controllers' strike by permanently replacing striking workers. Other corporations, including Greyhound, Phelps Dodge, and Eastern Airlines, began to use strikebreakers as permanent replacements for workers out on strike. Although the 1935 National Labor Relations Act (or Wagner Act) guarantees the right to strike, workers risk their jobs by doing so. In 1997, work stoppages (including strikes and lockouts) dropped to a record low of 29 ("Work Stoppages Drop to Record Low in 1997" 1998).

A study conducted by Kate Bronfenbrenner, director of labor education research at Cornell's School of Industrial and Labor Relations, found that employers threatened to close the plant in more than half of all union-organizing drives (Mokhiber and Weissman 1998). Where union organizing drives are successful, Bronfenbrenner found that employers carry out their threat and close their plant, in whole or in part, 15 percent of the time. This study also found that employers engaged in other ruthless and often illegal antiunion measures. In union-organizing drives from 1993 to 1995, more than a third of employers discharged workers for union activity, 38 percent gave bribes or special favors to employees who opposed the union, and 14 percent used electronic surveillance of union activists.

By 1983, only 20 percent of U.S. workers were union members (Bureau of Labor Statistics 1998a). Union representation continued to decline into the 1990s, except among government employees.

In addition to corporate hostility to unionization and probusiness government policies and actions, another reason for the decline in U.S. labor union membership is the reduction in industrial blue-collar jobs, where unions have drawn most of their membership. Caston (1998) suggests that another reason for union membership decline is "the increased willingness of corporate business to act more responsibly in providing for the welfare of employees, even if

> The Labor Movement: the folks who brought you the weekend.
>
> —From a bumper sticker, 1995

> The low wages at which women will work form the chief reason for employing them at all. . . . A woman's cheapness is, so to speak, her greatest economic asset. She can be used to keep down the cost of production where she is regularly employed. Where she has not been previously employed she can be introduced as a strike breaker to take the place of men seeking higher wages, or the threat of introducing her may be used to avert a strike. But the moment she organizes a union and seeks by organization to secure better wages she diminishes or destroys what is to the employer her chief value.
>
> —U.S. Bureau of Labor, *Report on Conditions of Women and Child Wage-Earners in the United States,* vol. 10, 1911

> I consider it important, indeed urgently necessary, for intellectual workers to get together, both to protect their own economic status and, also generally speaking, to secure their influence in the political field.
>
> —Albert Einstein (commenting on why he joined the American Federation of Teachers, AFL-CIO)

Although United Parcel Service striking workers were successful in achieving their demands, strikers risk losing their jobs.

only to make workers happy enough that they won't form unions or make new union demands" (p. 210).

National Data

In 1997, 14.1 percent of U.S. wage and salary workers were union members (down from 14.5 percent in 1996). Among government employees, 37 percent were unionized in 1997.

SOURCES: AFL-CIO 1998; Bureau of Labor Statistics 1998a

STRATEGIES FOR ACTION: RESPONSES TO WORKERS' CONCERNS

Government, private business, human rights organizations, and labor organizations play important roles in responding to the concerns of workers. Next we look at responses to child labor, health and safety regulations and work site programs, work-family policies and programs, efforts to strengthen labor, and workforce development programs.

Responses to Child Labor

In a global Agenda for Action, the 1997 International Conference on Child Labour urged all countries to eliminate child labor, giving urgent priority to the most extreme forms of child labor—slave or slavelike practices and forced or compulsory labor (including debt bondage), the use of children in the sex industry and drug trade, and other forms of dangerous or hazardous work that interferes with children's education ("Child Labour" 1997). Because child labor is often hidden, better methods of collecting and monitoring child labor are needed. Two important strategies in the campaign against child labor include the provision of free and compulsory education for all children and the establishment and *enforcement* of laws prohibiting oppressive and harmful child labor practices (UNICEF 1997).

> The appropriate societal response to exploitative child labor is straightforward: eliminate it and meet the economic needs of individuals forced to work under such conditions in other ways.
>
> —Letitia Davis
> Director, Occupational Health Surveillance
> Massachusetts Department of Public Health

In the United States, the first federal law governing child labor was passed in 1938—the Fair Labor Standards Act. Under this law, a child in the United States must be at least 14 years old to have a job (exceptions are made for newspaper delivery, farmwork, and acting). In response to the growing concern for child labor in other countries, a 1997 U.S. law was enacted prohibiting the U.S.

Customs Service from allowing the importation of any product that is made by "forced or indentured child labor" (International Labor Rights Fund 1997).

The United Nations Children's Fund also recommends that national and international corporations adopt codes of conduct guaranteeing that neither they nor their subcontractors will employ children in conditions that violate their rights (UNICEF 1997). Some industries, including rug and clothing manufacturers, use labels and logos to indicate that their products are not made by child laborers.

Human rights organizations such as the International Labour Organization, UNICEF, and the Child Labor Coalition, are active in the campaign against child labor. Another organization, the Bonded Labor Liberation Front (BLLF) has led the fight against bonded and child labor in Pakistan, freeing 30,000 adults and children from brick kilns, carpet factories, and farms, and placing 11,000 children in its own primary school system (Silvers 1996). However, employers in Pakistan have threatened workers with violence if they talk with "the abolitionists" or possess "illegal communist propaganda" (Silvers 1996). Human rights activists campaigning against child labor have also been victims of threats and violence.

Efforts to eliminate child labor cannot succeed unless the impoverished conditions that contribute to its practice are alleviated. A living minimum wage must be established for adult workers so they won't need to send their children to work. One labor rights advocate said, "We will not end child labor merely by attacking it in the export sector in poor nations. If children's parents in all countries around the world are not earning a living wage . . . , children will be driven into working in dangerous informal sector jobs. Labor rights for adults are essential if we truly want to eliminate child labor" (Global March against Child Labor 1998).

International Data

In most countries, child labor laws dictate that children must be at least 14 or 15 years old to work. In about 30 countries, the minimum age is 12 or 13.

SOURCE: Parker 1998

Efforts to Strengthen Labor

Labor unions have played an important role in fighting for fair wages and benefits, healthy and safe work environments, and other forms of worker advocacy. For example, the United Food and Commercial Workers (UFCW), the country's largest union representing poultry processing workers, was one of the most vocal advocates of a new Occupational Safety and Health Administration (OSHA) rule that establishes a federal workplace "potty" policy governing when employees can use the bathroom while on the job. According to UFCW international president Doug H. Dority, "For years workers in food processing industries have had to suffer the indignity of being denied the right to go to the bathroom when needed, just to maintain ever-increasing assembly-line speeds" ("New OSHA Policy Relieves Employees" 1998, 8). Dority claims that "poultry processors often have no other choice than to relieve themselves where they stand on the assembly line because their floor boss will not let them leave their workstation" (p. 8). The new OSHA rule mandates that employers must make toilet facilities available so that employees can use them when they need to. The employer may not impose unreasonable restrictions on employee use of the facilities to ensure that employees need not wait an unreasonably long time to use the bathroom.

Although efforts to strengthen labor are viewed as problematic to corporations and employers, such efforts have potential to remedy many of the problems facing workers. In an effort to strengthen their power, labor unions are merging with one another. Between 1992 and 1997, 18 unions affiliated with the AFL-CIO have combined—compared with only 10 mergers in the 5 years before

Labor unions have been instrumental in achieving a higher standard of living and in improving working conditions. They have helped to obtain safety and health measures against occupational risk; to achieve a larger degree of protection against child labor; to relieve the disabled, the sick, the unemployed; and to gain a more equitable share in the value of what they produce. These and other gains which labor unions have done much to win have reached far beyond their own membership and have benefited those who have not shared in the activity.

—The Presbyterian Church

that (National Center for Policy Analysis 1997). Labor union mergers result in higher membership numbers, thereby increasing the unions' financial resources, which are needed for successful recruiting and to withstand long strikes.

CONSIDERATION

Some workers, such as police and teachers, are legally prohibited from striking. An alternative tactic called a "sickout" involves mass numbers of workers calling in sick and not showing up for work. In 1999, American Airlines pilots staged a sickout, resulting in the cancellation of more than 6,000 flights. The travel plans of about 600,000 passengers were affected.

A woman's place is in her union.

—Coalition of Labor Union Women (CLUW)

Since 1960, the percentage of women in unions has nearly doubled from 20 to 39 percent ("Labor's 'Female Friendly' Agenda" 1998). The increasing percentages of women in labor unions is helping to strengthen labor unions' "female friendly" agenda. For example, a number of unions have been successful in bargaining for expanded family leave benefits, subsidized child care, elder care, and pay equity.

Labor unions are not the only groups advocating for worker rights. Numerous national and international human rights groups support the labor agenda. For example, Cleveland-based Nine to Five/National Association of Working Women combines education, lobbying, and organizing to deal with issues ranging from pay equity, work/family issues, and health and safety concerns.

I haven't seen as much raw anger as I see in the workplace today. One thing I've heard repeatedly around the country from unorganized workers is the following: "I never thought about joining a union, but for the first time I'm now thinking about it, because I need somebody to protect me."

—Robert Reich
Former U.S. Secretary of Labor

In June 1996, 1,400 elected delegates gathered in Cleveland, Ohio, to found the Labor Party. As an alternative to the Democratic and Republican Parties, the Labor Party was formed to unite union members and other prolabor activists for political action. The Labor Party is concerned with a number of key issues, including health care reform, increasing the minimum wage, preserving Social Security, fairness in international trade, and workers' rights. Labor Party leaders are striving to build the Labor Party into a large and powerful movement of working people.

Workforce Development and Job-Creation Programs

A variety of workforce development programs have been implemented since the first major employment programs were instituted during the Great Depression in the 1930s. Workforce development programs have provided a variety of services, including assessment to evaluate skills and needs, career counseling, job search assistance, basic education, occupational training (classroom and on-the-job), public employment, job placement, and stipends or other support services for child care and transportation assistance (Levitan, Mangum, and Mangum 1998). Workforce development programs of the last several decades include the 1962 Manpower Development and Training Act, the 1973 Comprehensive Employment and Training Act (CETA), the 1982 JTPA and the Jobs Corps program. These programs primarily assist youths, the handicapped, welfare recipients, displaced workers, the elderly, farmworkers, Native Americans, and veterans. Numerous studies have looked at the effectiveness of workforce development programs. In general, "evaluations indicate that employment and training programs enhance the earnings and employment of participants, although the effects vary by service population, are often modest because of brief training durations and the inherent difficulty of alleviating long-term deficiencies, and are not always cost effective" (Levitan, Mangum, and Mangum 1998, 199).

Efforts to prepare high school students for work include the establishment of technical and vocational high schools and high school programs and school-to-work programs. School-to-work programs involve partnerships between business, labor, government, education, and community organizations that help prepare high school students for jobs. Although school-to-work programs vary, in general, they allow high school students to explore different careers, and they provide job skill training and work-based learning experiences (Leonard 1996).

Unfortunately, funding for federal job training programs is insufficient to reach more than a small fraction of the workforce (Kenworthy 1995). Even those who complete the retraining do not always find new jobs at comparable wages. The rehiring of displaced workers also requires an improved economy that generates new jobs. Without job openings, the value of job retraining is limited.

One strategy for creating jobs involves local, state, and federal government providing benefits to corporations in the form of subsidies, tax breaks, real estate, and low-interest loans to corporations with the hope that this "corporate welfare" will result in new jobs (see also Chapters 10 and 13). However, most recent job creation in the United States is with small and medium-size companies. Although Fortune 500 companies are the biggest beneficiaries of corporate welfare, they have eliminated more jobs than they have created in the past decade (Barlett and Steele 1998). And, many of the jobs that are created are part-time or temporary jobs.

National Data

Employment in temporary agencies doubled between 1982 and 1989 and doubled again between 1989 and 1997.

SOURCE: Mishel, Bernstein, and Schmitt 1999

Efforts to create jobs must consider where the jobs are being created. The U.S. economy is described as a **split labor market** (or dual economy), because it is made up of two labor markets. The *primary labor market* refers to jobs that are stable, economically rewarding, and come with benefits. These jobs are usually occupied by the most educated and trained individuals (e.g., a corporate attorney, teacher, or accountant), most often white males. The *secondary labor market* refers to jobs that involve low pay, no security, few benefits, and little chance for advancement. Domestic servants, clerks, and food servers are examples of these jobs. Women and racial and ethnic minorities are disproportionately represented in the secondary labor market. These workers often have no union to protect them and are more likely to be dissatisfied with their job than workers in the primary labor market.

Responses to Worker Health and Safety Concerns

Over the last few decades, health and safety conditions in the workplace have improved as a result of media attention, demands by unions for change, more white-collar jobs, and regulations by OSHA. Through OSHA, the government develops, monitors, and enforces health and safety regulations in the workplace. But OSHA's more than 1,000 inspectors are able to visit only 2 percent of the 6 million workplaces in the United States each year (Kenworthy 1995). Because "the task of monitoring and enforcement simply cannot be effectively carried out by a government administrative agency," Kenworthy (1995) suggests that the United States follow the example of many other industrialized countries: turn over the bulk of responsibility for health and safety monitoring to the workforce (p. 114). Worker health and safety committees are a standard feature of companies in many other industrialized

What the public wants is called "politically unrealistic." Translated into English, that means power and privilege are opposed to it.

—Noam Chomsky

countries and are mandatory in most of Europe. These committees are authorized to inspect workplaces and cite employers for violations of health and safety regulations.

Due to the increase in cumulative trauma disorders, **ergonomics**—the designing or redesigning of the workplace to prevent and reduce cumulative trauma disorders—is becoming an accepted health and safety strategy in the workplace. The Occupational Safety and Health Administration has attempted to establish rules on workplace ergonomics, but efforts have been stalled in Congress (as of this writing). Table 11.3 lists recommendations for computer workstations designed to prevent cumulative trauma disorders associated with working on a computer. Does the work that you do on a computer conform to these guidelines?

Maximizing the health and safety of workers involves more than implementing, monitoring, and enforcing regulations. Increasingly, businesses and corporations are attempting to maximize worker's health (and corporate profits) by offering work-site health promotion programs. Work-site health promotion consists of health education, screening, and interventions designed to achieve better health among workers. Programs range from single interventions (such as screening for high blood pressure) to comprehensive health and fitness programs that may include hypertension screening, aerobic exercise and fitness, nutrition and weight control, stress management, smoking cessation, cancer-risk screening, drug and alcohol abuse prevention, accident prevention, and health information (Conrad 1999). Some workplaces have nap rooms, allowing workers to take naps (Bettelheim 1998). Many companies have employee assistance programs to help employees and their families with substance abuse, family discord, depression, and other mental health problems. Some companies assume total financial responsibility for health promotion programs; other companies share the expense with employees, who are asked to pay a nominal fee.

Health promotion programs benefit both employees and corporations. Corporations benefit because healthier employees have lower job absenteeism and turnover, file fewer health insurance claims (which results in lower health insurance costs for employers), file fewer workers' compensation claims, and exhibit higher morale and productivity. In addition, health promotion programs can enhance a company's image among workers and the community.

Table 11.3 Basic Ergonomic Recommendations for Computer Workstations

1. Keyboard and other input devices at elbow level
2. Video display screen 1 to 60 degrees below eye level
3. Screen display equipped or positioned to minimize glare
4. Rounded and/or padded hand, wrist, or forearm contact surfaces
5. Adequate space for performing task
6. Adequate leg room beneath work surface
7. Chair with appropriate back and arm support
8. Frequent short breaks or variation of tasks
9. Correct posture
10. Adjustments to work station to minimize awkward positions and applied force

SOURCE: Donald G. Downs. 1997. "Nonspecific Work-Related Upper Extremity Disorders." *American Family Physician* 55(4):1296–1302.

Employees enjoy the wide range of physical, mental, and social benefits that are associated with health promotion programs.

Work/Family Policies and Programs

The influx of women into the workforce has been accompanied by an increase in government and company policies designed to help women and men balance their work and family roles. In 1993, President Clinton signed into law the Family and Medical Leave Act, which requires all companies with 50 or more employees to provide each worker (who works at least 25 hours a week and has been working for at least a year) with up to 12 weeks of unpaid leave for reasons of family illness, birth, or adoption of a child. Yet, more than one-quarter (30.9 percent) of employed women and more than one-third (36.5 percent) of employed men are not covered by the Family and Medical Leave Act (Commission on Leave, U.S. Department of Labor 1996).

> In the long run, no work-family balance will ever fully take hold if the social conditions that might make it possible—men who are willing to share parenting and housework, communities that value work in the home as highly as work on the job, and policymakers and elected officials who are prepared to demand family-friendly reform—remain out of reach.
>
> –Arlie Hochschild
> Sociologist

Aside from government-mandated work/family policies, corporations and employers have begun to initiate policies and programs that address the family concerns of their employees. Employer-provided assistance with child care, assistance with elderly parent care, flexible work options, and job relocation assistance are becoming more common.

Offering employees more flexibility in their work hours helps parents balance their work and family demands. Flexible work arrangements, which benefit child-free workers as well as employed parents, include flextime, job sharing, a compressed workweek, and telecommuting. **Flextime** allows the employee to begin and end the workday at different times as long as 40 hours per week are maintained. For example, workers may begin at 6 A.M. and leave at 2 P.M. instead of the traditional 9 A.M. to 5 P.M. With **job sharing**, two workers share the responsibility of one job. A **compressed workweek** allows employees to condense their work into fewer days (e.g., four 10-hour days each week). **Telecommuting** allows employees to work part- or full-time at home or at a satellite office. The number of U.S. workers who telecommuted grew 30 percent from 1995, to more than 11 million (Koch 1998b). A study of U.S. companies found that the more women and minorities a company has in managerial positions, the more likely that company is to offer flexible work options (Galinsky and Bond 1998).

International Data

In Europe, women are granted 5 months' leave at full pay after giving birth. In addition, they are provided prenatal and obstetrical care, generous hospital stays, and baby equipment subsidies.

SOURCE: Hewlett and West 1998

Work/family policies benefit both employees and their families and corporations. For example, Corning Glass Works in upstate New York found that its turnover rate for female employees was twice as high as for male employees. This high turnover rate was costly—replacing each lost worker cost $40,000 (for search costs, on-the-job training costs, and the like). Corning conducted a survey and discovered that "family stress—particularly child-care problems—was the main reason so many women quit their jobs" (Hewlett 1992, 27). Corning decided to implement a family support package that included parenting leave, on-site child care, part-time work options, job sharing, and a parent resource center. The company's chairman, James P. Houghton, said that Corning's efforts "go way beyond simple justice; it's a matter of good business sense in a changing world . . . it's a matter of survival" (Hewlett 1992, 27).

> Our leaders talk as though they value families, but act as though families were a last priority.
>
> –Sylvia Hewlett
> Cornell West
> Family advocates

Fran Rodgers, president of Work/Family Directions explains the need for work/family policies:

For over 20 years we at Work/Family Directions have asked employees in all industries what it would take for them to contribute more at

> How can you ask your people to unselfishly support the needs of the corporation if the corporation will not support the needs of its people?
>
> —Roger Meade
> Scitor Corporation

work. Every study found the same thing: They need aid with their dependent care, more flexibility and control over the hours and conditions of work, and a corporate culture in which they are not punished because they have families. These are fundamental needs of our society and of every worker. (Galinsky et al. 1993, 51)

> If there ever is another war in this country, it will be between capital and labor.
>
> —Frank James
> (Jesse's brother)

UNDERSTANDING WORK AND UNEMPLOYMENT

To understand the social problems associated with work and unemployment, we must first recognize the power and influence of governments and corporations on the workplace. We must also be aware of the role that technological developments and postindustrialization have on what we produce, how we produce it, where we produce it, and who does the producing. In regard to what we produce, the United States is moving away from producing manufactured goods to producing services. In regard to production methods, the labor-intensive blue-collar assembly line is declining in importance, and information-intensive white-collar occupations are increasing. Due to increasing corporate multinationalization, U.S. jobs are being exported to foreign countries where labor is cheap, regulations are lax, and raw materials are available. Finally, in regard to who is producing, the workforce is becoming more diverse in terms of gender and racial and ethnic background and is including more contingent workers than in the past.

Decisions made by U.S. corporations about what and where to invest influence the quantity and quality of jobs available in the United States. As conflict theorists argue, such investment decisions are motivated by profit, which is part of a capitalist system. Profit is also a driving factor in deciding how and when technological devices will be used to replace workers and increase productivity. But if goods and services are produced too efficiently, workers are laid off and high unemployment results. When people have no money to buy products, sales slump, a recession ensues, and social welfare programs are needed to

Figure 11.2 From Layoff to Layoff: A Cycle

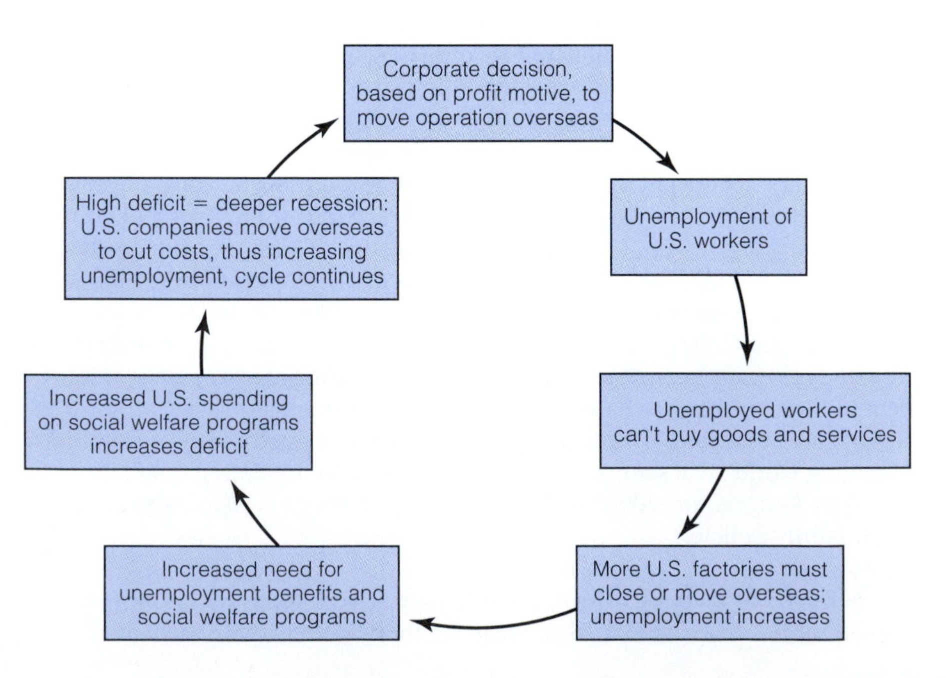

support the unemployed. When the government increases spending to pay for its social programs, it expands the deficit and increases the national debt. Deficit spending and a large national debt make it difficult to recover from the recession, and the cycle continues (see Figure 11.2).

What can be done to break the cycle? Those adhering to the classic view of capitalism argue for limited government intervention on the premise that business will regulate itself via an "invisible hand" or "market forces." For example, if corporations produce a desired product at a low price, people will buy it, which means workers will be hired to produce the product, and so on.

Ironically, those who support limited government intervention also sometimes advocate that the government intervene to bail out failed banks and lend money to troubled businesses such as Chrysler. Such government help benefits the powerful segments of our society. Yet, when economic policies hurt less powerful groups, such as minorities, there has been a collective hesitance to support or provide social welfare programs. It is also ironic that such bail-out programs, which contradict the ideals of capitalism, are needed because of capitalism. For example, the profit motive leads to multinationalization, which leads to unemployment, which leads to the need for government programs. The answers are as complex as the problems. The various forces transforming the American economy are interrelated—technology, globalization, capital flight through multinationalization, and the movement toward a service economy (Eitzen and Zinn 1990). For the individual worker, the concepts of work, job, and career have changed forever.

> I see in the near future a crisis approaching that unnerves me and causes me to tremble for the safety of my country. . . . corporations have been enthroned and an era of corruption in high places will follow, and the money power of the country will endeavor to prolong its reign by working upon the prejudices of the people until all wealth is aggregated in a few hands and the Republic is destroyed.
>
> —Abraham Lincoln

CRITICAL THINKING

1. The U.S. Department of Labor reported that in 1997, 20.2 percent of black men were union members (Bureau of Labor Statistics 1998a). With this rate, black men have the highest rate of union membership compared with any other demographic group. Why might this be so?
2. U.S. industries dominated by female workers have a higher "win rate" in organizing elections compared to industries dominated by men. Female-dominated industries have an average win rate of 60 percent, compared to 38 percent for industries where there are more males ("Labor's 'Female Friendly' Agenda" 1998). Why do think this is so?
3. In the late 1990s, a glut of media attention has been given to the positive economic indicators in the United States: low inflation rates, low unemployment rates, and surging stock market values. How might a conflict theorist view the media's portrayal of the U.S. economy?

KEY TERMS

alienation
automation
bonded labor
brain drain
capitalism
child labor
compressed workweek
contingent workers
convergence hypothesis
corporate downsizing
corporate multinationalism
cumulative trauma disorders
discriminatory unemployment
economic institution
ergonomics
flextime
global economy
industrialization

job burnout
job exportation
job sharing
postindustrialization
socialism
split labor market
structural unemployment
sweatshop
telecommuting
underemployment
union density
work-experience unemployment rate
work sectors (primary, secondary, tertiary)

INTERNET

You can find more information on labor unions, global economic issues, child labor, and worker health and safety issues at the *Understanding Social Problems* web site at **http://sociology.wadsworth.com.**

INFOTRAC COLLEGE EDITION

Either from the Wadsworth sociology resource center at **http://sociology.wadsworth.com** or directly from your web browser, you may access InfoTrac College Edition, an on-line university library that includes over 700 popular and scholarly journals in which you can find articles related to the topics in this chapter. Suggested articles and questions relating to these articles are listed below.

Pollin, Robert. 1998. "The Natural Rate of Unemployment." *Dollars and Sense,* September–October, 12–16.

1. What economic changes in the 1960s and 1990s have had significant effects on unemployment rates?
2. According to Marx, what is the function of unemployment?
3. What does the author see as the reason behind the sudden decrease in unemployment rates?

Sidanius, Che. 1998. "Immigrants in Europe: The Rise of a New Underclass." *Washington Quarterly* 21(4):5–9.

1. What was the political result of European countries' permitting open immigration?
2. What effects has unemployment had on immigrant families?
3. According to the article, what future social tensions might result from European governments' continuing to treat immigrants as noncitizens?

Wray, L. Randall. 1998. "Zero Unemployment and Stable Prices." *Journal of Economic Issues* 32(June):539–46.

1. What is the basis of the author's ELR program?
2. Assuming the author's ELR plan were initiated, what positive results for the government would occur?
3. What economic benefits does the author see in his ELR program?

CHAPTER TWELVE

Problems in Education

Is It True?

Is It True?

1. Due to court-ordered busing and a national emphasis on the equality of education, segregation in the public schools has been largely eliminated.
2. When a national sample of American adults were asked about education in America, over half replied that they were dissatisfied with the education children were receiving.
3. More than 1 out of every 10 Americans ages 18 to 24 has not completed high school.
4. More than half of people in developing nations are illiterate.
5. IQ is the best predictor of school success.

Answers: 1 = F; 2 = T; 3 = T; 4 = T; 5 = F

. . . in a democracy, all citizens have a stake in a well-educated populace and suffer from one that is ill educated. In a diverse and pluralistic country like ours, public schools are the glue that helps hold society together.

Albert Shanker
President
American
Federation
of Teachers

The U.S. educational system is often thought of as a source of national and global embarrassment. Students continue to graduate from high school who cannot read, work simple math problems, or write grammatically correct sentences. Graduates discover that they are ill prepared for corporations who demand literate, articulate, informed employees. Teachers leave the profession because of uncontrollable discipline problems, inadequate pay, and overcrowded classrooms. Students and teachers alike are "dumbing down," that is, lowering their standards, expectations, and role performances to fit increasingly undemanding and unresponsive systems of learning (Trout 1998; Leo 1998).

And yet it is education that is often claimed as a panacea—the cure-all for poverty and prejudice, drugs and violence, war and hatred, and the like. Can one institution, riddled with obstacles, be a solution for other social problems? In this chapter we focus on this question and what is being called one of the major sociopolitical issues of the next century—the educational crisis (Associated Press 1998). We begin with a look at education around the world.

The more you know, the more you know you don't know.

–Unknown

THE GLOBAL CONTEXT: CROSS-CULTURAL VARIATION IN EDUCATION

Looking only at the American educational system might lead one to conclude that most societies have developed some method of formal instruction for their members. After all, in the United States there are more than 80,000 schools, 5.6 million teachers, and 70 million students (U.S. Bureau of the Census 1998a). In reality, many societies have no formal mechanism for educating the masses. As a result, more than half of the people in the developing nations are illiterate. In India alone, more than 400 million people cannot read or write, and nearly 35 million children do not attend school (*Hindustan Times* 1999).

Education is an indispensable tool for the improvement of the quality of life.

–United Nations Conference on Population and Development

On the other end of the continuum are societies that emphasize the importance of formal education. In Japan, students attend school on Saturday, and the school calendar is 40 days longer than in the United States. Spain, China, Korea, Israel, Switzerland, Scotland, and Canada also have more mandatory school days than the United States (*Youth Indicators* 1996).

Some countries also empower professionals to organize and operate their school systems. Japan, for example, hires professionals to develop and implement a national curriculum and to administer nationwide financing for its schools. In contrast, school systems in the United States are often run at the local level by school boards and PTAs composed of lay people. In effect, local communities raise their own funds and develop their own policies for operating the school system in their area; the result is a lack of uniformity from district to district. Thus, parents who move from one state to another often find the quality of education available to their children differs radically. In societies with professionally operated and institutionally coordinated schools such as Japan, teaching is a prestigious and respected profession. Consequently, Japanese students are attentive and obedient to their teachers. In the United States, the phrase "Those that can, do; those that can't, teach" reflects the lack of esteem for the teaching profession. The insolence and defiance

International Data

An equal or higher number of 25- to 64-year-olds in the United States have completed high school or college compared with their counterparts in Japan, Germany, England, France, Italy, and Canada.

SOURCE: *Condition of Education* 1997, Indicator 23

Because teaching is a highly respected vocation in Japan, students treat their teachers with respect and obedience. Teaching is less well regarded in the United States, and the consequences are felt by our teachers every day.

among students in American classrooms are evidence of the disrespect that many American students have for their teacher.

National Data

Between 1987 and 1997 the percentage of SAT test takers with an A average rose from 28 percent to 37 percent; at the same time the combined SAT verbal and math scores of A average students dropped 14 points.

SOURCE: Marlein 1997

SOCIOLOGICAL THEORIES OF EDUCATION

The three major sociological perspectives—structural-functionalism, conflict theory, and symbolic interactionism—are important in explaining different aspects of American education.

Structural-Functionalist Perspective

According to structural-functionalism, the educational institution serves important tasks for society, including instruction, socialization, the provision of custodial care, and the sorting of individuals into various statuses. Many social problems, such as unemployment, crime and delinquency, and poverty, may be linked to the failure of the educational institution to fulfill these basic functions (see Chapters 4, 10, and 11). Structural-functionalists also examine the reciprocal influences of the educational institution and other social institutions, including the family, political institution, and economic institution.

Instruction A major function of education is to teach students the knowledge and skills that are necessary for future occupational roles, self-development, and social functioning. Although some parents teach their children basic knowledge and skills at home, most parents rely on schools to teach their children to read, spell, write, tell time, count money, and use computers. As discussed later, many U.S. students display a low level of academic achievement. The failure of schools to instruct students in basic knowledge and skills both causes and results from many other social problems.

> The beautiful thing about learning is that nobody can take it away from you.
>
> —B. B. King
> Musician

Socialization The socialization function of education involves teaching students to respect authority—behavior that is essential for social organization

(Merton 1968). Students learn to respond to authority by asking permission to leave the classroom, sitting quietly at their desks, and raising their hands before asking a question. Students who do not learn to respect and obey teachers may later disrespect and disobey employers, police officers, and judges.

The educational institution also socializes youth into the dominant culture. Schools attempt to instill and maintain the norms, values, traditions, and symbols of the culture in a variety of ways, such as celebrating holidays (Martin Luther King Jr. Day, Thanksgiving); requiring students to speak and write in standard English; displaying the American flag; and discouraging violence, drug use, and cheating.

As the number and size of racial and ethnic minority groups increase, American schools are faced with a dilemma: Should public schools promote only one common culture, or should they emphasize the cultural diversity reflected in the American population? In a recent survey, 75 percent of Americans responded that schools should do both—they should promote one common culture and emphasize diverse cultural traditions (Elam, Rose, and Gallup 1994). Banks and Banks (1993) suggest that **multicultural education** is necessary to "help students to develop the knowledge, attitudes, and skills needed to function within their own micro cultures, the U.S. macro culture, other micro cultures, and within the global community" (p. 25). Further, multicultural education is a necessary element in the battle against race, class, gender, and ethnic inequalities (Banks and Banks 1995).

National Data

In 1997, 82 percent of American adults age 25 and over had graduated from high school; 24 percent had a bachelor's degree or higher, and 8 percent held a graduate degree.

SOURCE: U.S. Bureau of the Census 1998a

CONSIDERATION

Recent research on fraternity members suggests that despite supporters who argue the benefits of the Greek system, the brotherhood has a negative effect on its participants. Several studies reported in Kuh, Pascarella, and Wechsler's "The Questionable Value of Fraternities" (1996) conclude that fraternity membership is associated with (1) lower intellectual development, (2) higher alcohol use, and (3) a reduced respect for human differences (e.g., race, ethnicity, sexual orientation).

> We need to give up the notion of a single ideal of the educated person and replace it with a multiplicity of models designed to accommodate the multiple capacities and interests of students.
>
> —Nel Noddings
> Educational reformer

Sorting Individuals into Statuses Schools sort individuals into statuses by providing credentials for individuals who achieve various levels of education, at various schools, within the system. These credentials sort people into different statuses—for example, "high school graduate," "Harvard alumni," and "English major." Further, schools sort individuals into professional statuses by awarding degrees in such fields as medicine, nursing, and law. The significance of such statuses lies in their association with occupational prestige and income. Table 12.1 documents the rewards of staying in school and the impact of sex on income differentials.

> There is a growing realization that completion of high school is the absolute minimal educational level necessary to prepare youngsters for the vast majority of jobs in the modern economy.
>
> —Gary Natriello
> Teachers College,
> Columbia University

Custodial Care The educational system also serves the function of providing custodial care (Merton 1968), which is particularly valuable to single-parent and dual-earner families, and the likely reason for the increase in enrollments of 3- and 4-year-olds. In 1966 and 1986, 13 percent and 39 percent, respectively, of 3- and 4-year-olds were enrolled in formal classes. Today, almost half of all 3- and 4-year-olds are enrolled in either nursery school or kindergarten (U.S. Bureau of the Census 1998a).

Table 12.1 **1996 Mean Earnings by Educational Attainment and Sex***

Educational Credentials	Male	Female
Less than ninth grade	$20,153	$15,150
Ninth to twelfth grade (no diploma)	25,283	17,313
High school graduates	32,521	21,289
Some college, no degree	38,491	25,889
Associate's degree	39,873	28,401
Bachelor's degree	52,354	36,555
Master's degree	70,859	44,471
Doctoral degree (Ph.D.)	86,436	62,169
Professional degree (M.D., D.D.S., J.D.)	112,873	90,711

*For year-round full-time workers 18 and over.

SOURCE: U.S. Bureau of the Census 1998b.

The school system provides supervision and care for children and adolescents until they are 16—12 years of school totaling almost 13,000 hours per pupil! Some school districts are experimenting with offering classes on a 12-month basis, Saturday classes, and/or longer school days. Providing more hours of supervision for children and adolescents may reduce juvenile delinquency and teenage pregnancy. Indeed, "lack of parental supervision is one of the strongest predictors of the development of conduct problems and delinquency" (Sautter 1995, K8).

Conflict Perspective

Conflict theorists emphasize that the educational institution solidifies the class positions of groups and allows the elite to control the masses. Although the official goal of education in society is to provide a universal mechanism for achievement, in reality educational opportunities and the quality of education are not equally distributed.

Conflict theorists point out that the socialization function of education is really indoctrination into a capitalist ideology. In essence, students are socialized to value the interests of the state and to function to sustain it. Such indoctrination begins in kindergarten. Rosabeth Moss Kanter (1972) coined the term "the organization child" to refer to the child in nursery school who is most comfortable with supervision, guidance, and adult control. Teachers cultivate the organization child by providing daily routines and rewarding those who conform. In essence, teachers train future bureaucrats to be obedient to authority.

Further, to conflict theorists, education serves as a mechanism for **cultural imperialism**, or the indoctrination into the dominant culture of a society. When cultural imperialism exists, the norms, values, traditions, and languages of minorities are systematically ignored. A Mexican-American student recalls his feelings about being required to speak English (Rodriquez 1990, 203):

> **When I became a student, I was literally "remade"; neither I nor my teachers considered anything I had known before as relevant. I had to forget most of what my culture had provided, because to remember it was a disadvantage. The past and its cultural values became detachable, like a piece of clothing grown heavy on a warm day and finally put away.**

National Data

In a recent poll, 36 percent of whites ranked public schools as either fair or poor; almost twice as many minorities, 64 percent of African Americans and 60 percent of Hispanic Americans, ranked public schools as either fair or poor.

SOURCE: Bolick 1998

Kozol contends that schools in poorer, mostly minority districts tend to have less funding and therefore less adequate facilities, books, materials, and personnel.

Finally, the conflict perspective focuses on what Kozol (1991) calls the "savage inequalities" in education that perpetuate racial inequality. Kozol documents gross inequities in the quality of education in poorer districts, largely composed of minorities, compared with districts that serve predominantly white middle-class and upper-middle-class families. Kozol reveals that schools in poor districts tend to receive less funding and have inadequate facilities, books, materials, equipment, and personnel (see Table 12.2).

Symbolic Interactionist Perspective

Whereas structural-functionalism and conflict theory focus on macro-level issues such as institutional influences and power relations, symbolic interactionism examines education from a micro perspective. This perspective is concerned with individual and small group issues, such as teacher-student interactions, the students' self-esteem, and the self-fulfilling prophecy.

Teacher-Student Interactions

Symbolic interactionists have examined the ways in which students and teachers view and relate to each other. For example, children from economically advantaged homes may be more likely to bring social and verbal skills into the classroom that elicit approval from teachers. From the teachers' point of view, middle-class children are easy and fun to teach: they grasp the material quickly, do their homework, and are more likely to "value" the educational process. Children from economically disadvantaged homes often bring fewer social and verbal skills to those same middle-class teachers, who may, inadvertently, hold up social mirrors of disapproval. Teacher disapproval contributes to the lower self-esteem among disadvantaged youth.

Self-Fulfilling Prophecy

The **self-fulfilling prophecy** occurs when people act in a manner consistent with the expectations of others. For example, a teacher who defines a student as a slow learner may be less likely to call on that student or to encourage the student to pursue difficult subjects. As a consequence of the teacher's behavior, the student is more likely to perform at a lower level.

A study by Rosenthal and Jacobson (1968) provides empirical evidence of the self-fulfilling prophecy in the public school system. Five elementary school students in a San Francisco school were selected at random and identified for their teachers as "spurters." Such a label implied that they had superior intelligence and academic ability. In reality, they were no different from the other students in their classes. At the end of the school year, however, these five stu-

Table 12.2 Public School Expenditures per Pupil (in 1997 Constant Dollars) by Function and School District Median Income

Median Household Income of District	Total	Instruction	Services	Capital Outlay	Total
Less than $20,000	$5,634	$3,052	$1,745	$407	$430
20,000–24,999	5,899	3,190	1,772	470	466
25,000–29,999	6,361	3,480	1,849	551	482
30,000–34,999	6,124	3,292	1,851	545	436
35,000 or more	7,027	3,806	2,174	606	441

SOURCE: U.S. Department of Education. 1998b. "School District Fiscal Data." National Center for Education Statistics. Washington, D.C.: U.S. Government Printing Office.

THE HUMAN SIDE

"THERE THEY THINK I'M DUMB"

The following excerpt poignantly recounts a conversation between author-ethnographer Thomas Cottle (1976) and Ollie, an 11-year-old boy, labeled as "dumb."

"You know what, Tom?" he said, looking down at his ice cream as though it suddenly had lost its flavor, "nobody, not even you or my Dad, can fix things now. The only thing that matters in my life is school, and there they think I'm dumb and always will be. I'm starting to think they're right. . . ."

"Even if I look around and know that I'm the smartest in my group, all that means is that I'm the smartest of the dumbest, so I haven't gotten anywhere at all, have I? I'm right where I always was. Every word those teachers tell me, even the ones I like most, I can hear in their voice that what they're really saying is, 'alright you dumb kids, I'll make it easy as I can, and if you don't get it then, then you'll never get it.' That's what I hear every day, man. From every one of them. Even the other kids talk to me that way too. . . ."

"I'll tell you something else," he was saying, unaware of the ice cream that was melting on his hand. "I used to think, man, that even if I wasn't so smart, that I could talk in any class in that school, if I did my studying, I mean, and have everybody in that class, all the kids and the teacher too, think I was alright. Maybe better than alright too. You know what I mean?"

"That you were intelligent," I said softly.

"Right. That I was intelligent like they were. I used to think that all the time, man. Had myself convinced that whenever I had to stand up and give a little speech, you know, about something, that I'd just be able to go do it." He tilted his head back and forth. "Just like that," he added excitedly.

"I'm sure you could too."

"I could have once, but not anymore."

"How do you know, Ollie?"

"I know."

"But how?" I persisted.

"Because last year just before they tested us and talked to us, you know, to see what we were like, I was in this one class and doing real good. As good as anybody else. Did everything they told me to do. Read what they said, wrote what they said, listened when they talked. . . ."

"Then they told me, like on a Friday, that today would be my last day in that class. That I should go to it today, you know, but that on Monday I had to switch to this other one. They just gave me a different room number, but I knew what they were doing. Like they were giving me one more day with the brains, and then I had to go to be with the dummies, where I was suppose to be. . . . So I went with the brains one more day, on that Friday like I said, in the afternoon. But the teacher didn't know I was moving, so she acted like I belonged there. Wasn't her fault. All the time I was just sitting there thinking this is the last day for me. This is the last time I'm ever going to learn anything, you know what I mean? Real learning."

He had not looked up at me even once since leaving the ice cream store. . . . "From now on," he was saying, "I knew I had to go back where they made me believe I belonged . . . then the teacher called on me, and this is how I know just how not smart I am. She called on me, like she

(continued)

THE HUMAN SIDE

"THERE THEY THINK I'M DUMB" *(continued)*

always did, like she'd call on anybody, and she asked me a question. I knew the answer, 'cause I'd read it the night before in my book. . . . So I began to speak, and suddenly I couldn't say nothing. Nothing, man. Not a word. Like my mind died in there. And everybody was looking at me, you know, like I was crazy or something. My heart was beating real fast, I knew the answer, man. And she was just waiting, and I couldn't say nothing. And you know what I did? I cried. I sat there and cried, man, 'cause I couldn't say nothing. That's how I know how smart I am. That's when I really learned at that school, how smart I was. I mean, how smart I thought I was. I had no business being there. Nobody smart's sitting in no class crying. That's the day I found out for real. That's the day that made me know for sure."

Ollie's voice had become so quiet and hoarse that I had to lean down to hear him. We were walking in silence, I was almost afraid to look at him. At last he turned toward me, and for the first time I saw the tears pouring from his eyes. His cheeks were bathed in them. Then he reached over and handed me his ice cream cone.

"I can't eat it now, man," he whispered. "I'll pay you back for it when I get some money."

SOURCE: Excerpted from Thomas J. Cottle. 1976. *Barred from School.* Washington, D.C.: The New Republic Book Company, Inc., pp. 138–40. Used by permission.

dents scored higher on their intelligence quotient (IQ) tests and made higher grades than their classmates who were not labeled as spurters. In addition, the teachers rated the spurters as more curious, interesting, and happy and more likely to succeed than the "nonspurters." Because the teachers expected the spurters to do well, they treated the students in a way that encouraged better school performance. This chapter's *The Human Side* illustrates how negative labeling affects a student's self-concept and performance.

WHO SUCCEEDS? THE INEQUALITY OF EDUCATIONAL ATTAINMENT

As noted earlier, conflict theory focuses on inequalities in the educational system. Educational inequality is based on social class and family background, race and ethnicity, and gender. Each of these factors influences who succeeds in school.

> Educational reform measures alone can have only modest success in raising the educational achievements of children from low-income families. The problems of poverty must be attacked directly.
>
> —Richard J. Murnane
> Graduate School of Education
> Harvard University

Social Class and Family Background

The best predictor of educational success and attainment is socioeconomic status (Lam 1997). Children whose families are in middle and upper socioeconomic brackets are more likely to perform better in school and to complete more years of education than children from lower socioeconomic families. Average SAT scores are positively correlated with family income—the higher the income, the higher the SAT score (College Board 1997). For example, on the average, students who come from well-off families, those with incomes of $100,000 a year or more, score 257 points higher than students whose family incomes are under $10,000 a year.

> The family plays a key role in their children's success in school.
>
> —Albert Bandura
> Social scientist

Families with low incomes have fewer resources to commit to educational purposes. Low-income families have less money to buy books, computers, tutors, and lessons in activities such as dance and music and are less likely to take their children to museums and zoos. Parents in low-income brackets are also less likely to expect their children to go to college, and their behavior may lead to a self-fulfilling prophecy. Disproportionately, children from low-income families do not go to college.

Low-income parents are also less involved in their children's education. Yet parental involvement is crucial to the academic success of the child. According to Barton (1992), the most powerful measure of school quality is a high parent-teacher ratio. Although working-class parents may value the education of their children, in contrast to middle- and upper-class parents, they are intimidated by their child's schools and teachers, don't attend teacher conferences, and have less access to educated people to consult about their child's education (Lareau 1989).

Because low-income parents are often themselves low academic achievers, their children are exposed to parents who have limited language and academic skills. Children learn the limited language skills of their parents, which restricts their ability to do well academically. Low-income parents are unable to help their children with their math, science, and English homework because they often do not have the academic skills to do the assignments. Interestingly, Call et al. (1997) report that even among impoverished youths, parental education is one of the best predictors of a child's academic success.

Children from poor families also have more health problems and nutritional deficiencies. Children cannot learn when they are hungry, malnourished, sick, or in pain. Kozol (1991) interviewed children in the South Bronx and observed:

> **Bleeding gums, impacted teeth and rotting teeth are routine matter for the children I have interviewed. . . . Children get used to feeling constant pain. They go to sleep with it. . . . Children live for months with pain that grown-ups would find unendurable. The gradual attrition of accepted pain erodes their energy and aspiration. I have seen children in New York with teeth that look like brownish, broken sticks. I have also seen teen-agers who were missing half their teeth. . . . Many teachers in urban schools have seen this. It is almost commonplace. (pp. 20–21)**

National Data

In 1992, 40 percent of high school seniors from the 25 percent poorest families did not attend college; only 10 percent of students from the 25 percent richest families did not attend college.

SOURCE: Weinstein 1999

CONSIDERATION

In 1965, Project Head Start began to help preschool children from disadvantaged families. It provided an integrated program of care emphasizing four primary areas: (1) education, (2) health care (medical, dental, nutritional, and mental health), (3) parental involvement, and (4) social services. According to the Select Committee on Children, Youth and Families, every dollar spent on such programs as Head Start saves taxpayers $6 in reduced costs for special education, truant officers, welfare benefits, and crime (Hewlett 1992, 67). Participation in Head Start is also associated with higher rates of high school graduation, employment, and college attendance (Svestka 1996; Barnett 1995; Zigler and Styfco 1996).

Lack of adequate funding for Head Start is part of a larger problem: children who live in lower socioeconomic conditions receive fewer public educational resources. Schools that serve low socioeconomic districts are largely overcrowded and understaffed; they lack adequate building space and learning materials.

The U.S. tradition of decentralized funding means that local schools depend upon local taxes, usually property taxes; about 45 percent of school funding comes from local sources. The amount of money available in each district varies by the socioeconomic status, or SES, of the district (see Table 12.2). This system of depending on local communities for financing has several consequences:

- Low SES school districts are poorer because less valuable housing means lower property values; in the inner city, houses are older and more dilapidated; less desirable neighborhoods are hurt by "white flight" with the result that there is a low tax base for local schools in deprived areas.
- Low SES school districts are less likely to have businesses or retail outlets where revenues are generated; such businesses have closed or moved away.
- Because of their proximity to the downtown area, low SES school districts are more likely to include hospitals, museums, and art galleries, all of which are tax-free facilities. These properties do not generate revenues.
- Low SES neighborhoods are often in need of the greatest share of city services; fire and police protection, sanitation, and public housing consume the bulk of the available revenues. Precious little is left for education in these districts.
- In low SES districts, a disproportionate amount of the money has to be spent on maintaining the school facilities, which are old and in need of repair, so less is available for the children themselves.

Although the state provides additional funding to supplement local taxes, it is not enough to lift schools in poorer districts to a level that even approximates the funding available to schools in wealthier districts.

Give me children from a two-parent Asian-American family. They will outscore almost everybody because they are taught at home that school is special, that teachers must be respected.

—William Bennett
Former Secretary of Education

Race and Ethnicity

Socioeconomic status interacts with race and ethnicity. Because race and ethnicity are so closely tied to socioeconomic status, it appears that race or ethnicity alone can determine school success. While race and ethnicity also have independent effects on educational achievement (Bankston and Caldas 1997; Jencks and Phillips 1998), their relationship is largely due to the association between race and ethnicity, and socioeconomic status. As Table 12.3 shows, educational attainment varies by race and ethnicity.

One reason why some minority students have academic difficulty is that they did not learn English as their native language. Indeed, there are over 2.3 million students with limited English proficiency (Executive Summary 1998).

To help American children who do not speak English as their native language, some educators advocate **bilingual education**—teaching children in both English and their non-English native language. Advocates claim that bilingual education results in better academic performance of minority students, enriches all students by exposing them to different languages and cultures, and enhances the self-esteem of minority students. Critics argue that bilingual education limits minority students and places them at a disadvantage when they compete outside the classroom, reduces the English skills of

International Data

In the fall of 1995, more than 450,000 foreign (nonimmigrant) students were enrolled in U.S. colleges and universities, an increase from 180,000 in 1975.

SOURCE: U.S. Bureau of the Census 1998a

Table 12.3 Educational Attainment by Race/Ethnicity, 1997 (Adults 25 or Older)

Educational Attainment	Race/Ethnic Background	Percentage
Four years of high school or more	Whites	83.0
	Blacks	74.9
	Hispanics	54.7
Four years of college or more	Whites	24.6
	Blacks	13.3
	Hispanics	10.3

SOURCE: *Statistical Abstract of the United States: 1998,* 118th ed. U.S. Bureau of the Census. Washington, D.C.: U.S. Government Printing Office, Table 260.

minorities, costs money, and leads to hostility with other minorities who are also competing for scarce resources.

Another factor that hurts minority students academically is the fact that many tests used to assess academic achievement and ability are biased against minorities. Questions on standardized tests often require students to have knowledge that is specific to the white middle-class majority culture, and students for whom English is not their native language are seriously disadvantaged. With recent White House recommendations to initiate national testing, it is not surprising that the Mexican-American Legal Defense and Educational Fund, the National Association for the Advancement of Colored People (NAACP) Legal Defense and Educational Fund, and the National Center for Fair and Open Testing have objected to English-only examinations (FairTest 1997).

In addition to being hindered by speaking a different language and being from a different cultural background, minority students in white school systems are also disadvantaged by overt racism and discrimination. Much of the educational inequality experienced by poor children is due to the fact that a high percentage of them are also nonwhite or Hispanic. Discrimination against minority students takes the form of unequal funding, as discussed earlier, and school segregation.

In 1954, the U.S. Supreme Court ruled in *Brown v. Board of Education* that segregated education was unconstitutional because it was inherently unequal. Despite this ruling, many schools today are racially segregated. In 1966, a landmark study entitled *Equality of Educational Opportunity* (Coleman et al. 1966) revealed the extent of segregation in U.S. schools. In this study of 570,000 students and 60,000 teachers in 4,000 schools, the researchers found that almost 80 percent of all schools attended by whites contained 10 percent or fewer blacks, and that whites outperformed minorities (excluding Asian Americans) on academic tests. Coleman and his colleagues emphasized that the only way of achieving quality education for all racial groups was to desegregate the schools. This recommendation, known as the **integration hypothesis**, advocated busing to achieve racial balance.

In spite of the Coleman report, court-ordered busing, and an emphasis on the equality of education, public schools in the 1990s and into the twenty-first century remain largely segregated. Research documents the harmful effects of such continued practices. After examining the reading and mathematics achievement levels of a nationally representative sample of high school students, Roscigno (1998, 1051) concludes that "school racial composition matters . . . in

> Reducing the black-white test score gap would probably do more to promote racial equality than any other strategy that commands broad political support.
>
> —Christopher Jencks
> Social scientist

National Data

Since 1984, the desegregation program in St. Louis has bused 14,000 African-American students to predominantly white suburban schools; participating students are two times as likely to graduate from high school and three times as likely to go to college as students not bused.

SOURCE: Wells and Crain 1997

the direction one would expect, even with class composition and other familial and educational attributes accounted for. Attending a black segregated school continues to have a negative influence on achievement." Nonetheless, for financial as well as political reasons, busing has essentially been abandoned (Orfield and Eaton 1997).

Gender

Worldwide, women receive less education than men. Two-thirds of the world's 920 million illiterate people are women (United Nations Population Fund 1999).

Historically, U.S. schools have discriminated against women. Prior to the 1830s, U.S. colleges accepted only male students. In 1833, Oberlin College in Ohio became the first college to admit women. Even so, in 1833, female students at Oberlin were required to wash male students' clothes, clean their rooms, and serve their meals and were forbidden to speak at public assemblies (Fletcher 1943; Flexner 1972).

In the 1960s, the women's movement sought to end sexism in education. Title IX of the Education Amendments of 1972 states that no person shall be discriminated against on the basis of sex in any educational program receiving federal funds. These guidelines were designed to end sexism in the hiring and promoting of teachers and administrators. Title IX also sought to end sex discrimination in granting admission to college and awarding financial aid. Finally, the guidelines called for an increase in opportunities for female athletes by making more funds available to their programs. The push toward equality in education for women has had a positive effect on the aspirations of women.

Traditional gender roles account for many of the differences in educational achievement and attainment between women and men. As noted in Chapter 7, schools, teachers, and educational materials reinforce traditional gender roles in several ways. Some evidence suggests, for example, that teachers provide less attention and encouragement to girls than to boys and that textbooks tend to stereotype females and males in traditional roles (Bailey 1993; Mensh and Lloyd 1997).

Studies of academic performance suggest that females tend to lag behind males in math and science. One explanation is that women experience workplace discrimination in these areas, which restricts their occupational and salary opportunities. The perception of restricted opportunities, in turn, negatively impacts academic motivation and performance among girls and women (Baker and Jones 1993).

International Data

A higher percentage of U.S. women complete an undergraduate college degree (23 percent) than their counterparts in other industrialized countries such as Germany (11 percent) and Japan (12 percent).

Source: U.S. Department of Education 1995

National Data

In 1996, women earned 40 percent of all doctorates, compared with 12 percent in 1966; however, they earned only 21 percent of Ph.D.s in the physical sciences and 12 percent in engineering; women earned 62 percent of doctoral degrees in education.

SOURCE: Summary Report 1998

CONSIDERATION

Educational reformers have recently suggested a return to single-sex schools for both males and females. For males the argument is that boys, particularly African-American boys, need the masculine environment provided by all-male academies where male teachers serve as positive adult role models. For females, the argument is that schools and their teachers are consistently biased against girls, and that sex segregation minimizes the stereotyping and discrimination inherent in mixed sex classrooms. However, researcher Maggie Ford concludes that the evidence indicates ". . . that separating by sex is not the solution to gender inequity in education. When elements of good education are present, girls and boys succeed" (AAUW 1998).

Most of the research on gender inequality in the schools focuses on how female students are shortchanged in the educational system. But what about male students? For example, schools fail to provide boys with adequate numbers of male teachers to serve as positive role models. To remedy this, some school systems actively recruit male teachers, especially in the elementary grades, where female teachers dominate.

The problems that boys bring to school may indeed require schools to devote more resources and attention to them. More than 70 percent of students with learning disabilities such as dyslexia, are male, as are about 75 percent of students identified as having serious emotional problems. Boys are also more likely than girls to have speech impairments and to be labeled as mentally retarded. Lastly, boys are more likely than girls to exhibit discipline problems (Bushweller 1995; Goldberg 1999).

PROBLEMS IN THE AMERICAN EDUCATIONAL SYSTEM

When a sample of persons was asked, "On the whole, would you say you were satisfied or dissatisfied with the education children are getting today," 57.3 percent responded dissatisfied (IRSS 1998). Such concerns are not misplaced—academic achievement is low, dropout rates remain high, teacher training is questionable, and school violence continues to rise. These and other problems contribute to the widespread concern over the quality of American education.

National Data

In a 1996 poll asking the major problems facing schools, a lack of quality education, school violence, students' use of drugs, a lack of discipline, and a lack of financial support were the most common responses.

SOURCE: Digest Of Education Statistics 1997

Illiteracy and Low Levels of Academic Achievement

The Educational Testing Service for the Department of Education conducted a National Adult Literacy Survey, which involved interviewing and testing 26,000 adults aged 16 and older. This survey required respondents to demonstrate that they could understand basic written information, complete a job application form, and balance a checkbook. The results were embarrassing. About half of all adults over the age of 16 functioned at only the lowest literacy level and were unfit to work in most workplaces (Simpson 1993).

Although literacy levels tend to be lowest among students who attend the poorest schools in the nation, apathy and ignorance may be found among students at more affluent schools as well. For example, an ABC News Special entitled "Burning Questions: America's Kids: Why They Flunk" began with the following interview with students from middle-class high schools:

> **Interviewer:** Do you know who's running for president?
>
> **First Student:** Who, run? Ooh. I don't watch the news.
>
> **Interviewer:** Do you know when the Vietnam War was?
>
> **Second Student:** Don't even ask me that. I don't know.
>
> **Interviewer:** Which side won the Civil War?
>
> **Third Student:** I have no idea.
>
> **Interviewer:** Do you know when the American Civil War was?
>
> **Fourth Student:** 1970.

Results of the National Assessment of Educational Progress (NAEP), a nationwide testing effort, are also disturbing. They indicate that the majority of

> Our society does not need to make its children first in the world in mathematics and science. It needs to care for its children—to reduce violence, to respect honest work of every kind, to reward excellence at every level, to ensure a place for every child and emerging adult in the economic and social world, to produce people who can care competently for their own families and contribute effectively to their communities.
>
> –Nel Noddings
> Educational reformer

National Data

Since 1967 the average scores on the math and verbal portions of the SAT have decreased: in 1967 the verbal and math scores were 543 and 516, respectively; in 1997 the average scores were 505 and 511.

SOURCE: *Statistical Abstract of the United States: 1998,* Table 290

International Data

The results of the 1997 Third International Mathematics and Science Study (TIMSS) indicate that American high school seniors ranked nineteenth out of 21 countries in math and science.

SOURCE: Pinkerton 1998

American children are behind desirable proficiency levels in history, mathematics, science, and reading (NCES 1997b). For example, although overall reading levels improved between 1971 and 1980, they declined between 1980 and 1990 and have remained stable since that time (Condition of Education 1998).

While some international comparisons are improving, American students are still outperformed by their foreign counterparts on many tests. The Third International Mathematics and Science Study (TIMSS) assesses the math and science performance of students worldwide. TIMSS indicates that U.S. fourth graders scored above the 26-nation average in both mathematics and science; eighth graders scored above the international average in science and below it in mathematics; and twelfth graders scored below the international average in both math and science (*Condition of Education* 1998, Indicator 20). It must be noted, however, that comparisons between U.S. students and students in other countries may be unfair—in many other countries students are channeled into specialty areas (e.g., math, science). Such specializations may account for achievement differences (Pinkerton 1998).

CONSIDERATION

A recent report by the National Science Foundation concludes that "the deficiency of America's average students is a major reason for the woeful U.S. performance" on the TIMSS examinations (Ratnesar 1998, 60). Why has the "average" student's performance decreased so dramatically over the years? The answer may lie in how school monies are allocated. In 1967, regular education, that is, education for the average student, received 80 percent of school funds. In 1996 that number was 59 percent (Ratnesar 1998). The increased number of special education students, in part, explains the change. In 1975, 800,000 students were called learning disabled; in 1997, that number was 2.6 million, costing over $9 billion dollars a year. Average children, then, may be "casualties of a spoils system in which every morsel of every school district's budget has a different interest group staking claim to it" (Ratnesar 1998, p. 60).

School Dropouts

More than 1 out of every 10 Americans aged 18 to 24 has not completed high school. The highest dropout rate is among Hispanic Americans (see Figure 12.1). Truancy and dropout rates are also highest among poor students living in school districts that are inadequately funded. Kozol (1991) examined some of the poorest schools in the nation and found dropout rates of up to 86 percent in some Chicago schools, 60 percent in some Cincinnati schools, and over 60 percent in Jersey City schools.

Why do students drop out of high school? An analysis of data from the National Educational Longitudinal Study lends some insight. Teachman, Paasch, and Carver (1997) conclude that staying in school between tenth and twelfth grade is a function of several variables. Children who attend Catholic schools, are from in-tact, two parent families, and come from families with greater income and educational levels have lower odds of dropping out of school. The authors conclude that social capital, that is, resources that are a consequence of specific social relationships, are predictive of school continuation.

The economic and social consequences of dropping out of school are significant. Dropouts are more likely than those who complete high school to be

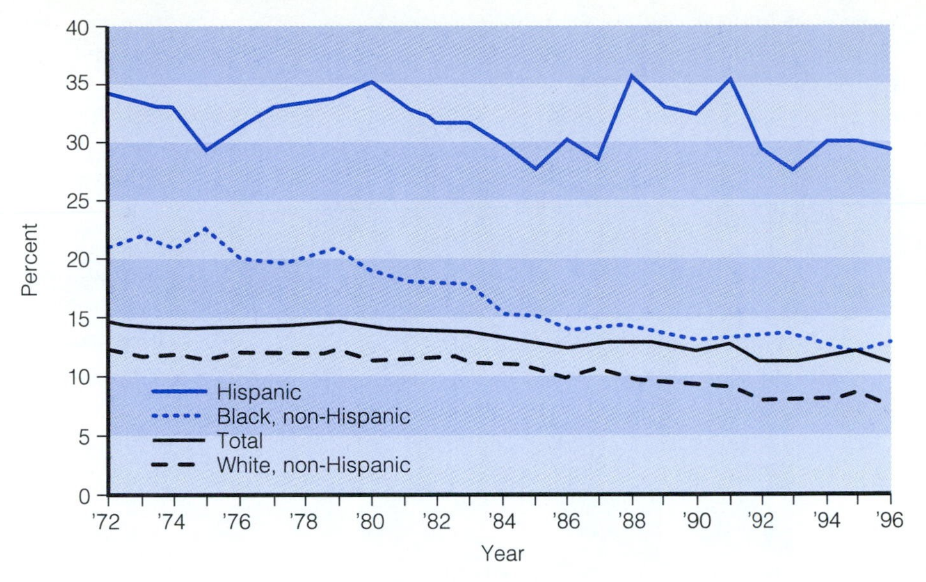

Figure 12.1 **Student Dropout Rate, Aged 16 to 24, by Race/Ethnicity, 1972–96 (A dropout has not completed high school and is not enrolled.)**

SOURCE: U.S. Department of Commerce, Bureau of Census. Current Population Survey, October (various years), unpublished data.

National Data

Between 1992 and 1996, parents who did not complete high school were twice as likely to have children who dropped out of school than parents with some college.

SOURCE: *Condition of Education* 1998, Indicator 6

unemployed and to earn less when they are employed. Individuals who do not complete high school are also more likely to engage in criminal activity, have poorer health and lower rates of political participation, and require more government services such as welfare and health care assistance (Rumberger 1987; Natriello 1995).

Student Violence

As we discussed in Chapter 4, the high rate of violence in U.S. society has infiltrated the nation's schools. A report of a White House Conference on School Violence (1998) found that (1) compared to 5 years ago, today's students are less likely to be victimized but more likely to feel unsafe, (2) school violence occurs more often in urban than rural or suburban schools, (3) serious crime and violence is in a small number of schools—only about 10 percent, (4) theft and fistfights are the most common school offenses, and (5) the overall school crime rate has actually decreased since 1993.

In response to violence, many schools throughout the country have police officers patrolling the halls, require students to pass through metal detectors before entering school, and conduct random locker searches. Video cameras set up in classrooms, cafeterias, halls, and buses purportedly deter some student violence. More than 2,000 schools nationwide conduct peer mediation and conflict resolution programs to help youth resolve conflict in nonviolent ways (see also Chapter 4). One successful antiviolence program, PeaceBuilders of Phoenix, Arizona, enlists the aid of parents, trained community volunteers, local media, and marketing campaigns to promote a "prosocial way of life" among elementary school students (Embry et al. 1996).

National Data

In 1995, approximately 13 percent of students aged 12 to 19 knew a student who brought a gun to school, and 1 in 12 high schoolers were threatened or injured with a weapon while in school. Five percent of students in high school report staying home out of fear of violence.

SOURCE: NCES 1997a; OJJDP 1997

Inadequate School Facilities and Personnel

In *Savage Inequalities*, Jonathan Kozol (1991) provides vivid and shocking descriptions of some of our nation's poorer schools. His descriptions reveal the gross inadequacies in school facilities and personnel that plague our school sys-

Many schools have responded to the problem of violence by taking drastic measures, such as installing metal detectors at entrances. According to the National Crime Victimization Survey, 14 percent of all violent crimes in 1997 took place on school property.

National Data

Over 57 percent of public school facilities are in need of major repair or renovation in such areas as plumbing, roofing, wiring, and heating. Those in most need of plumbing repairs include Washington, D.C. (64.9 percent), Delaware (49.6 percent), Oregon (40.8 percent), and Arizona (39.7 percent).

SOURCE: *USA Today* 1997

tem. Inadequate school facilities not only impede learning, but also contribute to behavioral problems and low self-esteem among students. Consider the following description:

> **At Irvington High School . . . the gym is used by up to seven classes at a time. To shoot one basketball, according to the coach, a student waits for 20 minutes. There are no working lockers. Children lack opportunities to bathe. They fight over items left in lockers they can't lock. They fight for their eight minutes on the floor. Again, the scarcity of things that other children take for granted in America—showers, lockers, space and time to exercise—creates the overheated mood that also causes trouble in the streets. The students perspire. They grow dirty and impatient. They dislike who they are and what they have become. (Kozol 1991, 159)**

According to a 1998 Government Accounting Office report, over 14 million students attend schools that are in need of extensive renovations or repairs. In response to such needs, President Clinton (1998b) established "school modernization tax credits", and a bill has been introduced into Congress that would establish state infrastructure banks. SIBs, as they are called, would act like private banks, making low interest loans to school districts with monies initially funded by federal grants (Tauscher 1997).

School districts with inadequate funding often have difficulty attracting qualified school personnel. Unfortunately, these poorer districts are in dire need of talented teachers who can meet the needs of children from diverse backgrounds and of varying abilities. Many schools have a shortage of minority teachers who can provide professional role models, have had similar life experiences, and have similar language and cultural backgrounds as minority students. Black and Hispanic male children may perceive that white female teachers are less understanding of what they experience than minority male teachers.

> These are not just cosmetic concerns. We are referring to conditions that are unsafe or even harmful.
>
> —Carol Moseley-Braun
> U.S. Senator

Undisciplined and violent students and inadequate teaching materials and school facilities contribute to low teacher morale. Low morale interferes with teaching effectiveness and drives some teachers out of the profession. The relatively low pay of teachers also contributes to morale problems and the questionable quality of instruction.

National Data

In 1997, the average salary of a public school teacher was $38,921; the average salary for a beginning teacher was $25,462.

SOURCE: NCES 1997b

Deficient Teachers

In a National Press Club speech, Department of Education Secretary Richard Riley reiterated the national education goal of "a talented, dedicated, well-prepared teacher in every classroom" by the year 2000. Given that over the next decade over 2 million additional teachers will be needed, the challenge is not an insignificant one (ED Initiatives 1998).

Individuals who enter and remain in the teaching profession are not necessarily competent and effective teachers. There is evidence that those who go into teaching, on the average, have lower college entrance exam scores than the average college student (Finsterbusch 1999). In response to such concerns, more than half of the states have implemented mandatory competency testing, which requires prospective teachers to pass the National Teacher's Examination or some other test of knowledge on the subject they intend to teach. In 1997, over 50 percent of prospective teachers in Massachusetts failed the state's competency exam, some unable to spell such basic words as "burned" and "abolished" (Leo 1998; Silber 1998). In response to such scandalous results, a proposed amendment to the Higher Education Act would cut federal funding to teachers' colleges where more than 30 percent of their graduates failed licensing exams (*USA Today* 1998).

National Data

In a 1998 national survey of 4,049 teachers, only one in five said they felt well prepared to "work in a modern classroom," 20 percent said they were confident using modern technology, and only 18 percent of those whose main assignment was math either majored or minored in it.

SOURCE: NCES 1999

Additionally, more than half of the states have adopted **alternative certification programs**, whereby college graduates with degrees in fields other than education may become certified if they have "life experience" in industry, the military, or other relevant jobs. Teach for America, originally conceived by a Princeton University student in an honor's thesis, is an alternative teacher education program aimed at recruiting liberal arts graduates into teaching positions in economically deprived and socially disadvantaged schools. After completing an 8-week training program, recruits are placed as full-time teachers in rural and inner city schools. Critics argue that these programs may place unprepared personnel in schools.

The actual success of a teacher depends in large measure on the capacity to state the subject matter of instruction in terms of the experience of the student.

—George Herbert Mead

STRATEGIES FOR ACTION: TRENDS AND INNOVATIONS IN AMERICAN EDUCATION

Since the 1983 publication of *A Nation at Risk*, which documented the dismal state of public education, American public schools have made attempts to improve. For example, 84 percent of the states have raised their graduation requirements, 70 percent of schools no longer permit students to participate in extracurricular activities if they are failing academic subjects, and 40 percent of schools have lengthened their school year (Celis 1993). Further, there is a nationwide movement to eliminate **social promotion**, the passing of students from grade to grade even if they are failing ("State of the Union" 1999). However, educational reformers are calling for changes that go beyond get-tough policies that maintain the status quo.

National Data

In 1996, 44 percent of teachers in colleges and universities were women, up from 36 percent in 1983; 74 percent of all other teachers were women, up from 71 percent in 1983.

SOURCE: U.S. Bureau of the Census 1998b

Moral and Interpersonal Education

Violence, divorce, teen pregnancy, and drug abuse suggest that more attention needs to be placed on moral education. Most school curricula neglect the human side of education—the moral and interpersonal aspects of developing as an individual and as a member of society. Some educational reformers oppose the current emphasis on increasing academic standards and recommend that the main goal of education should be "to encourage the growth of competent, caring, loving and lovable people" (Noddings 1995, 366). To achieve this, Noddings suggests that students work together on school projects, help younger students, contribute to the care of buildings and grounds, and do supervised volunteer work in the community. Such activities would likely reduce the high rates of alienation experienced by many of today's students (see this chapter's *Self & Society*). She recommends that:

> **All students should be engaged in a general education that guides them in caring for self, intimate others, global others, plants, animals, the environment, objects and instruments, and ideas. Moral life so defined should be frankly embraced as the main goal of education. (p. 368)**

A moral and interpersonal emphasis in education implies that schools should prepare students not only for the world of work, but also for parenting and for civic responsibility. Noddings (1995) notes that:

> **Almost all of us enter into intimate relationships, but schools largely ignore the centrality of such interests in our lives. And although most of us become parents, evidence suggests that we are not very good at parenting—and again the schools largely ignore this huge human task. (p. 367)**

International Data

As part of TIMSS, U.S. and Japanese math teachers were videotaped and their performances evaluated; U.S. teachers received 87 percent medium marks and 13 percent low marks; Japanese teachers received 13 percent low, 57 percent medium, and 30 percent high marks.

SOURCE: Pinkerton 1998

How does the public feel about moral education? In a national poll, the majority of public school parents approved of teaching courses on values and ethical behavior in the public schools. Although public schools are constitutionally prohibited from teaching any particular religion, about two-thirds of public school parents favor nondevotional instruction about various religions (Elam, Rose, and Gallup 1994).

Moral and interpersonal education occurs to some extent in schools that have peer mediation and conflict resolution programs (discussed earlier and in Chapter 4). Such programs teach the value of nonviolence, collaboration, and helping others as well as skills in interpersonal communication and conflict resolution. Moral and interpersonal education also occurs in micro-society schools as discussed in the following section.

> Education is the key to prosperity and the wisest investment we can make in our children's and our nation's future.
>
> —Richard Riley
> U.S. Secretary of Education

Micro-Society Schools

A **micro-society school** is a simulation of the "real" or nonschool world. In a micro-society school, the students—with the help and guidance of their teachers—design and run their own democratic, free-market society within the school (Sommerfeld 1998). At the first micro-society school, the City Magnet School in Lowell, Massachusetts, developed in the early 1980s, students set up their own government, complete with legislative, executive, and judicial

Student Alienation Scale

Indicate your agreement to each statement by selecting one of the responses provided:

1. It is hard to know what is right and wrong because the world is changing so fast.
   _____ Strongly agree _____ Agree _____ Disagree _____ Strongly disagree
2. I am pretty sure my life will work out the way I want it to.
   _____ Strongly agree _____ Agree _____ Disagree _____ Strongly disagree
3. I like the rules of my school because I know what to expect.
   _____ Strongly agree _____ Agree _____ Disagree _____ Strongly disagree
4. School is important in building social relationships.
   _____ Strongly agree _____ Agree _____ Disagree _____ Strongly disagree
5. School will get me a good job.
   _____ Strongly agree _____ Agree _____ Disagree _____ Strongly disagree
6. It is all right to break the law as long as you do not get caught.
   _____ Strongly agree _____ Agree _____ Disagree _____ Strongly disagree
7. I go to ball games and other sports activities at school.
   _____ Always _____ Most of the time _____ Some of the time _____ Never
8. School is teaching me what I want to learn.
   _____ Strongly agree _____ Agree _____ Disagree _____ Strongly disagree
9. I go to school parties, dances, and other school activities.
   _____ Always _____ Most of the time _____ Some of the time _____ Never
10. A student has the right to cheat if it will keep him or her from failing.
    _____ Strongly agree _____ Agree _____ Disagree _____ Strongly disagree
11. I feel like I do not have anyone to reach out to.
    _____ Always _____ Most of the time _____ Some of the time _____ Never
12. I feel that I am wasting my time in school.
    _____ Always _____ Most of the time _____ Some of the time _____ Never
13. I do not know anyone that I can confide in.
    _____ Strongly agree _____ Agree _____ Disagree _____ Strongly disagree
14. It is important to act and dress for the occasion.
    _____ Always _____ Most of the time _____ Some of the time _____ Never
15. It is no use to vote because one vote does not count very much.
    _____ Strongly agree _____ Agree _____ Disagree _____ Strongly disagree
16. When I am unhappy, there are people I can turn to for support.
    _____ Always _____ Most of the time _____ Some of the time _____ Never
17. School is helping me get ready for what I want to do after college.
    _____ Strongly agree _____ Agree _____ Disagree _____ Strongly disagree
18. When I am troubled, I keep things to myself.
    _____ Always _____ Most of the time _____ Some of the time _____ Never
19. I am not interested in adjusting to American society.
    _____ Strongly agree _____ Agree _____ Disagree _____ Strongly disagree
20. I feel close to my family.
    _____ Always _____ Most of the time _____ Some of the time _____ Never
21. Everything is relative and there just aren't any rules to live by.
    _____ Strongly agree _____ Agree _____ Disagree _____ Strongly disagree

(continued)

Student Alienation Scale *(continued)*

22. The problems of life are sometimes too big for me.
_____ Always _____ Most of the time _____ Some of the time _____ Never
23. I have lots of friends.
_____ Strongly agree _____ Agree _____ Disagree _____ Strongly disagree
24. I belong to different social groups.
_____ Strongly agree _____ Agree _____ Disagree _____ Strongly disagree

INTERPRETATION

This scale measures four aspects of alienation: powerlessness, or the sense that high goals (e.g., straight A's) are unattainable; meaninglessness, or lack of connectedness between the present (e.g., school) and the future (e.g., job); normlessness, or the feeling that socially disapproved behavior (e.g., cheating) is necessary to achieve goals (e.g., high grades); and social estrangement, or lack of connectedness to others (e.g., being a "loner"). For items 1, 6, 10, 11, 12, 13, 15, 18, 19, 21, and 22, the response indicating the greatest degree of alienation is "strongly agree" or "always." For all other items, the response indicating the greatest degree of alienation is "strongly disagree" or "never."

SOURCE: Rosalind Y. Mau. 1992. "The Validity and Devolution of a Concept: Student Alienation." *Adolescence* 27:107, pp. 739–40. Used by permission of Libra Publishers, Inc., 3089 Clairemont Drive, Suite 383, San Diego, California 92117.

branches. They write their own school constitution and elect a legislature to make their own school laws. The school has its own courts and system of justice, as well as its own police force, called the City School Crime Stoppers. Students also devised and implemented a tax system operated by their own internal revenue service.

The school has an economic system with its own currency and banks in which every student has an account. Students have started numerous retail businesses that sell such things as pencils and stationery. They also set up their own publishing business within the language arts program.

All these micro-society activities are real jobs for which students get paid. To learn all the basic skills necessary to hold a job and participate in the micro-society, students go to the school within the school—the City School Academy. Students pay tuition from the money they earn by holding jobs in the rest of the micro-society.

Through participation in the micro-society school, both students and their teachers "constantly face moral dilemmas that they must solve as they strive to build a 'good' society" says reformer/founder George Richmond (1989, 232):

> **Do you want a society with the extremes of poverty and wealth? Do you want a state of fear and violence? Should the microsociety's government assist or ignore children who may not be succeeding? Do you want a democracy or a totalitarian state? What liberties should children have? And what responsibilities should they shoulder? What kinds of activities should be taxed? When does one put the community's welfare ahead of the rights of the individual? What civil rights should children enjoy in their microsociety?**

National Data

The number of micro-society schools in the United States has increased dramatically in recent years with 24 presently in operation, 12 in planning stages, and several dozen others seeking funding.

SOURCE: Sommerfeld 1998

National Data

Of the 1998 incoming college class, 21.3 percent report that their high schools required community service for graduation.

SOURCE: Fact File 1999

National Educational Standards

During the Clinton administration, Congress passed Goals 2000: The Educate America Act. This national agenda identified eight national goals for education to be achieved by the year 2000. These goals include increasing the high school graduation rate to at least 90 percent, eliminating drugs and violence in schools, eliminating illiteracy among U.S. adults, and enabling U.S. students to rank first in the world in math and science. Parental involvement and teacher training are also important goals. To achieve the academic objectives in Goals 2000, the United States is moving toward a system of national standards that will define what students should know and be able to do at each grade level. To achieve national standards, "experts" are working on developing a national curriculum; to assess whether students are mastering the national curriculum objectives, educators are developing national tests. The administration's testing plans, however, suffered a serious blow in October of 1998 when Congress passed an amendment prohibiting "any new national tests that are not specifically and explicitly authorized by Congress" (Legislative Action 1998).

Opponents of national testing argue that it will lower the quality of education in the United States. How? According to FairTest, an educational testing watchdog group, national testing relies too heavily on multiple choice questions, thereby encouraging memorization and ignoring critical thinking skills. The reading portion of the national tests proposed is only in English, and there are concerns that those less proficient in English may be victimized by test results (e.g., tracking). There is also the question of how the test results will be used. Will single test scores be used to make important decisions about a child's educational future? (FairTest 1998)

Some even criticize the logic behind national testing—that is, how does national testing lead to higher educational standards as suggested by the Clinton administration? Critics argue that more tests may actually contribute to lower standards as poor test results lead to the "dumbing-down" of any proposed national curriculum. Finally, Apple (1995, 360) suggests that a "national curriculum is a mechanism for the political control of knowledge" and that the power to make decisions about what is taught and how it is taught should be in the hands of parents, teachers, and students in local school districts (see Table 12.4).

You don't fatten cattle by weighing them more often, and you don't improve student learning by giving more low quality tests.

–Dr. Monty Neill
Associate Director,
FairTest

CONSIDERATION

Imposing standards of test performance sometimes leads to cheating–not by students, but by teachers. Teachers cheat by giving students advance copies of tests, helping students during tests, teaching what is on the test, and altering students' responses. Toch (1992) reports that in a national survey of educators in 1990, 1 in 11 teachers reported pressure from administrators to cheat on standardized tests. Why are teachers and administrators motivated to cheat? School superintendents with high scores in their district may gain a national reputation and contract for larger salaries. Increasingly, states give funding bonuses to schools with the highest test scores. Teachers whose students perform well on standardized tests are also sometimes eligible to receive bonuses.

National Data

In a 1998 Harris poll of the college graduating class of 2001, 37 percent of the students responded that technology, including computers, e-mail, and the Internet, was the most important advantage they had over preceding generations.

SOURCE: Harris Poll 1998a

Computer Technology in Education

Computers in the classroom allow students to access large amounts of information (see this chapter's *Focus on Technology*). The proliferation of computers

Table 12.4 Public Opinion on a National Testing Plan, 1997*

1. Opinion of proposed voluntary testing of fourth and eighth graders:
   - Favored: 68 percent
   - Opposed: 28 percent
   - No opinion: 4 percent
2. National testing would:
   - Improve the quality of education: 44 percent
   - Make it worse: 7 percent
   - Make no difference: 45 percent
   - No opinion: 4 percent
3. National testing would give the federal government too much power.
   - Agree: 47 percent
   - Disagree: 49 percent
   - No opinion: 4 percent
4. National testing would hurt those who score poorly, including minorities and the poor.
   - Agree: 39 percent
   - Disagree: 57 percent
   - No opinion: 4 percent

*Random sample of adults over the age of 18.

SOURCE: Gallup Organization 1997

both in school and at home may mean that teachers will become facilitators and coaches rather than sole providers of information. Not only do computers enable students to access enormous information including that from the World Wide Web, but they also allow students to progress at their own pace. Unfortunately, computer technology is not equally accessible by all students. Students in poor school districts are less likely to have access to computers either in school or at home.

Interestingly, the conclusion of one of the largest studies of school computers found that students who use computers often scored lower on math tests than their low-use counterparts. The Educational Testing Service's study of 14,000 fourth and eighth graders concluded that how the computers were used—repetitive math drills versus real-life simulation—was responsible for the test variations. Also of note, students in classrooms where teachers were trained in computer use did better than students in classrooms where teachers were less skilled. Minority and low-income students were the least likely to have teachers highly skilled in computer technology.

The U.S. Department of Education has issued a "Technology Challenge" composed of four education goals: (1) teacher proficiency in computer technology, (2) modern computers in every classroom, (3) all classrooms connected to the Internet, and (4) effective use of software and online services as part of a school's curriculum (U.S. Department of Education 1998a). Progress is already being made. For example, in 1985 there was one computer for every 63 students; in 1998 there was one computer for every 7.4 students (U.S. Bureau of the Census 1998b; *Statistical Abstract of the United States: 1998,* Table 281).

> Every adult American has the right to vote, the right to decide where to work, where to live. It's time parents were free to choose the schools that their children attend. This approach will create the competitive climate that stimulates excellence. . . .
>
> —George Bush
> Former U.S. President

Alternative School Choices

Traditionally, children have gone to school in the district where they live. School vouchers, charter schools, home schooling, and private schools provide

DISTANCE LEARNING AND THE NEW EDUCATION

Imagine never having an eight o'clock class or walking in to the lecture room late. Imagine no room and board bills, or having to eat your roommate's cooking one more time. Imagine going to class when you want, even three o'clock in the morning. Imagine not worrying about parking! The future of higher education? Maybe. It's possible that the World Wide Web and other information technologies have so revolutionized education that the above scenarios are a *fait accompli*.

What is distance learning? Distance learning separates, by time or place, the teacher from the student. They are, however, linked by some communication technology: videoconferencing, satellite, computer, audiotape or videotape, real-time chat room, closed-circuit television, electronic mail, or the like.

Examples of distance learning abound. Washington State University offers on-line classes to over 2,300 students who might not otherwise attend college (Rudich 1998). Penn State christened its "World University" this fall; California's consortium of over 100 colleges and universities opened its "doors" with over 1,600 cyberclasses; and the University of Phoenix, a for-profit educational venture owned by the Apollo Group, has doubled its on-line student population in just 2 years (Bulkeley 1998; Arenson 1998).

One of the most ambitious campaigns began in 1995 when the Western Governors Association met to "explore ways to expand educational opportunities and access to higher education". The proposed "new education" was to be an independent, market-oriented, degree-granting institution that would emphasize quality, competency-based education (WGU 1998). The result of this summit, Western Governors University, began offering classes in September 1998 (see Figure 12.2).

Have a bachelor's degree? Looking for a graduate degree? Graduate and professional degrees can also be completed via the net. Kaplan Education Centers, known for their test preparation courses, opened the Internet's first law school in December of 1998—Concord School of Law. Although the school is not accredited by the American Bar Association, students who complete the four-year, $17,000 program can sit for the bar examination in California—and only California thus far (Bulkeley 1998).

The benefits of distance learning are clear. It provides a less expensive, accessible, and often more convenient way to complete a college degree. There are even pedagogical benefits. Research suggests that "students of all ages learn better when they are actively engaged in a process, whether that process comes in the form of a sophisticated multimedia package or a low-tech classroom debate on current events" (Carvin 1997). Distance education also benefits those who have historically been disadvantaged in the classroom. A review of research on gender differences suggests that females outperform males in distance learning environments (Koch 1998).

But all that glitters is not gold. There is evidence that students feel more estranged from their distance learning instructors than from teachers in conventional classrooms. The results of a study by Freitas, Meyers, and Avtgis (1998) indicate that college students in distance learning classes perceived significantly lower levels of teacher nonverbal immediacy (e.g., eye contact) when compared with students in conventional classes.

Additionally problematic is the proliferation of "virtual degrees." A Miami elementary school teacher enrolled in an online university to complete a master's degree in special education. After paying $800 of a total $2,000 bill, she was sent a book to summarize as part of her degree requirements. Shortly after returning her summary she was sent not only a master's degree, but a Ph.D. and transcripts of courses she had never taken with a recorded 3.9 grade point average (GPA) (Koeppel 1998)! Although not a new problem, (e.g., correspondence schools), there is the concern that the Internet lends itself to such fraudulent practices.

(continued)

FOCUS ON TECHNOLOGY

DISTANCE LEARNING AND THE NEW EDUCATION *(continued)*

Further, teachers, particularly in higher education, are concerned about the quality of distance education. Several regulatory bodies, including the Council of Graduate Schools, are presently establishing quality standards and guidelines (Guernsey 1998). And while a committee of the American Association of University Professors acknowledged that distance learning may be a "valuable pedagogical tool," it also questioned whether "academic quality, academic freedom, intellectual property rights and instructor's workloads and compensation" will be compromised (Arenson 1998, A14).

Over the objections of many, distance education continues to expand. Over 100 colleges and universities have joined the Internet2 project, which promises to deliver information from 100 to 1,000 times faster than the Internet (Rudich 1998). Its expansion may, in part, reflect the fact that distance education is a moneymaker—a $225 billion industry (Bulkeley 1998). New York University, for example, is planning to market an Internet subsidiary that is likely to earn millions of dollars for the university through private investors and stock offerings (Arenson 1998).

Although some would say that distance learning is only a fad, the growth of such private sector ventures would suggest otherwise. Thus, speed, availability, and accessibility are likely to increase as a response to market demands and, with them, use. Will distance learning solve all the problems facing education today? The answer is clearly no. While not the technological fix some are looking for, distance education, from digital libraries to cutting-edge technologies, does provide a provocative and financially lucrative alternative to traditional education providers.

I'd like parents to go into the school and be involved with the teacher, the principal, the PTA. . . .

—Richard Riley
U.S. Secretary of Education

parents with alternative school choices for their children. **School vouchers** are tax credits that are transferred to the public or private school the parents select for their child. Almost half of all states now offer such vouchers. Proponents of the voucher system argue that it reduces segregation and increases the quality of schools since they must compete for students to survive. Opponents argue that vouchers increase segregation because white parents use the vouchers to send their children to a private school with few minorities. Vouchers, opponents argue, are also unfair to economically disadvantaged students who are not able to attend private schools because of the high tuition. Further, they argue, the use of vouchers for religious schools violates the constitutional guarantee of separation of church and state. A Wisconsin Supreme Court decision, hailed as "the most significant school voucher ruling in the nation to date," held that taxpayers money can be used to send children to religious schools (Belluck 1998). The decision is being appealed to the U.S. Supreme Court.

Vouchers can also be used for **charter schools.** Charter schools are "public schools begun by parents, teachers and communities, open to all students regardless of background or ability, and given greater flexibility in exchange for higher levels of accountability" (Clinton 1998b). Charter schools are funded by foundations, universities, private benefactors, and entrepreneurs and are maintained by school tax dollars (Toch 1998). The Charter School Expansion Act of 1998 is likely to increase the number of charter schools from approximately 1,000 now to over 3,000 by the next century. Charter schools, like school vouchers, were designed to expand schooling options and increase the quality of education through competition.

National Data

A *Washington Post* poll found that more African Americans supported school choice (57 percent) than were opposed to it (38 percent), and that Hispanic Americans supported it in even greater numbers than blacks (65 percent in favor and 29 percent opposed). Those most in favor of school choice had school-aged children.

SOURCE: Bolick 1998

About WGU

Welcome to WGU

What We Are

Who's Involved

Vision, History & Mission

Press Releases

Newsletters

Administration

About Your Privacy

Glossary

FAQs

Western Governors University offers a totally new way of looking at higher education. We're a new type of university centered around you, the student.

What makes WGU so different and exciting is that we know that not everyone who wants a college degree or courses can live on, or near, the campus of their choice. We offer distance learning courses from dozens of colleges, universities, and corporations across the United States (and soon the world!). Courses offered through WGU (you'll find them in the Catalog) will come to you, wherever you are, not the other way around. These courses use both high-tech and low-tech ways – from the Internet to satellite to the Postal Service – to provide you with real options.

Through WGU, you will also be able to earn degrees by focusing only on the skills and knowledge areas that you need. We call this "competency-based" education, since it is not based on the number of credits you may have accumulated. We won't make you relearn what you already know. You can count skills and knowledge you've gained at other universities, on the job, or just through life, toward your WGU degree.

Our students have online access to our Catalog, directory of programs, advising services, and important resources, like our online library with over 60 full-text and comprehensive citation databases.

Come in, tour our online campus and resources, and discover the WGU advantage. You can be part of WGU as a student, an education provider, or a corporate partner. If you still have questions about WGU, call us at 877-HELP-WGU (877-435-7948). Wondering if distance education is right for you? Take this quiz and find out!

Sign up for the WGU mailing list!

Finally! Real options for higher education.

Home

About WGU | Academics | Catalog | Admission | Library | Bookstore | Union
Guide | Students | Educators | Corporate | Site Map | Contact

©1999, Western Governors University (WGU)
www.wgu.edu

Figure 12.2 **Western Governors University Homepage.**

SOURCE: From Western Governors University web site, **http://www.wgu.edu**. Reprinted by permission.

One form of school choice involves parents teaching their children at home. Some research has found that children taught at home were as successful in college and employment as those educated in the public schools.

Some parents are choosing not to send their children to school at all but to teach them at home. For some parents, **home schooling** is part of a fundamentalist movement to protect children from perceived non-Christian values in the public schools. Other parents are concerned about the quality of their children's education and their safety. How does being schooled at home instead of attending public school affect children? Webb (1989) found that children educated at home, either partly or completely, were equally successful in going to college and securing employment as those educated in the public school system.

National Data

In 1995, an estimated 14 percent of U.S. students at the elementary and secondary levels attended private schools.

SOURCE: *Statistical Abstract of the United States: 1998,* Table 255

Another choice parents may make is to send their children to a private school. The primary reason parents send their children to private schools is for religious instruction. The second most frequent reason is the belief that private schools are superior to public schools in terms of academic achievement. Research suggests, however, that when controlling for parents' education and income, there are few differences in private and pubic school educational outcomes (Shanker 1996; Ascher, Fruchter, and Berne 1997; Cohen 1998). Parents also choose private schools for their children in order to have greater control over school policy, to avoid busing, or to obtain a specific course of instruction such as dance or music.

Privatization

Privatization, or the use of private services in the public educational system, is a growing trend in education. Examples of privatization range from private practice teachers hired to teach selected courses to private management corporations that run entire public school systems. For example, schools in several cities are now run by Education Alternatives, Inc. and Sylvan Learning Systems. A recent review of privatization programs, however, indicates that they may fail to stimulate higher academic achievement, are more rather than less expensive, are unresponsive to parents' concerns, and lack accountability (Ascher, Fruchter, and Berne 1997).

We anticipate a continuing shift of public funds to private enterprise over the next ten years.

—Gerald Odening
Investment specialist

One form of privatization is the development of partnerships with businesses and private foundations, which provide schools with equipment, personnel, expertise, and money. Partnerships benefit schools by providing needed resources and help corporations by creating a better-educated workforce for the future. Corporations can contribute in other ways as well. In 1987, for example, Procter & Gamble adopted a Cincinnati school and established Project ASPIRE, enlisting 95 Procter & Gamble employees to work one-to-one with students at risk for dropping out.

Eugene Lang, a wealthy businessman, promised 62 sixth graders in one of the poorest school districts in New York City that he would personally finance their college education if they graduated from high school. Eight years later, all but a few had graduated, and over half were in college. Lang had given students with little academic future a vision of hope—and the certainty of a paid college education. Subsequently, Lang started the I Have a Dream Foundation in which 160 classes in 57 cities now participate. Lang recruits sponsors who, during the class's sixth grade, commit to paying college tuition for all who attain specified academic levels (IHAD 1999).

National Data

In 1992, almost half of all public schools participated in some partnership. In the same year, corporations and corporate foundations donated about $2 billion—or one-third of their total donations—for education-related projects.

SOURCE: Cited in Goldring and Sullivan 1995

UNDERSTANDING PROBLEMS IN EDUCATION

What can we conclude about the educational crisis in the United States? Any criticism of education must take into account that over a century ago the United States had no systematic public education system at all. Many American children did not receive even a primary school education. Instead, they worked in factories and on farms to help support their families. Whatever education they received came from the family or the religious institution. In the mid-1800s, educational reformer Horace Mann advocated at least 5 years of mandatory education for all U.S. children. Mann believed that mass education would function as the "balanced wheel of social machinery" to equalize social differences among members of an immigrant nation. His efforts resulted in the first compulsory education law in 1852, which required 12 weeks of attendance each year. By World War I, every state mandated primary school education, and by World War II, secondary education was compulsory as well.

In order for all children to enter elementary school prepared to learn, we will have to eliminate child poverty in this country and provide every child who needs them not only with adequate health and social services but also with an early childhood education program well beyond anything envisioned by Head Start.

—Evans Clinchy
Educational reformer

Public schools are supposed to provide all U.S. children with the academic and social foundations necessary to participate in society in a productive and meaningful way. But, as conflict theorists note, for many children the educational institution perpetuates an endless downward cycle of failure, alienation, and hopelessness.

Breaking the cycle requires providing adequate funding for teachers, school buildings, equipment, and educational materials. The public supports such spending. In a recent national survey of persons 18 and over, 66 percent responded that the government spends too little on public education (Harris Poll 1998b). Legislation such as the 1998 amendments to the Higher Education Act, which allocate $300 billion to teacher preparation and recruitment, reduce interest on college student loans, and increase the maximum in Pell grants, must be a priority (Burd 1998). But, even with such financial support, as functionalists argue, education alone cannot bear the burden of improving our schools.

Jobs must be provided for those who successfully complete their education. Students with little hope of school success will continue to experience low

National Data

The proposed 2000 federal budget includes $130 million for charter schools, $1.4 billion for 38,000 new teachers in grades 1–3, $600 million to create or expand after-school or summer programs, and an increase in the maximum Pell Grant to $3,250.

SOURCE: ED Initiatives 1999

motivation as long as job prospects are bleak and earnings in available jobs are low. Ray and Mickelson (1993) explain:

> **. . . School reforms of any kind are unlikely to succeed if non-college bound students cannot anticipate opportunity structures that reward diligent efforts in school. Employers are not apt to find highly disciplined and motivated young employees for jobs that are unstable and low paying. (pp. 14–15)**

Finally, "if we are to improve the skills and attitudes of future generations of workers, we must also focus attention and resources on the quality of the lives children lead outside the school" (Murnane 1994, 290). We must provide support to families so children grow up in healthy, safe, and nurturing environments. Children are the future of our nation and of the world. Whatever resources we provide to improve the lives and education of children are sure to be wise investments in our collective future.

CRITICAL THINKING

1. Clearly, micro-society schools have both advantages and disadvantages. After making a list of each, would you want your child to attend a micro-society school? Why or why not?
2. As discussed in Chapter 6, the proportion of elderly in the United States is increasing dramatically as we move into the twenty-first century. Since the elderly are unlikely to have children in public schools, how will the allocation of needed school funds be affected by this demographic trend?
3. Student violence is one of the most pressing problems in U.S. public schools. One response is defensive, that is, police patrolling halls, metal detectors, etc. Other than such defensive tactics, what violence prevention techniques should be instituted?
4. Students who drop out of school are often blamed for their lack of motivation. How may a teenager's dropping out of high school be explained as a failure of the educational system rather than as a "motivation" problem?

KEY TERMS

alternative certification programs
bilingual education
charter schools
cultural imperialism
home schooling
integration hypothesis
micro-society school
multicultural education
privatization
school vouchers
self-fulfilling prophecy
social promotion

INTERNET

You can find more information on charter schools, Title IX, distance learning, and Goals 2000: The Educate America Act at the *Understanding Social Problems* web site at **http://sociology.wadsworth.com.**

INFOTRAC COLLEGE EDITION

Either from the Wadsworth sociology resource center at **http://sociology.wadsworth.com** or directly from your web browser, you may access InfoTrac College Edition, an online university library that includes over 700 popular and scholarly journals in which you can find articles related to the topics in this chapter. Suggested articles and questions relating to these articles are listed below.

Cirillo, Kathleen J., B.E. Pruitt, et al. 1998. "School Violence: Prevalence and Intervention Strategies." *Adolescence* 33(130):319–31.

1. What are the four parts of Wodarski and Hendrick's model of violence prevention?
2. What is PACT, and how does it work?
3. What were the three findings of the survey detailed within this article?

Gavin, Tom. "Truancy: Not Just Kids' Stuff Anymore." 1997. *The FBI Law Enforcement Bulletin* 66(3):8–15.

1. Detail the links social reformers have identified between truancy and delinquency.
2. What are the functions of a truancy interdiction program?
3. What methods do law enforcement officers use to get parents involved in making sure their children attend school?

Rembolt, Carole. 1998. "Making Violence Unacceptable." *Educational Leadership* 56(1):32–39.

1. What causes does the article give for the recent rise in school violence?
2. In fighting school violence, which strategies have been the most and least effective?
3. How does the Respect and Protect program try to get students involved in violence prevention?

CHAPTER THIRTEEN

Cities in Crisis

- The Global Context: A World View of Urbanization
- Sociological Theories of Urbanization
- *Focus on Social Problems Research: Big Cities Health Inventory*
- The Effects of Urban Living on Social Relationships
- Cities and Social Problems
- *Self & Society: Road Rage Survey*
- Strategies for Action: Saving Our Cities
- *The Human Side: Trees of Life*
- Understanding Cities in Crisis

Is It True?

Is It True?

1. "Urban psychosis" has been used successfully as a defense for murder.
2. Most of the world's largest cities are in industrialized countries.
3. More than three-fourths of the U.S. population live in urban areas.
4. The entire state of New Jersey is composed of metropolitan areas.
5. More people die as a result of car accidents than of urban violent crime.

Answers: 1 = F; 2 = F; 3 = T; 4 = T; 5 =

> **Increasing urbanization has the potential for improving human life or increasing human misery. The cities can provide opportunities or frustrate their attainment; promote health or cause disease; empower people to realize their needs and desires or impose on them a simple struggle for basic survival.**
>
> *United Nations' 1996 State of the World Population Report*

Think back to the year 1996. What are some of the newsworthy events that occurred in that year? Some that may come to mind include the re-election of Bill Clinton to the U.S. presidency, the bombing at Atlanta's Centennial Park during the Olympics, the arrest of Ted Kaczynski, (the "Unabomber"), and the crash of a ValueJet DC9 in the Florida Everglades. But another very important historical event occurred in 1996—an event that went largely unnoticed. According to Clark (1998), 1996 was the year in which, for the first time in human history, half of the world's population lived in urban areas. "No longer are towns and cities exceptional settlement forms in predominantly rural societies. The world is an urban place" (Clark 1998, 85).

Many U.S. citizens are intrigued by cities and their diverse populations; their history; their architectural wonders; and their assorted offerings of cultural events, entertainment, sports, museums, conventions, universities, government centers, tourist attractions, restaurants, and innumerable consumer products and services. Many argue, however, that the problems of urban living outweigh its advantages. In this chapter, we look at the historical and social forces that have created our cities, the modern social forces that are shaping our cities today, the social problems that affect and are affected by cities in the United States and throughout the world, and strategies for revitalizing cities and improving the quality of life among urban dwellers.

International Data

In 1800, fewer than 3 percent of the world's population lived in urban places. In 1950, the figure was about 27 percent, increasing to 50 percent in 1996.

SOURCE: Clark 1998

THE GLOBAL CONTEXT: A WORLD VIEW OF URBANIZATION

As early as 5000 B.C., cities of 7,000 to 20,000 people existed along the Nile, Tigris-Euphrates, and Indus river valleys. But it was not until the Industrial Revolution in the nineteenth century that **urbanization**, the transformation of a society from a rural to an urban one, spread rapidly. According to the 1990 U.S. census definition, an **urban population** consists of persons living in cities or towns of 2,500 or more inhabitants. An **urbanized area** refers to one or more places and the adjacent densely populated surrounding territory that together have a minimum population of 50,000. Urbanization is expected to continue into the twenty-first century. Not only is the percentage of the world's population that is urban growing, so is the number of large cities.

U.S. citizens tend to think of New York, Los Angeles, and Chicago when they think of large cities, yet 11 of the world's 15 biggest cities are in developing countries (United Nations Population Fund 1996). Nearly all of the urban population growth is occurring in developing countries. Between 1970 and 2020, 93 percent of urban population growth will have occurred in the developing world (United Nations Population Fund 1996) (see Table 13.1). The most urbanized continent in the world is South America, with the population in all but one of its countries (Guyana) being more urban than rural (Clark 1998). Levels of urban development are low in most of Africa, South and East Asia.

International Data

Projections indicate that in 2010, 59 cities in the world will have populations over 5 million; 47 of these will be in developing countries.

SOURCE: Ingram 1998

Table 13.1 Ten Most Populated Cities in the World

	Projected Population in 2000
1. Tokyo-Yokohama, Japan	29,971,000
2. Mexico City, Mexico	27,872,000
3. Sao Paulo, Brazil	25,354,000
4. Seoul, South Korea	21,976,000
5. Bombay, India	15,357,000
6. New York City, USA	14,648,000
7. Osaka-Kobe-Kyoto, Japan	14,287,000
8. Tehran, Iran	14,251,000
9. Rio de Janeiro, Brazil	14,169,000
10. Calcutta, India	14,088,000

SOURCE: *Statistical Abstract of the United States: 1994*, 114th ed. U.S. Bureau of the Census. Washington, D.C.: U.S. Government Printing Office, Table 1355.

Tokyo is the most populated city in the world.

Several factors have contributed to the urbanization of the world. First, urbanization occurs as countries experience increased industrialization and economic growth (Ingram 1998). Corporate multinationalism has spurred foreign investment in foreign cities as profit-seeking corporations locate industry in developing countries to gain access to cheap labor, raw materials, and new markets for their products and services. Migration from rural to urban areas has also contributed to urbanization. Rural dwellers migrate to urban areas to flee from war or natural disasters or to find employment. As foreign corporate-controlled commercial agriculture displaces traditional subsistence farming in poor rural areas, peasant farmers flock to the city looking for employment. Some rural dwellers migrate to urban areas in search of a better job—one that has higher wages and better working conditions. Governments have also stimulated urban growth by spending more to improve urban infrastructures and services, while neglecting the needs of rural areas (Clark 1998). Lastly, urban populations in developing countries have increased due to high fertility rates.

History of Urbanization in the United States

Urbanization of the United States began as early as the 1700s, when most major industries were located in the most populated areas, including New York City, Philadelphia, and Boston. Unskilled laborers, seeking manufacturing jobs, moved into urban areas as industrialization accelerated in the nineteenth century. The "pull" of the city was not the only reason for urbanization, however. Technological advances were making it possible for fewer farmers to work the same amount of land. Thus, "push" factors were also involved—making a living as a farmer became more and more difficult as technology, even then, replaced workers.

Urban populations continued to multiply as a large influx of European immigrants in the late 1800s and early 1900s settled in U.S. cities. This was followed by a major migration of southern rural blacks to northern urban areas. People were lured to the cities by the promise of employment and better wages and such urban amenities as museums, libraries, and entertainment. Immigrants are also attracted to cities with large ethnic communities that provide a familiar cultural environment in which to live and work. Today, a large percentage of immigrants into inner cities are Asian or Hispanic, and they settle disproportionately in certain cities, including Miami, San Francisco, Los Angeles, and Chicago.

Figure 13.1 depicts the tremendous growth of the U.S. urban population over the last 200 years. Between 1800 and 1990, the percentage of the U.S. population that is urban grew from 6.1 percent to 77.5 percent.

Suburbanization

In the late nineteenth century, railroad and trolley lines enabled people to live outside the city and still commute into the city to work. As more and more people moved to the **suburbs**—urban areas surrounding central cities—America underwent **suburbanization.** As city residents left the city to live in the suburbs, cities lost population and experienced **deconcentration**, or the redistribution of the population from cities to suburbs and surrounding areas.

Many factors have contributed to suburbanization and deconcentration. Following World War II, many U.S. city dwellers moved to the suburbs out of concern for the declining quality of city life and the desire to own a home on a spacious lot. Suburbanization was also spurred by racial and ethnic prejudice, as the white majority moved away from cities that, due to immigration, were becoming increasingly diverse. Mass movement into suburbia was encouraged by the federal road building program, which promoted automobile use and by federal grants that have allowed the extensive construction of sewers. Due to the perceived threat of nuclear attack during the Cold War, federal, state, and local governments decentralized their offices and facilities, and defense contractors located new plants outside city centers (Kelbaugh 1997). In the 1950s, Veterans Administration (VA) and Federal Housing Administration (FHA) loans made housing more affordable, enabling many city dwellers to move to the suburbs. Suburb dwellers who worked in the central city could commute to work or work in a satellite branch in suburbia that was connected to the main downtown office. The new technologies of fax machines, home computers, cellular phones, pagers, and e-mail have also facilitated the movement to the suburbs. As increasing numbers of people moved to the suburbs,

As anyone who has even a moderate addiction to political punditry knows by now, "for the first time, the majority of voters live in suburbia." This purportedly epochal demographic shift is broadly interpreted as the death knell of the power and relevance of urbanism in America.

—Ruth Messinger and Andrew Breslau
Social scientists

National Data

In the United States, from one-third to one-half of large central cities have lost population over the past 25 years.

SOURCE: Ingram 1998

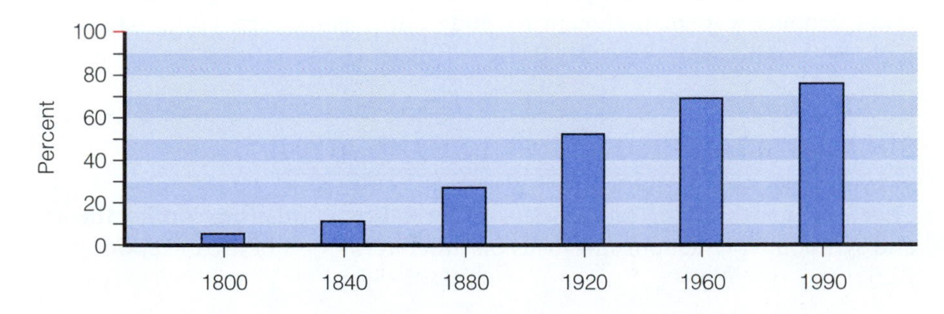

Figure 13.1 **Percentage of U.S. Population that is Urban, 1800–1990 (An urban population consists of persons living in cities or towns of 2,500 or more residents.)**

SOURCES: *Statistical Abstract of the United States: 1993,* 113th ed.; *Statistical Abstract of the United States: 1998,* 118th ed. U.S. Bureau of the Census. Washington, D.C.: U.S. Government Printing Office.

so did businesses and jobs. Without a strong economic base, city services and the quality of city public schools declined, which furthered the exodus from the city.

You can't keep spreading out. The cost to make roads and sewers gets to the point where it doesn't work.

—Mike Burton
Executive director
Portland's metro government

U.S. Metropolitan Growth and Urban Sprawl

Simply defined, a **metropolitan area** is a densely populated core area, together with adjacent communities. The largest city in each metropolitan area is designated as a "**central city.**" Another term for "metropolitan area" is ***metropolis,*** from the Greek meaning "mother city." Robert Geddes (1997) denotes cities and their surrounding regions as "city-regions."

Current standards require that each newly qualifying metropolitan area must include at least one city or urban area with at least 50,000 residents and a total metropolitan population of at least 100,000 (75,000 in New England). The most populated U.S. metropolitan area encompasses several cities in New York, New Jersey, Connecticut, and Pennsylvania.

Metropolises have grown rapidly in the United States. As new areas reach the minimum required city or urbanized area population, and as adjacent towns, cities, and counties satisfy the requirements for inclusion in metropolitan areas, both the number and size of metropolitan areas have grown. As of June 1998, there were 256 metropolitan areas in the United States, and 3 in Puerto Rico *(Statistical Abstract of the United States: 1998).* There is one U.S. state—New Jersey—that is entirely occupied by metropolitan areas as designated by the U.S. census (Geddes 1997).

National Data

In 1996, 79.8 percent of the U.S. population lived in metropolitan areas. In the same year, 19.9 percent of U.S. land area was designated as a metropolitan area.

SOURCE: *Statistical Abstract of the United States: 1998,* Table 40

Referring to the growth of metropolises, Robert Geddes (1997) observes, "a new form of human settlement has emerged in the twentieth century, radically different from the cities of the past" (p. 40). Those who find this new form of human settlement—the ever-increasing outward growth of urban areas—to be disturbing refer to it as "**urban sprawl.**" Critics of urban sprawl lament the loss of green, open spaces and cite traffic congestion and pollution among its liabilities. However, others view the outward growth of urban areas as the preferred lifestyle of many Americans (Gordon and Richardson 1998).

National Data

One in 20 Americans lives in a micropolitan area.

SOURCE: Heubusch 1998

Those who enjoy the conveniences and amenities of urban life, yet find large metropolitan areas undesirable, may choose to live in a **micropolitan area**—a small city located beyond congested metropolitan areas. These areas are large enough to attract jobs, restaurants, community organizations, and other benefits, yet small enough to elude traffic jams, high crime rates, and high costs of housing and other living expenses associated with larger cities.

SOCIOLOGICAL THEORIES OF URBANIZATION

The three main sociological theories—structural-functionalism, conflict theory, and symbolic interactionism—focus on different aspects of urbanization.

Structural-Functionalist Perspective

Structural-functionalists view the development of urban areas as functional for societal development. Although cities initially functioned as centers of production and distribution, today they are centers of finance, administration, education, and information.

The expansion of urban areas, although functional, also leads to increased rates of anomie, or normlessness, as the bonds between individuals and social groups become weak (see Chapter 1 and Chapter 4). While in rural areas social cohesion is based on shared values and beliefs, in urban areas social cohesion is a consequence of specialization and interdependence spurred by increased social diversity. Urbanization thus leads to higher rates of deviant behavior, including crime, drug addiction, and alcoholism. To functionalists, these increased rates are an indication of the social disorganization of the area.

I'd rather sleep in the middle of nowhere than in any city on earth.

—Steve McQueen
Actor

The structural-functionalist perspective is also concerned with the dysfunctions of urbanization. Historically, a major dysfunction of cities is that, by concentrating large numbers of people in small areas, they have facilitated the rapid spread of infectious disease (see this chapter's *Focus on Social Problems Research* feature). Overcrowding, poverty, and environmental destruction are also viewed as dysfunctions associated with urbanization.

Conflict Perspective

Conflict theorists emphasize the role of power, wealth, and the profit motive in the development and operations of urban areas. According to the conflict perspective, capitalism requires that the production and distribution of goods and services be centrally located thus, at least initially, leading to urbanization. Today, corporate multinationalism, in search of new markets, cheap labor, and raw materials, has largely spurred urbanization of the developing world. Capitalism also contributes to migration from rural areas into cities. Peasant farmers who have traditionally produced goods for local consumption are being displaced by commercial agriculture that is geared to producing fruits, flowers, and vegetables for export to the developed world. Displaced from their traditional occupations, peasant farmers are flocking to cities to find employment (Clark 1998).

National Data

In 1997, 36 percent of adults viewed cities as centers of business, culture, and progress, up from 30 percent in 1991. The percentage who viewed cities as centers of poverty, crime, and other social problems decreased from 42 percent in 1991 to 38 percent in 1997.

SOURCE: Fisher 1997

The conflict perspective also focuses on how policy makers and corporations base decisions that affect urban populations on economic considerations. The decision to build a mall or not and where to locate that mall are economic decisions based on the desire to maximize profits. The decision to tear down inner-city homes to make room for office complexes or parking decks is also profit motivated. The decision maker rarely considers that residents in low-income housing will be displaced or that 200-year-old oak trees will be cut down.

Symbolic Interactionist Perspective

The symbolic interaction perspective emphasizes how meanings, labels, and definitions affect behavior. One application of this perspective is found in efforts to change the public's negative definitions of cities. Cities are often viewed as dangerous and crime-ridden, decaying, dirty places. These negative definitions of cities discourage people from living and vacationing in cities and deter businesses from locating there. Negative definitions of cities may also contribute to a self-fulfilling prophecy. Detroit mayor Dennis Archer (1998) observed that "negative perceptions of the City of Detroit and low expectations for our future convinced many observers that Detroit was beyond hope. This made issues that many large cities face—unemployment—crime—flight of the middle class—dissatisfaction with public schools—much harder to handle in Detroit" (p. 341). Efforts to redefine cities in positive terms are

FOCUS ON SOCIAL PROBLEMS RESEARCH

Big Cities Health Inventory

Public health researchers Benbow, Wang, and Whitman (1998) were concerned about the lack of city-level health data in the United States. Employed by the Chicago Department of Public Health, the researchers had access to Chicago health statistics but wanted to compare their data with comparable cities. No such information was available. Thus, they initiated a research project that permitted inter-city health comparisons.

Sample and Methods

The researchers began by identifying important indicators of urban health. Twenty indicators were identified, including rates of AIDS, cancer, tuberculosis, homicide, sexually transmitted diseases, low birthweight, infant mortality rate, and heart disease. Data on each of the indicators were obtained from the 46 U.S. cities with populations greater than 350,000 based on the 1990 census (Benbow, Wang, and Whitman 1998, 471). Where available, data over several years were collected. Comparisons between cities and by gender, race, and ethnicity were made.

Findings

Selected findings suggest some interesting trends. Tuberculosis (TB) for example, a communicable disease, was highest in New York City, followed by San Francisco, Houston, and Atlanta. In no city were TB rates higher for women than for men, and in some cities female rates were less than half male rates.

Information on the health indicators over several years allowed comparisons over time. For example, of the 23 cities for which syphilis rates were available, 18 cities had decreased rates between 1992 and 1994. Five cities reported increased rates during the same time period (Baltimore, Cleveland, Denver, Indianapolis, and St. Louis).

Benbow, Wang, and Whitman (1998) argue that AIDS rates, homicide rates, and infant mortality rates constitute the three key public health issues in urban areas. The authors conclude that:

> the number of infant deaths had been declining steadily during this interval [1980–1995], the number of homicides had been fluctuating a bit but has remained rather constant over time, and the number of deaths due to AIDS had been steadily increasing, although there is some sign of a leveling off in recent years. (p. 50)

In 1994, in over half the cities (60 percent) AIDS killed more people than murder or infant deaths; in 20 percent of the cities infant mortality was responsible for more deaths than AIDS or homicide, and in the remaining 20 percent of the cities homicides took more lives than AIDS or infant mortality. The authors note the inconsistency between their findings and the media's portrayal of homicide as the leading cause of death in urban areas.

When examining mortality rates from all causes, Sacramento had the highest age-adjusted mortality rate, followed by Atlanta; San Jose, California, had the lowest mortality rate of all 46 cities. Racial and ethnic differences were also detected. Although Miami had the highest overall mortality rate for non-Hispanic blacks, Philadelphia had the highest mortality rate for Hispanics.

To conclude from this study that certain cities are more or less "healthy" would be premature. Mortality differentials are related to demographic characteristics of the city's population. However, it would also be premature to conclude that the ecology of the city is unrelated to mortality. Central to urban research and, specifically, urban epidemiological studies is the identification of urban variables and their relationship to the health of city residents.

SOURCE: Nanette Benbow, Yue Wang, and Steven Whitman. 1998. "The Big Cities Health Inventory, 1997." *Journal of Community Health* 23:471–82.

reflected in advertising campaigns sponsored by cities' convention and visitors bureaus. For example, an advertising slogan for Detroit indicates, "It's a Great Time in Detroit."

The symbolic interactionist perspective also focuses on how social forces influence individuals' self-concepts, values, and behaviors. People tend to identify with the place where they live, and this is especially true of urban dwellers. The distinctive cultures and lifestyles of cities shape the self-concepts, values, and behaviors of their residents. Although each city is unique, city dwellers tend to share a basic culture and lifestyle, known as **urbanism.** This urban culture and lifestyle is characterized by individualistic and cosmopolitan norms, values, and styles of behavior. In contrast to the slow-paced and traditional nature of rural life, life in the city is fast-paced and always on the "cutting edge" in terms of the latest technology, entertainment, fashions, food trends, and other forms of cultural change. Symbolic interactionists emphasize the effects of urbanization on interaction patterns and social relationships. We discuss these effects in the following section.

THE EFFECTS OF URBAN LIVING ON SOCIAL RELATIONSHIPS

Various theories of urbanism focus on how living in a city affects interaction patterns and social relationships. Here we examine both the classical and the modern theoretical views, as well as attempts to synthesize the two perspectives.

> Our cities are more pestilential than yellow fever to the morals, the health and the liberties of man.
>
> —Thomas Jefferson
> Former U.S. President

Classical Theoretical View

Cities have the reputation of being cold and impersonal. As early as 1902, George Simmel observed that urban living involved an overemphasis on punctuality, individuality, and a detached attitude toward interpersonal relationships (Wolff 1978). It is not difficult to find evidence to support Simmel's observations: witness New Yorkers pushing each other to get onto the subway during rush hour, hear motorists curse each other in Los Angeles traffic jams, watch Chicago residents ignore the homeless man asleep on the sidewalk.

Lewis Wirth (1938), a second-generation student of Simmel, argued that urban life is disruptive for both families and friendships. He believed that because of the heterogeneity, density, and size of urban populations, interactions become segmented and transitory, resulting in weakened social bonds. Those bonds that do develop are superficial and detract from the closeness of primary relationships. Wirth held that as social solidarity weakens, people exhibit loneliness, depression, stress, and antisocial behavior.

> "The city," an alien place where by definition middle class Americans refuse to live.
>
> —Jerry Adler
> Journalist

CONSIDERATION

The presumed effect of the urban environment on antisocial behavior has been used in court. In 1991, a 17-year-old youth with a history of victimization and neighborhood violence unsuccessfully used "urban psychosis" as a defense after shooting another teenager for a leather jacket (Shoop 1993). In a subsequent legal case, Daimion Osby contended that he shot two unarmed men to death because of "urban survival syndrome." (Davis 1994)

Wilson (1993) empirically assessed the validity of Wirth's classical view of cities by examining the effects of city size on a number of variables. Wilson hypothesized that urbanism inhibits family completeness in that people are less likely to marry; that it disrupts family relationships in terms of the number and frequency of contacts between relatives; and that it changes the functions served by kin—as community size increases, relatives are less likely to rely on one another for help.

Wilson's findings did not support Wirth's theory. Urbanism did not inhibit family completeness, and contact with and reliance on relatives were unaffected. He did find, however, that the larger the urban area, the greater the likelihood that individuals were separated or divorced. The fact that there are fewer

Recent immigrants may fail to assimilate socioeconomically, and the tendency to settle in enclaves means that intragroup solidarity increases, which further lowers the probability of assimilation.

informal social controls and that attitudes are generally more liberal in an urban environment may contribute to the higher divorce rate. Further, contrary to Wirth's theory, the prevalence of psychiatric disorders does not differ significantly between urban and rural areas (Kessler et al. 1994).

Modern Theoretical View

A community must simultaneously nurture both a respect for group values and a tolerance for individuality . . .

–Douglas Kelbaugh
University of Washington

In contrast to Wirth's pessimistic view of urban areas, Herbert Gans ([1962] 1984) argued that cities do not interfere with the development and maintenance of functional and positive interpersonal relationships. Among other communities, he studied an Italian urban neighborhood in Boston called the West End and found such neighborhoods to be community oriented and marked by close interpersonal ties. Rather than finding the social disorganization described by Wirth, Gans observed that kinship and ethnicity helped bind people together. These enclaves were characterized by intimate small groups with strong social bonds. Thus, Gans saw the city as a patchwork quilt of different neighborhoods or **urban villages**, each of which helped individuals deal with the pressures of urban living.

A Theoretical Synthesis

Fisher (1982) interviewed more than 1,000 respondents in various urban areas and found evidence for both the classical and the modern theoretical views of urbanism. From the classical perspective, he found that heterogeneity (the diversity among urban residents) does make community integration and consensus difficult—there is less community cohesion. Ties that do exist are less often kin-related than in nonurban areas and are more often based on work relationships, memberships in voluntary and professional organizations, and proximity to neighbors.

Fisher also found, however, that the diversity of urban populations facilitates the development of subcultures that have a sense of community ties. For example, large urban areas include such diverse groups as gays, ethnic and racial minorities, and artists. These individuals "find each other" and develop their own unique subcultures.

Another researcher (Tittle 1989) empirically examined each of the three theories of the effects of urbanism—classical, modern, and Fisher's theoretical synthesis. Holding such demographic variables as race, education, age, length of residence, and marital status constant, Tittle examined the effects of community size on a respondent's social bonds, anonymity, tolerance, alienation, and deviant behavior.

Tittle (1989) found some support for both the classical and modern view and, thus, for Fisher's synthesis. As predicted by the classical view, the larger the size of the community, the weaker a respondent's social bonds and the higher his or her anonymity, tolerance, alienation, and reported incidence of deviant behavior. He also found that racial and ethnic groups created their own sense of community, lending support to the modern theory of urbanism.

CITIES AND SOCIAL PROBLEMS

Given that urban areas comprise high concentrations of people, it is not surprising that they also experience high concentrations of various social problems. Urban areas tend to be plagued by poverty, poor housing, and inadequate schools. Other urban problems include HIV/AIDS, addiction, and crime; and transportation and traffic problems. While many urban problems in the United States are exacerbated by the loss of inner-city residents to the suburbs, in developing countries, urban problems are compounded by rapid urban population growth.

Urban Poverty

The number of urban poor is growing rapidly worldwide. As discussed in Chapter 2, poverty is a main determinant of health problems. Infant mortality, a widely accepted indicator of the health of a country's population, has declined sharply in towns and rural areas since the 1970s. However, in large cities of Latin America, infant mortality rates have declined by a modest 7 percent, and in the largest cities of sub-Saharan Africa, less than 5 percent. Some cities in tropical Africa have actually experienced an *increase* in infant mortality rates. Cities that have grown the fastest have higher infant mortality rates than cities that have grown at a slower pace (Brockerhoff and Brennan 1997).

Poverty is concentrated in central cities of the metropolitan areas of the United States. In some U.S. central cities, nearly one-third of the population is poor (Pack 1998). Half of Maryland's poor live in the city of Baltimore (Schmoke 1998). Increasingly, as more affluent city residents move to the outer suburbs, the older inner suburbs surrounding central cities are also characterized by high poverty rates. Between 1970 and 1990, neighborhood poverty expanded in the majority of U.S. metropolitan areas (Jargowsky and Wilson 1997).

International Data

Over one-quarter of the developing world's urban population lives below the official poverty line. In some of the world's poorest developing countries, half of the urban population live in conditions of extreme deprivation.

SOURCES: United Nations Population Fund 1996; World Health Organization and United Nations Joint Programme on HIV/AIDS 1998

> **National Data**
>
> In 1994 and 1995, about 45 percent of the nation's poor lived in central cities.
>
> SOURCE: Accordino 1998

Urban poverty is reflected in the high numbers of homeless and street children in urban areas. In Los Angeles, for example, there are an estimated 84,000 people homeless on any given night and only 11,000 beds in homeless shelters (Wolch 1998) (see also Chapter 10). Children who live and/or work on the streets (sometimes referred to as "street children") are found in virtually every city throughout the world. The dramatic increase in the number of street children in developing countries has been linked to societal stress associated with rapid industrialization and urbanization (Le Roux and Smith 1998). In industrialized countries, the epidemic of street children is linked to inner-city decay and chronic unemployment.

> **International Data**
>
> There are as many as 100 million street children in the world.
>
> SOURCE: Le Roux and Smith 1998

In the United States and other industrialized countries, urban unemployment and poverty is related to **deindustrialization**, or the loss and/or relocation of manufacturing industries (McNulty 1993). Since the 1970s, many urban factories have closed or relocated, forcing blue-collar workers into unemployment. Flint, Michigan, was once the site of a large General Motors plant boasting 70,000 manufacturing employees in 1980. When the plant closed, Flint's manufacturing workforce dropped by almost half, to 40,000, and by 1987 the city was ranked last out of 300 U.S. cities in a quality-of-life survey (Lord and Price 1992). Loss of jobs has been a main cause of unemployment for many U.S. cities. New York City had 200,000 fewer jobs in 1996 than it had in 1990; Los Angeles County had 300,000 fewer jobs; Baltimore, 64,000; and Philadelphia, 57,000 (Goozner 1998).

> We will neglect our cities to our peril, for in neglecting them we neglect the nation.
>
> —John F. Kennedy
> Former U.S. President

CONSIDERATION

Welfare recipients tend to be concentrated in central cities. Due to the welfare reform legislation of 1996, poor parents are eligible to receive public assistance (called Temporary Aid for Needy Families or TANF) for only 5 years over the course of their lifetimes, and numerous other benefits, such as food stamps, have been cut (see Chapter 10). This welfare reform law is likely to have an overall negative impact on the economies of central cities. Accordino (1998) forewarns that "welfare reform may hit cities with a double whammy" (p. 14). First, welfare recipients who do not find jobs lose their income, deepening urban poverty. Second, consumer spending declines (as the jobless have no spending money), which hurts small businesses that employ many of the working poor, who end up losing their jobs as well.

Urban Housing Problems

> **International Data**
>
> In a study of eight megacities, the International Labor Organization estimated slum populations at between 12 percent (Seoul) and 84 percent (Cairo).
>
> SOURCE: United Nations Population Fund 1996

Urban housing problems include lack of affordable housing, substandard housing, and housing segregation. Many cities are experiencing a housing crisis. The Center for Budget Policy Priorities reported that in the past 25 years the number of low-income renters increased by 70 percent, while the number of low-cost rental units has dropped ("Crisis in Low-Income Rental Housing" 1998). Housing that is available and affordable is often substandard, characterized by outdated plumbing and wiring, overcrowding, rat infestations, and fire hazards (see also Chapter 10). Older housing units often have some lead-based paint, which may cause lead poisoning.

Concentrated areas of poor housing and squalor in heavily populated urban areas are called **slums**. A slum section of a city occupied primarily by a minority group is referred to as a **ghetto.** U.S. minorities, who are disproportionately represented among the poor, tend to be segregated in concentrated areas of low-income housing (see also Chapter 8). The fact that many minorities do not have the resources to move out of inner-city neighborhoods results in involuntary segregation—whites move out, leaving nonwhites behind. Segregation is particularly evident in public housing (Bickford and Massey 1991).

Low-income housing tends to be concentrated in inner-city areas of concentrated poverty. But jobs are increasingly moving to the suburbs, and central-city residents often lack transportation to get to these jobs. The more affluent suburbs restrict development of affordable housing to keep out "undesirables" and maintain their high property values. Suburban zoning regulations that require large lot sizes, minimum room sizes, and single-family dwellings serve as barriers to low-income development in suburban areas (Orfield 1997).

> Cities have been ignored because they are blacker, browner, poorer, and more female than the rest of the nation.
>
> —Julianne Malveaux
> Writer/Scholar

Inadequate Schools

When jobs and middle-class residents left the city for the suburbs in mass exodus, local revenues decreased, and the quality of city schools, particularly in inner-city neighborhoods, declined. This decline in school quality contributed further to residents leaving the city—at least those who could afford to.

Although public schools in other countries are often supported at the national level, U.S. schools depend primarily on local funding. The result is dramatic differences between inner-city and suburban schools. The former are highly segregated with an estimated 90 percent of the school population being nonwhite (Kozol 1991). Further, as discussed in Chapter 12, inner city schools are often the oldest and most deteriorated. Consequently, they lack adequate facilities, such as computer laboratories, and have higher teacher turnover because of the often dangerous working conditions. As educational quality continues to deteriorate, those who can afford to move out do so, leaving the poorest residents behind. As a result, property values continue to decline, the tax base decreases, schools further erode, and the cycle repeats itself.

The Urban "Three-Headed Monster": AIDS, Addiction, and Crime

In a speech delivered at the Conference on the Successful City of the 21st Century, Baltimore's mayor, Kurt Schmoke (1998), used the term "three-headed monster" to refer to three problems that plague cities: HIV/AIDS, addiction, and crime.

HIV/AIDS and Drug Addiction Rates of HIV and AIDS are higher in urban areas than in rural areas because individuals who live in cities have higher rates of risk behavior (see Chapter 2). For example, rates of drug addiction are higher among city dwellers than the rest of the population. Baltimore has an estimated 59,000 drug addicts (Schmoke 1998). Drug use impairs judgment and interferes with safer sex practices, and some addicts resort to the high-risk behavior of prostitution to pay for their drugs. Most importantly, drug use that involves needle injection accounts for a significant percentage of HIV transmission and the disease that often follows: AIDS.

National Data

In 1996, there were 715 violent crimes reported to the police per 100,000 population in metropolitan areas compared to 222 in rural areas. The property crime rate was 4,798 in metropolitan areas compared to 1,828 in rural areas.

SOURCE: *Statistical Abstract of the United States: 1998*

Urban Crime In general, the rates of violent crime and property crime are higher in metropolitan areas than in rural areas. High rates of urban crime are linked to high rates of poverty, single-parent households where children have little supervision, inadequate schools that fail to channel students into productive roles, high rates of drug addiction, and lack of positive role models. Orfield (1997) explains:

> **In neighborhoods lacking successful middle-class role models, gang leaders, drug dealers, and other antisocial figures are often the only local residents with money and status. Tightly knit gangs replace nonexistent family structures. These factors interact with anger, frustration, isolation, boredom, and hopelessness, and create a synergism of disproportionate levels of crime, violence, and other antisocial behavior. (p. 19)**

CONSIDERATION

Far more people are killed and injured in automobile accidents than by violent crime. Therefore, it has been argued that despite higher crime rates in the inner city, the suburbs are the more dangerous place to live because suburbanites "drive three times as much, and twice as fast, as urban dwellers. All told, city dwellers are much safer." (Durning 1996, 24)

Transportation and Traffic Problems

The United States is a country where private automobiles, often carrying only one person, are the dominant mode of travel. Cars and light trucks are the largest source of urban air pollution (Union of Concerned Scientists 1999). Air pollution and traffic congestion have been major forces driving residents and

Due to the high volume of traffic in Los Angeles, this city has the worst air quality in the United States.

Road Rage Survey

Increased congestion of urban highways has led to a phenomenon called road rage. Please answer the questions below to assess your level of road rage and then compare your answers to a national sample of respondents.

1. Which of the following do you ever do?

Category I (Check all that apply)

- Going 5 to 10 mph over the speed limit ☐
- Making rolling stops ☐
- Making illegal turns ☐
- Lane hopping without signaling ☐
- Following very close as a habit ☐
- Going through red lights ☐
- Denying right of way (failure to yield) ☐
- Swearing, cursing, name calling ☐

Combined, how regularly do you do the things in Category I (circle one)?
Never 1 2 3 4 5 6 7 8 9 10 Quite regularly

Category II (check all that apply)

- Going 15 to 25 mph over the speed limit ☐
- Yelling at another driver ☐
- Honking in protest ☐
- Revving the engine ☐
- Making an insulting gesture ☐
- Tailgating dangerously ☐
- Shining the headlights to retaliate ☐
- Cruising in the passing lane (forcing others to pass on the right) ☐

Combined, how regularly do you do the things in Category II (circle one)?
Never 1 2 3 4 5 6 7 8 9 10 Quite regularly

Category III (check all that apply)

- Braking suddenly to punish a tailgater ☐
- Deliberately cutting someone off to retaliate ☐
- Using the car to block the way ☐
- Using the car as a weapon to threaten someone ☐
- Chasing another car in pursuit ☐
- Getting into a physical fight with another driver ☐
- Other things ☐

Combined, how regularly do you do the things in Category III (circle one)?
Never 1 2 3 4 5 6 7 8 9 10 Quite regularly

2. Which of the following emotions do you experience when driving?

Anger or rage behind the wheel
Never 1 2 3 4 5 6 7 8 9 10 Regularly

Enjoying fantasies of violence
Never 1 2 3 4 5 6 7 8 9 10 Regularly

(continued)

SELF & SOCIETY

Road Rage Survey *(continued)*

Experiencing fear for self or family in the car
Never 1 2 3 4 5 6 7 8 9 10 Regularly
Feeling compassion for another driver
Never 1 2 3 4 5 6 7 8 9 10 Regularly
Feeling competitive with other drivers
Never 1 2 3 4 5 6 7 8 9 10 Regularly
Feeling impatient and the constant urge to rush
Never 1 2 3 4 5 6 7 8 9 10 Regularly
Wanting to drive dangerously
Never 1 2 3 4 5 6 7 8 9 10 Regularly
Feeling levelheaded and calm
Never 1 2 3 4 5 6 7 8 9 10 Regularly

Below are the results of a national survey of 761 U.S. drivers. Numbers indicate the percentage of people who have engaged in the behavior ∓3 percent.

Category I	*Percentage*
Going 5 to 10 mph over the speed limit	92
Making rolling stops	50
Making illegal turns	15
Lane hopping without signaling	25
Following very close as a habit	14
Going through red lights	10
Denying right of way (failure to yield)	8
Swearing, cursing, name calling	58
Mean regularity rating	6.0

Category II	*Percentage*
Going 15 to 25 mph over the speed limit	39
Yelling at another driver	32
Honking in protest	35
Revving the engine	11
Making an insulting gesture	24
Tailgating dangerously	12
Shining the headlights to retaliate	18
Cruising in the passing lane (forcing others to pass on the right)	11
Mean regularity rating	3.9

Category III	*Percentage*
Braking suddenly to punish a tailgater	32
Deliberately cutting someone off to retaliate	15
Using the car to block the way	18
Using the car as a weapon to threaten someone	2
Chasing another car in pursuit	8
Getting into a physical fight with another driver	2
Other things	8
Mean regularity rating	2.7

Road Rage Survey *(continued)*

Emotions	*Mean Regularity Rating*
Anger or rage behind the wheel	4.5
Enjoying fantasies of violence	2.4
Experiencing fear for self or family in the car	4.0
Feeling compassion for another driver	4.9
Feeling competitive with other drivers	4.9
Feeling impatient and the constant urge to rush	5.7
Wanting to drive dangerously	2.7
Feeling levelheaded and calm	6.7

SOURCE: Leon James. 1998. "World Road Rage Survey" by Leon James, Ph.D. as appeared on **http://www.drdriving.org**. Reprinted by permission.

businesses away from densely populated urban areas (Warren 1998). Many public roads in urban areas are afflicted with what some call *autosclerosis*—defined as "clogged vehicular arteries that slow rush hour traffic to a crawl or a stop, even when there are no accidents or construction crews ahead" ("The Bridge to the 21st Century Leads to Gridlock in and around Decaying Cities" 1997, 1). In the United States, Washington, D.C., and Los Angeles are known for their chronic traffic jams, but foreign cities such as Sao Paulo, Brazil; Bangkok, Thailand; and Cairo, Egypt, are even worse. According to one report, "it sometimes takes so long to reach the Bangkok airport from downtown—from 3 to 6 traffic-paralyzed hours—that roadside entrepreneurs sell minitoilet kits for use by desperate riders in traffic-jammed cars" ("The Bridge to the 21st Century Leads to Gridlock in and around Decaying Cities" 1997, 1). Traffic congestion creates stress on drivers, which sometimes leads to aggressive driving and violent reactions to other drivers—a phenomenon known as **road rage** (see this chapter's *Self & Society* feature).

Dependency on cars has been encouraged by a number of factors: unwillingness to tax gasoline commensurate with its cost to society, free and tax-free parking typically provided by corporations to their employees outside of the city, the glamorization of automobiles—largely perpetuated by the automobile industry—and federal subsidies that favor the building of highways over investments in public transit. The conversion of the United States to an automobile-based system of transportation was heavily influenced by industries that profit from automobile use. In the 1930s, National City Lines, a company backed by the three major automakers, major oil companies, tire manufacturers, and the trucking and construction industries, succeeded in systematically buying and closing down more than 100 electric trolley lines in 45 U.S. cities. Although National City Lines was convicted of this conspiracy in 1949, the dismantling of the rail transit system had already been accomplished (Warren 1998).

National Data

In a 1997 survey of U.S. adults, 51 percent said crime is a major disadvantage of city living, followed by traffic congestion and crowds, at 41 percent. Nine percent said it costs too much to live in cities, 8 percent said the pace of city life is too hectic, and 7 percent consider pollution and environmental concerns a negative of city life.

SOURCE: Fisher 1997

National Data

American consumers would pay \$5.60 to \$15.14 per gallon of gasoline if the environmental and social impacts of producing it were reflected in the price at the pump.

SOURCE: "News Briefs" 1999

STRATEGIES FOR ACTION: SAVING OUR CITIES

In the film *Field of Dreams* Kevin Costner was told by unseen voices to "build it and they will come"; the Republican approach to cities seems to be "ignore them and they will go away."

—Ruth Messinger and Andrew Breslau
Social scientists

Numerous federal, state, and private policies and programs have been implemented to attempt to revitalize the cities and address the various urban problems discussed in this chapter. Here we summarize strategies to alleviate poverty and stimulate economic development in inner cities, tame the urban "three-headed monster" (AIDS, addiction, and crime), improve transportation and alleviate traffic congestion, and curb urban growth in developing countries.

Urban Economic Development and Revitalization

A number of strategies have been proposed and implemented to restore prosperity to U.S. cities and well-being to their residents, businesses, and workers, including strategies to attract new businesses, create jobs, and repopulate cities. Strategies to improve inner-city schools are found in Chapter 12, and housing is discussed in Chapter 10. Although alleviating other urban problems, such as HIV/AIDS, addiction, crime, and traffic and transportation problems also plays a key role in the economic development and revitalization of cities, these issues are addressed in later sections of this chapter.

We simply can't allow our cities to become nothing but massive shelters for the poor.

—Kurt L. Schmoke
Mayor of Baltimore

Empowerment Zone/Enterprise Community Program One strategy to stimulate economic development in impoverished areas (many of which are urban areas) is the Empowerment Zone/Enterprise Community Initiative, or EZ/EC program. This program provides tax incentives and performance grants and loans to create jobs for residents living within the designated zone or community. Thus, **empowerment zones** and enterprise communities stimulate the local economy, provide jobs for the unemployed, and increase city revenues. They also provide funds for job training, child care, and transportation to enable impoverished adults to work.

Some cities have attributed significant overall economic gains and improvements to the EZ/EC program. Detroit mayor Dennis Archer (1998) claimed that,

> **Since winning our Empowerment Zone designation in December 1994, it has indeed become a powerful contributor to the city's economic momentum. One hundred million dollars in federal money has been provided to 49 Detroit community organizations for construction projects and social programs. Motivated by tax breaks for employment within the Empowerment Zone, businesses have invested at least $3.9 billion in the Zone since early 1995. At least 2,750 new jobs—with 400 more on the way—have been created within the 18.3-square mile area during that period. (pp. 341–42)**

National Data

In December 1994, President Clinton and Vice President Gore designated 72 urban areas and 33 rural communities as empowerment zones or enterprise communities.

SOURCE: "Empowerment Zone/Enterprise Community Initiative" 1998

There are no quick fixes or dazzling new policy options for the problems that beleaguer America's cities.

—Ruth Messinger and Andrew Breslau
Social scientists

Public/Private Partnerships Some efforts to develop and revitalize urban areas involve city governments teaming up with private for-profit and nonprofit businesses and organizations. Increasingly, the resources and expertise of the private sector are being applied toward solving the needs of cities and towns. Public-private partnerships are being used to provide public services; to develop, finance, and implement infrastructure and community facilities; and to retain business and industry (McDonough 1998). For example, Baltimore's business community, in partnership with government, created Baltimore's Inner Harbor. Mayor Kurt Schmoke (1998) remarked, "What had once been a neglected area of dilapidated buildings, roadways, and railroad tracks was transformed into a festive waterfront attraction featuring museums, shopping and dining pavilions, hotels, promenades, and marinas" (p. 111).

Public-sector involvement in providing city services is also increasing. In some cities, services such as garbage collection and transportation services are being contracted out to private firms. Case studies indicate that privatization of city services tends to be less costly and more efficient than having such work done by public employees (Ingram 1998).

Urban neighborhood revitalization efforts include painting over graffiti.

Community-Based Urban Renewal Efforts Some residents of deteriorating urban areas have begun grassroots programs to improve living conditions in their neighborhoods. Community-based development programs involve small-scale developers and volunteers working with a small professional staff who are concerned with improving the community. For example, 26,000 volunteers spent a Saturday cleaning Detroit streets, parks, and playgrounds during the fourth annual spring Clean Sweep campaign (Archer 1998). In Detroit's annual Paint the Town event, individuals, community organizations, and corporate volunteers fix and paint the homes of the poor and elderly.

Gentrification and Incumbent Upgrading **Gentrification** is a type of neighborhood revitalization in which middle- and upper-income persons buy and rehabilitate older homes in a depressed neighborhood. They may live there or sell or rent to others. The city provides tax incentives for investing in old housing with the goal of attracting wealthier residents back into these neighborhoods and increasing the tax base. However, low-income residents are often forced into substandard housing as less and less affordable housing is available. In effect, gentrification often displaces the poor and the elderly (Johnson 1997).

An alternative to gentrification is **incumbent upgrading**, in which aid programs help residents of depressed neighborhoods buy or improve their homes and stay in the community. Both gentrification and incumbent upgrading improve decaying neighborhoods, which attracts residents as well as businesses.

Taming the Urban "Three-Headed Monster": Strategies to Reduce AIDS, Addiction, and Crime

Strategies to reduce AIDS, addiction, and crime are discussed in other chapters in this text (see Chapters 2, 3, and 4). Here, we highlight how some U.S. cities have responded to these problems.

Cities Respond to AIDS and Drug Addiction As noted earlier, HIV is often transmitted through sharing infected needles used to inject drugs. In Baltimore, 85 percent of new HIV infections are attributed to drug addicts using infected needles (Schmoke 1998). Some cities have succeeded in reducing HIV transmission by setting up needle-exchange programs whereby drug users can exchange used (and perhaps infected) needles for free clean ones. The largest needle-exchange program operated by a local government is in Baltimore. An evaluation of the program found that enrolled addicts lowered their risk of contracting HIV by almost 40 percent (Schmoke 1998). Other studies have found that HIV infection rates are over three times lower in drug injectors who participate in needle-exchange programs than in those who do not (World Health Organization and United Nations Joint Programme on HIV/AIDS 1998). Needle-exchange programs represent a public health approach to drug addiction, as they not only reduce the spread of HIV and AIDS, but also serve as a stepping-stone to drug treatment.

Cities Respond to Crime One strategy to reduce urban crime is to design housing, public spaces, and roads in such a way as to reduce crime. A national program, Crime Prevention through Environmental Design, emphasizes the security effects of walkways, building entries, lighting, parking areas, and landscaping (Johnson 1997). For example, "proper placement of lights, vegetation, and access doors will reduce a building's vulnerability to burglary and vandalism" (p. 153). Regarding traffic, some cities have implemented traffic diverters at intersections that force drivers to make a right-hand turn, making alleys into cul-de-sacs to discourage easy escape from crime scenes.

Other cities have taken a "get tough on crime" approach as state governments have lengthened sentences and built more prisons. However, Orfield (1997) notes that "these very large expenditures and their small effects strongly suggest that it is time to seek solutions to crime by reducing the concentration of its breeding ground, the poverty core" (p. 24). Orfield argues that providing poor inner-city residents with access to opportunity is likely to reduce crime. Opportunities for poor city residents increase by improving city schools, expanding job training programs, and integrating low-income residents among the middle-class through mixed-income housing.

Improving Transportation and Alleviating Traffic Congestion

A number of strategies to improve transportation and alleviate traffic congestion have been implemented and proposed. Some states have established high-occupancy vehicle (HOV) lanes designed to reduce traffic congestion (and pollution-causing car emissions) by inducing more commuters to carpool. Cars carrying two or more persons are permitted to drive in less crowded HOV lanes. However, efforts to encourage carpooling, such as HOV lanes and park-and-ride lots, generally have not been successful (Nelessen 1997).

The **New Urbanists**, a growing group of planners, architects, developers, and traffic engineers support neighborhood designs that create a strong sense of community by incorporating features of traditional small towns. New Urbanists advocate designing **mixed-use neighborhoods** that combine residential and commercial elements along with public and private facilities such as schools, recreation centers, and places of worship. The idea of mixed-use neighborhoods is to provide suburban residents with convenient access to jobs, stores and service providers, schools, and other facilities, thus reducing driving distances. In many cases, mixed-use neighborhoods are designed to allow residents to walk or bike to various destinations, minimizing the need to drive. New Urbanists envision pedestrian communities in which it is possible to walk to schools, parks, recreation, jobs, and transit stops.

Some cities have turned to high-tech solutions to ease traffic congestion. Thanks to the Integrated Traveler Information Sharing System, commuters in some California counties, may use their car radios and cell phones to access traffic advisory information regarding accidents, alternative routes, and lane closures ("Technology Smooths the Ride for Santa Ana Commuters" 1998).

National Data

In the United States, about one-fifth of person-trips are for work-related purposes, one-fifth are for shopping, and three-fifths are for "social" reasons (including "school/church," "visit friends or relatives," and "other social or recreational" purposes).

SOURCE: Gordon and Richardson 1998

Another strategy for reducing traffic congestion involves improving public transportation systems. In 1998, President Clinton signed the Transportation Equity Act for the 21st Century (TEA21). This provided $169.5 billion for highway improvements and $42 billion for mass transit over 6 years. In addition to large public buses, trolleys, subways, and rails, neighborhood transit has emerged as a form of public transportation designed to serve suburban areas. Also referred to as "dial-a-ride" systems, **neighborhood transit** provides transportation on small buses on demand. Nelessen (1997) describes how the system might work: you dial a 1-800 number, and a computer asks, "What pedestrian precinct are you in?" and you punch it in. The computer then asks, "To which pedestrian precinct would you like to go?" You then punch in your response. The computer answers, "Six minutes, happy to serve you." In 6 minutes, a small bus arrives at your neighborhood transit stop. When the bus arrives, you pass your credit card or bank card through the reader inside the door of the bus and are billed monthly.

Responding to Urban Sprawl: Growth Boundaries and Smart Growth

Some cities have tried to manage urban sprawl by establishing growth boundaries. Since 1973, Oregon has required each city to draw a growth boundary based on its estimate of economic development and community needs in the next 20 years (Geddes 1997). A number of cities in California have also enacted urban growth boundaries (Froehlich 1998). Rather than simply put a limit on urban growth, another approach to managing urban sprawl is to pursue what is called "smart growth." Rather than take a hard stand either for or against urban growth, advocates of **smart growth** encourage development that serves the economic, environmental, and social needs of communities. A smart growth urban development plan entails the following principles (Froehlich 1998; Smart Growth Network 1999):

- Mixed-use land use, whereby residences, jobs, schools, grocery stores, etc. are located within close proximity of each other; ample sidewalks for a walkable community
- Compact building design
- Housing and transportation choices
- Distinctive and attractive community design
- The preservation of open space, farmland, natural beauty, and critical environmental areas (see this chapter's *The Human Side*)
- The redevelopment of existing communities, rather than letting them decay and building new communities around them
- Regional planning and collaboration between business, private residents, community groups, and policy makers on development/redevelopment issues

National Data

In January 1999, President Clinton signed the Lands Legacy Initiative, which provides $600 million to help U.S. communities save open spaces, develop outdoor recreation areas, and set aside new coastal and wildlife preserves.

However, most local zoning codes prohibit smart growth design by mandating large housing setbacks, wide streets, and separation of residential and commercial areas (Pelley 1999).

Regionalism

Addressing the various social problems facing urban areas may best be achieved through **regionalism**—a form of collaboration among central cities

THE HUMAN SIDE

TREES OF LIFE

I never thought much about the huge oak growing between the sidewalk and curb in front of my house in Washington, D.C., even though it was the biggest tree on the block. Then one day not long ago, I came home from work and it was gone; apparently the tree was diseased and city authorities removed it. Now there's a hole in the sky where the oak once stood and a surprising void in my life. I didn't realize how important that tree was to my home until it was gone. Will the city ever replace it? If it does, I promise never to take another street tree for granted.

That's a lesson perhaps we all ought to learn. In this age of mounting urban budget deficits, tree programs are among the first municipal services to get the axe. As a result, street trees are disappearing from American cities at an alarming rate. One study found that in the country's 20 largest metropolitan areas, an average of four trees die or are uprooted for every new one that is planted. Now, experts assert, more than half of the tree-planting spaces along the nation's city streets are empty—an unfortunate situation, considering the valuable role trees play in the hostile urban environment.

According to a recent report by the U.S. Forest Service, trees save city governments and homeowners considerable amounts of money over the long run. In the study, which used Chicago as its model, researchers found that if officials planted 95,000 trees and tended them for 30 years, the city would save $38 million—or an average of $402 per tree. The reason: Trees help reduce energy use and pollution damage.

In a single year, observes Forest Service researcher Gregory McPherson, trees in the Chicago area perform $9.2 million worth of air-pollution removal. "They scrub ozone, sulfur dioxide and other pollutants out of the air," he says. And by blocking out summer sun and winter wind, the trees reduce air conditioning and heating costs. McPherson and his associates found that properly placed trees can save homeowners nearly $100 a year in energy bills.

I think about that fact now that warm weather is approaching and I see the afternoon sun beating directly onto the front of my house. Last year, the nonprofit group American Forests reported that, since 1972, temperatures in downtown Atlanta have increased by 6 to 9 degrees more than those of the surrounding countryside. Why? Because the city center has lost 65 percent of its tree cover, creating what scientists call an "urban heat island." Is that the sort of effect I can expect this summer around my own tiny urban island?

While I wait for the answer, I do know one thing: The intangible benefits of street trees go far beyond dollars and cents. "Older trees are a sign of strength and endurance that help define the character of a neighborhood," says another U.S. Forest Service researcher, John Dwyer. "Their aesthetic qualities may be their most valuable contributions of all to the urban landscape."

In studying apartment dwellers not long ago, University of Illinois researchers discovered that people whose units are surrounded by trees have better relationships with their neighbors and less violence in their

THE HUMAN SIDE

TREES OF LIFE *(continued)*

homes than people whose living spaces are bordered only by concrete. "Perhaps," observes Roger S. Ulrich, a Texas A&M University environmental psychologist, "even small pockets of green space provide people with a better sense of community."

In recent years, Ulrich has devoted considerable time to examining the effects of city trees on human health. Among other things, he has found that the sight of trees alone can quickly lower a person's blood pressure and relax muscle tension and brain wave patterns, which indicate reduced stress. He also discovered that patients who had views of trees through their hospital windows recovered faster from surgery than patients who viewed only a brick wall.

"Even just a few minutes of visual contact with trees has proven to be enough to produce significant recovery from stress," says Ulrich. "And so when we remove trees from our environment and never replace them, we may be incurring some very high, unintended costs in terms of our health and emotional well-being." As I look out on the big hole in front of my house, I can understand what he is saying.

SOURCE: Adapted from Mark Wexler. 1998. "Money Does Grow on Trees—and So Does Better Health and Happiness." *National Wildlife*, April/May. **http://www.nwf.org/nwf/natlwild/1998/urbanam8.html.** Reprinted by permission.

and suburbs that encourages local governments to share common responsibility for common problems. Central cities, declining inner suburbs, and developing suburbs are often in conflict over the distribution of government-funded resources, zoning and land use plans, transportation and transit reform, and development plans. Rather than compete with each other, regional government provides a mechanism for achieving the interests of an entire region. A metropolitan-wide government would handle the inequities and concerns of both suburban and urban areas. As might be expected, suburban officials resist regionalization because they believe it will hurt their neighborhoods economically by draining off money for the cities. But some regions have had success with regionalism. Minnesota state representative Myron Orfield (1997) successfully formed a political coalition between legislators from Minneapolis-St. Paul, their declining blue-collar suburbs, and more affluent developing suburbs.

Strategies for Reducing Urban Growth in Developing Countries

In developing countries, limiting population growth is essential for alleviating social problems associated with rapidly growing urban populations. Strategies for lowering the birthrate and reducing population growth are discussed in detail in Chapter 15.

Another strategy for minimizing urban growth in less developed countries involves redistributing the population from urban to rural areas. Such redistribution strategies include the following (United Nations 1994): (1) promote agricultural development in rural areas, (2) provide incentives to industries and businesses to relocate from urban to rural areas, (3) provide incentives to encourage new businesses and industries to develop in rural areas, (4) develop the infrastructure of rural areas, including transportation and communication systems, clean water supplies, sanitary waste disposal systems, and social services. Of course, these strategies require economic and material resources, which are in short supply in less developed countries.

UNDERSTANDING CITIES IN CRISIS

> America and its cities are at a crossroads. If they choose to make their separate ways into the future, decline and decay almost surely lie ahead for both. But if they stride together as partners into the twenty-first century, the nation and its great urban areas will compete and brilliantly succeed in the new global economy.
>
> –Final Report of the 82nd American Assembly

What can we conclude from our analysis of urban life? First, attention to urban problems and issues is increasingly important, as the United States and the rest of the world are becoming increasingly and rapidly urbanized. Second, the social forces affecting urbanization in industrialized countries are different from those in developing countries. Countries such as the United States have experienced urban decline due to deindustrialization and deconcentration. At the same time, developing countries have experienced rapid urban growth due to industrialization fueled, in part, by a global economy in which corporate multinationals locate industry in developing countries to gain access to cheap labor, raw materials, and new markets.

While U.S. cities struggle to repopulate, cities in developing countries grapple with the problem of limiting population growth. In both cases, problems affecting urban residents include high rates of HIV/AIDS, addiction, and crime (also referred to as the "three-headed monster"), poverty and unemployment, inadequate schools, inferior or unaffordable housing, pollution, and traffic and transportation problems.

People living in rural areas are also affected by urban processes. In developing countries, peasant farmers are being displaced by the growth of large-scale commercial agriculture geared toward producing products for export rather than for local consumption. Displaced rural dwellers migrate to cities looking for work, but often find that urban conditions are not as promising as they had hoped. In the United States and other developed countries, rural and suburban dwellers who leave or avoid cities find increasingly that the city is coming after them as urban sprawl replaces green open places with concrete. Also, as central cities deteriorate, the surrounding suburbs tend to do so as well.

Urban problems are multifaceted and interrelated. For example, urban poverty leads to underfunded and inadequate schools, which results in a poorly trained workforce, leading to loss of jobs (or discouraging new jobs from locating in the city), which leads to more poverty, which causes high rates of crime, drug addiction, and HIV/AIDS (and other health problems). All of these problems are confounded by racial and ethnic prejudice, the rise in single-parent households, and increasing economic inequalities in society. Each problem facing cities seems to reinforce other problems. It is difficult to know what causes what. For example, did suburbanization contribute to the deterioration of the inner cities or vice versa? Is poverty a cause of urban ills or the result? It is difficult to determine the ordering of the events and, therefore, what to do to resolve the problems.

Although the number and scope of problems facing cities is daunting, the progress that some U.S. cities have made is encouraging. Many formerly decaying cities are bustling with new waterfront development, sports arenas, and beautification projects. And there is optimism for the future. The fourteenth annual *State of the Cities* survey of municipal officials in cities with populations above 10,000 found improvement in most indicators of municipal conditions. Among the 30 indicators, 23 (or 77 percent) were ranked by more officials as improved rather than worsened (National League of Cities 1998). The majority of municipal officials in this survey reported improved overall economic conditions and less unemployment during the past year, improved recreation services, and increased range of city services. Nearly 9 out of 10 are optimistic about the direction in which their city appears to be heading and rated local services as good or very good in relation to community needs. In another survey of 21 large U.S. cities, all but one expect their center-city populations to increase in the next decade (Friedman 1998). However, given the growing inequality between the haves and the have-nots and the projected increase in poverty and hardship due to welfare reform (see Chapter 10), cities may see their poor population increase as well. In the fourteenth annual *State of the Cities* survey, nearly one-third of municipal officials reported an increase in the need for "survival services" such as food, shelter, clothing, and health care over the past year (National League of Cities 1998). About one in five said their city's finances and infrastructure needs worsened, 18 percent said homelessness worsened, 34 percent said youth violence and crime worsened, 30 percent said drug and alcohol abuse worsened, and 25 percent said family stability worsened. Donald J. Borut, Executive Director of the National League of Cities, summed up the fourteenth annual *State of the Cities* survey as follows:

> **This report describes a remarkable record of sustained progress and continuing recovery in cities and towns all across America, but it's not a one-dimensional picture. There are shadows that linger in each area of gleaming success. There are needs that remain even as each measure of progress is recorded. (National League of Cities 1998, 2)**

CRITICAL THINKING

1. Today, one of the most significant forces in the United States and throughout the developed world is the aging of the population. As the population ages, what issues must cities confront in order to accommodate the needs of their elderly residents?
2. Develop a 5-year plan to recommend to the members of the city council of Decaysville. You should consider not only the effects of the policies and regulations you recommend (e.g., toll to enter the city limits to raise money), but also the consequences of each action taken (e.g., reduces number of visitors to city).
3. How might the growing trend of telecommuting affect cities?
4. As discussed in Chapter 14, there is a growing trend toward e-commerce. One part of e-commerce is the increased tendency for goods and services to be purchased over the Internet. How might this trend affect urban life?

5. Develop a list of common interests suburbanites and urbanites share in "saving our cities." Do you think such commonalities will lead to regional government? Why or why not?

KEY TERMS

central city
deconcentration
deindustrialization
empowerment zones
gentrification
ghetto
incumbent upgrading
metropolis
metropolitan area
micropolitan area
mixed-use neighborhods
neighborhood transit
New Urbanists
regionalism
road rage
slum
smart growth
suburbanization
suburbs
urbanism
urbanization
urban population
urbanized area
urban sprawl
urban villages

INTERNET

You can find more information on smart growth, New Urbanism, sustainable communities, and road rage at the *Understanding Social Problems* web site at **http://sociology.wadsworth.com**.

INFOTRAC COLLEGE EDITION

Either from the Wadsworth sociology resource center at **http://sociology.wadsworth.com** or directly from your web browser, you may access InfoTrac College Edition, an online university library that includes over 700 popular and scholarly journals in which you can find articles related to the topics in this chapter. Suggested articles and questions relating to these articles are listed below.

Clark, David. 1998. "Interdependent Urbanization in an Urban World." *The Geographical Journal* 164(1):85–97.

1. Where are the highest areas of urbanization in the world?
2. What are some of the links the author describes between capitalism and urbanization?
3. What is transnationalization of production?

Gordon, Peter, and Harry W. Richardson. 1998. "Prove It: the Costs and Benefits of Sprawl." *Brookings Review* 16(4):23–26.

1. What U.S. city is considered to be the "sprawl capital" and why?
2. What problems does the author see with collective transportation?
3. How is the public's demand for "community" being met in modern society?

Marcuse, Peter. 1997. "The Ghetto of Exclusion and the Fortified Enclave." *American Behavioral Scientist* 41(3):311–27.

1. What three developments characterize urban patterns since the 1970s?
2. What is "ghettoization," and what is it based upon?
3. What is the difference between a ghetto and an enclave?

Section 4
Problems of Modernization

Section 4 focuses on problems of modernization—the cultural and structural changes that occur as a consequence of society changing from traditional to modern. Both Durkheim and Marx were concerned with the impact of modernization. Each theorized that as societies moved from "mechanical" to "organic solidarity" (in Durkheimian terms) or from "production for use" to "production for exchange" (in Marxian terms), fundamental changes in social organization would lead to increased social problems. Although modernization has contributed to many of the social problems we have already discussed, it is more directly related to the four problems we examine in this section—technology, population, the environment, and global conflict.

One of the difficulties in understanding social problems involves sorting out the numerous social forces that contribute to social problems. Every social problem is related, in some way, to many other social problems. Nowhere is this more apparent than in the final three chapters. For example, even though scientific and technological advances are designed and implemented to enhance the quality and conditions of social life, they contribute to other social problems. Science and technology (Chapter 14) have extended life through var-

ious medical advances and have successfully lowered the infant mortality rate in many developing nations. However, these two "successes" (fewer babies dying and an increased life expectancy), when coupled with a relatively high fertility rate, lead to expanding populations. Many nations struggle to feed, clothe, house, or provide safe drinking water and medical care for their increased numbers. Further, in responding to problems created by overpopulation, science and technology have contributed to the growing environmental crisis. For example, many developing countries use hazardous pesticides to increase food production for their growing populations, overuse land, which leads to desertification, and, out of economic necessity, agree to deforestation by foreign investors.

Developed countries also contribute to the environmental crisis. Indeed modernization itself, independent of population problems, exacerbates environmental concerns as the fragile ecosystem is overburdened with the by-products of affluent societies and scientific and technological triumphs: air pollution from the burning of fossil fuels, groundwater contamination from chemical runoff, nuclear waste disposal, and destruction of the ozone layer by chlorofluorocarbons. Thus, science and technology, as well as population and the environment (Chapter 15), are inextricably related.

While population patterns and the resultant increased scarcity of resources provide a motivation for global conflict (Chapter 16), science and technology provide more efficient means of worldwide destruction. Conversely, global conflict has devastating effects on the environment (e.g., nuclear winter) and has inspired much scientific and technological research and development such as laser-based and nuclear technologies. Thus, each of the chapter topics to follow is both an independent and a dependent variable in a complex web of cause and effect.

CHAPTER FOURTEEN

Science and Technology

Is It True?

Is It True?

1. In 1995, most U.S. federal research and development funds were spent on health, space research and technology, and energy.
2. In 1994, a defective Pentium chip was discovered to exist in over 2 million computers in aerospace, medical, scientific, and financial institutions, as well as schools and government agencies.
3. Over 30 million people in 160 countries use the Internet.
4. The clock was invented by capitalistic industrialists as a means of controlling the time workers spent on the job.
5. About 40 percent of American homes have a personal computer.

Answers: 1 = F; 2 = T; 3 = T; 4 = F; 5 =

Most of the consequences of technology that are causing concern at the present time—pollution of the environment, potential damage to the ecology of the planet, occupational and social dislocations, threats to the privacy and political significance of the individual, social and psychological malaise . . . are with us in large measure because it has not been anybody's explicit business to foresee and anticipate them.

Emmanuel Mesthene, Former Director of the Harvard Program on Technology and Society

Virtual reality, cloning, and teleportation are no longer just the stuff of popular sci-fi shows such as *Star Trek*, *Millennium*, and *The X-Files*. Virtual reality is now used to train workers in occupations as diverse as medicine, engineering, and professional football. The ability to genetically replicate embryos has sparked worldwide debate over the ethics of reproduction, and California Institute of Technology scientists have beamed a ray of light from one location to another (Fox 1998). Just as the telephone, the automobile, the television, and countless other technological innovations have forever altered social life, so will more recent technologies.

Science and technology go hand in hand. **Science** is the process of discovering, explaining, and predicting natural or social phenomena. A scientific approach to understanding AIDS, for example, might include investigating the molecular structure of the virus, the means by which it is transmitted, and public attitudes about AIDS. **Technology**, as "a form of human cultural activity that applies the principles of science and mechanics to the solution of problems," is intended to accomplish a specific task—in this case, the development of an AIDS vaccine.

Societies differ in their level of technological sophistication and development. In agricultural societies, which emphasize the production of raw materials, **mechanization**, or the use of tools to accomplish tasks previously done by hand, dominates. As societies move toward industrialization and become more concerned with the mass production of goods, automation prevails. **Automation** involves the use of self-operating machines, as in an automated factory where autonomous robots assemble automobiles. Finally, as a society moves toward postindustrialization, it emphasizes service and information professions (Bell 1973). At this stage, technology shifts toward **cybernation**, whereby machines control machines—making production decisions, programming robots, and monitoring assembly performance.

What are the effects of science and technology on humans and their social world? How do science and technology help to remedy social problems and how do they contribute to social problems? Is technology, as Postman (1992) suggests, both a friend and a foe to humankind? This chapter addresses each of these questions.

International Data

According to a Commerce Department report, it took 38 years for radio use to reach 50 million people; it took television only 13 years to reach the same number; the Internet reached 50 million people in 4 years.

SOURCE: Leslie 1998

If religion was formerly the opiate of the masses, then surely technology is the opiate of the educated public today.

—John McDermott
State University of New York

THE GLOBAL CONTEXT: THE TECHNOLOGICAL REVOLUTION

Less than 50 years ago, traveling across state lines was an arduous task, a long-distance phone call was a memorable event, and mail carriers brought belated news of friends and relatives from far away. Today, travelers journey between continents in a matter of hours, and for many, e-mail, faxes, videoconferencing, and electronic fund transfers have replaced conventional means of communication.

The world is a much smaller place than it used to be and will become even smaller as the technological revolution continues. The Internet is projected to have a quarter of a billion users in over 100 countries by the year 2000 (GIP 1998a). Although over half of Internet users are English speaking, non-English

The world was made a smaller place in the mid- to late 1800s by the Pony Express. And it gets smaller all the time, particularly now that much of the world is connected to the Internet.

> Science is committed to the universal. A sign of this is that the more successful a science becomes, the broader the agreement about its basic concepts; there is not a separate Chinese or American or Soviet thermodynamics, for example, there is simply thermodynamics.
>
> —O. B. Hardison Jr.
> Humanities professor

National Data

Eighty percent of American households have a car, 65 percent a stereo, 37 percent video games, and 14 percent coffee makers.

SOURCE: Hafner 1999

speakers constitute the fastest growing language group on the Internet. Of the 67 million non-English-speaking Internet users, 20 percent speak Spanish, 20 percent German, 18 percent Japanese, 9 percent French, and 6 percent Chinese (Global Statistics 1998). The Global Internet Project, begun by software and computer industry executives ". . . believes that the Internet must be viewed as a global medium transcending geographical differences that is transforming not only how commerce is conducted, but education, health care, and society in general" (GIP 1998b).

The movement toward globalization of technology is, of course, not limited to the use and expansion of the Internet. The world robot market continues to expand (IFR 1997); Latin America's computer industry is dominated by Texas-based manufacturer Compaq; 60 percent of Microsoft's Internet platform and support products are sold overseas (GIP 1998c); and "globe-trotting scientists" collect skin and blood samples from remote islanders for genetic research (Shand 1998).

To achieve such scientific and technological innovations, sometimes called research and development (R&D), countries need material and economic resources. Research entails the pursuit of knowledge; development refers to the production of needed materials, systems, processes, or devices directed toward the solution of a practical problem. In 1996, the United States spent over $140 million dollars on R&D. As in most other countries, 1996 U.S. funding sources were primarily from four sectors: private industry (65 percent), the federal government (31 percent), colleges and universities (3 percent), and other nonprofit organizations such as research institutes (2 percent) (*Statistical Abstract of the United States: 1998,* Table 962).

Scientific discoveries and technological developments also require the support of a country's citizens and political leaders. For example, although abortion has been technically possible for years, 24 percent of the world's population live in countries where abortion is either prohibited or permitted only when the life of the mother is in danger (United Nations Population Fund 1991). Thus, the degree to which science and technology are considered good or bad, desirable or undesirable, is to a large extent socially constructed.

Postmodernism and the Technological Fix

Many Americans believe that social problems can be resolved through a **"technological fix"** (Weinberg 1966) rather than through social engineering. For example, a social engineer might approach the problem of water shortages by persuading people to change their lifestyle: use less water, take shorter showers, and wear clothes more than once before washing. A technologist would avoid the challenge of changing people's habits and motivations and instead concentrate on the development of new technologies that would increase the water supply. Social problems may be tackled through both social engineering and a "technological fix." In recent years, for example, social engineering efforts to reduce drunk driving have included imposing stiffer penalties for drunk driving and disseminating public service announcements such as "Friends Don't Let Friends Drive Drunk." An example of a "technological fix" for the same problem is the development of car air bags, which reduce injuries and deaths due to car accidents.

Not all individuals, however, agree that science and technology are good for society. **Postmodernism**, an emerging world view, holds that rational thinking and the scientific perspective have fallen short in providing the "truths" they were once presumed to hold. During the industrial era, science, rationality, and technological innovations were thought to pave the way to a better, safer, and more humane world. Today, postmodernists question the validity of the scientific enterprise, often pointing to the unforeseen and unwanted consequences of resulting technologies. Automobiles, for example, began to be mass-produced in the 1930s in response to consumer demands. But the proliferation of automobiles also led to increased air pollution and the deterioration of cities as suburbs developed, and today, traffic fatalities are the number one cause of death from all accidents. Examine Figure 14.1 and consider the negative consequences of each of these modern-day technologies.

International Data

Worldwide, the number of scientists and engineers per million population varies dramatically: 3,548 in Japan, 2,685 in the United States, 1,632 in Europe, 209 in the Arab states, and 53 in Africa.

SOURCE: Kennedy 1998

> I've got gigabytes. I've got megabytes. I'm voice-mailed. I'm e-mailed. I surf the net. I'm on the Web. I am Cyber-Man. So how come I feel so out of touch?
>
> —Volkswagen television commercial

CONSIDERATION

Postman (1992) suggests that technology is both a friend and an enemy:

> Technology . . . makes life easier, cleaner, and longer. Can anyone ask more of a friend? . . . It is the kind of friend that asks for trust and obedience, which most people are inclined to give because its gifts are truly bountiful. But of course, there is a dark side to this friend. Its gifts are not without a heavy cost. . . . The uncontrolled growth of technology destroys the vital sources of our humanity. It creates a culture without a moral foundation. It undermines certain mental processes and social relations that make human life worth living. Technology, in sum, is both friend and enemy. (p. xii)

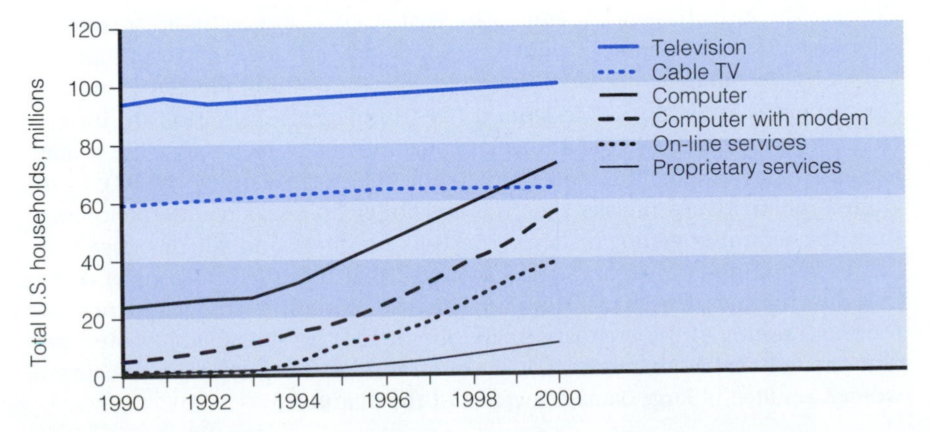

Figure 14.1 Technology in U.S. Households, 1990–2000

SOURCE: FAS. 1998a. From Cyberstrategy Project, **http://www.fas.org/cp/house.gif**. Reprinted by permission.

SOCIOLOGICAL THEORIES OF SCIENCE AND TECHNOLOGY

National Data

In a 1993 General Social Survey of U.S. adults, only 17 percent agreed or strongly agreed with the statement "overall, modern science does more harm than good"; 62 percent disagreed or strongly disagreed; and 21 percent neither agreed nor disagreed.

SOURCE: General Social Survey 1993

Each of the three major sociological frameworks helps us to better understand the nature of science and technology in society.

Structural-Functionalist Perspective

Functionalists view science and technology as emerging in response to societal needs—that "[science] was born indicates that society needed it" (Durkheim [1925] 1973). As societies become more complex and heterogeneous, finding a common and agreed-upon knowledge base becomes more difficult. Science fulfills the need for an assumed objective measure of "truth" and provides a basis for making intelligent and rational decisions. In this regard, science and the resulting technologies are functional for society.

If society changes too rapidly as a result of science and technology, however, problems may emerge. When the material part of culture (i.e., its physical elements) changes at a faster rate than the nonmaterial (i.e., its beliefs and values) a **cultural lag** may develop (Ogburn 1957). For example, the typewriter, the conveyor belt, and the computer expanded opportunities for women to work outside the home. With the potential for economic independence, women were able to remain single or to leave unsatisfactory relationships and/or establish careers. But while new technologies have created new opportunities for women, beliefs about women's roles, expectations of female behavior, and values concerning equality, marriage, and divorce have "lagged" behind.

Science and technology are . . . essential for strengthening the economy, creating high quality jobs, protecting the environment, improving our health care and education systems, and maintaining our national security. This country must sustain world leadership in science, mathematics, and engineering if we are to meet the challenges of today . . . and tomorrow.

—Bill Clinton
U.S. President

Robert Merton (1973), a functionalist and founder of the subdiscipline sociology of science, also argued that scientific discoveries or technological innovations may be dysfunctional for society and create instability in the social system. For example, the development of time-saving machines increases production, but also displaces workers and contributes to higher rates of employee alienation. Defective technology can have disastrous effects on society. In 1994, a defective Pentium chip was discovered to exist in over 2 million computers in aerospace, medical, scientific, and financial institutions, as well as schools and government agencies. Replacing the defective chip was a massive undertaking, but was necessary to avoid thousands of inaccurate computations and organizational catastrophe.

Conflict Perspective

Conflict theorists, in general, emphasize that science and technology benefit a select few. For some conflict theorists, technological advances occur primarily as a response to capitalist needs for increased efficiency and productivity and thus are motivated by profit. As McDermott (1993) notes, most decisions to increase technology are made by "the immediate practitioners of technology, their managerial cronies, and for the profits accruing to their corporations" (p. 93). In the United States, private industry spends more money on research and development than the federal government does. The Dalkon Shield and silicone breast implants are examples of technological advances that promised millions of dollars in profits for their developers. However, the rush to market took precedence over thorough testing of the products' safety. Subsequent lawsuits filed by consumers who argued that both products had compromised the physical well-being of women resulted in large damage awards for the plaintiffs.

Many of the things that we like to think of as mere tools or instruments now function as virtual members of society.

—Langdon Winner
Social scientist

Science and technology also further the interests of dominant groups to the detriment of others. The need for scientific research on AIDS was evident in the early 1980s, but the required large-scale funding was not made available as long as the virus was thought to be specific to homosexuals and intravenous drug users. Only when the virus became a threat to mainstream Americans were millions of dollars made available for AIDS research. Hence, conflict theorists argue that granting agencies act as gatekeepers to scientific discoveries and technological innovations. These agencies are influenced by powerful interest groups and the marketability of the product rather than by the needs of society.

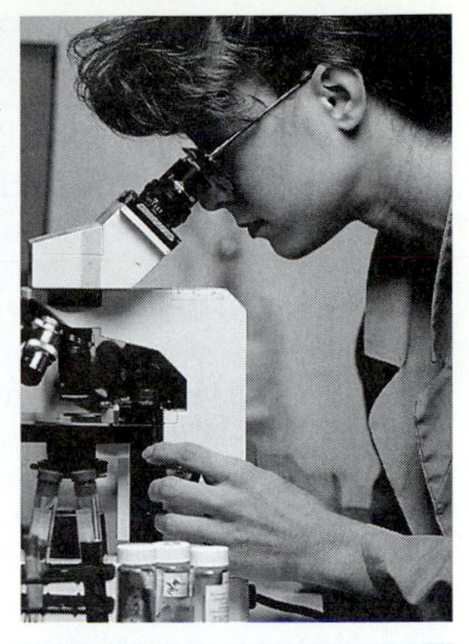

Motivated by profit, private industry spends more money on research and development than the federal government does.

CONSIDERATION

Some feminists argue that technology is an extension of the patriarchal nature of society that promotes the interest of men and ignores the needs and interests of women. For example, washing machines, although time-saving devices, disrupted the communal telling of stories and the resulting friendships among women who gathered together to do their chores. Bush (1993) observes:

> Technology always enters into the present culture, accepting and exacerbating the existing norms and values. In a society characterized by a sex-role division of labor, any tool or technique . . . will have dramatically different effects on men than on women. (p. 204)

Other feminists acknowledge that technological innovations have improved the lives of women by balancing employment opportunities, especially in new, skilled, blue-collar and technical jobs such as computer repair specialist and telecommunications operator. Walshok (1993) studied women in these occupations and found that they reported high levels of job satisfaction and a strong sense of job security.

> While gene hunters are obliged to obtain informed consent from their donors, they are not required to tell them that tissue donation could result in patented products. And donors are rarely offered a cut of the profits.
>
> –Hope Shand
> Rural Advancement
> Foundation International

Symbolic Interactionist Perspective

Knowledge is relative. It changes over time, over circumstances, and between societies. We no longer believe that the world is flat or that the earth is the center of the universe, but such beliefs once determined behavior as individuals responded to what they thought to be true. The scientific process is a social process in that "truths"—socially constructed "truths"—result from the interactions between scientists, researchers, and the lay public.

Kuhn (1973) argues that the process of scientific discovery begins with assumptions about a particular phenomenon (e.g., the world is flat). Since there are always unanswered questions about a topic (e.g., why don't the oceans drain?), science works to fill these gaps. When new information suggests that the initial assumptions were incorrect (e.g., the world is not flat), a new set of assumptions or framework emerges to replace the old one (e.g., the world is round). It then becomes the dominant belief or paradigm.

> This new regimen is simpler and potentially allows greater privacy than any other abortion method.
>
> –Dr. Etienne-Emile Raulieu
> Inventor of RU486

Symbolic interactionists emphasize the importance of this process and the impact social forces have on it. Conrad (1997), for example, describes the media's contribution in framing societal beliefs that alcoholism, homosexuality, and racial inequality are genetically determined. Technological innovations are also affected by social forces, and their success is, in part, dependent upon the social meaning assigned to any particular product. If a product is defined as impractical, cumbersome, inefficient, or immoral, it is unlikely to gain public acceptance.

FOCUS ON SOCIAL PROBLEMS RESEARCH

The Social Construction of the Hacking Community

Sample and Methods

Jordan and Taylor (1998) researched computer hackers and the hacking community through 80 semistructured interviews, 200 questionnaires, and an examination of existing data on the topic. As is often the case in crime, illicit drug use, and other similarly difficult research areas, a random sample of hackers was not possible. Nonetheless, the authors lend insight into this increasingly costly social problem and the symbolic interactionsist notion of "social construction"—in this case, of an on-line community.

Findings

Computer hacking, or "unauthorized computer intrusion," is an increasingly serious problem, particularly in a society dominated by information technologies. Unlawful entry into computer networks or databases can be achieved by several means including (1) guessing someone's password, (2) tricking a computer about the identity of another computer (called "IP spoofing"), or (3) "social engineering," a slang term referring to getting important access information by stealing documents, looking over someone's shoulder, going through their garbage, etc.

Hacking carries with it certain norms and values for, according to Jordan and Taylor, the hacking community can be thought of as a culture within a culture. The two researchers identify six elements of this socially constructed community:

- *Technology.* The core of the hacking community is the technology that allows it to occur. As one professor interviewed stated, the young today have ". . . lived with computers virtually from the cradle, and therefore have no trace of fear, not even a trace of reverence."
- *Secrecy.* The hacking community must, on the one hand, commit secret acts since their "hacks" are illegal. On the other hand, much of the motivation for hacking requires publicity in order to achieve the notoriary often sought. Further, hacking is often a group activity that bonds members together. As one hacker stated, hacking ". . . can give you a real kick some time. But it can give you a lot more satisfaction and recognition if you share your experiences with others. . . ."
- *Anonymity.* While secrecy refers to the hacking act, anonymity refers to the importance of the hacker's identity remaining anonymous. Thus, for example, hackers and hacking groups take on names such as Legion of Doom, the Inner Circle I, Mercury, and Kaos, Inc.
- *Membership Fluidity.* Membership is fluid rather than static, often characterized by high turnover rates, in part, as a response to law enforcement pressures. Unlike more structured organizations, there are no formal rules or regulations.
- *Male Dominance.* Hacking is defined as a male activity and, consequently, there are few female hackers. Jordan and Taylor also note, after recounting an incidence of sexual harrassment, that ". . . the collective identity hackers share and construct . . . is in part misogynist" (p. 768).
- *Motivation.* Contributing to the articulation of the hacking communities' boundaries are the agreed-upon definitions of acceptable hacking motivations, including: (1) addiction to computers, (2) curiosity, (3) excitement, (4) power, (5) acceptance and recognition, and (6) community service through the identification of security risks.

Finally, Jordan and Taylor (1998, 770) note that hackers also maintain group boundaries by distinguishing between their community and other social groups, including "an antagonistic bond to the computer security industry (CSI)". Ironically, hackers admit a desire to be hired by the CSI, which would not only legitimize their activities but give them a steady income as well.

The authors conclude that the general fear of computers and of those who understand them underlies the common although inaccurate portrayal of hackers as pathological, obsessed computer "geeks." When journalist Jon Littman asked hacker Kevin Mitnick if he was being demonized because of increased dependence on and fear of information technologies, Mitnick replied, "Yeah . . . That's why they're instilling fear of the unknown. That's why they're scared of me. Not because of what I've done, but because I have the capability to wreak havoc" (Jordan and Taylor 1998, 776).

SOURCE: Tim Jordan and Paul Taylor. 1998. "A Sociology of Hackers." *The Sociological Review* (November): 757–78.

Such is the case with RU486, an oral contraceptive, which is widely used in France, Great Britain, and China, but is rarely prescribed in the United States.

Not only are technological innovations subject to social meaning, but who becomes involved in what aspects of science and technology is also socially defined. Men, for example, outnumber women three to one in earning computer science degrees. They also score higher on measures of computer aptitude and report higher computer use than women (Lewin 1998). Societal definitions of men as being rational, mathematical, and scientifically minded and having greater mechanical aptitude than women are, in part, responsible for these differences. This chapter's *Focus on Social Problems Research* highlights one of the consequences of the masculinization of technology, as well as the ways in which computer hacker identities and communities are socially constructed.

New technologies alter the structure of our interests: the things we think about. They alter the character of our symbols: the things we think with. And they alter the nature of community: the arena in which thoughts develop.

–Neil Postman
New York University

TECHNOLOGY AND THE TRANSFORMATION OF SOCIETY

A number of modern technologies are considerably more sophisticated than technological innovations of the past. Nevertheless, older technologies have influenced the social world as profoundly as the most mind-boggling modern inventions. Indeed, without older technological innovations, such as the clock and printing press, most modern technology would not have been possible.

Technology has far-reaching effects on every aspect of social life. As noted earlier, technology has altered the very concept of society. New transportation and communication systems have created interconnections between previously separated societies and thus led to the development of a global society. The following sections discuss other societal transformations due to various modern technologies, including workplace technology, computers, the information highway, and science and biotechnology.

Technology in the Workplace

All workplaces, from doctors' offices to factories and from supermarkets to real estate corporations, have felt the impact of technology. The Office of Technology Assessment of the U.S. Congress estimates that over 7 million U.S. workers are under some type of computer surveillance, which lessens the need for supervisors and makes control by employers easier. Further, technology can make workers more accountable by gathering information about their performance. Through such time-saving devices as personal digital assistants and battery-powered store shelf labels, technology can enhance workers' efficiency. Technology is also changing the location of work. Over 3 million employees telecommute that is, complete all or part of their work away from the workplace. Fifty-one percent of corporations now allow at least some of their employees to work from home (Carey 1998).

International Data

With an estimated 1.18 million people on-line and 90 percent of computer networks not secure, the Chinese government has assigned over 200 full-time "Internet security guards" to protect the systems from hackers.

SOURCE: Fang 1998

Information technologies are also changing the nature of work. Lilly Pharmaceutical employees communicate via their own "intranet" on which all work-related notices are posted. Federal Express not only created a FedEx network for their 30,000 employees, but allowed customers to enter their package-tracking database, saving the Memphis-based company $2 million a year. It is estimated that one-fifth of all corporations are now using such telecommunication devices, leading to the potential of a paperless workplace (GIP 1998c).

Robotic technology, sometimes called computer-aided manufacturing (CAM), has also revolutionized work, particularly in heavy industry such as automobile manufacturing (see Chapter 11). An employer's decision to use robotics depends on direct (e.g., initial investment) and indirect (e.g., unemployment compensation) costs, the feasibility and availability of robots performing the desired tasks, and the increased rate of productivity. Use of robotics may also depend on whether there is union resistance to the replacement of workers by machines.

The time is fast approaching when anything an entire company can do, its every single employee can do. From his [her] desk. Anywhere on earth.

–Global Internet Project

National Data

Sales of software needed to operate corporate intranets totaled more than $476 million in 1975; in 1997, approximately $4 billion; in 1998, an estimated $8 billion.

SOURCE: GIP 1998c

International Data

In 1997, the total stock of "operational industrial robots" was estimated to be over 700,000 units; Japan accounts for 58 percent of the robot inventory, the United States 11 percent, and Germany 9 percent.

SOURCE: IFR 1997

I think there is a world market for maybe five computers.

—Thomas Watson
Chairman of IBM,
speaking in 1943

National Data

It is estimated that 30 million U.S. households have computers and that by the year 2000, between 40 and 70 million households will have computers, over half of which will have modems for accessing the Internet.

SOURCE: FAS 1998b

CONSIDERATION

Postman (1992) describes how the clock—a relatively simple innovation that is taken for granted in today's world—profoundly influenced not only the workplace but the larger economic institution:

> The clock had its origin in the Benedictine monasteries of the twelfth and thirteenth centuries. The impetus behind the invention was to provide a more or less precise regularity to the routines of the monasteries, which required, among other things, seven periods of devotion during the course of the day. The bells of the monastery were to be rung to signal the canonical hours; the mechanical clock was the technology that could provide precision to these rituals of devotion. . . . What the monks did not foresee was that the clock is a means not merely of keeping track of the hours but also of synchronizing and controlling the actions of men. And thus, by the middle of the fourteenth century, the clock had moved outside the walls of the monastery, and brought a new and precise regularity to the life of the workman and the merchant. . . . In short, without the clock, capitalism would have been quite impossible. The paradox . . . [is] that the clock was invented by men who wanted to devote themselves more rigorously to God; it ended as the technology of greatest use to men who wished to devote themselves to the accumulation of money. (pp. 14–15)

The Computer Revolution

Early computers were much larger than the small machines we have today and were thought to have only esoteric uses among members of the scientific and military communities. In 1951, only about half a dozen computers existed (Ceruzzi 1993). The development of the silicon chip and sophisticated microelectronic technology allowed tens of thousands of components to be imprinted on a single chip smaller than a dime. The silicon chip led to the development of laptop computers, mini-television sets, cellular phones, electronic keyboards, and singing birthday cards. The silicon chip also made computers affordable. Although the first personal computer (PC) was developed only 20 years ago, today about 40 percent of American homes have one (Kate 1998).

Today, computers form the backbone of universities, government agencies, corporations, and businesses. Between 1984 and 1993, the percentage of U.S. public school students using computers increased from 27 percent to 59 percent (*Statistical Abstract of the United States: 1998,* Table 283).

Computers are big business, and the United States is one of the most successful producers of computer technology, boasting the three top-selling desktops in the world—Packard Bell, Compaq, and Gateway. Retail sales of computers exceeded $18 billion in 1997 with an average home computer cost of $1,745 (Scout Report 1998). By the year 2006, spending on home computers is predicted to grow tenfold as consumers increasingly define PCs as a necessity rather than a luxury (Klein 1998).

Computer education has also mushroomed in the last two decades. In 1971, 2,388 U.S. college students earned a bachelor's degree in computer and information sciences; by 1995 that number had increased to 24,404 (*Statistical Abstract of the United States: 1998,* Table 325). Universities are moving toward mandatory laptop policies for their students, and college and university spending on hardware and software needs are at an all-time high—$2.8 billion in the 1998–99 school year ("College Technology" 1998).

As we move into the twenty-first century there are concerns that computers programmed with double-digit years (i.e., 97, 98, 99) will interpret 00 as 1900, 01 as 1901, and so forth rather than the years 2000, 2001, etc. Known as the year 2000 **(Y2K)** problem, countries vary dramatically in their preparation for such an event. While the United States has invested billions of dollars in preparing for the glitch, many other nations have virtually ignored the problem. For example, in a U.S. State Department survey of 113 countries, telecommunication companies in 33 countries were working on the problem but with little success, and 29 countries were unaware of the problem or had not begun to address it (Bajak 1998).

Given the global economy, it is likely that some Y2K effects will be felt regardless of U.S. efforts. Telecommunications may be disrupted, affecting global fund transfers; international airlines may be grounded as computer-guided air traffic jams; payrolls around the world may not be met; and once-taken-for-granted goods (e.g., electricity) and services (e.g., medical care) to millions of people may be disrupted. Computer experts are divided on the impact of the Y2K problem, some predicting doom and gloom and others mild inconveniences. No one knows, however, exactly what the result of the "millennium computer bug" will be (Bajak 1998).

International Data

A recent survey of 6,000 companies in 47 countries found that half will not have critical computer systems fixed in time for the year 2000.

SOURCE: Bajak 1998

The Information Highway

Information technology, or **IT** for short, refers to any technology that carries information. Most information technologies were developed within a hundred-year span: photography and telegraphy (1830s), rotary power printing (1840s), the typewriter (1860s), transatlantic cable (1866), telephone (1876), motion pictures (1894), wireless telegraphy (1895), magnetic tape recording (1899), radio (1906), and television (1923) (Beniger 1993). The concept of an "information society" dates back to the 1950s when an economist identified a work sector he called "the production and distribution of knowledge." In 1958, 31 percent of the labor force were employed in this sector—today more than 50 percent are. When this figure is combined with those in service occupations, more than 75 percent of the labor force are involved in the information society (Beniger 1993).

The information superhighway may be mostly hype today, but it is an understatement about tomorrow. It will exist beyond people's wildest predictions.

—Nicholas Negroponte
Professor of Media Technology, M.I.T.

The Clinton administration embraces the development of a national information infrastructure (NII) as outlined in the Communications Act of 1994. An information infrastructure performs three functions (Kahin 1993). First, it carries information, just as a transportation system carries passengers. Second, it collects data in digital form that can be understood and used by people. Finally, it permits people to communicate with one another by sharing, monitoring, and exchanging information based on common standards and networks. In short, an information infrastructure facilitates telecommunications, knowledge, and community integration.

The **Internet** is an international information infrastructure—a network of networks—available through universities, research institutes, government agencies, and businesses. Today over one-fifth of Americans use the Internet. U.S. users are more likely to be male (55 percent), white (85 percent), and between the ages of 12 and 17, although in 1998, 18 percent of 2- to 12-year-olds were also on-line (Headcount 1998).

The First Amendment shouldn't end where the Internet begins.

—Marc Rotenberg
Electronic Privacy Information Center

> The question pornography poses in cyberspace is the same one it poses everywhere else: whether anything will be done about it.
>
> –Catharine MacKinnon
> Law professor

With a personal computer, a modem, and a telephone line, users log into locations around the world to access information, transfer files, and send and receive e-mail. The Internet also provides access to thousands of discussion groups, databases, bulletin boards, videos, and reservation systems from around the global village. Commercial access to the Internet is available through such services as Prodigy, America On-line, and Compuserve.

As Internet use has expanded, so has **e-commerce**, or the buying and selling of goods and services over the Internet. For example, France expects 10 percent of all retail revenues to be "Internet generated" by the year 2001; Great Britain expects 12 percent, and the Scandinavian countries 15 percent. Not surprisingly, advertising on the Internet has increased dramatically over the last several years as has *spam,* the slang term for Internet junk mail.

International Data

In 1992 there were just over 100 World Wide Web sites; today there are over 200,000 sites containing 11 million pages of information; 300,000 Internet pages are added every day.

SOURCE: GIP 1998a

National Data

In 1995, 7 percent of all U.S. adults had a computer with access to the Internet; in 1996, 20 percent; 1997, 28 percent, and in 1998, 36 percent. Three-quarters of Internet users also have e-mail accounts.

SOURCE: Harris Poll 1998a

Table 14.1 Results of 1998 Harris Poll*

Began to use Internet/WWW/on-line services?	
Past 6 months	33.1
Last year	30.8
Last 2 years or more	36.0
Use computer to access the Internet?	
Yes	60.1
No	39.9
Use computer to access the World Wide Web?	
Yes	40.3
No	59.7
Check phone bill on the Internet?	
Very or somewhat interested	45.9
Not very or not at all interested	53.9
On-line purchases in last year?	
Yes	32.4
No	67.6
Access Internet once a month or more?	
Yes	65.0
No	34.4
Days in last week using Internet?	
None to 3	81.2
4 to 7	17.8
Best government approach to Internet privacy?	
Voluntary standards	18.8
Recommend privacy standards	22.5
Pass laws now	52.8
None of the above	2.3
Don't know/refused	3.4
Concerned about e-mail privacy?	
Very or somewhat concerned	88.0
Not very or not at all concerned	12.0

*National sample of U.S. adults 18 years and older

SOURCE: Harris Poll. 1998. "A 21st Century Juxtaposition: Grandma, Grandpa and High Technology." *The Harris Poll* 12: February 26.

Science and Biotechnology

While recent computer innovations and the establishment of an information highway have led to significant cultural and structural changes, science and its resulting biotechnologies have led to not only dramatic changes, but also hotly contested issues. Here we will look at some of the issues raised by developments in genetics and reproductive technology.

Genetics Molecular biology has led to a greater understanding of the genetic material found in all cells—DNA (deoxyribonucleic acid)—and with it the ability for **genetic screening.** Currently, researchers are trying to complete genetic maps that will link DNA to particular traits. Already, specific strands of DNA have been identified as carrying such physical traits as eye color and height, as well as such diseases as breast cancer, cystic fibrosis, prostate cancer, and Alzheimer's. By the year 2000, "routine tests will detect predispositions to dozens of diseases as well as indicate a wide range of normal human traits" (Weinberg 1993, 319).

The Human Genome Project, a 15-year effort to map human DNA, "will have decoded all 3 billion chemical letters that spell out our 70,000 or so genes" by the year 2003 (Begley 1998, 62). The hope is that if a defective or missing gene can be identified, it may be possible to get a healthy duplicate and transplant it to the affected cell. This is known as **gene therapy.** Alternatively, viruses have their own genes that can be targeted for removal. Experiments are now under way to accomplish these biotechnological feats.

Genetic engineering is the ability to manipulate the genes of an organism in such a way that the natural outcome is altered. Genetic engineering is accomplished by splicing the DNA from one organism into the genes of another. Often, however, there are unwanted consequences. For example, through genetic engineering some plants are now self-insecticiding, that is, the plant itself produces an insect-repelling substance. Ironically, the continual plant production of the insecticide, in contrast to only sporadic application by farmers, is leading to insecticide-resistant pests (Ehrenfeld 1998).

CONSIDERATION

Gregory Carey, a geneticist at the University of Colorado, poses an interesting question. Suppose scientists discovered a genetic marker that would permit accurate prediction of violent behavior before a child's birth. If we knew that this child was, say, nine times more likely to be arrested and convicted of a violent act, should that child be aborted? What if the child had Down's syndrome or was physically handicapped, an alcoholic, or gay? What if the child had a known life expectancy of 35? Which, if any, of these fetuses should be aborted? Carey reminds us that we have already identified the gene that predicts violent behavior—males are nine times more likely to be arrested and convicted of a violent crime than females (Elmer-Dewitt 1994).

Reproductive Technologies The evolution of "reproductive science" has been furthered by scientific developments in biology, medicine, and agriculture. At the same time, however, its development has been hindered by the stigma associated with sexuality and reproduction, its link with unpopular social

International Data

In a recent survey of Australians, a clear majority were in favor of genetic research; more than 90 percent of the sample were in favor of genetically engineered products such as "a treatment for blood cancer, a drug that lowers blood pressure, and cotton that resists insects. . . ."

SOURCE: Executive Summary 1998

It is now a matter of a handful of years before biologists will be able to irreversibly change the evolutionary wisdom of billions of years with the creation of new plants, new animals, and new forms of human and post human beings.

—Ted Howard
Jeremy Rifkin
Biotechnology critics

International Data

It is estimated that in 2000, over $3.2 billion will be spent on gene therapy products; cancer-related items will account for 60 percent of this total; therapies for cystic fibrosis and AIDS will also be marketed.

SOURCE: Find/SVP 1997

Although the first successful attempt at in-vitro fertilization occurred in 1944, it was 1978 before the first test tube baby was born. Today over 300 fertility clinics in the United States provide this procedure. This photo depicts children born by this method and their families.

movements (e.g., contraception), and the feeling that such innovations challenge the natural order (Clarke 1990). Nevertheless, new reproductive technologies have been and continue to be developed.

In **in-vitro fertilization** (IVF), an egg and a sperm are united in an artificial setting such as a laboratory dish or test tube. Although the first successful attempt at IVF occurred in 1944, it was not until 1978 that the first test-tube baby, Louise Brown, was born. Today, more than 300 fertility clinics in the United States provide this procedure, resulting in about 10,000 live births. Criticisms of IVF are often based on traditional definitions of the family and the legal complications created when a child can have as many as five potential parental ties—egg donor, sperm donor, surrogate mother, and the two people who raise the child (depending on the situation, IVF may not involve donors and/or a surrogate). Litigation over who are the "real" parents has already occurred.

Perhaps more than any other biotechnology, **abortion** epitomizes the potentially explosive consequences of new technologies. Abortion is the removal of an embryo or fetus from a woman's uterus before it can survive on its own. Since the U.S. Supreme Court's ruling in *Roe v. Wade* in 1973, abortion has been legal in the United States. However, recent Supreme Court decisions have limited the *Roe v. Wade* decision. In *Planned Parenthood of Southeastern Pennsylvania v. Casey,* the Court ruled that a state may restrict the conditions under which an abortion is granted, such as by requiring a 24-hour waiting period or parental consent for minors.

Most recent debates concern intact dilation and extraction (D&X) abortions. Opponents refer to such abortions as partial birth abortions because the limbs and the torso are typically delivered before the fetus has expired. D&X abortions are performed because the fetus has a serious defect, the woman's health is jeopardized by the pregnancy, or both. Legislation banning partial birth abortions was passed by Congress in 1995 but was vetoed by President Clinton. In 1996, the House voted to override the veto, but the Senate failed to get the necessary two-thirds majority needed to do so. In 1997, and again in the fall of 1998, the Senate began talks of resurrecting the ban. President Clinton

International Data

Globally, U.S. abortion rates are relatively low at 21 per 1,000 women aged 15 to 44. Countries with over twice the U.S. rate include Cuba (57), Bulgaria (77), Vietnam (84), Russian Federation (119), and Romania (172).

SOURCE: "World Abortion Policies 1994" 1997

What most Americans want to do with abortion is to permit it but also to discourage its use.

—Roger Rosenblatt
Author/social commentator

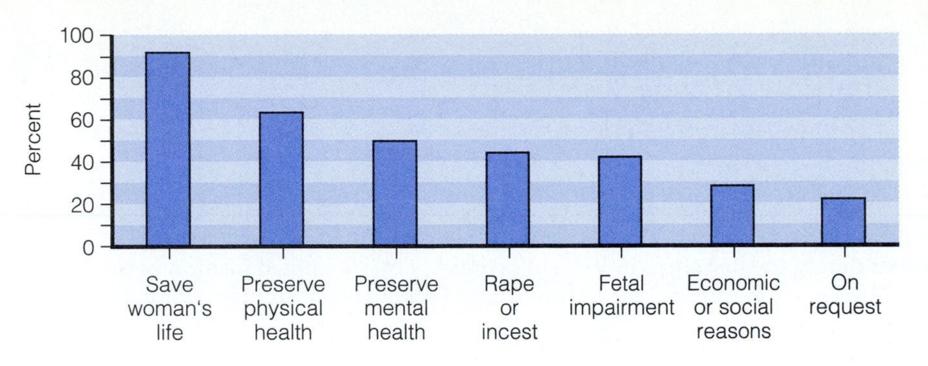

Figure 14.2 Percentages of Countries Permitting Abortions on Specified Grounds

SOURCE: "World Abortion Policies 1994" 1997

has said that he would sign the ban if there was an exemption for when the mother's life was in danger (Reuters 1998).

Abortion is a complex issue for everyone, but especially for women, whose lives are most affected by pregnancy and childbearing. Women who have abortions are disproportionately poor, unmarried minorities who say they intend to have children in the future. Abortion is also a complex issue for societies, which must respond to the pressures of conflicting attitudes toward abortion and the reality of high rates of unintended and unwanted pregnancy. Figure 14.2 portrays the percentage of countries allowing abortions on various grounds.

Attitudes toward abortion tend to be polarized between two opposing groups of abortion activists—prochoice and prolife. Advocates of the prochoice movement hold that freedom of choice is a central human value, that procreation choices must be free of government interference, and that since the woman must bear the burden of moral choices, she should have the right to make such decisions. Alternatively, prolifers hold that the unborn fetus has a right to live and be protected, that abortion is immoral, and that alternative means of resolving an unwanted pregnancy should be found.

In July of 1996, scientist Ian Wilmut of Scotland successfully cloned an adult sheep named Dolly. He did so by placing an udder cell from a 6-year-old sheep with an immature egg cell from another sheep and implanting the resulting embryo in a third sheep. This technological breakthrough has caused worldwide concern about the possibility of human cloning, which prompted quick European action that resulted in a ban on cloning. President Clinton ordered a national commission to study the legal and ethical implications of cloning, which concluded that creating a child through cloning is morally unacceptable.

One argument in favor of developing human cloning technology is its medical value; it may potentially allow everyone to have "their own reserve of therapeutic cells that would increase their chance of being cured of various diseases, such as cancer, degenerative disorders and viral or inflammatory diseases" (Kahn 1997, 54). Human cloning could also provide an alternative reproductive route for couples who are infertile and for those in which one partner is at risk for transmitting a genetic disease.

Arguments against cloning are largely based on moral and ethical considerations. Critics of human cloning suggest that, whether used for medical therapeutic purposes or as a means of reproduction, human cloning is a threat to human dignity. For example, cloned humans would be deprived of their individuality and, as Kahn (1997, 119) points out, "creating human life for the sole purpose of preparing therapeutic material would clearly not be for the dignity of the life created." Despite the current moratorium on human cloning,

National Data

As of 1996, 26 states enforced parental consent or notification laws for minors seeking abortions.

SOURCE: *Facts in Brief* 1996

International Data

Each year, an estimated 50 million abortions occur worldwide. Of these, 30 million are obtained legally, and 20 million illegally.

SOURCE: *Facts in Brief* 1996

National Data

Fifty-seven percent of college and university first-year students in the United States agree that abortion should be legal.

SOURCE: American Council on Education and University of California 1997

Abortion Attitude Scale

This is not a test. There are no wrong or right answers to any of the statements, so just answer as honestly as you can. The statements ask you to tell how you feel about legal abortion (the voluntary removal of a human fetus from the mother during the first three months of pregnancy by a qualified medical person). Tell how you feel about each statement by circling one of the choices beside each sentence. Respond to each statement and circle only one response.

Strongly Agree	Agree	Slightly Agree	Slightly Disagree	Disagree	Strongly Disagree
5	4	3	2	1	0

1. The Supreme Court should strike down legal abortions in the United States.	5	4	3	2	1	0
2. Abortion is a good way of solving an unwanted pregnancy.	5	4	3	2	1	0
3. A mother should feel obligated to bear a child she has conceived.	5	4	3	2	1	0
4. Abortion is wrong no matter what the circumstances are.	5	4	3	2	1	0
5. A fetus is not a person until it can live outside its mother's body.	5	4	3	2	1	0
6. The decision to have an abortion should be the pregnant mother's.	5	4	3	2	1	0
7. Every conceived child has the right to be born.	5	4	3	2	1	0
8. A pregnant female not wanting to have a child should be encouraged to have an abortion.	5	4	3	2	1	0
9. Abortion should be considered killing a person.	5	4	3	2	1	0
10. People should not look down on those who choose to have abortions.	5	4	3	2	1	0
11. Abortion should be an available alternative for unmarried, pregnant teenagers.	5	4	3	2	1	0
12. Persons should not have the power over the life or death of a fetus.	5	4	3	2	1	0
13. Unwanted children should not be brought into the world.	5	4	3	2	1	0
14. A fetus should be considered a person at the moment of conception.	5	4	3	2	1	0

SCORING AND INTERPRETATION

As its name indicates, this scale was developed to measure attitudes toward abortion. It was developed by Sloan (1983) for use with high school and college students. To compute your score, first reverse the point scale for Items 1, 3, 4, 7, 9, 12, and 14. Total the point responses for all items. Sloan provided the following categories for interpreting the results:

70–56 Strong proabortion
55–44 Moderate proabortion
43–27 Unsure
26–16 Moderate prolife
15–0 Strong prolife

RELIABILITY AND VALIDITY

The Abortion Attitude Scale was administered to high school and college students, Right to Life group members, and abortion service personnel. Sloan (1983) reported a high total test estimate of reliability (0.92). Construct validity was supported in that Right to Life members' mean scores were 16.2; abortion service personnel mean scores were 55.6, and other groups' scores fell between these values.

SOURCE: "Abortion Attitude Scale" by L.A. Sloan. Reprinted with permission from the *Journal of Health Education* Vol. 14. No. 3, May/June 1983. The *Journal of Health Education* is a publication of the American Allegiance for Health, Physical Education, Recreation and Dance. 1900 Association Drive, Reston, Virginia 20191.

With the cloning of Dolly, the possibilities of biotechnology expand exponentially. Most Americans are opposed to cloning not only of humans but of animals as well.

according to Arthur Caplan, director of the Center on Bioethics at the University of Pennsylvania, the first infant clone could be created within the next 7 years (Kluger 1997).

Despite what appears to be a universal race to the future and the indisputable benefits of such scientific discoveries as the workings of DNA and the technology of IVF, some people are concerned about the duality of science and technology. Science and the resulting technological innovations are often life assisting and life giving; they are also potentially destructive and life threatening. The same scientific knowledge that led to the discovery of nuclear fission, for example, led to the development of both nuclear power plants and the potential for nuclear destruction. Thus, we now turn our attention to the problems associated with science and technology.

National Data

When a sample of adults were asked whether they favored cloning for purposes of creating a duplicate of an individual, 8 percent were in favor, 84 percent were opposed, and the remainder were unsure. When those opposed to cloning for purposes of replication were asked about cloning to allow infertile couples to have a child, 79 percent were opposed, 12 percent in favor, and the remainder unsure.

SOURCE: IRSS 1998

SOCIETAL CONSEQUENCES OF SCIENCE AND TECHNOLOGY

Scientific discoveries and technological innovations have implications for all social actors and social groups. As such, they also have consequences for society as a whole.

Alienation, Deskilling, and Upskilling

As technology continues to play an important role in the workplace, workers may feel there is no creativity in what they do—they feel alienated (see Chapter 11). The movement from mechanization, to automation, to cybernation increasingly removes individuals from the production process, often relegating them to flipping a switch, staring at a computer monitor, or entering data at a keyboard. For example, many low-paid employees, often women, sit at computer terminals for hours entering data and keeping records for thousands of businesses, corporations, and agencies. The work is monotonous and solitary and provides little autonomy.

Men [and women] have become the tools of their tools.

—Henry David Thoreau
Author/social activist

Not only are these activities routine, boring, and meaningless, they promote **deskilling**, that is, "labor requires less thought than before and gives them [workers] fewer decisions to make" (Perrolle 1990, 338). Deskilling stifles development of alternative skills and limits opportunities for advancement and creativity as old skill sets become obsolete. To conflict theorists, deskilling also provides the basis for increased inequality for "[T]hroughout the world, those who control the means of producing information products are also able to determine the social organization of the 'mental labor' which produces them" (Perrolle 1990, 337).

Technology in some work environments, however, may lead to **upskilling.** Unlike deskilling, upskilling reduces alienation as employees find their work more rather than less meaningful, and have greater decision-making powers as information becomes decentralized. Futurists argue that upskilling in the workplace could lead to a "horizontal" work environment where "employees do not so much what they are told to do, but what their expansive knowledge of the entire enterprise suggests to them needs doing." (GIP 1998c)

Social Relationships and Social Interaction

Online life is rich and rewarding, but it's no substitute for face-to-face interaction.

–Jacques Leslie
Journalist/writer

Technology affects social relationships and the nature of social interaction. The development of telephones has led to fewer visits with friends and relatives; with the coming of VCRs and cable television, the number of places where social life occurs (e.g., movie theaters) has declined. Even the nature of dating has changed as computer networks facilitate cyberdates and "private" chat rooms. As technology increases, social relationships and human interaction are transformed.

Technology also makes it easier for individuals to live in a cocoon—to be self-sufficient in terms of finances (e.g., Quicken), entertainment (e.g., pay-per-view movies), work (e.g., telecommuting), recreation (e.g., virtual reality), shopping (e.g., amazon.com), communication (e.g., Internet), and many other aspects of social life. Ironically, although technology can bring people together, it can also isolate them from each other, leading the Amish to ban any technology that "is seen as a threat to the cohesion of the community or might contaminate the culture's values" (Hafner 1999, 1). For example, some technological innovations replace social roles—an answering machine may replace a secretary, a computer-operated vending machine may replace a waitperson, an automatic teller machine may replace a banker, and closed circuit television, a teacher (Johnson 1988; Winner 1993). These technologies may improve efficiency, but they also reduce human contact.

National Data

In a Gallup survey, 38 percent of the respondents feared that computer technology would result in their losing face-to-face contact with other people.

SOURCE: Wiseman and Enrico 1994

Loss of Privacy and Security

People aren't aware that mouse clicks can be traced, packaged, and sold.

–Larry Irving
Commerce Department

When Massachusetts Institute of Technology (M.I.T). professor of media technology Nicholas Negroponte was asked "what do you fear most in thinking about the digital future?," his response was "privacy and security."

> **When I send you a message in the future, I want you to be sure it is from me; that when that message goes from me to you, nobody is listening in; and when it lives on your desk, nobody is snooping later. (Negroponte 1995, 88)**

Schools, employers, and the government are increasingly using technology to monitor individuals' performance and behavior. Today, through cybernation, machines monitor behavior by counting a telephone operator's minutes on-line, videotaping a citizen walking down a city street, or tracking the whereabouts of a needed employee (Quick 1998).

Employers and schools may subject individuals to drug testing technology (see Chapter 3). Through computers, individuals can obtain access to phone bills, tax returns, medical reports, credit histories, bank account balances, and driving records. Some companies sell such personal information. Unauthorized disclosure is potentially devastating. If a person's medical records indicate that he or she is HIV positive, for example, that person could be in danger of losing his or her job or health benefits. DNA testing of hair, blood, or skin samples can be used to deny individuals insurance, employment, or medical care. In response to the possibility of such consequences Brin (1998), author of *The Transparent Society,* argues that since it is impossible to prevent such intrusions, "reciprocal transparency," or complete openness, should prevail. If organizations can collect the information, then citizens should have access to it and to its uses.

Technology has created threats not only to the privacy of individuals, but also to the security of entire nations. Computers and modems can be used (or misused) in terrorism and warfare to cripple the infrastructure of a society and tamper with military information and communication operations.

To protect unauthorized access to information stored in or transmitted through computers and communication technologies, groups, organizations, and corporations may code information using a technique called encryption. Information that is encrypted is unintelligible without the code or "keys" to unscramble it. The U.S. government is opposed to exporting encryption technologies in the interest of national security. Encryption industries, however, can export their products if recovery keys are made available to law enforcement agencies. Currently, there are no restrictions on the sale and use of encryption systems in the United States (Lash 1998).

National Data

Over 75 million PC users say they are unlikely to use the Internet in the next year; reasons most often given concern privacy of information (44 percent) and security of on-line financial transactions (40 percent).

SOURCE: "Influencing Net Use" 1998

Unemployment

Some technologies replace human workers—robots replace factory workers, word processors displace secretaries and typists, and computer-assisted diagnostics reduce the need for automobile mechanics. It is estimated, for example, that 52,000 branch banks will be replaced by automatic teller machines in the next decade. The cost of a teller transaction is $1.07; an electronic transaction, 7¢ (Fix 1994).

Increases in unemployment due to technology are likely to be worldwide according to activist Jeremy Rifkin, whose book, *The End of Work* (1996), predicts a global reduction in service-sector employees. Unlike previous decades, according to Rifkin, where uprooted workers moved from farms to factories to offices, today's technologically displaced workers have no place to go. It should be noted that all-time low unemployment rates seem to contradict Rifkin's dire forecasts.

There is little doubt, however, that technology changes the nature of work and the types of jobs available. For example, fewer semiskilled workers are needed since many of these jobs have been replaced by machines. The jobs that remain, often white-collar jobs, require more education and technological skills. Technology thereby contributes to the split labor market as the pay gulf between skilled and unskilled workers continues to grow (Pascal 1996).

Human skills are subject to obsolescence at a rate perhaps unprecedented in American history.

—Alan Greenspan
Chairman, Federal Reserve

National Data

By some estimates, 18 million American workers—20 percent of the labor force—are at risk of losing their job because of technology.

SOURCE: Pascal 1996

The Haves and the Have-Nots

One of the most significant social problems associated with science and technology is the increased division between the classes. As Welter (1997, 2) notes,

> **. . . it is a fundamental truth that people who ultimately gain access to, and who can manipulate, the prevalent technology are enfranchised and flourish. Those individuals (or cultures) that are denied access to the new technologies, or can not master and pass them on to the largest number of their offspring, suffer and perish.**

The fear that technology will produce a "virtual elite" (Levy 1997) is not uncommon. Several theorists hypothesize that as technology displaces workers, most notably the unskilled and uneducated, certain classes of people will be irreparably disadvantaged—the poor, minorities, and women (Hayes 1998; Welter 1997). There is even concern that biotechnologies will lead to a "genetic stratification," whereby genetic screening, gene therapy, and other types of genetic enhancements are available only to the rich (Mehlman and Botkin 1998).

The wealthier the family, for example, the more likely the family is to have a computer. Of American families with an income of $75,000 a year or more, 74 percent have at least one PC. However, only about 40 percent of all U.S. homes have a computer (Lohr 1999). Further, white families are three times more likely to have a computer than black or Hispanic families (Hancock 1995, 50–51).

Additionally, schools in poorer districts are less likely to have Internet access and have fewer instructional rooms with Internet access if it is available (*Statistical Abstract of the United States: 1998,* Table 282). Inner-city neighborhoods, disproportionately populated by racial and ethnic minorities, are simply less likely to be "wired," that is, to have the telecommunications hardware necessary for schools to access on-line services. In fact, cable and telephone companies are less likely to lay fiber optics in these areas—a practice called "information apartheid" or "electronic redlining." Students who live in such neighborhoods are technologically disadvantaged and may never catch up to their middle-class counterparts (Welter 1997).

The cost of equalizing such differences is enormous, but the cost of not equalizing them may be even greater. Employees who are technologically skilled earn up to 15 percent higher pay than those who are not, and the gap is widening (Pascal 1996; Hancock 1995). Further, technological disparities exacerbate the structural inequities perpetuated by the split labor force and the existence of primary and secondary labor markets (see Chapter 11).

National Data

Unemployment among older information technology workers, those over 50, is 17 percent compared to only 3 percent for all workers 50 and older, and less than 5 percent for the general population.

SOURCE: Hayes 1998

National Data

According to a recent survey of first-year college students, technology is perceived as the single "most important advantage" over preceding generations (32 percent)—other advantages listed did not come close to the significance of technology.

SOURCE: Harris Poll 1998b

CONSIDERATION

As symbolic interactionists note, language itself perpetuates inequality. Until the mid-1980s, biologists described human conception in terms of gender-specific qualities. Such words as "velocity" and "penetrating" were used in reference to the male's sperm. The female egg, however, was described as docile, passively wandering about, as the determined sperm approaches. More recently scientists, in part motivated by feminist thinking, conducted further research that revealed that sperm are not quite the swimmers once thought. In fact, it appears as though sperm are rather inefficient swimmers, often losing direction as they approach the egg. It also appears that, contrary to traditional theorizing, the egg plays a more active role in conception, moving closer to the sperm as it nears (Begley 1997).

Mental and Physical Health

Some new technologies have unknown risks. Biotechnology, for example, has promised and, to some extent, has delivered everything from life-saving drugs to heartier pest-free tomatoes. Biotechnologies have also, however, created **technology-induced diseases** such as those experienced by Chellis Glendinning (1990). Glendinning, after using the "Pill" and, later, the Dalkon Shield IUD, became seriously ill.

> **Despite my efforts to get help, medical professionals did not seem to know the root of my condition lay in immune dysfunction caused by ingesting artificial hormones and worsened by chronic inflammation. In all, my life was disrupted by illness for twenty years, including six years spent in bed. . . . For most of the years of illness, I lived in isolation with my problem. Doctors and manufacturers of birth control technologies never acknowledged it or its sources.**

The history of the "rocky road of progress" is made up of apparent technological wonders that turned out to be techno-threats or outright disasters.

—Abigail Trafford
Andrea Gabor
Journalists

National Data

According to the Bureau of Labor Statistics, cumulative trauma disorders (carpal tunnel syndrome, low-back pain, tendinitis) cost American industries over $100 billion a year.

SOURCE: WRC 1998

Other technologies that pose a clear risk to a large number of people include nuclear power plants, DDT, automobiles, x rays, food coloring, and breast implants.

The production of new technologies may also place manufacturing employees in jeopardy. For example, the electronics industry uses thousands of hazardous chemicals, including freon, acetone, and sulfuric and nitric acids:

> **[Semiconductor] workers are expected to dip wafer-thin silicon that has been painted with photoresist into acid baths. The wafers are then heated in gas-filled ovens, where the gas chemically reacts with the photosensitive chemicals. After drying, the chips are then bonded to ceramic frames, wires are attached to contacts and the chip is encapsulated with epoxy. These integrated circuits are then soldered onto boards, and the whole device is cleaned with solvents. Gases and silicon lead to respiratory and lung diseases, acids to burning and blood vessel damage, and solvents to liver damage. (Hosmer 1986, 2)**

Finally, technological innovations are, for many, a cause of anguish and stress. Nearly 60 percent of workers report being "technophobes," that is, fearful of technology (Boles and Sunoo 1998). Says Dr. Michelle Weil, a clinical psychologist, the key to dealing with technophobia is to decide "which tools make sense in a person's life and will give them more control, and, ultimately, more enjoyment" (Kelly 1997, 4). This chapter's *The Human Side,* "Data Smog Removal Techniques," deals directly with this issue.

THE HUMAN SIDE

DATA SMOG REMOVAL TECHNIQUES

The following excerpt from David Shenk's *Data Smog: Surviving the Information Glut* (1997), outlines techniques for simplifying your life and reducing the stress associated with a "high-tech" lifestyle.

> Most of us have excess information in our lives, distracting us, pulling us away from our priorities and from a much-desired tranquility. If we stop just for a moment to look (and listen) around us, we will begin to notice a series of data streams that we'd be better off without, including some distractions we pay handsomely for.
>
> **Turn the television off.** There is no quicker way to regain control of the pace of your life, the peace of your home, and the content of your thinking than to turn off the appliance that supplies for all-too-many of us the ambiance of our lives. . . .
>
> . . . It is not enough to simply turn the TV off; one must also make it somewhat difficult to turn it on again. Many TV-free participants have liked the results so much that they have gotten rid of their televisions altogether. My own alternative to this radical gesture has been to move the offending item from our kitchen/living room into the closet. There it stays except for a few select hours per week, when I lug it out, plug it in, and turn it on.
>
> **Leave the pager and cell-phone behind.** It is thrilling to be in touch with the world at all times, but it's also draining and interfering. Are wireless communicators instruments of liberation, freeing people to be more mobile with their lives—or are they more like electronic leashes, keeping people more plugged-into their work and their info-glutted lives than is necessary and healthy?
>
> **Limit your e-mail.** As one spends more and more time online, e-mail quickly changes from being a stimulating novelty to a time-consuming burden, with dozens of messages to read and answer every day from colleagues, friends, family, news group posts, and unsolicited sales pitches. . . .
>
> . . . If we're spending too much time each day reading and answering e-mail that has virtually no value, we must take steps to control it. Ask people (nicely) not to forward trivia indiscriminately. "Unsubscribe" to the news groups that you're no longer really interested in. Tell spammers that you have no interest in their product, and ask them to remove you from their customer list.
>
> **Say no to dataveillance.** With some determination and a small amount of effort, one can also greatly reduce the amount of junk mail and unsolicited sales phone calls. It involves writing just a few letters, requesting to have your name put on "do-not-disturb" lists, which some 75 percent of direct marketers honor.
>
> **Resist advertising.** We read, watch, and listen to advertisements all day. Must we also wear them on our clothes?
>
> **Resist upgrade mania.** Remember: Upgrades are designed to be sales tools, not to give customers what they've been clamoring for.

THE HUMAN SIDE

DATA SMOG REMOVAL TECHNIQUES *(continued)*

Be your own "smart agent." A new class of robotic "smart agent" software is becoming available to help consumers automate their information-filtering needs—software such as IBM's InfoSage . . . delivers a custom set of news stories according to programmed preferences. Other programs weed out e-mail not written by a select list of people.

But . . . smart agents are not the answer to the information glut. Regardless of how efficient they become, they will never be adequate substitutes for our own manual filtering. Instead, we must become our own smart agents.

Cleanse your system with "data-fasts." As your own smart agent, you are also your own data dietitian. Take some time to examine your daily intake and consider whether or not your info diet needs some fine-tuning. Take some data naps in the afternoon, during which you stay away from electronic information for a prescribed period. You could also consider limiting yourself to no more than a certain number of hours on the Internet each week, or at least balancing the amount of time spent online with an equal amount of time reading books. . . .

[P]eriodic fasts of a week or month have a remarkably rejuvenating effect. One sure way to gauge the value of something, after all, is to go without it for a while.

SOURCE: Shenk, David. 1997. *Data Smog: Surviving the Information Glut.* San Francisco: HarperEdge, pp. 184–89. Reprinted by permission.

The Challenge to Traditional Values and Beliefs

Technological innovations and scientific discoveries often challenge traditionally held values and beliefs, in part because they enable people to achieve goals that were previously unobtainable. Before recent advances in reproductive technology, for example, women could not conceive and give birth after menopause. Technology that allows postmenopausal women to give birth challenges societal beliefs about childbearing and the role of older women. Macklin (1991) notes that the techniques of egg retrieval, in vitro fertilization, and gamete intrafallopian transfer (GIFT) make it possible for two different women to each make a biological contribution to the creation of a new life. Such technology requires society to reexamine its beliefs about what a family is and what a mother is. Should family be defined by custom, law, or the intentions of the parties involved?

> Technological systems are both socially constructed and society shaping.
>
> —Thomas Hughes
> Social scientist

Medical technologies that sustain life lead us to rethink the issue of when life should end. The increasing use of computers throughout society challenges the traditional value of privacy. New weapons systems challenge the traditional idea of war as something that can be survived and even won. And cloning challenges our traditional notions of family, parenthood, and individuality. Toffler (1970) coined the term **future shock** to describe the confusion resulting from rapid scientific and technological changes that unravel our traditional values and beliefs.

STRATEGIES FOR ACTION: CONTROLLING SCIENCE AND TECHNOLOGY

As technology increases, so does the need for social responsibility. Nuclear power, genetic engineering, cloning, and computer surveillance all increase the need for social responsibility: ". . . technological change has the effect of enhancing the importance of public decision making in society, because technology is continually creating new possibilities for social action as well as new problems that have to be dealt with" (Mesthene 1993, 85). In the following section, various aspects of the public debate are addressed, including science, ethics, and the law, the role of corporate America, and government policy.

Science, Ethics, and the Law

> Prohibiting scientific and medical activities would also raise troubling enforcement issues . . . Would they [FBI] raid research laboratories and universities? Seize and read the private medical records of infertility patients? Burst into operating rooms with their guns drawn? Grill new mothers about how their babies were conceived?
>
> —Mark Eibert
> Attorney

Science and its resulting technologies alter the culture of society through the challenging of traditional values. Public debate and ethical controversies, however, have led to structural alterations in society as the legal system responds to calls for action. The Institute of Science, Law and Technology at the Illinois Institute of Technology, for example, recently recommended that minimum standards for fertility treatments be set. In an effort to reduce the number of multiple births, the group called for a legal limit of four transferred embryos for each IVF attempt. Ethicist George Annas also called for states to pass laws "defining legal parentage to avoid confusion when disputes occur among the parties who helped create a child. . . ." ("Group Seeks Standards for Infertility Services" 1998, 12).

Are such regulations necessary? In a society characterized by rapid technological and thus social change—a society where custody of frozen embryos is part of the divorce agreement—many would say yes. Cloning, for example, is one of the most hotly debated technologies in recent years. Bioethicists and the public vehemently debate the various costs and benefits of this scientific technique. Despite such controversy, however, the chairman of the National Bioethics Advisory Commission recently stated that human cloning will be "very difficult to stop" (McFarling 1998). Such comments have fueled state legislative action. As of January 1, 1999, California became the first state to outlaw human cloning (Eibert 1998).

Should the choices that we make, as a society, be dependent upon what we *can do,* or what we *should do*? While scientists and the agencies and corporations who fund them often determine the former, who should determine the latter? Although such decisions are likely to have a strong legal component, that is, they must be consistent with the rule of law and the constitutional right of scientific inquiry (Eibert 1998), legality or the lack thereof often fails to answer the question, what *should* be done? *Roe v. Wade* (1973) did little to squash the public debate over abortion and, more specifically, the question of when life begins. Thus, it is likely that the issues surrounding the most controversial of technologies will continue into the twenty-first century and with no easy answers.

International Data

Over 30 public and private livestock centers in Japan have received $2.2 million from the Japanese government to conduct cattle cloning research; thus far, 400 cows have been cloned and sent to market.

SOURCE: Associated Press 1998

Technology and Corporate America

As philosopher Jean-Francois Lyotard notes, knowledge is increasingly produced to be sold (Powers 1998). The development of genetically altered crops, the commodification of women as egg donors, and the harvesting of regenerated organ tissues are all examples of potentially market-driven technologies. Like the corporate pursuit of computer technology, profit-motivated biotechnology creates several concerns.

First and foremost is the concern that only the rich will have access to such life-saving technologies as genetic screening and cloned organs. Such fears are justified. Several "companies with enigmatic names such as Progenitor, Millennium Pharmaceuticals, and Darwin Molecular have been pinpointing and patenting human life with the help of $4.5 billion in investments from pharmaceuticals companies" (Shand 1998, 46). Millennium Pharmaceutical holds the patent on the melanoma gene and the obesity gene; Darwin Molecular controls the premature aging gene, and Progenitor the gene for schizophrenia.

These patents result in **gene monopolies**, which could lead to astronomical patient costs for genetic screening and treatment. One company's corporate literature candidly states that its patent of the breast cancer gene will limit competition and lead to huge profits (Shand 1998, 47). The biotechnology industry argues that such patents are the only way to recoup research costs which, in turn, lead to further innovations.

The commercialization of technology causes several other concerns, including the tendency for discoveries to remain closely guarded secrets rather than collaborative efforts, and issues of quality control (Rabino 1998; Lemonick and Thompson 1999). Further, industry involvement has made government control more difficult as researchers depend less and less on federal funding. In 1996, private biotechnology industries spent $19 million on research and development (Burstein 1998).

Finally, although there is little doubt that profit acts as a catalyst for some scientific discoveries, other less commercially profitable but equally important projects may be ignored. As biologist Isaac Rabino states, "[I]magine if early chemists had thrown their energies into developing profitable household products before the periodic table was discovered. . . ." (Rabino 1998, 112).

International Data

In 1998 the Swiss pharmaceutical company Roche Holding paid Decode Genetics $200 million for research in Iceland; with a homogeneous population of 270,000, Icelanders provide a much-needed "narrow gene pool" for the identification of genes carrying diseases.

SOURCE: Marshall 1998

Runaway Science and Government Policy

Science and technology raise many public policy issues. Policy decisions, for example, address concerns about the safety of nuclear power plants, the privacy of electronic mail, the hazards of chemical warfare, and the legality of surrogacy. In creating science and technology, have we created a monster that has begun to control us rather than the reverse? What controls, if any, should be placed on science and technology? And are such controls constitutionally permitted? In the hopes of regulating sexually oriented material on the Internet, the Communications Decency Act of 1996 was signed into law. In 1997, the U.S. Supreme Court declared the act unconstitutional, in part, based on First Amendment considerations. The 1998 Child Online Protection Act has also been challenged by civil liberties groups as violating free speech rights (Weil 1998; Cohen 1999).

The government, often through such regulatory agencies and departments as the Civil Aeronautics Board, the Federal Communications Commission, and the Department of Transportation, prohibits the use of some technologies (e.g., assisted-suicide devices) and requires others (e.g., seat belts). For example, in 1998 the Food and Drug Administration announced

Science has liberated the ideas of those who read and reflect, and the American example had kindled feelings of right in the people. An insurrection has consequently begun, of science, talents and courage, against rank and birth, which have fallen into contempt. . . . Science is progressive.

—Thomas Jefferson
Former U.S. President

that under its "statutory authority over biological products . . ." it will begin regulating cloning (Eibert 1998, 52).

In addition to regulatory control, all proposed research and development projects seeking federal funds must be accompanied by an environmental impact statement. The Office of Technology Assessment, established in 1973, also provides information to Congress about the environmental impact of proposed projects. Additionally, the National Institute of Standards and Technology, a part of the Commerce Department's Technology Administration, promotes economic growth through working with industry in developing new technologies.

Restricting expression that can cause direct, provable harm is one thing. Regulating expression because it might cause "immoral" behavior is something else entirely.

—Marjorie Heins
Author and social commentator

In 1993, President Clinton, by executive order, established the National Science and Technology Council (NSTC), which includes the vice president and the secretaries of defense, energy, and commerce, among others. The major function of the NSTC is to coordinate the interagency science and technology policy-making process and to implement and integrate the president's science and technology agenda (NSTC 1994). President Clinton also created the President's Committee of Advisers on Science and Technology (PCAST), which advises him on issues of science and technology and assists the NSTC in securing private sector cooperation. Finally, as of this writing, Congress is considering a variety of technologically relevant legislation, including the Kennedy-Feinstein bill, which would ban human cloning for 10 years, and impose a $1 million fine on violators (Eibert 1998).

International Data

In 1998, a government ban prohibiting the exportation of any "personal information" about individuals without their permission went into effect in 15 European nations.

SOURCE: Clayton 1998

UNDERSTANDING **SCIENCE AND TECHNOLOGY**

What are we to understand about science and technology from this chapter? As functionalists argue, science and technology evolve as a social process and are a natural part of the evolution of society. As society's needs change, scientific discoveries and technological innovations emerge to meet these needs, thereby serving the functions of the whole. Consistent with conflict theory, however, science and technology also meet the needs of select groups and are characterized by political components. As Winner (1993) notes, the structure of science and technology conveys political messages including "[P]ower is centralized," "[T]here are barriers between social classes," "[T]he world is hierarchically structured," and "[T]he good things are distributed unequally" (Winner 1993, 288).

The scientific discoveries and technological innovations that are embraced by society as truth itself are socially determined. Research indicates that science and the resulting technologies have both negative and positive consequences—a technological dualism. Technology saves lives and time and money; it also leads to death, unemployment, alienation, and estrangement. Weighing the costs and benefits of technology poses ethical dilemmas as does science itself. Ethics, however, "is not only concerned with individual choices and acts. It is also and, perhaps, above all concerned with the cultural shifts and trends of which acts are but the symptoms" (McCormick 1994, 16).

It would be nice if the scientist and the engineer were cognizant of, and deeply concerned about, the potential risks, secondary consequences, and other costs of their creations.

—Allan Mazur
Sociologist

Thus, society makes a choice by the very direction it follows. Such choices should be made on the basis of guiding principles that are both fair and just (Winner 1993; Goodman 1993; Eibert 1998):

1. Science and technology should be prudent. Adequate testing, safeguards, and impact studies are essential. Impact assessment should include an evaluation of the social, political, environmental, and economic factors.

2. No technology should be developed unless all groups, and particularly those who will be most affected by the technology, have at least some representation "at a very early stage in defining what that technology will be" (Winner 1993, 291). Traditionally, the structure of the scientific process and the development of technologies has been centralized (that is, decisions have been in the hands of a few scientists and engineers); decentralization of the process would increase representation.
3. There should be no means without ends. Each new innovation should be directed toward fulfilling a societal need rather than the more typical pattern in which a technology is developed first (e.g., high-definition television) and then a market is created (e.g., "you'll never watch a regular TV again!"). Indeed, from the space program to research on artificial intelligence, the vested interests of scientists and engineers, whose discoveries and innovations build careers, should be tempered by the demands of society.

What the twenty-first century will hold, as the technological transformation continues, may be beyond the imagination of most of society's members. Technology empowers; it increases efficiency and productivity, extends life, controls the environment, and expands individual capabilities. But, as Steven Levy (1995) notes, there is a question as to whether society can accommodate such empowerment (p. 26).

As this century closes and we enter the first computational millennium, one of the great concerns of civilization will be the attempt to reorder society, culture, and government in a manner that exploits the digital bonanza, yet prevents it from running roughshod over the checks and balances so delicately constructed in those simpler precomputer years.

CRITICAL THINKING

1. Use of the Internet by neo-Nazi and white supremacist groups has recently increased. Despite such increases, the U.S. Supreme Court has strengthened First Amendment protections of Internet material (Whine 1997). Should such groups have the right to disseminate information about their organizations and recruit members through the Internet?
2. In 1996, President Clinton signed a bill that requires TV manufacturers to equip future television sets with the "V-chip"—a technological device designed to prevent children from watching programs their parents find objectional. Hollywood executives opposed the "V-chip" on the grounds that it is intrusive and violates constitutional guarantees of freedom of expression. Others have been critical, pointing to the difficulty of establishing a universal definition of violence, and the impact on advertising revenues (Makris 1996). How might a conflict theorist explain opposition to the "V-chip"?
3. What currently existing technologies have had more negative than positive consequences for individuals and for society?
4. Some research suggests that productivity actually declines with the use of computers (Rosenberg 1998). Assuming this "paradox of productivity" is accurate, what do you think causes the reduction in efficiency?

KEY TERMS

abortion
automation
cultural lag
cybernation
deskilling
e-commerce
future shock
gene monopolies
gene therapy
genetic engineering
genetic screening
Internet
in-vitro fertilization
IT
mechanization
postmodernism
science
technological fix
technology
technology-induced diseases
upskilling
Y2K

INTERNET

You can find more information on the Human Genome Project, cloning, genetic engineering, and government regulation of technology at the *Understanding Social Problems* web site at **http://sociology.wadsworth.com.**

INFOTRAC COLLEGE EDITION

Either from the Wadsworth sociology resource center at **http://sociology.wadsworth.com** or directly from your web browser, you may access InfoTrac College Edition, an on-line university library that includes over 700 popular and scholarly journals in which you can find articles related to the topics in this chapter. Suggested articles and questions relating to these articles are listed below.

Rabino, Isaac. 1998. "The Biotech future." *American Scientist* 86(2):110–13.

1. What differences does the author see between European and American scientists' views of public attention being given to scientific research?
2. What problems do scientists see in "commercialization" of their research?
3. What societal obstacles stand between scientists and possible medical breakthroughs?

Brown, Brian A. 1998. "Cloning: Where's the Outrage?" *The Human Life Review* 24(2):107–8.

1. What arguments does the author make in favor of or in opposition to cloning?
2. What are the differences between pragmatic and moral arguments, and which are more enduring?
3. Is this article from a conservative or liberal perspective? Why?

Morell, Virginia. 1998. "A Clone of One's Own." *Discover* 19(5):82–96.

1. According to the article, what is the next step in experimentation prior to human cloning?
2. Who are "Raelians" and what is Valiant Venture, Ltd.?
3. What procedures were used in producing Neti and Ditto?

CHAPTER FIFTEEN

Population and Environmental Problems

Is It True?

Is It True?

1. In some countries, a woman must obtain her husband's consent in order to obtain contraception.
2. Most of the growth in world population is occurring in wealthy, industrialized countries.
3. Evidence of global warming includes the fact that 1998 was the hottest year on record.
4. At least 1,000 species of life are lost each year to extinction.
5. Three states—North Dakota, South Dakota, and Texas—have enough harnessable wind energy to supply the energy needs of the entire United States.

Answers: 1 = T; 2 = F; 3 = T; 4 = T; 5 =

One defining characteristic of a civilized society is a sense of responsibility to the next generation. If we do not assume that responsibility, environmental deterioration leading to economic decline and social disintegration could threaten the survival of civilization as we know it.

Lester R. Brown
Jennifer Mitchell
of the World Watch Institute

If all the people in the world stood shoulder to shoulder they would fill:

a. The island of Bali, Indonesia
b. The state of California
c. The entire United States
d. All of Asia

You may be surprised that the answer is *(a)*—all the people on Earth (as of 1998) could fit on the small island of Bali, with each person occupying less than one square meter—about the size of half a bathtub (Stiefel 1998). Does this mean that there is plenty of room for the world's growing population? The real question, according to Professor Joel Cohen, is not how many people can the Earth support, but how many people can the Earth support with what quality of life (Livernash and Rodenburg 1998)? In 1992, the U.S. National Academy of Sciences and the Royal Society of London issued a report that warned:

> **If current predictions of population growth prove accurate and patterns of human activity on the planet remain unchanged, science and technology may not be able to prevent either irreversible degradation of the environment or continued poverty for much of the world. (cited in Brown 1995, 411)**

Population may be the key to all the issues that will shape the future: economic growth; environmental security; and the health and well-being of countries, communities, and families.

—Nafis Sadik
Executive Director
UN Population Fund

In this chapter, we discuss population growth and the concomitant environmental implications. These concerns reflect two branches of social science: demography and human ecology. **Demography** is the study of the size, distribution, movement, and composition of human populations. **Human ecology** is the study of the relationship between human populations and their natural environment. After discussing population growth in the world and in the United States, we view population and environmental problems through the lens of structural-functionalism, conflict theory, and symbolic interactionism. We also explore how population growth contributes to a variety of social problems and examine strategies for limiting population growth. The second half of the chapter examines the extent, social causes, and solutions to environmental problems. Before reading further, you may want to take the "Eco-Quiz" in this chapter's *Self & Society* feature.

THE GLOBAL CONTEXT: A WORLD VIEW OF POPULATION GROWTH

For thousands of years, the world's population grew at a relatively slow rate. During 99 percent of human history, the size of hunting and gathering societies was restricted by disease and limited food supplies. Around 8000 B.C., the development of agriculture and the domestication of animals led to increased food supplies and population growth, but even then harsh living conditions and disease still put limits on the rate of growth. This pattern continued until the mid-eighteenth century when the Industrial Revolution improved the standard of living for much of the world's population. The improvements included better food, cleaner drinking water, and improved housing, as well as advances

SELF & SOCIETY

Eco-Quiz

This quiz is designed to see how environmentally conscious you are. For each question, select the correct answer from the choices provided. When complete, calculate the number of correct answers you have and check your level of eco-awareness.

1. If you saved up all the unwanted junk mail for 1 year, the United States would save _____ trees in a year.
   a. 100,000 b. 100 million c. 1 billion
2. If just 25 percent of U.S. families used 10 fewer plastic bags a month, we would save over _____ bags a year.
   a. 250,000 b. 2.5 million c. 2.5 billion
3. Leaves alone account for _____ of our solid wastes in the fall season.
   a. 25 percent b. 75 percent c. 90 percent
4. Every ton of recycled office paper saves _____ gallons of oil.
   a. 12 b. 135 c. 380
5. Energy saved from one recycled aluminum can will operate a TV set for _____ .
   a. 3 hours b. 5 hours c. 5 days
6. Americans consume more than _____ trees a year through commercial paper.
   a. 150 million b. 250 million c. 850 million
7. Insulating your attic reduces the amount of energy loss in most houses by up to _____ .
   a. 5 percent b. 20 percent c. 50 percent
8. Every day _____ children die from preventable diseases somewhere in the world.
   a. 40,000 b. 4,000 c. 400
9. The human population of the world is expected to have nearly _____ by the year 2100.
   a. doubled b. tripled c. quadrupled
10. The uncontrolled fishing that is allowed has reduced the amount of some commercial species up to _____ of their original population.
    a. one-tenth b. one-half c. three-fourths
11. Every day _____ species of plants and animals become extinct as their habitat and human influences destroy them.
    a. 10–25 b. 25–50 c. 50–100
12. A study has shown that there are possibly over _____ species of insects dwelling in the canopies of tropical forests.
    a. 3 million b. 30 million c. 300,000
13. _____ square miles of rain forests are being destroyed each year.
    a. 13,000 b. 43,000 c. 63,000
14. Medicine produced in tropical forests brings in _____ dollars a year commercially.
    a. 30,000 b. 3 million c. 300 billion
15. Grasslands cover _____ of the land on Earth.
    a. one-sixteenth b. one-eighth c. one-fifth
16. Every year approximately _____ tons of carbon accumulates in the air.
    a. 400 billion b. 400 million c. 400,000
17. Batteries contain heavy metals, which cause contamination in dump sites and in the air when incinerated. This is why you should _____ your batteries.
    a. throw away b. recharge c. recycle
18. Life in the oceans provides the earth with most of our _____ , a most precious chemical.
    a. oxygen b. nitrogen c. fresh water

(continued)

19. Instead of using an incandescent light bulb, use a ______ light bulb, which uses one-fourth of the energy and lasts longer. Yet, it still gives off the same amount of light.
    a. fluorescent b. burnt-out c. black
20. Toxic chemicals that you dump down the storm drain may end up in ______ .
    a. your town's water treatment plant b. your dishwasher c. the ocean

ANSWERS

Question 1: The correct answer is (b); 100 million trees would be saved if all unwanted junk mail was recycled.

Question 2: The correct answer is (c); 2.5 billion would be saved!

Question 3: The correct answer is (b); nearly three-quarters of our waste during the autumn is made up of leaves, which should be composted!

Question 4: The correct answer is (c); now, that's a lot of oil!

Question 5: The correct answer is (a); now you see why recycling can save a lot of energy!

Question 6: The correct answer is (c); now, that's a lot of trees!

Question 7: The correct answer is (b); you should insulate your attic to save energy and save money!

Question 8: The correct answer is (a); isn't it sad that so many children die unnecessarily? The cost of medical treatment for these children is equal to just 2 percent of the Third World's expenditure on weapons!

Question 9: The correct answer is (b); just imagine how crowded the earth will be if our population triples!

Question 10: The correct answer is (a); uncontrolled fishing is wreaking havoc upon marine ecosystems and has resulted in the reduction of one-tenth of our commercial species!

Question 11: The correct answer is (c); now, that's a lot of lost plants and animals!

Question 12: The correct answer is (c); insects have been one of the most successful organisms to conquer this planet!

Question 13: The correct answer is (c); our rain forests are rapidly disappearing, and this needs to stop!

Question 14: The correct answer is (c); rain forest destruction not only harms the environment, but also eliminates possible cures for deadly diseases!

Question 15: The correct answer is (c); this biome sure covers a lot of our planet!

Question 16: The correct answer is (a); ever since the Industrial Revolution, we have been dumping tons of carbon dioxide into the atmosphere!

Question 17: The correct answer is (b); batteries contain hazardous chemicals, so it is best to use rechargeable batteries.

Question 18: The correct answer is (a); due to the high amounts of seaweed and phytoplankton, our oceans release most of the world's oxygen.

Question 19: The correct answer is (a); fluorescent lights save energy and money!

Question 20: The correct answer is (c); if you don't want to swim in it, don't dump it!

Score	*Rating*
20	EcoNaut
17–19	Eco-whiz
15–16	Eco-star
12–14	Eco-conscious
9–11	Eco-disaster
6–8	Eco-calamity
0–5	Eco-hazard

SOURCE: Adapted from "Eco-Quiz." 1998. *Save Our Earth and Make a Difference*. Think Quest. **http://library.advanced.org/11353/text/quiz.html**. Reprinted by permission.

in medical technology such as vaccinations against infectious diseases; all contributed to rapid increases in population.

World Population Growth

In the year A.D. 1, the world's population was about 250 million. The **doubling time**—or time it takes for a population to double in size from any base year—decreases as the population grows. Although the population in A.D. 1 took 1,650 years to double, the second doubling took only 200 years, the third 80 years, and the fourth 45 years. In 1998, the doubling time for the world's population was 49 years (Population Reference Bureau 1998). Figure 15.1 illustrates world population growth from 1950 projected until 2050.

International Data

Sixty-five countries are expected to double their populations in 30 years or less, and 14 countries will triple or nearly triple their populations by 2050.

SOURCE: Population Institute 1998

CONSIDERATION

If you were born before 1950, you are a member of the first generation ever to witness a doubling of the world's population. There has been more growth in population since 1950 than during the 4 million years since our early ancestors first walked on earth (Brown, Gardner, and Halweil 1998).

International Data

As many as 58 countries—43 of them in Africa and 12 in Asia—have fertility rates of five or more children per woman. Half of these countries, 29, have fertility rates of six or more children per woman.

SOURCE: Population Institute 1998

More than 90 percent of population growth is occurring in less developed countries (Livernash and Rodenburg 1998). Compared to the developed countries, these nations are characterized by greater illiteracy, higher infant mortality rates, less industrialization, shorter life expectancies, more poverty, lower-quality health care, and higher **fertility rates**, or average number of births per woman.

Falling Fertility Rates and the "Birth Dearth"

While many countries, especially in the developing world, are experiencing continued population growth, other countries are declining in population. In a growing number of countries births are falling below 2.1 children per woman—the **replacement level** below which the population begins to decline. Declining birthrates in some countries are thought to result from high unemployment and a high cost of living.

Spain has the lowest birthrate in the world: each woman has an average of 1.15 children in her lifetime. Falling population levels are also a concern in France, Germany, Greece, Italy, Russia, and Japan. According to Steve Mosher,

International Data

In about 61 countries the average number of children born to each woman has fallen below 2.1—the replacement level required to maintain the population.

SOURCE: United Nations 1999

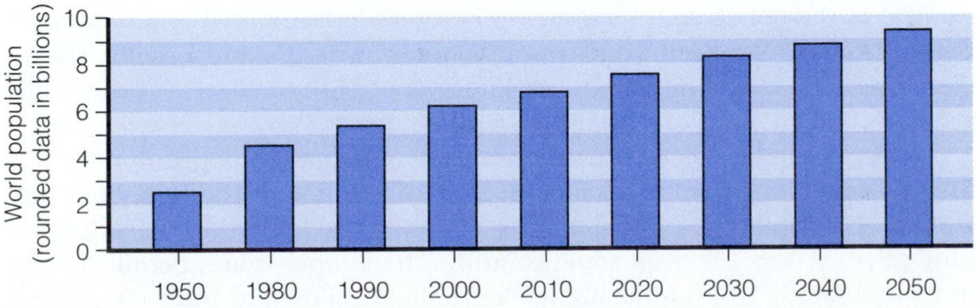

Figure 15.1 World Population Growth and Projections, 1950–2050

SOURCES: Lester R. Brown, Gary Gardner, and Brian Halweil. 1998. *Beyond Malthus: Sixteen Dimensions of the Population Problem.* World Watch Paper 143. Washington, D.C.: World Watch Institute; *Statistical Abstract of the United States: 1998,* 118th ed. U.S. Bureau of the Census. Washington, D.C.: U.S. Government Printing Office, Table 1340.

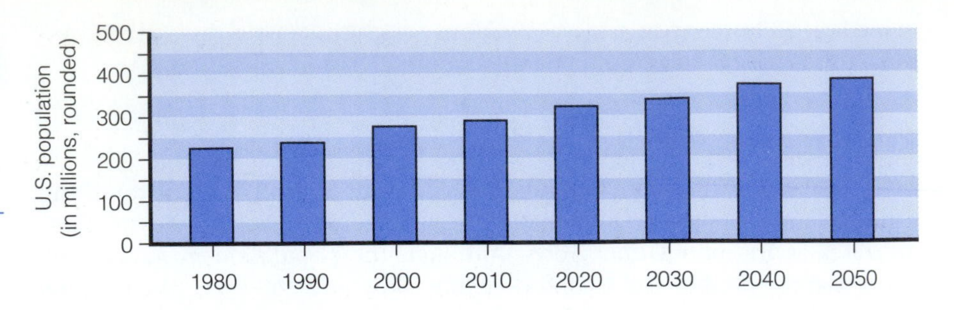

Figure 15.2 U.S. Population Growth and Projections, 1980–2050

SOURCE: *Statistical Abstract of the United States: 1998,* 118th ed. U.S. Bureau of the Census. Washington, D.C.: U.S. Government Printing Office, Table 4

president of the Population Research Institute, "humanity's long-term problem is not too many children being born but too few . . . Over time, the demographic collapse will extinguish entire cultures" (Cooper 1998, 612).

United States Population Growth

National Data

Of all the world's industrialized countries, the United States has the most population growth. About one-third of U.S. growth results from immigration.

SOURCE: "Ninety-Eight Percent of Growth in Developing Countries" 1997

During the colonial era (c. 1650), the U.S. population included about 50,000 colonists and 750,000 Native Americans. By 1859, disease and warfare had reduced the American Indian population to 250,000, but the European population had increased to 23 million. The U.S. population continued to increase into the 1900s. From the mid-1940s to the late 1960s the U.S. birthrate increased significantly. This period of high birthrates, commonly referred to as the baby boom, peaked in 1957 when the crude birthrate (number of live births per 1,000 people) reached 25.3. Although the birthrate dropped to 15.3 in 1986, the U.S. population continues to increase. As Figure 15.2 shows, the U.S. population is expected to be over 390 million by the year 2050.

SOCIOLOGICAL THEORIES OF POPULATION AND ENVIRONMENTAL PROBLEMS

The three main sociological perspectives—structural-functionalism, conflict theory, and symbolic interactionism—may be applied to the study of population and environmental problems.

Structural-Functionalist Perspective

Structural-functionalism focuses on how changes in one aspect of the social system affect other aspects of society. For example, the **demographic transition theory** of population describes how industrialization has affected population growth. According to this theory, in traditional agricultural societies, high fertility rates are viewed as necessary to offset high mortality and to ensure continued survival of the population. As a society becomes industrialized and urbanized, improved sanitation, health, and education lead to a decline in mortality. The increased survival rate of infants and children, along with the declining economic value of children, leads to a decline in fertility rates.

Other changes in social structure and culture that affect population include the shift in values toward individualism and self-fulfillment. The availability and cultural acceptability of postnatal forms of family size limitation also affect fertility rates (Mason 1997). In some countries, traditional values permit parents to control family size postnatally by "returning" children at birth (i.e., killing them), selling them to families in need of a child, sending them into bonded labor or prostitution, or marrying them off in early childhood.

Structural-functionalism emphasizes the interdependence between human beings and the physical environment. From this perspective, human actions, social arrangements, and cultural values contribute to environmental problems, which in turn affect social life. For example, war, industrialization, urbanization, and materialistic values contribute to destruction of the environment and depletion of natural resources. Population growth also affects the environment; however, its impact varies tremendously according to a society's lifestyle and economic activities (Livernash and Rodenburg 1998).

Conflict Perspective

The conflict perspective focuses on how wealth, power, or the lack thereof affect population and environmental problems. In 1798, Thomas Malthus predicted that population would grow faster than the food supply and that masses of people were destined to be poor and hungry. According to **Malthusian theory**, food shortages would lead to war, disease, and starvation that would eventually slow population growth. However, conflict theorists argue that food shortages result primarily from inequitable distribution of power and resources. The world's farmers produce enough food for the world's nearly 6 billion people, yet about 800 million people are malnourished, primarily in sub-Saharan Africa and South Asia. "Poverty and the lack of political power are the primary causes of their hunger, not insufficient food" (Livernash and Rodenburg 1998, 4).

Conflict theorists also note that population growth results from pervasive poverty and the subordinate position of women in many less developed countries. Poor countries have high infant and child mortality rates. Hence, women in many poor countries feel compelled to have many children to increase the chances that some will survive into adulthood. The subordinate position of women prevents many women from limiting their fertility. For example, in 14 countries around the world, a woman must get her husband's consent before she can receive any contraceptive services (United Nations Population Fund 1997). Thus, according to conflict theorists, population problems result from continued inequality within and between nations.

The conflict perspective also emphasizes how wealth, power, and the pursuit of profit underlie many environmental problems. Wealth is related to consumption patterns that cause environmental problems. Wealthy nations have higher per capita consumption of petroleum, wood, metals, cement, and other commodities that deplete the earth's resources, emit pollutants, and generate large volumes of waste.

The capitalistic pursuit of profit encourages making money from industry regardless of the damage done to the environment. Further, to maximize sales, manufacturers design products intended to become obsolete. As a result of this **planned obsolescence,** consumers continually throw away used products and purchase replacements. Industry profits at the expense of the environment, which must sustain the constant production and absorb ever-increasing amounts of waste. Industries also use their power and wealth to resist environmental policies that would hurt the industries' profits. For example, there is growing consensus among scientists that global warming is occurring and that the burning of fossil fuels is a major cause of the problem. To fight the attack on fossil-fuel use, petroleum, automobile, coal, and other industries that profit from fossil-fuel consumption created and fund the Global Climate Coalition, an industry-front group that proclaims global warming a myth and characterizes hard evidence of global climate change as "junk science" (Jensen 1999).

Symbolic Interactionist Perspective

The symbolic interaction perspective focuses on how meanings, labels, and definitions learned through interaction affect population and environmental problems. For example, in some societies, women learn through interaction with others that deliberate control of fertility is defined as deviant and socially unacceptable. Once some women learn new definitions of fertility control, they become role models and influence the attitudes and behaviors of others in their personal networks (Bongaarts and Watkins 1996). The level of development of a country is relevant in that it facilitates or hinders social interaction. The more developed a country is, the more likely women will be exposed to new meanings of fertility control through interaction in educational settings and through media and information technologies.

Meanings, labels, and definitions learned through interaction and the media also affect environmental problems. Whether or not an individual recycles, carpools, or joins an environmental activist group is influenced by the meanings and definitions of these behaviors that the individual learns through interaction with others.

Increasingly, businesses and industries are attempting to increase profits through a strategy called **greenwashing**, which essentially involves redefining their corporate image and products as being "environmentally friendly" or socially responsible. For example, many companies publicly emphasize the steps they have taken to help the environment. Another greenwashing strategy is to retool, repackage, or relabel a company's product. For example, in 1990, McDonald's announced it was phasing out foam packaging and switching to a new, paper-based packaging that is partially degradable. Many products have labels that indicate the packaging is made from recycled material. Proctor and Gamble reformulated its liquid detergent, changing from a 64-ounce bottle to a 50-ounce concentrated formula that can be refilled from another container made from 50 percent recycled plastic. ACX Technologies, a subsidiary of Adolph Coors Co., has patented Heplon, a plastic polymer made from plant sources such as corn and beets ("A New Way to Make a Six-Pack Disappear" 1998). Beer cans made from this plastic are biodegradable and can be thrown in a compost pile.

Switzer (1997) explains that greenwashing is commonly used by public relations firms that specialize in damage control for clients whose reputations and profits have been hurt by poor environmental practices. DuPont, the biggest private generator of toxic waste in the United States, attempted to project a "green" image by producing a TV ad showing seals clapping, whales and dolphins jumping, and flamingos flying. In an Earth Day event on the National Mall in Washington, D.C., the National Association of Manufacturers (NAM) highlighted renewable and energy-efficient technologies. Yet, members of NAM have spent millions of dollars lobbying against use of these very same technologies (Karliner 1998). In another greenwashing attempt, a leading ant- and roach-killing product included a label that read: "Made with Pyrethrins: Pyrethrin Insecticide Is Made from Flowers." Yet, the fine print reveals that pyrethrins make up only 0.08 percent of the product, while the neurotoxic pesticide Dursban makes up three times as much of the product (0.24 percent) (Karliner 1998). Although greenwashing often involves manipulation of public perception to maximize profits, some firms make genuine and legitimate efforts to improve their operations, packaging, or overall sense of corporate responsibility toward the environment (Switzer 1997).

SOCIAL PROBLEMS RELATED TO POPULATION GROWTH

Some of the most urgent social problems today are related to population growth. They include poor maternal and infant health, shortages of food and water, environmental degradation, overcrowded cities, and conflict within and between countries.

Poor Maternal and Infant Health

As noted in Chapter 2, maternal deaths (deaths due to causes related to pregnancy and childbirth) are the leading cause of mortality for reproductive-age women in the developing world. Having several children at short intervals increases the chances of premature birth, infectious disease, and death for the mother or the baby. Childbearing at young ages has been associated with anemia and hemorrhage, obstructed and prolonged labor, infection, and higher rates of infant mortality (Zabin and Kiragu 1998). In developing countries, one in four children are born unwanted, increasing the risk of neglect and abuse. In addition, the more children a woman has, the fewer parental resources (parental income and time, and maternal nutrition) and social resources (health care and education) are available to each child (Catley-Carlson and Outlaw 1998). The adverse health effects of high fertility on women and children are, in themselves, compelling reasons for providing women with family planning services. "Reproductive health and choice are often the key to a woman's ability to stay alive, to protect the health of her children and to provide for herself and her family" (Catley-Carlson and Outlaw 1998, 241).

> The United Nations' Food and Agriculture Organization now lists virtually every major commercial species of fish as depleted, fully exploited or overexploited.
>
> —Melissa Healy
> Journalist

Food Shortages, Malnourishment, and Disease

Countries with large populations, few resources, and limited land are particularly vulnerable to food shortages. Food shortages lead to malnourishment, disease, and premature death.

In 1950, 500 million people (20 percent of the world's population) were considered malnourished; in the late 1990s, more than 3 billion people (one-half of the world's population) suffered from malnutrition (Pimentel et al. 1998). In many countries, shortages of vitamin A cause blindness and death. Deficiencies of iron cause anemia and death, and iodine deficiencies cause iodine-deficiency disease—a leading cause of brain damage in children and infants.

International Data

The World Health Organization reports that 19,000 people, mostly infants and children, die *each day* from hunger and malnutrition.

SOURCE: Brown, Gardner, and Halweil 1998

CONSIDERATION

Another factor contributing to food shortages is the increase in meat consumption in affluent countries. Reid (1998) explains that "a meat-centered diet is an inefficient use of resources because you have to feed the animal before it's fed to people" (p. 74). For example, a pig consumes about 600 pounds of corn and 100 pounds of soybean meal before it is sent to slaughter at 240 pounds. The meat from that pig would provide a person with a minimum daily caloric intake for 49 days. Eating the corn and soybean meal directly would provide the same person with enough food for more than 500 days (Reid 1998).

The greatest challenge of the coming century is the maintenance of growth in global food production to match or exceed the projected doubling (at least) of the human population.

—Paul Ehrlich,
Anne H. Ehrlich, and
Gretchen C. Daily
Demographers/biologist

Water Shortages and Depletion of Other Natural Resources

Authors of the *State of the World 1998* suggest that "one of the most underrated issues facing the world as it enters the third millennium is spreading water scarcity" (Brown, Flavin, and French 1998, 5). About 40 percent of the world population faces water shortages at some time during the year (Zwingle 1998). Because 70 percent of all water pumped from underground or drawn from rivers is used for irrigation, water scarcity also threatens food supplies.

Water shortages are exacerbating international conflict. Jordon, Israel, and Syria compete for the waters of the Jordan River basin. Jordon's King Hussein declared that water was the only issue that could lead him to declare war on Israel (Mitchell 1998). Speaking of the water shortage, General Federico Mayor, director of UNESCO, warned, "As it becomes increasingly rare, it becomes coveted, capable of unleashing conflicts. More than petrol or land, it is over water that the most bitter conflicts of the near future may be fought" ("Water Wars Forecast if Solutions Not Found" 1999).

Population growth also contributes to the depletion of other natural resources such as forests, oil, gas, coal, and certain minerals. Whether our planet will be able to sustain the world's population depends not only on how many people there are to sustain, but also on how these people make use of the resources that sustain them.

International Data

About 33 countries are expected to have chronic water shortages by 2025.

SOURCE: "Water Wars Forecast if Solutions Not Found" 1999

Urban Crowding and High Population Density

Population growth contributes to urban crowding and high population density. Without economic and material resources to provide for basic living needs, urban populations in developing countries often live in severe poverty. Urban poverty in turn produces environmental problems such as unsanitary disposal of waste. In Nigerian urban ghettos, for example, "mounds of refuse (including human wastes) that litter everywhere—gutters, schools, roads, market places and town squares—have been accepted as part of the way of life" (Nzeako, quoted in Agbese 1995). Urbanization also diverts water from irrigation and food production to industrial and residential uses.

Densely populated urban areas facilitate the spread of disease among people. Infectious diseases cause more than one-third of all deaths worldwide. Crowded conditions in urban areas provide the ideal environment for the culture and spread of diseases such as cholera and tuberculosis (Pimentel et al. 1998).

International Data

The World Health Organization estimates that half the people in the world do not have access to a decent toilet. Unsanitary disposal of human waste contaminates water supplies. Half the people in the developing world suffer from disease caused by poor sanitation. One of these diseases, diarrhea, is the leading killer of children today.

SOURCE: Gardner 1998

CONSIDERATION

India has one-third the land area of the United States, but more than three times the population. Imagine tripling the U.S. population. Then imagine that this tripled population all lived in the eastern third of the United States. That will give you an idea of how crowded countries like India are.

STRATEGIES FOR ACTION: SLOWING POPULATION GROWTH

Both governments and nongovernmental organizations (NGOs) have attempted to slow world population growth by reducing fertility levels. Strategies for slowing population growth include providing access to birth control methods, improving the status of women, increasing economic development and improving health status, and imposing governmental regulations and policies.

International Data

In developing countries, at least 120 million married women—and a large number of unmarried women—want more control over their pregnancies, but cannot obtain family planning services.

SOURCE: Mitchell 1998

CONSIDERATION

In some European countries with below-replacement-level birthrates, government policies have attempted to encourage rather than discourage childbearing. For example, Italy, Germany, and France have implemented generous child subsidies, in the form of tax credits for every child born, extended maternal leave with full pay, guaranteed employment upon returning to work, and free child care (Cooper 1998).

Provide Access to Birth Control Methods

Although researchers have developed a wide range of contraceptives, many of these methods are not widely available or are prohibitively expensive for individuals with few economic resources. In some countries, methods of birth control are provided only to married women. Family planning personnel often refuse, or are forbidden by law or policy, to make referrals for contraceptive and abortion services for unmarried women. Finally, many women throughout the world do not have access to legal, safe abortion. Without access to contraceptives, many women who experience unwanted pregnancy resort to abortion—even under illegal and unsafe conditions (see also Chapter 2).

International Data

Over the past 30 years, the global percentage of couples using some form of contraception has increased dramatically—from less than 10 to more than 50 percent.

SOURCE: Mitchell 1998

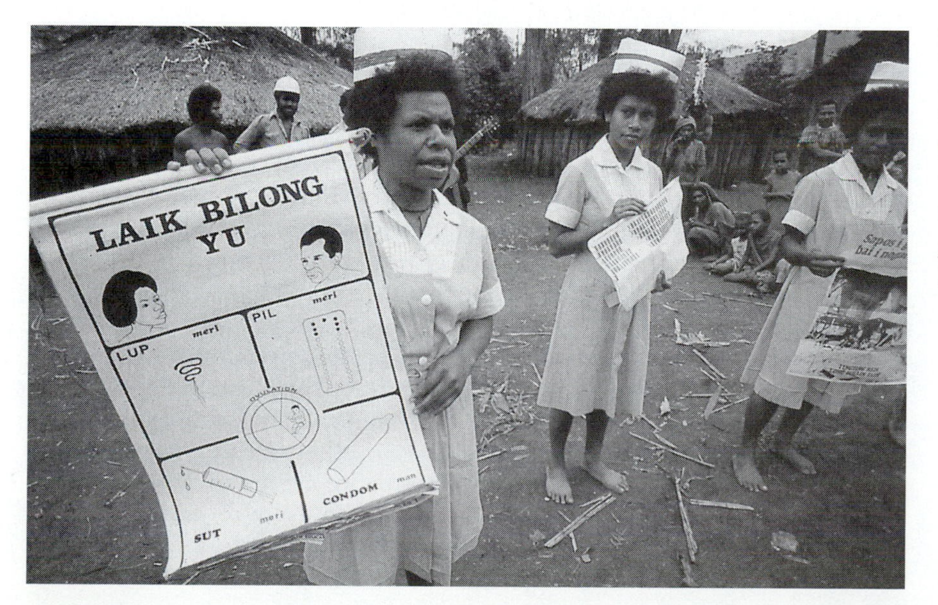

Information about birth control and contraceptives is provided more freely in some countries than in others. Here nurses provide contraceptive information to local people at a family planning clinic in Papua, New Guinea.

> The question that society must answer is this: Shall family limitation be achieved through birth control or abortion? Shall normal, safe, effective contraceptives be employed, or shall we continue to force women to the abnormal, often dangerous surgical operation? Contraceptives or Abortion—which shall it be?
>
> –Margaret Sanger
> Birth control advocate

Contraceptive use has increased in many countries throughout the world. Yet many countries have low rates of contraceptive use. In Ethiopia, less than 5 percent of women use contraception; in Niger and Chad, only 4 percent use contraception. In Nigeria and Pakistan, the figures are 15 percent and 18 percent (Population Institute 1998).

In 1994, delegates from 180 countries met in Cairo, Egypt, for the International Conference on Population and Development and adopted a 20-year Program of Action aimed at reducing population growth. The Cairo conference determined that about $17 billion would be needed annually to meet global family planning programs—and industrialized countries pledged to provide one-third of this amount. However, family planning assistance funds from developed countries have fallen far short of the amount promised. Cuts in U.S. assistance to international family planning programs are largely due to opposition to abortion practices in some countries. According to the United Nations Population Fund, funding cuts in international family planning programs means that "870,000 women will be deprived of effective contraception that would cost only $29 per woman. This will result in 500,000 unwanted pregnancies and 200,000 abortions—the same unfortunate outcome some members of Congress have been professing steadfastly to prevent" (Population Institute 1998, 10).

International Data

The amount of money needed each year to provide reproductive health care for all women in developing countries is $12 billion—the same amount that is spent annually on perfumes in England and the United States.

SOURCE: "Matters of Scale" 1999

CONSIDERATION

Making contraceptives available does not reduce fertility unless women and men want to use birth control methods to prevent childbearing. In many societies, **pronatalism**–a cultural value that promotes having children–contributes to high fertility rates. In traditional Chinese culture, having many children was considered a blessing and a way of honoring one's ancestors. Confucius, the ancient Chinese philosopher, said that the greatest sin is to die without an heir. Throughout history, many religions have worshiped fertility and recognized it as being necessary for the continuation of the human race. In many countries, religions prohibit or discourage birth control, contraceptives, and abortion.

Improve the Status of Women

Throughout the developing world, the status of women is primarily restricted to that of wife and mother. Women in developing countries traditionally have not been encouraged to seek education or employment, but rather to marry early and have children. In some countries, a woman must obtain the consent of her husband before she can receive contraceptive services. This means that women who are never-married, divorced, or widowed, or who cannot persuade their husbands to grant them permission to use contraceptives, are denied family planning services.

> In my opinion, if we are to make genuine progress in economic and social development, if we are to make progress in achieving population goals, women increasingly must have greater freedom of choice in determining their roles in society.
>
> –Jay Rockefeller
> U.S. Senator

Improving the status of women is vital to curbing population growth. Nafis Sadik, executive director of the United Nations Population Fund, argues that population growth cannot be slowed without gender equality and the social empowerment of women to control their lives, especially their reproductive lives. Education plays a key role in improving the status of women and reducing fertility rates. Educated women are more likely to delay their first pregnancies, to use safe and effective contraception, and to limit and space their children (United Nations Population Fund 1998).

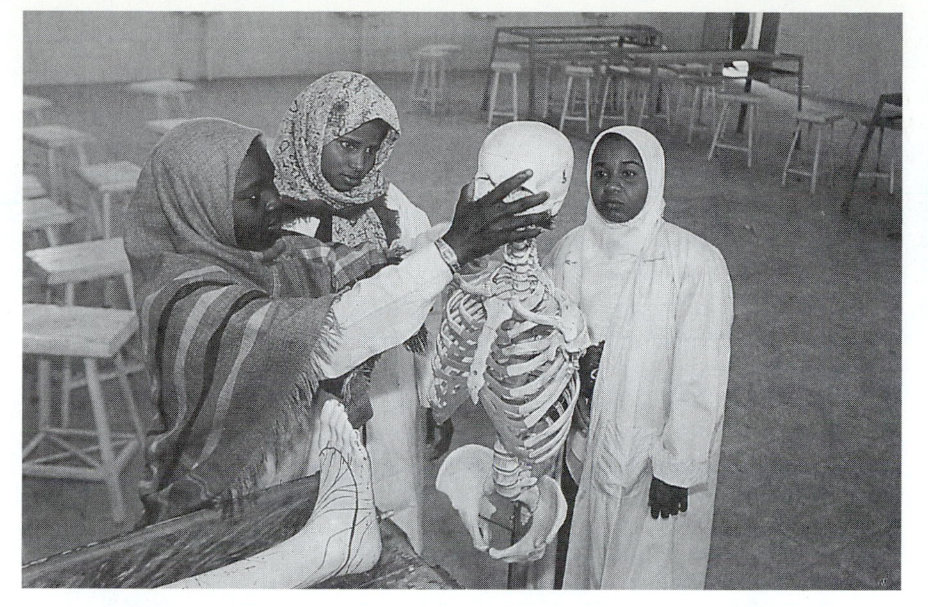

Researchers hypothesize that more education for women results in reduced fertility in developing countries because it delays marriage and reduces desired family size by stimulating aspirations for a higher standard of living.

Increase Economic Development and Health Status

Although fertility reduction may be achieved without industrialization, economic development may play an important role in slowing population growth. Families in poor countries often rely on having many children to provide needed labor and income to support the family. Economic development decreases the economic value of children. Economic development is also associated with more education for women and greater gender equality. As previously noted, women's education and status are related to fertility levels. Finally, economic development tends to result in improved health status of populations. Reductions in infant and child mortality are important for fertility decline, as couples no longer need to have many pregnancies to insure that some children survive into adulthood.

The increase in education of women and girls contributes to greater empowerment of women, to a postponement of the age of marriage and to a reduction in the size of families.

—Program of Action of the 1994 International Conference on Population and Development (Chapter XI)

Impose Government Regulations and Policies

Some countries have imposed strict governmental regulations on reproduction. In the 1970s, India established mass sterilization camps and made the salaries of public servants contingent on their recruiting a quota of citizens who would accept sterilization. The Indian state of Maharashtra enacted compulsory sterilization legislation and in a 6-month period sterilized millions of people, many of them against their will (Boland, Rao, and Zeidenstein 1994). This policy has since been rescinded.

Governmental population regulations in China are less extreme, but still controversial. In 1979, China developed a one-child family policy whereby parents get special privileges for limiting their family size. Specifically, parents who pledge to have only one child receive an allowance for health care for 5 years that effectively increases their annual income by 10 percent. Parents who have only one child also have priority access to hospitals and receive a pension when they reach old age. In addition, their only child is given priority enrollment in nursery school and is exempted from school tuition. Chinese couples who have three or

The greatest single obstacle to the economic and social advancement of the majority of the peoples of the underdeveloped world is rapid population growth.

—Robert McNamara
Former Secretary of Defense

In China, parents who have only one child have priority access to hospitals and receive a pension when they reach old age. Additionally, their only child is given priority enrollment in nursery school and exempted from school tuition.

more children lose these privileges and must pay a special tax that effectively lowers their income for the first 14 years of the child's life (Zhang and Strum 1994).

China also implemented mandatory gynecologic exams. Twice each year health care workers visit each village and, theoretically, check every woman of reproductive age. Under the guise of reproductive health care, each woman is checked for gynecologic problems, has an IUD inserted if she has already had a child, and is subjected to an abortion if she already has a child and her new pregnancy has not been approved by her village. The Chinese government also promotes sterilization (Greenhalgh, Zhu Chujuzhu, and Li Nan 1994).

Critics of China's population policy argue that government policies that are intended to directly influence fertility are coercive and abusive of women's right to choose the number and timing of their children (McIntosh and Finkle 1995). The 1994 International Conference on Population Development advanced the agenda to replace coercive population programs with those that empower women, improve their health, and raise their status in the family and community.

CONSIDERATION

Even if countries around the world achieved replacement-level fertility rates (an average of 2.1 births per woman), populations would continue to grow for several decades because of the large numbers of people now entering their reproductive years. This population momentum is expected to account for up to two-thirds of the projected growth of world population (United Nations Population Fund 1998).

ENVIRONMENTAL PROBLEMS

What good is a house if you don't have a decent planet to put it on?

—Henry D. Thoreau
Author

An expanding population is one of many factors that contribute to environmental problems such as land, water, and air pollution and depletion of natural resources. Each of these environmental problems poses a growing threat to the physical, economic, and social well-being of all people throughout the world.

Air Pollution

Transportation vehicles, fuel combustion, industrial processes (such as the burning of coal and wood), and solid waste disposal have contributed to the growing levels of air pollutants, including carbon monoxide, sulfur dioxide, nitrogen dioxides, and lead. Air pollution levels are highest in areas with both heavy industry and traffic congestion, such as Los Angeles and Mexico City. In China, heavy reliance on coal has resulted in some of the world's most polluted air. In 1996, the National Environmental Protection Agency of China reported 3 million deaths in cities during the preceding 2 years from chronic bronchitis resulting from urban air pollution (Brown 1998a).

National and International Data

In the United States, an estimated 80 million people are exposed to polluted air. Globally, 1.4 billion of the world's 6 billion people are exposed to air that impairs health.

SOURCE: Kemps 1998

Indoor Air Pollution When we hear the phrase "air pollution," we typically think of smokestacks and vehicle exhausts pouring gray streams of chemical matter into the air. But much air pollution is invisible to the eye and exists where we least expect it—in our homes, schools, workplaces, and public buildings.

Sick building syndrome (SBS) is a situation in which occupants of a building experience symptoms that seem to be linked to time spent in a building, but no specific illness or cause can be identified (Lindauer 1999). Building occupants complain of symptoms such as headaches; eye, nose, and throat irritation; a dry cough; dry or itchy skin; dizziness and nausea; difficulty in concentrating; fatigue; and sensitivity to odors. While specific causes of SBS remain unknown, contributing factors may include polluted outdoor air that enters a building through poorly located air intake vents, windows, and other openings. Most indoor air pollution comes from sources inside the building, such as adhesives, upholstery, carpeting, copy machines, manufactured wood products, cleaning agents, and pesticides.

International Data

As many as 1 billion people, mostly women and children in developing countries, are exposed to high levels of indoor air pollution.

SOURCE: World Health Organization 1997

In poor countries fumes from cooking and heating are major contributors to indoor air pollution. Globally, 4 million people die each year from acute respiratory problems caused by breathing fumes from cooking and heating (Kemps 1998).

In developed countries, common household, personal, and commercial products contribute to indoor pollution. For some individuals, exposure to these products can result in a variety of temporary acute symptoms such as drowsiness, disorientation, headache, dizziness, nausea, fatigue, shortness of breath, cramps, diarrhea, and irritation of the eyes, nose, throat, and lungs. Long-term exposure can affect the nervous system, reproductive system, liver, kidneys, heart, and blood. Some solvents found in household, commercial, and workplace products cause cancer; others, birth defects ("The Delicate Balance" 1994). Some of the most common indoor pollutants include carpeting (which emits nearly 100 different chemical gases), mattresses (which may emit formaldehyde and aldehydes), drain cleaners, oven cleaners, spot removers, shoe polish, dry-cleaned clothes, paints, varnishes, furniture polish, potpourri, mothballs, fabric softener, and caulking compounds. Air fresheners, deodorizers, and disinfectants emit the pesticide paradichlorobenzene. Potentially harmful organic solvents are present in numerous office supplies, including glue, correction fluid, printing ink, carbonless paper, and felt-tip markers.

Many consumer products contain fragrances. The air in public and commercial buildings often contains deodorizers, disinfectants, and fragrances that are emitted through the building's heating, ventilation, and air conditioning (HVAC) system. Some fragrance fumes produce various combinations of sensory irritation, pulmonary irritation, decreases in expiratory airflow velocity, and possible neurotoxic effects (Fisher 1998). Fragrance products can cause skin sensitivity, rashes, headache, sneezing, watery eyes, sinus problems, nausea, wheezing, shortness of breath, inability to concentrate, dizziness, sore throat, cough, hyperactivity, fatigue, and drowsiness (DesJardins 1997). More and more businesses are voluntarily limiting fragrances in the workplace or banning them altogether to accommodate employees who experience ill effects from them.

Indoor air pollution is particularly problematic for sufferers of a controversial condition known as "**multiple chemical sensitivity**" (MCS). People suffering from MCS say that after one or more acute or traumatic exposures to a chemical or group of chemicals, they began to experience adverse effects from low levels of chemical exposure that do not produce symptoms in the general public. But even individuals who are not diagnosed as having MCS may experience adverse health effects from indoor air pollution.

International Data

In 1997, the ozone hole over the Antarctic was more than twice the size of Europe. An ozone hole over the Arctic, observed for the first time in 1995, is a new source of concern.

SOURCE: O'Meara 1998a

Destruction of the Ozone Layer The use of human-made chlorofluorocarbons (CFCs), which are used in refrigerators, cleaning computer chips, hospital sterilization, solvents, dry cleaning, and aerosols, has damaged the ozone layer of the

International Data

According to the National Oceanic and Atmospheric Administration (1999), 1998 was the hottest year on record and the twentieth consecutive year that a global mean temperature exceeded the long-term average.

The rising atmospheric concentration of greenhouse gases is potentially the most economically disruptive and costly change that has been set in motion by our modern industrial society.

—Lester R. Brown
Worldwatch Institute

earth's atmosphere. The depletion of the ozone layer allows hazardous levels of ultraviolet rays to reach the earth's surface. Ultraviolet light has been linked to increases in skin cancer and cataracts, declining food crops, rising sea levels, and global warming.

Greenhouse Effect and Global Warming Various gases collect in the atmosphere and act like the glass in a greenhouse, holding heat from the sun close to the earth and preventing it from rising back into space. Greenhouse gases include carbon dioxide, CFCs, and methane. The most prevalent, carbon dioxide, results from energy production, transportation, and deforestation (World Resources Institute 1998b). Evidence for the **greenhouse effect** lies in the increasing temperatures recorded around the world. Heat and drought caused by the greenhouse effect threaten crops and food supplies and accelerate the extinction of numerous plant and animal species.

The greenhouse effect also threatens to melt polar ice caps and other glaciers, resulting in a rise in sea level. Some island countries would become uninhabitable. Low-lying deltas, such as in Egypt and Bangladesh, would be flooded, displacing millions of people. Brown (1995) warns that "rising seas in a warming world would be not only economically costly, but politically disruptive as well" (p. 415).

Acid Rain Air pollutants, such as sulfur dioxide and nitrogen oxide, mix with precipitation to form **acid rain.** Polluted rain, snow, and fog contaminate crops, forests, lakes, and rivers.

Land Pollution

About 30 percent of the world's surface is land, which provides soil to grow the food we eat. Increasingly, humans are polluting the land with toxic and nuclear waste, solid waste, and pesticides.

Toxic and Nuclear Waste The global community generates more than a million tons of hazardous waste each day (Brown 1995).

Decades of nuclear weapons production has polluted over 2.3 million acres of land in at least 24 states; cleanup will extend well into the twenty-first century at a cost of over $200 billion (Black 1999). Radioactive waste from nuclear power plants and major weapons production sites is associated with cancer and genetic defects. Radioactive plutonium, used in both nuclear power and weapons production, has a half-life (the time it takes for its radioactivity to be reduced by half) of 24,000 years (Mead 1998). The disposal of nuclear waste, is particularly problematic. The U.S. Nuclear Regulatory Commission licenses nuclear reactors for 40 years. When nuclear reactors reach the end of their 40-year licenses, the radioactive waste must be stored somewhere. About 28,000 metric tons of nuclear waste are already stored around the United States. This figure is expected to grow to 48,000 metric tons by 2003 and 87,000 metric tons by 2030 (not including waste from military nuclear operations) (Stover 1995). The United States has also promised to accept nuclear waste from about 40 nations with nuclear research reactors.

The cost of storing nuclear waste and cleaning other hazardous waste sites is immense. Cleaning up hazardous waste sites in the United States would require an estimated $750 billion, roughly three-fourths of the U.S. federal budget in 1990 (Brown 1995). But not cleaning these sites is also costly. Brown

We have found the sources of hazardous waste and they are us.

—U.S. EPA booklet, *Everybody's Problem: Hazardous Waste* 1980

(1995) remarks that "one way or another, society will pay—either in clean up bills or in rising health care costs" (p. 416).

Under the Nuclear Waste Policy Act, which was passed in 1982, beginning in 1998, the Department of Energy will handle the storage of nuclear waste from commercial nuclear power plants. Scientists have proposed several options for disposing of nuclear waste, including burying it in rock formations deep below the earth's surface or under Antarctic ice, injecting it into the ocean floor, or hurling it into outer space. Each of these options is risky and costly.

Solid Waste In 1960, each U.S. citizen generated 2.7 pounds of garbage on average every day. By 1996, this figure had increased to 4.3 pounds (*Statistical Abstract of the United States: 1995,* Table 380; *Statistical Abstract of the United States: 1998,* Table 402). In the United States, we discard nearly 160 million tons of solid waste each year, enough to bury 2,700 football fields in a layer 10 stories high (National Solid Wastes Management Association 1991). Some of this waste is converted into energy by incinerators; more than half is taken to landfills. The availability of landfill space is limited, however. Some states have passed laws that limit the amount of solid waste that can be disposed of; instead, they require that bottles and cans be returned for a deposit or that lawn clippings be used in a community composting program.

Pesticides Pesticides are used worldwide in the growing of crops and gardens; outdoor mosquito control; the care of lawns, parks, and golf courses; and indoor pest control.

Pesticides contaminate food, water, and air and can be absorbed through the skin, swallowed, or inhaled. Many common pesticides are considered potential carcinogens and neurotoxins. Even when a pesticide is found to be hazardous and is banned in the United States, other countries may continue to use it. Although DDT is banned in the United States, the pesticide is used in many other countries from which we import food. About one-third of the foods purchased by U.S. consumers have detectable levels of pesticides; from 1 to 3 percent of the foods have pesticide levels above the legal level (Pimentel and Greiner 1997).

Water Pollution

Our water is being polluted by a number of harmful substances, including pesticides, industrial waste, acid rain, and oil spills. In 1993, there were more than 9,000 oil spills in and around U.S. waters alone, totaling more than one and one-half million gallons of spilled oil (*Statistical Abstract of the United States: 1998,* Table 396). However, worldwide, more disease and death are caused by water contaminated from feces, not chemicals (Kemps 1998).

About 1.2 billion people lack access to clean water. In developing nations, as much as 95 percent of untreated sewage is dumped directly into rivers, lakes, and seas that are also used for drinking and bathing (Pimentel et al. 1998). Approximately 35,000 river miles in the United States are currently polluted by runoff from livestock operations ("New Rules for Feedlots" 1998).

Depletion of Natural Resources

Overpopulation contributes to the depletion of natural resources, such as coal, oil, and forests. The demand for new land, fuel, and raw materials has resulted

National Data

In 1997 there were 1,231 hazardous waste sites in the United States. States with the highest number of such waste sites include New Jersey (110), Pennsylvania (99), California (94), New York (80), Michigan (74), and Florida (54).

SOURCE: *Statistical Abstract of the United States: 1998,* Table 407

International Data

Worldwide, over 2.5 million tons of pesticides are applied each year. An estimated 3 million severe human pesticide poisonings occur in the world each year, of which about 220,000 are fatal.

SOURCE: Pimentel and Greiner 1997

National Data

In the United States, about 500,000 tons of 600 different types of pesticides are used annually.

SOURCE: Pimentel and Greiner 1997

> The faster populations grow, the more environmental problems are aggravated.
>
> –Werner Fornos
> President, Population Institute

> Do not live with a vocation that is harmful to humans and nature. Do not invest in companies that deprive others of their chance to live.
>
> –Thich Nhat Hanh
> Buddhist monk and teacher

International Data

Over the last century, the world has lost nearly half of its original forest area.

SOURCE: Brown 1998a

in **deforestation**—the destruction of the earth's rain forests. The major cause of deforestation is commercial logging.

Trees naturally absorb carbon dioxide—one of the main gases that contribute to the "greenhouse effect" and global warming—from the air. Deforestation is responsible for about 30 percent of the atmospheric buildup of carbon dioxide (World Resources Institute 1998b). Deforestation also displaces people and wild species from their lands. About half of the world's approximately 14 million species live in tropical forests (World Resources Institute 1998b). Once the trees are gone, these species disappear forever. Finally, soil erosion caused by deforestation may lead to flooding.

A related concern is **desertification**, which occurs when semiarid land on the margins of deserts is overused. Overgrazing by cattle and other herd animals and the cutting of brush and trees for firewood leave the land denuded of vegetation and allow the desert to expand. By the year 2000 the world will have lost 25 million acres of cultivated land.

Environmental Injustice

Although environmental pollution and degradation and depletion of natural resources affect us all, some groups are more affected than others. **Environmental injustice**, also referred to as **environmental racism**, refers to the tendency for socially and politically marginalized groups to bear the brunt of environmental ills. In the United States, polluting industries, industrial and waste facilities, and transportation arteries (that generate vehicle emissions pollution) are often located in minority communities (Bullard 1996; Bullard and Johnson 1997). U.S. communities that are predominantly African American, Hispanic, or Native American are disproportionately affected by industrial toxins, contaminated air and drinking water, and the location of hazardous waste treatment and storage facilities (see this chapter's *Focus on Social Problems Research* feature). One study found, for example, that hog industries—and the associated environmental and health risks associated with hog waste—in eastern North Carolina tend to be

This sugar cane factory in central Mexico has been closed twice but has reopened again and continues to emit toxic fumes so that some local residents wear gauze over their noses.

FOCUS ON SOCIAL PROBLEMS RESEARCH

Environmental Injustice

Stretsky and Hogan's empirical investigation of environmental injustice (1998) addresses three research questions. First, do environmental injustices, that is, the tendency for hazardous waste sites to be located in minority neighborhoods, exist? Second, if they exist, can such injustices be explained by race and ethnicity or, alternatively, are they a function of socioeconomic variables such as income, poverty, and unemployment? Finally, assuming that environmental injustices against racial and ethnic minorities exist, do these injustices result from discrimination against minorities or from other social factors?

Methods and Sample

The researchers identified census tract locations of 53 Florida hazardous waste sites listed on the Environmental Protection Agency's Superfund list. Independent variables thought to be related to hazardous waste site locations were then selected: (1) the percent of blacks and Hispanics living in a census tract, (2) the median household income of a census tract, (3) the percent of individuals living below poverty in a census tract, and (4) the percent of unemployed persons in a census tract.

Findings

First, and most importantly, the researchers found that "Blacks and Hispanics are more likely than Whites to live near Superfund sites in Florida" (p. 284). This relationship holds true regardless of income level, unemployment rate, and level of urbanization. Somewhat surprisingly, Superfund sites in Florida are equally likely to be located in urban as rural areas. Not surprisingly, they are more likely to be located in areas with lower-than-average property values.

Not only are race and ethnicity the single best predictors of the dependent variable, location of a Superfund site, examining the variables over time permitted the researchers to assess whether the detected environmental injustices were abating or getting worse over time. As Stretsky and Hogan (1998, 284) note:

> . . . we find evidence that suggests that this form of injustice may be intensifying. For example, in Florida census tracts containing Superfund sites the percentage of Blacks and Hispanics has increased between 1970 and 1990. In addition, race and ethnicity are much stronger predictors of the location of Superfund sites in 1990 [than in 1970].

The authors have thus addressed two of their three research questions. Environmental injustices exist and, at least in Florida, are related more to race and ethnicity rather than to socioeconomic variables. Are such injustices a consequence of overt prejudice or structural forces over which no individual has control? The authors conclude that environmental injustices, although there are variations over time, are unlikely to be the result of bigoted individuals but rather reflect housing patterns.

For example, in 1970, Superfund census tracts in Florida were 14.7 percent black; by 1990, they were 19.7 percent black—a statistically significant increase. Over time, black residents are more rather than less likely to live in census tracts containing hazardous waste sites. The authors suggest that this *indirect discrimination* may be a consequence of plummeting housing costs. Property values in census tracts with hazardous waste sites, understandably, decreased over the years, making housing in those tracts more affordable for the poor who are disproportionately racial and ethnic minorities. Thus, there is evidence that rather than hazardous waste sites being located in minority neighborhoods, minorities are located in neighborhoods with hazardous waste sites.

Unfortunately, the results of this study do little to alter the environmental injustices that exist. If direct discrimination had been detected, prejudiced individuals or biased public policies could have been addressed. However, Stretsky and Hogan's research revealed that environmental injustice is part of the larger structural injustices that exist in society, including income inequality and segregated housing patterns. As the authors conclude, "[I]n the end the EPA and state environmental agencies may need to look beyond the direct forms of discrimination as the cause of environmental injustice and focus their policies on larger social justice issues" (pp. 168, 284).

SOURCE: Paul Stretsky and Michael Hogan. 1998. "Environmental Justice: An Analysis of Superfund Sites in Florida." *Social Problems* 45:268–87.

located in communities with high black populations, low voter registration, and low incomes (Edwards and Ladd 1998). Not only are minority communities more likely to be polluted than others, but environmental regulations are also less likely to be enforced in these areas (Stephens 1998).

Environmental injustice affects marginalized populations around the world, including minority groups, indigenous peoples, and other vulnerable and impoverished communities such as peasants and nomadic tribes (Renner 1996). These groups are often powerless to fight against government and corporate powers that sustain environmentally damaging industries. For example, in the early 1970s, a huge copper mine began operations on the South Pacific island of Bougainville. While profits of the copper mine benefitted the central government and foreign investors, the lives of the island's inhabitants were being destroyed. Farming and traditional hunting and gathering suffered as mine pollutants covered vast areas of land, destroying local crops of cocoa and bananas, contaminating rivers and their fish. Indigenous groups in Nigeria, such as the Urhobo, Isoko, Kalabare, and Ogoni, are facing environmental threats caused by oil production operations run by multinational corporations. Oil spills, natural gas fires, and leaks from toxic waste pits have polluted the soil, water, and air and compromised the health of various local tribes. "Formerly lush agricultural land is now covered by oil slicks, and much vegetation and wildlife has been destroyed. Many Ogoni suffer from respiratory diseases and cancer, and birth defects are frequent" (Renner 1996, 57). The environmental injustices experienced by the Bougainville and Ogoni reflect only the tip of the iceberg. Renner (1996) warns that "minority populations and indigenous peoples around the globe are facing massive degradation of their environments that threatens to irreversibly alter, indeed destroy, their ways of life and cultures" (p. 59).

Environmental Illness

International Data

According to the World Health Organization (1997), poor environmental quality is directly responsible for about 25 percent of all preventable ill health in the world today.

Exposure to pollution, toxic substances, and other environmental hazards is associated with numerous illnesses and health problems. The smoke of cooking fires fueled by wood, coal, and other organic material contains harmful particulate matter as well as carcinogenic chemicals (such as benzene and formaldehyde). According to one World Bank study, fuelwood cooking smoke causes the death of approximately 4 million children worldwide each year (Pimentel et al. 1998).

By one estimate, every 1 percent decrease in the ozone layer increases ultraviolet B (UVB) radiation by 1.4 percent, and the incidence of skin cancer is increasing accordingly. The number of new cases of skin cancer in the United States has quadrupled over the past 20 years, resulting in nearly 10,000 deaths per year (Pimentel et al. 1998). Destruction of the ozone layer also results in increased exposure of the eyes to radiation in sunlight, which increases the risk of cataracts (Bergman 1998).

International Data

Environmental factors, including various chemicals, radiation, and tobacco smoke, are estimated to cause roughly 80 percent of all cancers.

SOURCE: Murray and Lopez 1996

Environmental illness also results from exposure to toxic chemicals, such as benzene, lead, pesticides, and cyanides. Of the approximately 80,000 chemicals in use today, 10 percent are recognized as carcinogens (cancer causing) (Pimentel et al. 1998). Between 1989 and 1996, the number of pesticide poisonings reported in the United States nearly doubled.

Persistent organic pollutants (POPs) accumulate in the food chain, and persist in the environment, taking centuries to degrade. While the effects of POPs on human health are unclear, many researchers believe that long-term exposure contributes to increasing rates of birth defects, fertility problems, greater susceptibility to disease, diminished intelligence, and certain cancers (Fisher 1999). The United Nations Environment Programme has identified 12 POPs that require urgent regulatory attention, 9 of which are pesticides.

International Data

Worldwide, about 1 million deaths and chronic illnesses each year are due to pesticide poisoning.

SOURCE: Pimentel et al. 1998

Threats to Biodiversity

Biodiversity refers to the great variety of life forms on Earth. In recent decades, we have witnessed mass extinction rates of diverse life forms. Most estimates

Table 15.1 Threatened Species Worldwide

Species	Total Share of Species Threatened with Extinction
Birds	11%
Mammals	25%
Reptiles	20%
Amphibians	25%
Fish	34%

SOURCE: Lester R. Brown, Gary Gardner, and Brian Halweil. 1998. *Beyond Malthus: Sixteen Dimensions of the Population Problem*. World Watch Paper 143. Washington, D.C.: World Watch Institute. Used by permission.

suggest that at least 1,000 species of life are lost per year (Tuxill 1998). Unlike the extinction of the dinosaurs millions of years ago, humans are the primary cause of disappearing species today. Air, water, and land pollution; deforestation; disruption of native habitats, and overexploitation of species for their meat, hides, horns, or medicinal or entertainment value threaten biodiversity and the delicate balance of nature. Table 15.1 lists the percentages of various species that are currently classified as threatened.

The loss of biodiversity affects everyone. The diverse forms of life on Earth provide humanity with food, fibers, and many other products and "natural services." For example, about 25 percent of drugs prescribed in the United States include chemical compounds derived from wild species (Tuxill 1998). Insects, birds, and bats provide pollination services that enable us to feed ourselves. Frogs, fish, and birds provide natural pest control. Various aquatic organisms filter and cleanse our water and plants and microorganisms enrich and renew our soil.

> We are triggering the greatest extinction of plant and animal species since the dinosaurs disappeared. As our numbers go up, their numbers go down.
>
> —Lester R. Brown, Gary Gardner, and Brian Halweil
> World Watch Institute

SOCIAL CAUSES OF ENVIRONMENTAL PROBLEMS

In addition to population growth, various other structural and cultural factors have also played a role in environmental problems.

Industrialization and Economic Development

Many of the environmental problems confronting the world have been caused by industrialization and economic development. Industrialized countries, for example, consume more energy and natural resources than developing countries. Environmental problems of developed countries are due to industrialization, overconsumption of natural resources, and the demand for increasing quantities of goods and services. Industrialization and economic development have been the primary cause of global environmental problems such as the ozone hole and global warming (Koenig 1995). Conflict theorists argue that governments pursue economic development and industrialization at the expense of environmental conditions. In less developed countries, environmental problems are largely due to poverty and the priority of economic survival over environmental concerns. Vajpeyi (1995) explains:

> Modern industrial civilization, as presently organized, is colliding violently with our planet's ecological system. . . . We must make the rescue of the environment the central organizing principle for civilization.
>
> —Al Gore
> U.S. Vice President

The earth has enough for everyone's need but not enough for everyone's greed.

–Mahatma Gandhi
Peace activist

Policymakers in the Third World are often in conflict with the ever-increasing demands to satisfy basic human needs—clean air, water, adequate food, shelter, education—and to safeguard the environmental quality. Given the scarce economic and technical resources at their disposal, most of these policymakers have ignored long-range environmental concerns and opted for short-range economic and political gains. (p. 24)

Cultural Values and Attitudes

Cultural values and attitudes that contribute to environmental problems include individualism, capitalism, and materialism. Individualism, which is a characteristic of U.S. culture, puts individual interests over collective welfare. When a Gallup poll asked U.S. adults if they would be willing to have the United States take steps to reduce global warming if costs for gasoline or electricity went up a great deal, nearly half (48 percent) said no (Gallup and Saad 1997). Even though recycling is good for our collective environment, many individuals do not recycle because of the personal inconvenience involved in washing and sorting recyclable items. Similarly, individuals often indulge in countless behaviors that provide enjoyment and convenience at the expense of the environment: long showers, use of dishwashing machines, recreational boating, meat eating, and use of air conditioning, to name just a few.

Finally, the influence of materialism, or the emphasis on worldly possessions, also encourages individuals to continually purchase new items and throw away old ones. The media bombard us daily with advertisements that tell us life will be better if we purchase a particular product. Materialism contributes to pollution and environmental degradation by supporting industry and contributing to garbage and waste.

The cultural value of militarism also contributes to environmental degradation. This issue is discussed in detail in Chapter 16.

STRATEGIES FOR ACTION: RESPONDING TO ENVIRONMENTAL PROBLEMS

Solving environmental problems is difficult and costly. Lowering fertility, as discussed earlier, helps reduce population pressure on the environment. Environmentalist groups, modifications in consumer behavior, government regulations, and the development of alternative sources of energy can also alleviate environmental problems.

Thanks to an activist campaign which NRDC helped spearhead, airborne lead has been reduced 95% from the uncontrolled levels of the 1970s.

–Eric A. Goldstein
NRDC Attorney

Environmentalist Groups

Environmentalist groups exert pressure on the government to initiate or intensify actions related to environmental protection. For example, lawsuits filed in the 1970s by the Natural Resources Defense Council (NRDC) resulted in the phase-out of leaded gasoline. Lobbying by the same organization led to a 1993 White House order directing that the federal government use recycled paper. Environmentalist groups also design and implement their own projects, and disseminate information to the public about environmental issues.

In the United States, environmentalist groups date back to 1892 with the establishment of the Sierra Club, followed by the Audubon Society in 1905. Other environmental groups include the National Wildlife Federation, World Wildlife Fund, Environmental Defense Fund, Friends of the Earth, Union of Concerned Scientists, Greenpeace, Environmental Action, Natural Resources Defense Council, World Watch Institute, National Recycling Coalition, World Resources Institute, Rainforest Alliance, Global Climate Coalition, and Mothers & Others, to name a few. A recent development in the environmental movement is the emergence of **ecofeminism**—a synthesis of feminism, environmentalism, and antimilitarism.

In the mid-1980s, membership in the 10 largest environmental organizations was about 4 million. That figure doubled by 1990, when it began a gradual decline (Switzer 1997).

Modifications in Consumer Behavior

Public opinion survey data from the 1990s show that the majority of Americans are deeply concerned about the environment and identify themselves as environmentally supportive (Switzer 1997). Increasingly, consumers are making choices in their behavior and purchases that reflect environmental awareness. For example, between 1988 and 1997, global sales of energy-efficient compact fluorescent bulbs increased eightfold (O'Meara 1998b). The estimated 980 million compact fluorescent bulbs in use today save the equivalent of roughly 100 coal-fired power plants. Recycling has also increased. In the 1970s and early 1980s, between 22 and 27 percent of wastepaper was recycled, increasing to 45 percent by 1995 (Mattoon 1998).

> That's not the way to save the planet!
>
> —Four-year-old Benjamin Lewis, who yelled at a man who threw a soda can out of his car window

Sales of organic foods (that don't use pesticides) have increased, as have other products and services that are environmentally friendly. From credit card and long-distance telephone companies that donate a percentage of their profits to environmental causes, to socially responsible investment services, consumers are increasingly voting with their dollars to support environmentally responsible practices. This chapter's *The Human Side* feature offers examples of how consumers can resist materialistic urges and reduce waste and overconsumption.

> In their gas-guzzling lives full of electric gadgetry and plastic and chemical waste dumping, [many citizens] can't imagine that the freedom of all to breathe air everywhere is threatened by their pollutants. . . . Can they come to realize that so-called primitive tribal ways are more intelligently in tune with Earth, in terms of saving Earth's resources, than those of obscenely exploitative nation-states which call themselves "civilized"?
>
> —Daniela Gioseffi
> Poet, novelist, editor, journalist, performer, and activist

Alternative Sources of Energy

Generating energy from coal and oil contributes to the world's growing level of pollution. In addition, the earth's supplies of coal and oil are finite and are being depleted. Nuclear power produces harmful radioactive wastes and involves potential harm caused by accidents, such as the 1986 explosion at the Chernobyl nuclear reactor in the Soviet Ukraine. One solution to the problems associated with the use of coal and oil for energy is to use alternative sources of energy, such as solar power, geothermal power, wind power, ocean thermal power, nuclear power, tidal power, and power from converting corn, wood, or garbage.

The fastest developing alternative source of energy is wind power (see Table 15.2). While new energy technologies are promising, the World Resources Institute (1998a) suggests that "technology alone is not a panacea, and changes in consumer behavior would be necessary to achieve rapid rates of improvement in the energy intensity of the global economy" (p. 15).

THE HUMAN SIDE

"ESCAPE FROM AFFLUENZA"

It's not that consumption is bad—we have to consume to live. It's that we've gone way, way overboard, at enormous cost to ourselves and the earth.

–Donella H. Meadows
Adjunct professor of environmental studies at Dartmouth

A 1997 PBS television program, "Affluenza," "explores the high social and environmental costs of materialism and overconsumption." In a 1998 sequel, "Escape from Affluenza," people working to reduce waste and overconsumption and live their lives in balance with the environment are profiled. Below is a partial list, by category, of some of the ways **you** can "Escape from Affluenza" (KCTS 1998).

Shopping

- Biggest shopping trap–it was on sale. If it wasn't something you identified as a need, you didn't SAVE money, you SPENT money. Learn this mantra: It's not a bargain if I don't need it.
- Mail-order shopping is a great way to save time, fuel and money. But avoid the companies that make up the cost of glossy catalogs by charging twice as much. About 94% of these catalogs go unused into the waste stream: don't be one of the 6% who pay more to make this practice profitable . . . !
- The average American family of four spends an estimated 10% of its annual income on clothing. Using principles you already know (only buying what you need, buying secondhand items that are as good as new, for instance), you can easily cut your clothing expenditures in half . . .

Waste and Clutter

- A poorly insulated house can easily waste 30% to 50% of the energy poured into it . . . Simple changes around the house: run the dishwasher half as often as you do now, and save 50% in water, energy and time. Wash only full loads of clothing, and since 90% of the cost of washing clothes is to heat the water, avoid using hot water . . .
- Half of the average family's household energy goes to heating and/or cooling, at a cost of about $450 a year. Put a 15% dent in this expense by keeping the thermostat set at 65 F in the daytime, and 55 or 60 at night.
- Planting a large tree to shade your home can save you an estimated $73 a year in air-conditioning bills. If every household did this, the country would save more than $4 billion in energy costs . . .
- An estimated 50% of the waste stream in the U.S. is discarded packaging. Be very aware of packaging excess, and when all else is equal, choose the least-packaged. Help raise awareness of this issue by being a packaging activist: ask your grocers to phase out pre-wrapped fruits and vegetables.

Home and Hearth

- When home-hunting, pick the smallest amount of space in which you are comfortable. This will limit the amount of stuff you can accumulate, and take far less of your time and resources to furnish, clean, maintain, insure and pay for.
- Eighty percent of the dirt in your home is brought in on shoes. Save time and cleaning expenses by starting a no-shoes policy. . . .
- Start a neighborhood swap of seldom-used tools. Why should a street of 10 houses have 10 lawnmowers, 10 paint-sprayers, and 10 band saws (or 55 Disney videos, for that matter)? . . .

THE HUMAN SIDE

"ESCAPE FROM AFFLUENZA" *(continued)*

Food

- Eat a local diet as much as possible. This creates jobs in the region, reduces transportation costs and energy consumption, ensures higher nutritional value and encourages local small-scale agriculture that protects land from development. Ask your grocers to mark foods local. Hint—If it's out of season (e.g., strawberries in December in Seattle), you know it's not local.
- To dramatically reduce your ecological footprint, save about 50% in food costs and maximize your prospects for a longer, healthier life, become a vegetarian . . .

Health and Happiness

- Prevention is the cheapest and best health insurance. Get an annual checkup (even frugal fanatics agree this is worth the $$). Take care of your body by eating and sleeping right, and stop smoking.
- Stop equating the amount of fun and pleasure you get with the amount of money you spend to get it. Sit down and make a list of 25 things you like to do that cost little or no money, and keep it where you can see it every day.
- Instead of spending your time and energy buying and taking care of stuff, volunteer to feed the hungry, care for the suffering, visit the lonely, or mentor a child.
- Be happy with what you have. If you make a habit of thinking in terms of what you have, rather than what you don't, you may well find that you've got enough.

SOURCE: From "100 Ways to Escape from Affluenza" as appears on **www.pbs.org/kcts/affluenza/escape** by Vivia Boe. Reprinted by permission. Films are available from BULLFROG FILMS, 800-543-FROG or **www.bullfrogfilms.com**.

Government Regulations and Funding

Through regulations and the provision of funds, governments play a vital role in protecting and restoring the environment. Governments have responded to concerns related to air and water pollution, ozone depletion, and global warming by imposing regulations affecting the production of pollutants. Governments have also been involved in the preservation of deserts, forests, reefs, wetlands, and endangered plant and animal species.

For example, to reduce air pollution caused by automobiles, the U.S. government requires cars to have catalytic converters. In 1990, Congress amended the Clean Air Act to provide marketable pollution rights for sulfur dioxide emissions by electric companies. The Environmental Protection Agency issued more than 5 million marketable permits, each allowing the emission of 1 ton of sulfur dioxide annually, to 110 companies, which must have permits for all of their sulfur dioxide emissions or face large fines. Companies that want to emit more sulfur dioxide can buy permits from other companies, while those that reduce their own pollution can profit from selling their permits (Holcombe 1995).

City governments also play a role in environmental issues. In a landmark city ordinance, San Francisco banned the use of pesticides by all city departments, with the exception of schools and public housing. The city has mandated

. . . as long as 70 million Americans live in communities where the air is dangerous to breathe; as long as half our rivers, our lakes and our streams are too polluted for fishing and swimming; as long as people in our poorest communities face terrible hazards from lead paint to toxic waste dumps . . . our journey is far from finished.

–Bill Clinton
U.S. President

National Data

A study by the U.S. Department of Energy concluded that North Dakota, South Dakota, and Texas had enough harnessable wind energy to meet all U.S. electricity needs.

SOURCE: Brown 1998b

Table 15.2 **Trends in Energy Use, by Source, 1990–97**

Energy Source	Annual Rate of Growth
Wind power	25.7
Solar power	16.8
Geothermal power	3.0
Natural gas	2.1
Hydroelectric power	1.6
Oil	1.4
Coal	1.2
Nuclear power	0.6

SOURCE: Lester R. Brown. 1998. "Overview: New Records, New Stresses." From *Vital Signs 1993: The Trends That Are Shaping Our Future* by Lester R. Brown, Hal Kane, and Ed Ayres. Copyright © 1993 by the Worldwatch Institute. Reprinted by permission of W. W. Norton & Company, Inc.

that integrated pest management methods, which utilize least-toxic alternatives, be used in public parks, golf courses, and office buildings ("San Francisco Bans Pesticides" 1997).

Some environmentalists advocate using taxes to discourage polluting practices. Governments now rely on personal and corporate income taxes for revenue, while taxes on environmentally destructive activities are either negligible or nonexistent. If government heavily taxed destructive activities such as carbon emissions, sulfur emission, the generation of toxic waste, the use of pesticides, and the use of virgin raw material, these activities would decrease (Brown and Mitchell 1998).

International Data

Since 1991, six national governments in western Europe have made small cuts in income taxes and raised taxes on environmentally destructive practices.

SOURCE: Roodman 1998

CONSIDERATION

Brown and Mitchell note that "not only are we not taxing many environmentally destructive activities, some of these efforts are actually being subsidized" (p. 182). For example, more than $600 billion a year of taxpayers' money is spent to subsidize deforestation, overfishing, the burning of fossil fuels, the use of virgin raw materials, and other environmentally destructive activities. The Earth Council remarked, "There's something unbelievable about the world spending hundreds of billions of dollars annually to subsidize its own destruction." (Brown and Mitchell 1998, 182)

International Cooperation

Global environmental concerns such as global warming and climate change require international governmental cooperation. In 1997, delegates from 160 nations met in Kyoto, Japan, and forged the Kyoto Protocol—the first international agreement to place legally binding limits on greenhouse gas emissions from developed countries.

The 1972 Stockholm Conference on the Human Environment, the United Nations Environment Program (UNEP) since the 1970s, and the 1983 World Commission on Environment and Development represent efforts to address environmental concerns at the international level. The 1992 Earth Summit in Rio de Janeiro—a 12-day event—brought together heads of states, delegates from more than 170 nations, nongovernmental organizations, representatives

> Protection of the environment is a goal that virtually everyone believes is worthwhile and would enhance the quality of life. The disagreement comes in deciding how to implement policies to achieve the goal.
>
> —Randall G. Holcombe
> Florida State University

of indigenous people, and the media to discuss an international agenda for both economic development and the environment. International cooperative efforts have resulted in, for example, the Global Environmental Facility (GEF), located at the World Bank in Washington, and the United Nations' International Environmental Technology Center in Osaska and Shiga, Japan. The United Nations Development Program supports countries in environmental management. The 1992 Earth Summit in Rio resulted in the Rio Declaration—"a nonbinding statement of broad principles to guide environmental policy, vaguely committing its signatories not to damage the environment of other nations by activities within their borders and to acknowledge environmental protection as an integral part of development" (Koenig 1995, 15).

Another example of international cooperation on environmental issues is the agreement made by 70 nations to curb the production of CFCs, which contribute to ozone depletion and global warming. The largest consumers of CFCs have established a fund to help developing countries acquire alternatives to CFCs (Koenig 1995).

Because industrialized countries have more economic and technological resources, they bear primary responsibility for leading the nations of the world toward environmental cooperation. Jan (1995) emphasizes the importance of international environmental cooperation and the role of developed countries in this endeavor:

> **Advanced countries must be willing to sacrifice their own economic well-being to help improve the environment of the poor, developing countries. Failing to do this will lead to irreparable damage to our global environment. Environmental protection is no longer the affair of any one country. It has become an urgent global issue. Environmental pollution recognizes no national boundaries. No country, especially a poor country, can solve this problem alone. (pp. 82–83)**

Sustainable Economic Development: An International Agenda

Achieving global cooperation on environmental issues is difficult, in part, because developed countries (primarily in the Northern Hemisphere) have different economic agendas from developing countries (primarily in the Southern Hemisphere). The northern agenda emphasizes preserving wealth, affluent lifestyles, and the welfare state while the southern agenda focuses on overcoming mass poverty and achieving a higher quality of life (Koenig 1995). Southern countries are concerned that northern industrialized countries—having already achieved economic wealth—will impose international environmental policies that restrict the economic growth of developing countries just as they are beginning to industrialize. Global strategies to preserve the environment must address both wasteful lifestyles in some nations and the need to overcome overpopulation and widespread poverty in others.

Development involves more than economic growth, it involves sustainability—the long-term environmental, social, and economic health of societies. **Sustainable development** involves meeting the needs of the present world without endangering the ability of future generations to meet their own needs. Achieving sustainable development has become a primary goal of governments throughout the world.

> It will not be possible for the community of nations to achieve any of its major goals—not peace, not environmental protection, not human rights or democratization, not fertility reduction, not social integration—except in the context of sustainable development.
>
> *–1994 Human Development Report*

> The world today is demographically divided. There is an aging world, mainly comprised of industrialized nations, where population growth has slowed down, stabalized or become negative. There is another world—a less developed, overcrowded world, in Africa, Asia and Latin America, where people are struggling to balance their resources and environment with their runaway population growth. People of these countries need the knowledge and resources to acquire family planning services to reduce their family sizes and ensure better lives for themselves.
>
> –Population Institute

UNDERSTANDING POPULATION AND ENVIRONMENTAL PROBLEMS

Because population increases exponentially, the size of the world's population has grown and will continue to grow at a staggering rate. Given the problems associated with population growth—deteriorating socioeconomic conditions, depletion of natural resources, and urban crowding—most governments recognize the value of controlling population size. Efforts to control population must go beyond providing safe, effective, and affordable methods of birth control. Slowing population growth necessitates interventions that change the cultural and structural bases for high fertility rates. These interventions include increasing economic development and improving the status of women, which includes raising their levels of education, their economic position, and their (and their children's) health.

Many countries have reported declining fertility rates in recent years: Bangladesh, from 6.2 children per woman to 3.4; India, from 4.5 to 3.4; Pakistan from 6.5 to 5.5; Turkey, from 4.1 to 2.7; Syria from 7.4 to 4.7, and Kenya, from 7.5 to 5.4 ("Fertility Declines Reported" 1997). Such accomplishments have helped to slow population growth. Unfortunately, high mortality rates due to HIV/AIDS and war have also contributed to the slowing of population growth. Although birthrates have decreased in most countries throughout the world, population continues to grow; because of past high fertility rates, more young women than ever are currently entering their childbearing years. In addition, low contraception use and high fertility rates persist in many less developed countries. These factors combined—population momentum and continued high fertility rates in some countries—contribute to escalating world population. As noted earlier, 65 countries are expected to double their populations in 30 years or less, and 14 countries will triple or nearly triple their populations by 2050. "The impact of population growth in these fast-growing countries will more than offset the gains of low fertility in 80 other countries" (Population Institute 1998, 3).

Rapid and dramatic population growth, along with expanding world industrialization and patterns of consumption, has contributed to environmental problems. The greater numbers of people make increased demands on natural resources and generate excessive levels of pollutants. Environmental problems are due not only to large populations, but also to the ways in which these populations live and work. As conflict theorists argue, individuals, private enterprises, and governments have tended to choose economic and political interests over environmental concerns.

Many Americans believe in a "technological fix" for the environment—that science and technology will solve environmental problems. Paradoxically, the same environmental problems that have been caused by technological progress may be solved by technological innovations designed to clean up pollution and preserve natural resources. While technological innovation is important in resolving environmental concerns, other social changes are needed as well. Addressing the values that guide choices, the economic contexts in which the choices are made, and the governmental policies that encourage various choices is critical to resolving environmental and population problems.

> Man is here for only a limited time, and he borrows the natural resources of water, land and air from his children who carry on his cultural heritage to the end of time. . . . One must hand over the stewardship of his natural resources to the future generations in the same condition, if not as close to the one that existed when his generation was entrusted to be the caretaker.
>
> –Delano Saluskin
> Yakima Indian Nation

Global cooperation is also vital to resolving environmental concerns, but is difficult to achieve because rich and poor countries have different economic development agendas: developing poor countries struggle to survive and provide for the basic needs of their citizens; developed wealthy countries struggle to maintain their wealth and relatively high standard of living. Can both agendas be achieved without further pollution and destruction of the environment? Is sustainable economic development an attainable goal? The answer must be yes.

CRITICAL THINKING

1. One writer observed that "on a certain November Day an obscure woman in Iowa gives birth to seven babies; we marvel and rejoice. On the same day an obscure woman in Nigeria gives birth to her seventh child in a row; we are distressed and appalled" (Zwingle 1998, 38). Why might reactions to these two events be so different?
2. In many developing countries that have strict laws against abortion, the use of child labor is common. Do you think there may a connection between laws prohibiting abortion and child labor in developing countries? How could you find out if there is a connection?
3. Do you think governments should regulate reproductive behavior to control population growth? If so, how?
4. Babies eat more, drink more, and breathe more, proportionally, than adults—which means babies are more susceptible to environmental toxins than are adults (Dionis 1999). Yet, federal environmental standards are largely set at levels designed to protect adults. What social forces do you think discourage the government from setting environmental standards based on what is safe for infants and young children?

KEY TERMS

acid rain
biodiversity
deforestation
demographic transition theory
demography
desertification
doubling time
ecofeminism
environmental injustice
environmental racism
fertility rate
greenhouse effect
greenwashing
human ecology
Malthusian theory
multiple chemical sensitivity
planned obsolescence
pronatalism
replacement level
sick building syndrome
sustainable development

INTERNET

You can find more information on population growth, the Environmental Protection Agency, indoor air pollution, multiple chemical sensitivity, and environmental groups at the *Understanding Social Problems* web site at **http://sociology.wadsworth.com.**

INFOTRAC COLLEGE EDITION

Either from the Wadsworth sociology resource center at **http://sociology.wadsworth.com** or directly from your web browser, you may access InfoTrac College Edition, an on-line university library that includes over 700 popular and scholarly journals in which you can find articles related to the topics in this chapter. Suggested articles and questions relating to these articles are listed below.

Daily, Gretchen, and Partha Dasgupta. 1998. "Food Production, Population Growth and the Environment." *Science* 281(5381):1291–93.

1. When is the next "world population boom" predicted, and where is it predicted to occur most drastically?
2. What four obstacles seem to be key in preventing current food production trends from keeping pace with rapid population growth?
3. What institutional reforms are needed to help prepare for the "world population boom"?

Mitchell, Jennifer D. 1998. "Before the Next Doubling." *World Watch* 11(1):21–29.

1. What are some of the obstacles to universal family planning?
2. What was the problem with contraception in Bangladesh, and what did the government do to avoid this problem?
3. What methods does the author suggest could be useful in publicizing the need for family planning and contraception in nondeveloped or developing countries?

Pimentel, David, et al. 1998. "Ecology of Increasing Disease." *Bioscience* 48(10):817–27.

1. What public health issues are associated with crowded urban environments?
2. What links do the authors draw between global warming and the spread of infectious diseases?
3. What links do the authors draw between infectious disease and the need to clear land to house a growing population?

CHAPTER SIXTEEN

Conflict around the World

Is It True?

Is It True?

1. Military spending has declined in recent years, averaging a 4.5 percent decrease over the last 10 years.
2. The resolution of conflict between nations also tends to result in the resolution of conflict within nations.
3. Although the military causes a great deal of environmental damage during wartime, during times of peace military forces are geared to help clean up the environment.
4. Military spending in the world as a whole has far exceeded spending for health care and education, especially in poor countries.
5. In 1998, the combined total of U.S. and Russian warheads was estimated to be 34,000. Only 500 to 2,000 are needed to induce a nuclear winter and destroy most of life on earth.

Answers: 1 = T; 2 = F; 3 = F; 4 = T; 5 = T

Every gun that is made, every warship launched, every rocket fired, signifies in the final sense a theft from those who hunger and are not fed, those who are cold and not clothed. The world in arms is not spending money alone. It is spending the sweat of its laborers, the genius of its scientists, and the hopes of its children.

Dwight D. Eisenhower, Former U.S. President/Military Leader

The history of the world is a history of conflict. Never in recorded history has there been a time when conflict did not exist between and within groups. The most violent form of conflict—**war**—refers to organized armed violence aimed at a social group in pursuit of an objective. Wars have existed throughout human history and continue in the contemporary world (see Figure 16.1).

War is one of the great paradoxes of human history. It both protects and annihilates. It creates and defends nations, but also destroys them. Whether war is just or unjust, defensive or offensive, it involves the most horrendous atrocities known to humankind. This chapter focuses on the causes and consequences of global conflict and war. Along with population and environmental problems, war and global conflict are among the most serious of all social problems in their threat to the human race and life on earth.

THE GLOBAL CONTEXT: CONFLICT IN A CHANGING WORLD

As societies have evolved and changed throughout history, the nature of war has also changed. Before industrialization and the sophisticated technology that resulted, war occurred primarily between neighboring groups on a relatively small scale. In the modern world, war can be waged between nations that are separated by thousands of miles, as well as between neighboring nations.

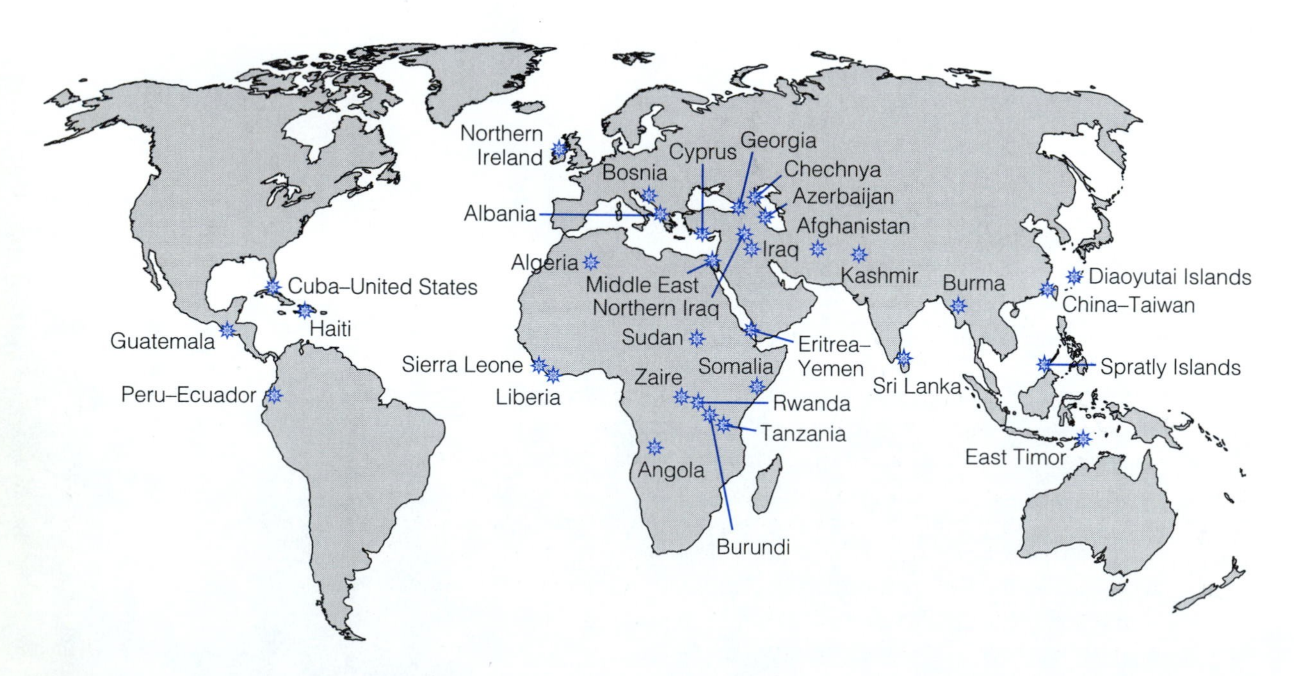

Figure 16.1 **As of 1999, conflict existed in the world at over 33 different locations.**

SOURCE: Map design by Aaron Bradley. **www.cfcsc.dnd.ca/links/wars/index.html** Copyright©1996–1998, Information Resource Centre, Canadian Forces College. Reprinted by permission.

In the following sections, we examine how war has changed our social world and how our changing social world has affected the nature of war in the industrial and postindustrial information age.

War and the Development of Civilization

The very act that now threatens modern civilization—war—is largely responsible for creating the advanced civilization in which we live. Before large political states existed, people lived in small groups and villages. War broke the barriers of autonomy between local groups and permitted small villages to be incorporated into larger political units known as "chiefdoms." Centuries of warfare between chiefdoms culminated in the development of the state. The **state** is "an apparatus of power, a set of institutions—the central government, the armed forces, the regulatory and police agencies—whose most important functions involve the use of force, the control of territory and the maintenance of internal order" (Porter 1994, 5–6). The creation of the state in turn led to other profound social and cultural changes:

> **And once the state emerged, the gates were flung open to enormous cultural advances, advances undreamed of during—and impossible under—a regimen of small autonomous villages. . . . Only in large political units, far removed in structure from the small autonomous communities from which they sprang, was it possible for great advances to be made in the arts and sciences, in economics and technology, and indeed in every field of culture central to the great industrial civilizations of the world. (Carneiro 1994, 14–15)**

> The significance of wars is not just that they lead to major changes during the period of hostilities and immediately after. They produced transformations which have turned out to be of enduring significance.
>
> –Anthony Giddens
> Sociologist

Thus, war, in a sense, gave rise to the state. Interestingly, the development of the state reduced the amount of lethal conflict (i.e., death through war, execution, homicide, or rebellion) in a society by providing alternative means of dispute resolution (Cooney 1997).

The Influence of Industrialization on War

Industrialization and technology could not have developed in the small social groups that existed before military action consolidated them into larger states. Thus, war contributed indirectly to the industrialization and technological sophistication that characterize the modern world.

Industrialization, in turn, has had two major influences on war. Cohen (1986) calculated the number of wars fought per decade in industrial and preindustrial nations and concluded that "as societies become more industrialized, their proneness to warfare decreases" (p. 265). Cohen summarized the evidence for this conclusion: the preindustrial nations had an overall mean of 10.6 wars per decade, while the industrial nations averaged 2.7 wars per decade. Perhaps industrialized nations have more to lose, so they avoid war and the risk of defeat. Indeed, Gentry (1998) notes that U.S. leaders "are increasingly reluctant to use [military forces] in roles that could lead to casualties in combat." The result—a "national aversion to danger" (pp. 179, 180).

Although industrialization may decrease a society's propensity to war, it also increases the potential destruction of war. With industrialization, military technology became more sophisticated and more lethal. Rifles and cannons

Industrialization may decrease a society's propensity to war, but, at the same time, it increases the potential destructiveness of war because with industrialization, warfare technology becomes more sophisticated and lethal.

replaced the clubs, arrows, and swords used in more primitive warfare and in turn were replaced by tanks, bombers, and nuclear warheads. This chapter's *Focus on Technology* looks at how modern computer and information technology is transforming warfare capabilities.

International Data

The Middle East and East Asia are the exceptions to the trend in decreased military spending; in the last 10 years military spending in the Middle East has increased 9 percent and in East Asia, 25 percent.

SOURCE: SIPRI 1998

The Economics of Military Spending

The increasing sophistication of military technology has commanded a large share of resources totaling, worldwide, $740 billion in 1997. Military spending has, however, declined in recent years, averaging a 4.5 percent decrease in the last decade. The largest cuts have occurred in Russia, the United States, Africa, and Central America (SIPRI 1998). Military spending includes expenditures for salaries of military personnel, research and development, weapons, veterans' benefits, and other defense-related expenses.

The U.S. government not only spends money on its own military, but also sells military equipment to other countries either directly or by helping U.S. companies sell weapons abroad (see Figure 16.2). Although the purchasing countries may use these weapons to defend themselves from hostile attack,

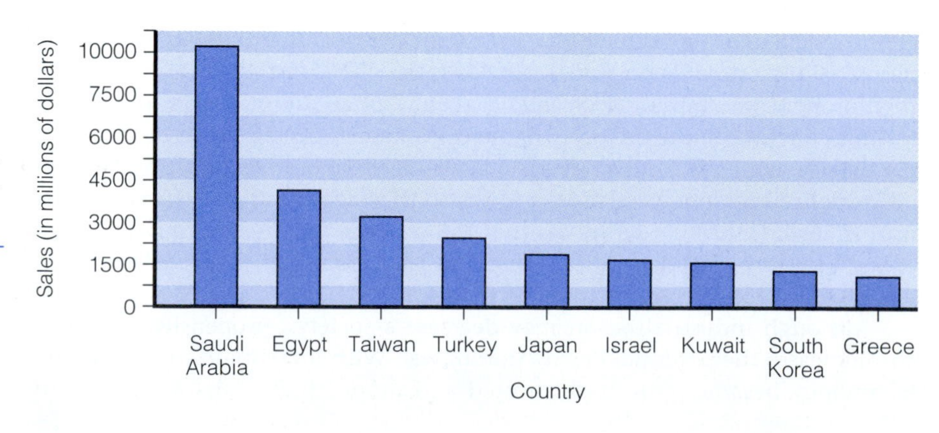

Figure 16.2 Arms Supplied by the United States, Cumulative Value for 1993–95

SOURCE: *Statistical Abstract of the United States: 1998,* 118th ed. U.S. Bureau of the Census. Washington, D.C.: U.S. Government Printing Office, Table 576.

FOCUS ON TECHNOLOGY

INFORMATION WARFARE

In the postindustrial information age, computer technology has revolutionized the nature of warfare and future warfare capabilities. Today, a "whole range of new technologies are offered for the next generations of weapons and military operations" (BICC 1998, 3) including the use of high-performance sensors, information processors, directed-energy technologies, and precision-guided munitions (O'Prey 1995; BICC 1998). With the increasing proliferation and power of computer technology, military strategists and political leaders are exploring the horizons of "information warfare," or **infowar.** Essentially, infowar utilizes technology to attack or manipulate the military and civilian infrastructure and information systems of an enemy. For example, infowar capabilities include the following (Waller 1995):

- Breaking into the communications equipment of the enemy army and disseminating incorrect information to enemy military leaders.
- Inserting computer viruses into the computer systems that control the phone system, electrical power, banking and stock exchanges, and air-traffic control of an enemy country.
- Jamming signals on the enemy's government television station and inserting a contrived TV program that depicts enemy leaders making unpopular statements that will alienate their people.
- Incapacitating the enemy's military computer systems, such as the one that operates the weapons system.

By 2010, the U.S. Army hopes to "digitize the battlefield" by linking every soldier and weapons systems electronically. Further, a prototype of the equipment to be used by the infantry in the twenty-first century is currently being developed:

> His helmet will be fitted with microphones and earphones for communications, night-vision goggles and thermal-imaging sensors to see in the dark, along with a heads-up display in front of his eyes to show him where he is on the ground and give him constant intelligence reports. (Waller 1995, 41)

The U.S. Army, Navy, and Air Force are setting up infowar offices. In 1995, the first 16 infowar officers graduated from the National Defense University in Washington after being specially trained in everything from defending against computer attacks to using virtual reality computer technology to plan battle maneuvers.

There are, however, two concerns with the use of infowar technologies. Although the United States leads the pack, other nations are catching up and in some cases threatening to overcome our position as the infowar leader (Bernstein and Libicki 1998). For example, France now leads in digitizing three-dimensional battles, a process in which information technologies are used to "acquire, exchange and employ timely information throughout the battlespace." (Cook 1996, 602). Clearly, if other nations develop information technologies that rival or even surpass American capabilities, U.S. military preparedness could be jeopardized.

Another problem is that infowar is relatively inexpensive and readily available. With a computer, a modem, and some rudimentary knowledge of computers, anyone could initiate an information attack (Broder 1999). Nation-states are not the only entities to be feared. Any individual or group with a cause could potentially become an "infowarrior". As a matter of fact, the U.S. government's Joint Security Commission described U.S. vulnerability to infowar as "the major security challenge of this decade and possibly the next century" (quoted in Waller 1995, 40). President Clinton's 2000 budget allocates $1.4 billion to fight "cyberattacks" as well as other unconventional threats to national security (Broder 1999).

SOURCES: Alvin Bernstein and Martin Libicki. 1998. "High-Tech: The Future of War." *Commentary* 105: 28–37; BICC (Bonn International Center for Conversion). 1998. "Chapter Six." *Conversion Survey, 1998.* Bonn, Germany: BICC; John Broder. 1999. "President Steps up War on New Terrorism." *New York Times,* January 23, 14; Nick Cook. 1996. "Battlespace 2000." *Interavia Business and Technology* 51:43–46; Kevin P. O'Prey. 1995. *The Arms Export Challenge: Cooperative Approaches to Export Management and Defense Conversion.* Washington, D.C.: The Brookings Institute; Douglas Waller. 1995. "Onward Cyber Soldiers." *Time,* August 21, 38–44.

foreign military sales may pose a threat to the United States by arming potential antagonists. For example, the United States, which provides more than half of the world's arms exports, supplied weapons to Iraq to use against Iran. These same weapons were then used against Americans in the Gulf War.

The **Cold War**, the state of political tension and military rivalry that existed between the United States and the former Soviet Union, provided justification for large expenditures on military preparedness. The end of the Cold War, along with the rising national debt, has resulted in cutbacks in the U.S. military budget as noted earlier.

The economic impact of these defense cutbacks, and cutbacks around the world, has been mixed. On the negative side, defense cutbacks result in job losses and the closure of military bases, facilities, factories and plants in defense industries. Certain industries and the countries, regions, and states where they are located suffer disproportionately as a result of defense cutbacks. Worldwide, 8.3 million jobs were lost in the defense industry between 1987 and 1996—47 percent of the 1987 total (BICC 1998).

National Data

U.S. defense spending decreased each year between 1992 and 1996.

SOURCE: *Statistical Abstract of the United States: 1998,* Table 567

On the positive side, defense cutbacks provide a **peace dividend**, in that resources previously spent on military purposes can be used for private or public investment or consumption. For example, the peace dividend could be used for new manufacturing plants and machinery (private investment) or for public education, transportation, housing, health, and the environment. The peace dividend could also be used to lower the national debt and/or taxes.

International Data

In 1995, military expenditures by developed countries equaled $668 billion; by developing countries, $197 billion and by NATO countries, $471 billion.

SOURCE: *Statistical Abstract of the United States: 1998,* Table 574

Achieving a peace dividend requires the reallocation of resources from military forces and defense industries to other sectors of the economy—a process referred to as **economic conversion.** For example, military bases could be converted to civil airports, prisons, housing developments, or shopping centers, and workers in the defense industry could be employed in the civil sector. In 1993 nearly $20 billion was committed to the "Defense Reinvestment and Conversion Initiative." However, in 1994, with the election of a Republican-dominated Congress, there were pressures "toward the dismantlement of conversion projects . . ." (BICC 1998, 13).

SOCIOLOGICAL THEORIES OF CONFLICT AND WAR

Sociological perspectives can help us understand various aspects of war. In this section, we describe how structural-functionalism, conflict theory, and symbolic interactionism may be applied to the study of conflict and war.

Structural-Functionalist Perspective

Structural-functionalism focuses on the functions war serves and suggests that war would not exist unless it had positive outcomes for society. It has already been noted that war has served to consolidate small autonomous social groups into larger political states. An estimated 600,000 autonomous political units existed in the world around the year 1000 B.C. Today, that number has dwindled to less than 200 (Carneiro 1994).

> Intersocietal conflicts have furthered the development of social structures.
>
> —Herbert Spencer
> Sociologist

Another major function of war is that it produces social cohesion and unity among societal members by giving them a "common cause" and a common enemy. Unless a war is extremely unpopular, military conflict promotes economic and political cooperation. Internal domestic conflicts between political parties, minority groups, and special interest groups dissolve as they unite

A major function of war is that it produces unity among societal members. They have a common cause and a common enemy. They feel a sense of cohesion and work together to defeat the enemy.

to fight the common enemy. During World War II, U.S. citizens worked together as a nation to defeat Germany and Japan.

In the short term, war also increases employment and stimulates the economy. The increased production needed to fight World War II helped pull the United States out of the Great Depression. The investments in the manufacturing sector during World War II also had a long-term impact on the U.S. economy. Hooks and Bloomquist (1992) studied the effect of the war on the U.S. economy between 1947 and 1972 and concluded that the U.S. government "directed, and in large measure, paid for a 65% expansion of the total investment in plant and equipment" (p. 304).

Another function of war is the inspiration of scientific and technological developments that are useful to civilians. Research on laser-based defense systems led to laser surgery, for example, and research in nuclear fission and fusion facilitated the development of nuclear power. The airline industry owes much of its technology to the development of air power by the Department of Defense, and the Internet was created by the Pentagon for military purposes. Presently, researchers are developing insectlike robots that will act as scouts in detecting the presence of enemy forces. The 2-inch spies will also be made available to police SWAT teams and other emergency personnel (Koerner 1998).

Finally, war serves to encourage social reform. After a major war, members of society have a sense of shared sacrifice and a desire to heal wounds and rebuild normal patterns of life. They put political pressure on the state to care for war victims, improve social and political conditions, and reward those who have sacrificed lives, family members, and property in battle. As Porter (1994) explains:

> **Only the promise of a better world can give meaning to a terrible conflict. Since . . . the lower economic strata usually contribute more of their blood in battle than the wealthier classes, war often gives impetus to social welfare reforms. (p. 19)**

National Data

Confidence in the leaders of America's institutions has steadily decreased since 1966. However, in 1998 44 percent of a national sample of adults responded they had "a great deal of confidence in the military," an increase of 7 percent from the preceding year.

SOURCE: Harris Poll 1998

> As war becomes more sophisticated, it continuously increases governmental authority and decreases the power of the people.
>
> –J. C. L. Simonde de Sismondi
> Swiss economist

Conflict Perspective

Conflict theorists emphasize that the roots of war often stem from antagonisms that emerge whenever two or more ethnic groups (e.g., Bosnians and Serbs), countries (United States and Vietnam), or regions within countries (the U.S. North and South) struggle for control of resources or have different political, economic, or religious ideologies. Further, conflict theory suggests that war benefits the corporate, military, and political elites. Corporate elites benefit because war often results in the victor taking control of the raw materials of the losing nations, thereby creating a bigger supply of raw materials for its own industries. Indeed, many corporations profit from defense spending. Under the Pentagon's bid and proposal program, for example, corporations can charge the cost of preparing proposals for new weapons as overhead on their Defense Department contracts. Also, Pentagon contracts often guarantee a profit to the developing corporations. Even if the project's cost exceeds initial estimates, called a cost overrun, the corporation still receives the agreed-upon profit. In the late 1950s, President Dwight D. Eisenhower referred to this close association between the military and the defense industry as the **military-industrial complex.**

The military elite benefit because war and the preparations for it provide prestige and employment for military officials. For example, Military Professional Resources, Inc. (MPRI), in Virginia, is described in its brochure as "a corporation of former military leaders organized to provide a wide range of professional and technical services" (Harris 1996, 12). The company employs over 2,000 retired military professionals, has clients around the world, and represents a movement toward the privatization of the military.

War also benefits the political elite by giving government officials more power. Porter (1994) observed that "throughout modern history, war has been the lever by which . . . governments have imposed increasingly larger tax burdens on increasingly broader segments of society, thus enabling ever-higher levels of spending to be sustained, even in peacetime" (p. 14). Political leaders who lead their country to a military victory also benefit from the prestige and hero status conferred on them.

National Data

As of January 1, 1999, there were over 1.3 million U.S. military personnel, a decrease of 37,000 from the previous year.

SOURCE: DefenseLink 1999

Symbolic Interactionist Perspective

The symbolic interactionist perspective focuses on how meanings and definitions influence attitudes and behaviors regarding conflict and war. The development of attitudes and behaviors that support war begins in childhood. American children learn to glorify and celebrate the Revolutionary War, which created our nation. Movies romanticize war, children play war games with toy weapons, and various video and computer games glorify heroes conquering villains. Indeed, from 1938 to 1942 a series of "Horrors of Wars" cards were manufactured and distributed in the United States and collected by millions of American youth much like baseball cards (Nelson 1997).

> Pilgrim settlers had to imagine Native Americans as subhuman savages in order to kill and rob them of their very sustenance and rightful use of their own lands and hunting grounds.
>
> –Daniela Gioseffi
> Activist

CONSIDERATION

Arguing that "political cultures operate as historical systems of meaning," Olick and Levy (1997, 934) investigate the impact of collective definitions of the Holocaust on German political claim making. The authors, using official records of the Federal Republic of Germany, trace post–World War II efforts to redefine Germany's negative image as a function of evolving definitions of the causes and consequences of the Nazi regime. Consistent with a symbolic interactionist perspective, the authors conclude Germany's "ongoing work to define who it is, what it can do, and what it should do" is dependent on a "continuous negotiation between past and present" (p. 934).

Symbolic interactionism helps to explain how military recruits and civilians develop a mind-set for war by defining war and its consequences as acceptable and necessary. The word "war" has achieved a positive connotation through its use in various phrases—the "war on drugs," the "war on poverty," and the "war on crime." Positive labels and favorable definitions of military personnel facilitate military recruitment and public support of armed forces. Military personnel wear uniforms that command public respect and earn badges and medals that convey their status as "heroes."

Many government and military officials convince the masses that the way to ensure world peace is to be prepared for war. Most world governments preach peace through strength rather than strength through peace. Governments may use propaganda and appeals to patriotism to generate support for war efforts and motivate individuals to join armed forces.

To legitimize war, the act of killing in war is not regarded as "murder." Deaths that result from war are referred to as "casualties." Bombing military and civilian targets appears more acceptable when nuclear missiles are "peacekeepers" that are equipped with multiple "peace heads." Finally, killing the enemy is more acceptable when derogatory and dehumanizing labels such as Gook, Jap, Chink, and Kraut convey the attitude that the enemy is less than human.

> The issue of war is always uncertain.
>
> –Cicero
> Roman orator/statesman

CAUSES OF WAR

The causes of war are numerous and complex. Most wars involve more than one cause. The immediate cause of a war may be a border dispute, for example, but religious tensions that have existed between the two countries for decades may also contribute to the war. The following section reviews various causes of war.

> We are destined to live together on the same soil in the same land. We, the soldiers who have returned from battles stained with blood; we who have seen our relatives and friends killed before our eyes . . . We who have fought against you, the Palestinians, we say to you today in a loud and clear voice: Enough of blood and tears. Enough!
>
> –Yitzhak Rabin
> Former Prime Minister of Israel

Conflict over Land and Other Natural Resources

Nations often go to war in an attempt to acquire or maintain control over natural resources, such as land, water, and oil. Disputed borders are one of the most common motives for war. Conflicts are most likely to arise when borders are physically easy to cross and are not clearly delineated by natural boundaries, such as major rivers, oceans, or mountain ranges.

Water is another valuable resource that has led to wars. At various times the empires of Egypt, Mesopotamia, India, and China all went to war over irrigation rights. Recently, Serbia's desire for access to the Adriatic Sea has contributed to the conflict between the Bosnian Serbs and the Muslims of Bosnia and Herzegovina.

Not only do the oil-rich countries in the Middle East present a tempting target in themselves, but war in the region can threaten other nations that are dependent on Middle Eastern oil. Thus, when Iraq seized Kuwait and threatened the supply of oil from the Persian Gulf, the United States and many other nations reacted militarily in the Gulf War. In a document prepared for the Center for Strategic and International Studies, Starr and Stoll (1989) warn:

> **By the year 2000, water, not oil will be the dominant resource issue of the Middle East. According to World Watch Institute, "despite modern technology and feats of engineering, a secure water future for much of the world remains elusive." The prognosis for Egypt, Jordan, Israel, the West Bank, the Gaza Strip, Syria, and Iraq is especially alarming. If present consumption patterns continue, emerging water shortages, combined with a deterioration in water quality, will lead to more competition and conflict. (p. 1)**

International Data

There are currently over 430 million people living in countries with water shortages; by 2050, that number will increase threefold to fivefold.

SOURCE: Gardner-Outlaw and Engelman 1997

Conflict over Values and Ideologies

Many countries initiate war not over resources, but over beliefs. World War II was largely a war over differing political ideologies: democracy versus fascism. The Cold War involved the clashing of opposing economic ideologies: capitalism versus communism. Wars over differing religious beliefs have led to some of the worst episodes of bloodshed in history, in part, because some religions are partial to martyrdom—the idea that dying for one's beliefs leads to eternal salvation. The Shiites (one of the two main sects within Islam) in the Middle East represent a classic example of holy warriors who feel divine inspiration to kill the enemy.

Conflicts over values or ideologies are not easily resolved. The conflict between secularism and Islam has lasted for 14 centuries (Lewis 1990). According to Brown (1994), wars fought over values and ideologies are less likely to end in compromise or negotiation because they are fueled by people's convictions and their desire to spread their way of life.

If ideological differences can contribute to war, do ideological similarities discourage war? The answer seems to be yes; in general, countries with similar ideologies are less likely to engage in war with each other than countries with differing ideological values (Dixon 1994). Democratic nations are particularly disinclined to wage war against one another (Doyle 1986).

International Data

Following the 1998 Northern Ireland–UK peace plan, a car bomb exploded in Omagh, Northern Ireland, killing 29 people and wounding more than 200.

SOURCE: "Light of Day in Bloody Eire" 1998

Racial and Ethnic Hostilities

Ethnic groups vary in their cultural and religious beliefs, values, and traditions. Thus, conflicts between ethnic groups often stem from conflicting values and ideologies. Racial and ethnic hostilities are also fueled by competition over land and other natural and economic resources. Gioseffi (1993) notes that "experts agree that the depleted world economy, wasted on war efforts, is in great measure the reason for renewed ethnic and religious strife. 'Haves' fight with 'have-nots' for the smaller piece of the pie that must go around" (xviii). As noted in Chapter 8, racial and ethnic hostilities are also perpetuated by the wealthy majority to divert attention away from their exploitations and to maintain their own position of power.

Gioseffi (1993) conveys the idiocy and irony of racial and ethnic hostilities:

> **At this dangerous juncture, as we near the year 2000, after more than eighty centuries of art and human creativity, philosophy, music, poetry, social and biological science—we humans, considered the**

The most important divide in this hard world is not over bloodlines or ways of worship. It's between those who pursue reason and those who dismiss it. It's between those who build and those who bomb.

—Ellen Goodman
Columnist

paragon of animals in our ability to think, named Homo sapiens, meaning wise or knowing animal, persist, brutishly, in hating and killing each other for the colors of our skin, the shapes of our features, our places of origin on our common terra firma, our styles of culture or language, and most ironically of all our "religious" beliefs—despite the fact that all the great religions of the Earth teach the same basic golden rule: "Do unto others as you would have them do unto you." (xlix)

My Brothers, my Sisters, one and all, please forgive me. While I was trying to build my life, yours was forcefully being taken away from you, and all I did was say, "God, give our leaders Wisdom."

–Anonymous
South African citizen

CONSIDERATION

As described by Paul (1998), sociologist Daniel Chirot argues that the recent worldwide increase in ethnic hostilities is a consequence of "retribalization," that is, the tendency for groups, lost in a globalized culture, to seek solace in the "extended family of an ethnic group" (p. 56). Chirot identifies five levels of ethnic conflict: (1) multiethnic societies without serious conflict (e.g., Switzerland), (2) multiethnic societies with controlled conflict (e.g., United States, Canada), (3) societies with ethnic conflict that has been resolved (e.g., South Africa), (4) societies with serious ethnic conflict leading to warfare (e.g., Sri Lanka), and (5) societies with genocidal ethnic conflict, including "ethnic cleansing" (e.g., Germany, Yugoslavia).

Defense against Hostile Attacks

The threat or fear of being attacked may cause the leaders of a country to declare war on the nation that poses the threat. The threat may come from a foreign country or from a group within the country. After Germany invaded Poland in 1939, Britain and France declared war on Germany out of fear that they would be Germany's next victims. Germany attacked Russia in World War I, in part out of fear that Russia had entered the arms race and would use its weapons against Germany. Japan bombed Pearl Harbor hoping to avoid a later confrontation with the U.S. Pacific fleet, which posed a threat to the Japanese military. In 1981, the Israelis conducted an air raid against Iraq's nuclear facilities in an attempt to disarm Iraq and remove its threat to Israel (Brown 1994).

Revolution

Revolutions involve citizens warring against their own government. A revolution may occur when a government is not responsive to the concerns and demands of its citizens and when there are strong leaders willing to mount opposition to the government (Brown 1994).

Those who make peaceful evolution impossible, make violent revolution inevitable.

–John F. Kennedy
Former U.S. President

The birth of the United States resulted from colonists revolting against British control. Contemporary examples of civil war include Sri Lanka, where the Tamils, a separatist group living in the northern region of the country, have been at war with the Sri Lankan government for over 10 years. According to Mylvaganam (1998) the Liberation of Tamil Tigers (LTTE) has successfully fought off the better prepared Sri Lankan government army through the "use of paramilitary and guerrilla warfare, coupled with their expert knowledge of the terrain . . ." (p. 1). Additionally, civil wars have erupted in newly independent republics created by the collapse of communism in Eastern Europe, as well as in Rwanda, Liberia, Guatemala, Chile, and Uganda.

> Lift your head. Be proud to be German.
>
> —Nazi War Criminal Judgment at Nuremberg

Nationalism

Some countries engage in war in an effort to maintain or restore their national pride. For example, Scheff (1994) argues that "Hitler's rise to power was laid by the treatment Germany received at the end of World War I at the hands of the victors" (p. 121). Excluded from the League of Nations, punished by the Treaty of Versailles, and ostracized by the world community, Germany turned to nationalism as a reaction to material and symbolic exclusion.

In the late 1970s, Iranian fundamentalist groups took hostages from the American Embassy in Iran. President Carter's attempt to use military forces to free the hostages was not successful. That failure intensified doubts about America's ability to effectively use military power to achieve its goals. The hostages in Iran were eventually released after President Reagan took office. But doubts about the strength and effectiveness of America's military still called into question America's status as a world power. Subsequently, U.S. military forces invaded the small island of Grenada because the government of Grenada was building an airfield large enough to accommodate major military armaments. U.S. officials feared that this airfield would be used by countries in hostile attacks on the United States. From one point of view, the large scale and "successful" attack on Grenada functioned to restore faith in the power and effectiveness of the American military.

SOCIAL PROBLEMS ASSOCIATED WITH WAR AND MILITARISM

Social problems associated with war and militarism include death and disability; rape, forced prostitution, and displacement of women and children; disruption of social-psychological comfort; diversion of economic resources; and destruction of the environment.

Death and Disability

> The object of war is not to die for your country but to make the other bastard die for his.
>
> —Gen. George Patton

Thus far in the twentieth century, war has taken the lives of more than 100 million persons—more than the total number of deaths in all previous wars or massacres in human history combined (Porter 1994). Many American lives have been lost in wars, including over 53,000 in World War I, 292,000 in World War II, 34,000 in Korea, and 47,000 in Vietnam (see Table 16.1). In our modern world, sophisticated weapons technology combined with increased population density has made it easier to kill large numbers of people in a short amount of time. When the atomic bomb was dropped on the Japanese cities of Hiroshima and Nagasaki during World War II, 250,000 civilians were killed. More recently, in the civil war in Rwanda, over half a million people died within 3 months (Gibbs 1994).

> Wars kill people. That is what makes them different from all other forms of human enterprise.
>
> —Gen. Colin L. Powell

War's impact extends far beyond those who are killed. Many of those who survive war incur disabling injuries as well as diseases. For example, Gulf War syndrome, a mysterious disease of unknown origin, impacts an estimated 80,000 to 100,000 U.S. veterans (National Gulf War Resource Center 1997). This chapter's *The Human Side* poignantly describes the destructive course of this multisymptom disease (Coker et al. 1999).

War-related deaths and disabilities also deplete the labor force, create orphans and single-parent families, and burden taxpayers who must pay for the care of orphans and disabled war veterans.

Table 16.1 **Armed Forces Personnel: Summary of Major Conflicts**

Item	Civil	Spanish-American	World War I	World War II	Korean	Vietnam
			War or Conflict			
Personnel serving (1,000)	2,213	307	4,735	16,113	5,720	8,744
Average length of service (months)	20	8	12	33	19	23
Battle deaths (1,000)	140[a]	<.5	53	292	34	47
Wounds not mortal (1,000)	282	2	204	671	103	153

[a]Union only; estimates of Confederate deaths are from 600,000 to 1.5 million.

SOURCE: *Statistical Abstract of the United States: 1998,* Table 585.

Persons who participate in experiments for military research may also suffer physical harm. Representative Edward Markey of Massachusetts identified 31 experiments dating back to 1945 in which U.S. citizens were subjected to harm from participation in military experiments. Markey charged that many of the experiments used human subjects who were captive audiences or populations considered "expendable," such as the elderly, prisoners, and hospital patients. Eda Charlton of New York was injected with plutonium in 1945. She and 17 other patients did not learn of their poisoning until 30 years later. Her son, Fred Shultz, said of his deceased mother:

> **I was over there fighting the Germans who were conducting these horrific medical experiments . . . at the same time my own country was conducting them on my own mother. (Miller 1993, 17)**

Rape, Forced Prostitution, and Displacement of Women and Children

Half a century ago, the Geneva Conventions prohibited rape and forced prostitution in war. Nevertheless, both continue to occur in modern conflicts.

Before and during World War II, the Japanese military forced an estimated 100,000 to 200,000 women and teenage girls into prostitution as military "comfort women" (Amnesty International 1995). These women were forced to have sex with dozens of soldiers every day in "comfort stations." Many of the women died as a result of untreated sexually transmitted disease, harsh punishment, or indiscriminate acts of torture.

More recently, armed forces in Bosnia-Herzegovina have raped women civilians and prisoners. Most of the victims have been Muslim women raped by Bosnian Serbian soldiers:

> **In one such case, a 17-year-old Muslim girl was taken by Serbs from her village to huts in woods nearby. . . . She was held there for three months. . . . She was among 12 women who were raped repeatedly in the hut in front of the other women—when they tried to defend her they were beaten off by the soldiers. (Amnesty International 1995, 19)**

International Data

Over 110 million land mines are located around the world. At the rate at which they are being located, it would take over 1,000 years to find and destroy them all. Every year, 30,000 individuals, often civilians, are maimed or killed by land mines.

SOURCE: "Fast New Gizmos Can Unearth Them Safely" 1998

The use of rape in conflict reflects the inequalities women face in their everyday lives in peacetime. . . . Women are raped because their bodies are seen as the legitimate spoils of war.

—Amnesty International

Some countries have seen the use of systematic sexual violence against women as a weapon of war to degrade and humiliate entire populations. Rape is the most despicable crime against women; mass rape is an abomination.

—Boutros Boutros-Ghali
UN Secretary-General

THE HUMAN SIDE

THE DEATH OF A GULF WAR VETERAN

The following letter to *American Legion Magazine* is from the parents of a Gulf War veteran.

> Our son may be unique. We really don't know. We're desperate for information that is extremely hard to come by. We are very frustrated that what happened to us may have already happened to other families—or will—before something is done.
>
> This is for Scott, our son, who said, "Go for it Mom", when I told him I'd never quit trying to find out what made him so ill.
>
> He was sent to the Persian Gulf with the 1133rd Transportation Co., National Guard, of Mason City, Iowa. He left for Saudi as a very healthy young man and he returned in much the same physical condition. Just tired, but not unusual considering where he'd been and what he'd seen.
>
> . . . two years after he returned from home in the summer of 1993, he developed a rash on his torso. It would erupt, disappear and then come back again. It didn't always look the same. We thought it was a heat rash, something he ate, new laundry soap, etc.
>
> In the fall of 1993 he developed sores in his mouth. He could eat very little and quickly lost about 40 pounds. He went from specialist to specialist, who sent tests many places to try and find a cause. He was put on steroids and depending on the dosage, it would get a little better and then flare up again. By winter he could only eat pureed foods and liquids. He could not use a straw as it hurt too much. Many doctors, many tests, many different medications. Nothing helped very much. And no concrete diagnosis.
>
> In May 1994, he was examined by the VA hospital in Des Moines. They were not able to find the cause for the terrible sores in his mouth and the rash that now also affected his feet, hands and arms. He made many trips to Des Moines—a three hour drive.
>
> In early August 1994, the VA Hospital in Des Moines came up with a diagnosis of lupus. My heart just broke when I heard those words, but he was thrilled to know there was finally a name and treatment for his symptoms.
>
> By mid-August he was hardly able to walk, his feet swollen and so extremely sore from the rash that seemed to get worse by the day. He went to the VA Hospital on Friday, August 19 . . . and was admitted. The rash had become blisters about the size of a 50-cent piece and were breaking and bleeding. He was running a fever . . .
>
> He was transferred to University [hospital] in Iowa City and taken to surgery . . . [where] they removed all his skin and replaced it with what is called pig skin. Out of surgery, bandaged from head to toe, he was given a five percent chance of survival.
>
> Our family gathered together to give him all the love and support we could. Ten days later the pig skin was removed. Infection had been found. It was too risky to take him back to

THE HUMAN SIDE

THE DEATH OF A GULF WAR VETERAN *(continued)*

surgery, so it was removed in a sterile room close to his room.

The next seven weeks are a blur. We had many ups—a good day for Scott—and many downs—a bad day for him. He had to endure burn baths every day. Water jets removed sloughed skin. He'd grow a tiny patch of skin only to lose it in a day or so later. He was fed through a tube. Many antibiotics were given in the hopes of warding off infection; morphine for the tremendous pain. We read to him—he'd correctly pronounce words we missed. His sense of humor never left. He worried about all of us.

Scott's life ended at 3 A.M. on October 15, 1994 . . .

We are convinced beyond anything we've ever felt as parents that Persian Gulf Syndrome killed our only son. We want someone to tell us the truth. We won't quit until we have believable answers to our question.

What do we tell his oldest son, now 11, who has nothing but pictures of his daddy . . . ?

What do we say to a three-year-old whose wish is to build a rocket so he can go get daddy and make everyone happy?

What do you say to a son who was born two weeks after his daddy died . . . ?

What do you say to a wife who longs only for her husband's love, strength and support?

If chemicals and gases were used over there, tell us the truth. It kills innocent people and destroys innocent families. We are real people with real feelings and we deserve the truth. This was not easy to write. We did it in hopes it may help someone. Only then will Scott's death make sense to us. He would have wanted it that way. He was that special.

Ardie and Rollie Siefken
Plainfield, Iowa

SOURCE: "The Sad Death of a Gulf War Veteran." *American Legion Magazine,* August 6, 1996. Copyright lifted for distribution. **http://www.ascension-research.org/part-5.html**

In 1994, militia forces roamed through the town of Kibuye, Rwanda, burning houses and killing civilians. "Women found sheltering in the parish church were raped, then pieces of wood were thrust into their vaginas, and they were left to die slowly" (Amnesty International 1995, 17). Feminist analysis of wartime rape emphasizes that the practice reflects not only a military strategy, but ethnic and gender dominance as well (Card 1997).

War also displaces women and children from their homes, forcing them to flee to other countries or other regions of their homeland. Refugee women and female children are particularly vulnerable to sexual abuse and exploitation by locals, members of security forces, border guards, or other refugees. In refugee camps, women and children may also be subjected to sexual violence from officials and male refugees.

International Data

More than 80 percent of refugees are women and children.

SOURCE: Amnesty International 1995

I have seen enough of one war never to wish to see another.

—Thomas Jefferson
Former U.S. President

Disruption of Social-Psychological Comfort

War and living under the threat of war interfere with social-psychological well-being. In a study of 269 Israeli adolescents, Klingman and Goldstein (1994) found a significant level of anxiety and fear, particularly among younger females, in regard to the possibility of nuclear and chemical warfare. Similarly, female scientists when compared to their male counterparts perceived significantly higher nuclear-related risks (Barke, Jenkins-Smith, and Slovic 1997).

Civilians who are victimized by war and military personnel who engage in combat may experience a form of psychological distress known as **post-traumatic stress disorder** (PTSD), a clinical term referring to a set of symptoms that may result from any traumatic experience, including crime victimization, rape, or war. Symptoms of PTSD include sleep disturbances, recurring nightmares, flashbacks, and poor concentration (Novac 1998). PTSD is also associated with other personal problems, such as alcoholism, family violence, divorce, and suicide.

National Data

About 8 percent of World War I veterans suffered from "post-combat psychiatric disorder"; an estimated 6 percent of World War II veterans had "war-related psychological disorders"; the Korean War saw 3 percent of its veterans diagnosed as "psychological casualties"; and 12 percent of Vietnam soldiers were lost due to "psychiatric reasons."

SOURCE: Danitz 1997

One study estimates that about 30 percent of male veterans of the Vietnam War have experienced PTSD, and about 15 percent continue to experience it (Hayman and Scaturo 1993). In another study of 215 Army National Guard and Army Reserve troops who served in the Gulf War and who did not seek mental health services upon return to the states, 16 to 24 percent exhibited symptoms of PTSD (Sutker et al. 1993). Finally, research on PTSD in children reveals that females are generally more symptomatic than males and that PTSD may disrupt the normal functioning and psychological development of children (Cauffman et al. 1998; Pfefferbaum 1997).

Diversion of Economic Resources

As discussed earlier, maintaining the military and engaging in warfare require enormous financial capital and human support. In 1997, an estimated $740 billion was spent worldwide on military research and development (SIPRI 1998). This amount exceeds the combined government research expenditures on developing new energy technologies, improving human health, raising agricultural productivity, and controlling pollution.

National Data

In 1998, an estimated $35.8 billion was spent on national defense research and development; in 1980, that number was $13.1 billion.

SOURCE: *Statistical Abstract of the United States: 1998*, Table 568

Money that is spent for military purposes could be allocated for social programs. The decision to spend $1.4 billion for one Trident submarine, equal in cost to a 5-year immunization program that would prevent nearly 1 million childhood deaths annually (Renner 1993b), is a political choice. Similarly, world leaders could choose to allocate the $774 billion needed to reverse environmental damage in four priority areas: reforesting the earth, raising energy efficiency, protecting croplands from erosion of topsoil, and developing renewable sources of energy (Renner 1993a). Although the end of the Cold War has resulted in a relative decrease in defense spending, the 1998 through 2003 U.S. budgets project spending more money on national defense than on education, the justice system, the environment, and transportation combined (Office of Management and Budget 1998) (see Figure 16.3).

Destruction of the Environment

Traditional definitions of and approaches to national security have assumed that political states or groups constitute the principal threat to national security and welfare. This assumption implies that national defense and security are best

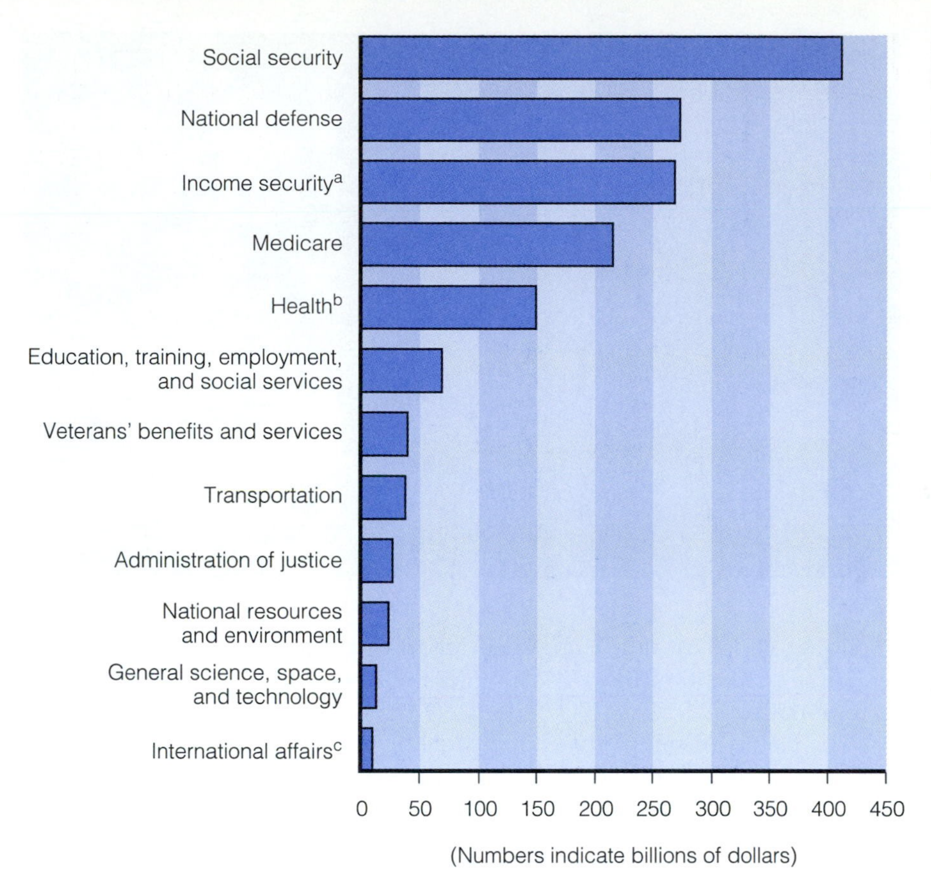

Figure 16.3 Selected U.S. Federal Outlays, 2000 (estimated)

SOURCE: Office of Management and Budget. 1998. **http://www.access.gpo.gov**

served by being prepared for war against other states. The environmental costs of military preparedness are often overlooked or minimized in such discussions. For example, in 1984, when a citizen complained to a Virginia base-commander about toxic contaminations from the facility, he responded, "We're in the business of protecting the nation, not the environment" (Calhoun 1996, 60). But environmental security is also vital to national and global security, and achieving one at the expense of the other is like taking one step forward and two steps back.

> My God . . . what have we done?
>
> –Robert Lewis
> Co-pilot of the Enola Gay after dropping the atomic bomb on Hiroshima

Destruction of the Environment during War The environmental damage that occurs during war continues to devastate human populations long after war ceases. In Vietnam, 13 million tons of bombs left 25 million bomb craters. Nineteen million gallons of herbicides, including Agent Orange, were spread over the countryside. An estimated 80 percent of Vietnam's forests and swamplands were destroyed by bulldozing or bombing (Funke 1994).

The Gulf War also illustrates how war destroys the environment (Funke 1994; Renner 1993a). In 6 weeks, 1,000 air missions were flown, dropping 6,000 bombs. In 1991, Iraqi troops set 650 oil wells on fire, releasing oil, which still covers the surface of the Kuwaiti desert and seeps into the ground, threatening to poison underground water supplies. The estimated 6–8 million barrels

Oil smoke from the 650 burning oil wells left in the wake of the Gulf War contains soot, sulfur dioxide, and nitrogen oxides, the major components of acid rain, and a variety of toxic and potentially carcinogenic chemicals and heavy metals.

of oil that spilled into the Gulf waters are threatening marine life. This spill is by far the largest in history—25–30 times the size of the 1989 *Exxon Valdez* oil spill in Alaska.

The clouds of smoke that hung over the Gulf region for 8 months contained soot, sulfur dioxide, and nitrogen oxides—the major components of acid rain—and a variety of toxic and potentially carcinogenic chemicals and heavy metals. The U.S. Environmental Protection Agency estimates that in March 1991 about 10 times as much air pollution was being emitted in Kuwait as by all U.S. industrial and power-generating plants combined. Acid rain destroys forests and harms crops. It also activates several dangerous metals normally found in soil, including aluminum, cadmium, and mercury.

The ultimate environmental catastrophe facing the planet is a massive exchange thermonuclear war. Aside from the immediate human casualties, poisoned air, poisoned crops, and radioactive rain, many scientists agree that the dust storms and concentrations of particles would block vital sunlight and lower temperatures in the Northern Hemisphere, creating a **nuclear winter.** In the event of large-scale nuclear war, most living things on earth would die. The fear of nuclear war has greatly contributed to the military and arms buildup, which, ironically, also causes environmental destruction even in times of peace.

National Data

A Pentagon report to Congress identified over 10,000 suspected hazardous waste sites on active military facilities in the United States.

SOURCE: Calhoun 1996

CONSIDERATION

Several researchers argue that the continued high-alert status of U.S. and Russian missiles is conducive to accidental nuclear war (Forrow et al. 1998). Of particular concern is the strategy of "launch on warning"–the policy that allows only minutes for detection of an approaching nuclear warhead, decision making, authorization of a launch, and a response. A launch response based on a false warning is the most likely scenario leading to an accidental nuclear war.

In assessing the consequences of such an event, Forrow et al. (1998) conclude that eight U.S. urban areas (Atlanta, Boston, Chicago, New York, Pittsburgh, San Francisco, Seattle, and Washington, D.C.)–four hit with four warheads and four hit with eight warheads–would, conservatively, result in over 6 million immediate deaths. The number of secondary deaths, that is, those not as a direct consequence of the bombing, would likely exceed that number. The authors also acknowledge that destruction to health infrastructures would be so great that medical care of the injured would be nearly impossible.

Destruction of the Environment in Peacetime Even when military forces are not engaged in active warfare, military activities assault the environment. For example, modern military maneuvers require large amounts of land. The use of land for military purposes prevents other uses, such as agriculture, habitat protection, recreation, and housing. More importantly, military use of land often harms the land. In practicing military maneuvers, the armed forces demolish natural vegetation, disturb wildlife habitats, erode soil, silt up streams, and cause flooding. Military bombing and shooting ranges leave the land pockmarked with craters and contaminate the soil and groundwater with lead and other toxic residues.

Further, the production, maintenance, and storage of weapons and military equipment poison the environment and its human, plant, and animal inhabitants. As Calhoun (1996) notes, "[D]ecades of improper and unsafe handling, storage and disposal of hazardous materials while building and maintaining the world's most powerful fighting force have severely polluted America's air, water and soil" (p. 60). The military is the largest producer of hazardous material in the United States (Calhoun 1996).

Bombs exploded during peacetime leak radiation into the atmosphere and groundwater. From 1945 to 1990, 1,908 bombs were tested—that is, exploded—at more than 35 sites around the world. Although underground testing has reduced radiation, some still escapes into the atmosphere and is suspected of seeping into groundwater (Renner 1993a).

Finally, although arms-control and disarmament treaties of the last decade have called for the disposal of huge stockpiles of weapons, there are no completely safe means of disposing of weapons and ammunition. Greenpeace has called for placing weapons in storage until safe disposal methods are found (Renner 1993b). But the longer weapons are stored, the more they deteriorate, increasing the likelihood of dangerous leakage.

National Data

Over 100 Department of Defense installations have been placed on the nation's "Superfund" list of the most contaminated sites in the United States; the Environmental Protection Agency estimates the number to be over 200.

SOURCE: Calhoun 1996

National Data

In a 1998 survey of U.S. adults, 36 percent responded that the development of the atomic bomb was a "good thing," 61 percent a "bad thing," and 3 percent had "no opinion." In 1945, those numbers were 69 percent, 17 percent, and 14 percent, respectively.

SOURCE: Gallup Poll 1998

CONFLICT AND POLITICS IN THE POST–COLD WAR ERA

One of the most significant recent events affecting global conflict has been the end of the Cold War due to the collapse of the communist regime of the former Soviet Union. Prospects for world peace in the post–Cold War era are examined next.

Internal Conflict in the Post–Cold War Era

In the discussion of the structural-functionalist view of global conflict and war, we noted that war functions as a catalyst for social cohesion. The corollary to the cohesive effect of war is that without a common enemy to fight, internal strife is likely to occur. Porter (1994) identifies a historical pattern in which "the end of an era of international rivalry and conflict has marked the beginning of internal conflict and disarray almost everywhere" (p. 300). According to Porter,

Perhaps for the first time, world leaders can move from responding to the Cold War to shaping environmentally healthy societies. The environment can then move to the center of the economic decision making, where it belongs.

—Worldwatch Institute

International Data

In 1999, escalating violence in the civil war in Kosovo, a province of Yugoslavia, led to the mass murder of 45 ethnic Albanians.
SOURCE: Eddy 1999

the post–Cold War era is likely to be an era of political turmoil and divisiveness among social, racial, religious, and class groups:

> **Given the growing diversity of American culture and values today, our society, no longer united by foreign threats, might find that its own internal cleavages are greater than anyone realized. . . . We can expect growing public disdain for the political process, rising unrest in the inner cities, proposals for radical constitutional change, third-party movements, one-term presidents, and a serious national identity crisis over what it means to be an American. (Porter 1994, 295)**

Terrorism and Guerrilla Warfare: A Growing Threat

If you ask me what is the obstacle to the implementation [of the peace plan], it is terrorism. We face a unique kind of terrorism—the suicidal terror mission. There is no deterrent to a person who goes with high explosives in his car in his bag and explodes himself.

–Yitzhak Rabin
Former Prime Minister of Israel

Terrorism is the premeditated use, or threatened use, of violence by an individual or group to gain a political or social objective (INTERPOL 1998). Terrorism may be used to publicize a cause, promote an ideology, achieve religious freedom, attain the release of a political prisoner, or rebel against a government. Terrorists use a variety of tactics, including assassinations, skyjackings, armed attacks, kidnaping and hostage taking, threats, and various forms of bombing. And, according to Laqueur (1998), terrorism has become "far more indiscriminate in its choice of targets" (p. 169).

Terrorism can be either domestic or transnational. Transnational terrorism occurs when a terrorist act in one country involves victims, targets, institutions, governments, or citizens of another country. The 1993 bombing of the World Trade Center in New York exemplifies transnational terrorism. Four Muslim extremists were convicted of the bombing, which killed six people and injured a thousand. In 1998, Ramzi Ahmed Yousef, one of the convicted, was sentenced to 240 years in solitary confinement for his role in the bombing (USIS 1998). Domestic terrorism is exemplified by the 1995 truck bombing of a nine-story federal office building in Oklahoma City, resulting in 168 deaths and more than 200 injured. Gulf War veteran Timothy McVeigh, who, along with Terry Nichols, was convicted of the crime, is reported to have been a member of a paramilitary group that opposes the U.S. government.

Another example of terrorism occurred in 1995, when the prime minister of Israel, Yitzhak Rabin, was assassinated by Yigal Amir, a fellow Jew. Amir was part of an extremist Jewish group that insisted that lands seized by Israel during the Six-Day War in 1967 (the Sinai, West Bank, and Gaza Strip) must remain as part of Israel's biblical birthright. Amir and other extremists opposed Rabin's plan to trade land for peace with the Palestinians. Amir assassinated Rabin in an attempt to thwart the peace process.

The August 7, 1998, bombings of American embassies in Kenya and Tanzania resulted in U.S. air strikes on Afghanistan and Sudan on August 20, 1998.

A government can use both defensive and offensive strategies to fight terrorism. Defensive strategies include using metal detectors at airports and strengthening security at potential targets, such as embassies and military command posts. Offensive strategies include retaliatory raids such as the U.S. bombing of terrorists facilities in Afghanistan, group infiltration, and preemptive strikes. Unfortunately, efforts to stop one kind of terrorism may result in an increase in other types of terrorist acts. For example, after the use of metal detectors in airports increased, the incidence of skyjackings decreased, but the incidence of assassinations increased (Enders and Sandler 1993).

Following the Oklahoma City bombing, President Clinton called for new powers against domestic terrorists, including (1) laws to establish an FBI-run counterterrorism center, (2) creation of a fund for infiltrating suspected domestic terrorist organizations, and (3) legislation to give the FBI more authority to search hotel registers, phone logs, and credit card records (Hagan 1997 168, 175). Further, the 1998 International Crime Control Act calls for actions that will "prevent acts of international crime planned abroad, including terrorist acts, before they occur" (see Chapter 4) (ICCA 1998, 12). In 1999, President Clinton announced he would ask for increased federal funding to combat the terrorist threat of "weapons of mass destruction," that is, chemical, biological, and nuclear weapons (Broder 1999). This chapter's *Self & Society* deals with American attitudes and beliefs about terrorism and one weapon of mass destruction—the nuclear threat.

Since the early 1980s, the Southern Poverty Law Center's Militia Task Force has been instrumental in fighting antigovernment militia groups. For example, in 1986, the Militia Task Force brought a court action that exposed the North Carolina White Patriot Party's thousand-member militia unit's plans to blow up federal facilities with stolen military explosives (Militia Task Force 1995). Over a 15-year period, the Militia Task Force has built extensive computerized files on militias and hate groups. The files include more than 11,000 photographs, information on 200 web sites ("The Patriot Movement" 1998), records of over 61,000 hate activities, reports on 14,000 individuals who have committed hate acts or are affiliated with hate groups, and intelligence on more than 3,200 hate group and militia organizations.

Unlike terrorist activity, which targets civilians and may be committed by lone individuals, **guerrilla warfare** is committed by organized groups opposing a domestic or foreign government and its military forces. Guerrilla warfare often involves small groups who use elaborate camouflage and underground tunnels to hide until they are ready to execute a surprise attack. Since 1945, more than 120 armed guerrilla conflicts have occurred, resulting in the death of over 20 million people (Perdue 1993). Most of these conflicts occurred in developing countries. Fidel Castro's guerrillas in Cuba and the Vietcong guerrillas in Vietnam are examples.

Two theories have been advanced to explain why guerrilla warfare occurs most often in less developed countries (Moaddel 1994). First, as societies change from a traditional to a modern industrialized society, they experience institutional instability, which leads to conflict as various groups compete for control and resources. A second theory holds that the conflicts arise from external international relations. Less developed countries are dependent on more developed countries, which exploit their labor forces and resources. The growing inequality between less developed and more developed countries creates conflict.

Politics in the Post–Cold War Era: Three Possible Scenarios

Porter (1994) outlines three possible scenarios for the future of global security. In the first scenario, civil strife grows out of the extreme dissatisfaction of ethnic communities living in a state dominated by other ethnic groups. Conflict between ethnic groups may ultimately result in each group forming its own independent political community. In Europe, which currently consists of 1,000 or more possible political communities, this scenario would result in the devolution of state structures. Small independent polities would replace centralized states, and Europe would essentially be remedievalized. "Post-modernity," suggests Porter, "would loom as a return to the past" (p. 301).

National Data

It took 18 years, 200 suspects, 20,000 toll-free phone calls, and thousands of interviews to catch confessed Unabomber Theodore Kaczynski.

SOURCE: CNN 1997

International Data

In 1997, there were 304 acts of transnational terrorism leading to 221 deaths and 693 wounded; 7 who died and 21 who were injured were U.S. citizens. Bombs were the most frequently used method of attack.

SOURCE: USIS 1998

National Data

All 50 states now have active antigovernment groups; there are "858 identifiable groups, including 380 armed ones" throughout the United States.

SOURCE: Tharp and Holstein 1997

Nuclear Terrorism

Answer the following questions. This is not a test, and there are no right or wrong answers. When complete, compare your responses to those below from a national sample of U.S. adults.

Statement	Likely	Unlikely
1. The U.S. will get into nuclear war in the next 10 years.	______	
2. Terrorists will use nuclear weapons in the U.S. in the next 10 years.	______	
3. Other countries will use nuclear weapons against each other in the next 10 years.	______	
4. The development of nuclear weapons is a good thing.	______	
Question	**Yes**	**No**
5. Does nuclear proliferation threaten U.S. security?	______	
6. Does nuclear proliferation threaten world peace?	______	
7. Do these countries' nuclear weapons threaten the U.S.?		
China	______	
Russia	______	
Pakistan	______	
India	______	
Britain	______	
8. Would these countries' nuclear weapons threaten the U.S.?		
Iraq	______	
Iran	______	
Israel	______	
Brazil	______	
9. Do these countries' nuclear weapons threaten world peace?		
Pakistan	______	
China	______	
Russia	______	
India	______	
Britain	______	
10. Would these countries' nuclear weapons threaten world peace?		
Iraq	______	
Iran	______	
Israel	______	
Brazil	______	

Results of the 1998 *CNN/USA Today* Gallup Poll

Statement	*Percentage* Likely	*Percentage* Unlikely
1. The U.S. will get into nuclear war in the next 10 years.	37	61
2. Terrorists will use nuclear weapons in the U.S. in the next 10 years.	50	47
3. Other countries will use nuclear weapons against each other in the next 10 years.	71	27
4. The development of nuclear weapons is a good thing.	36	61
5. Does nuclear proliferation threaten U.S. security?	41	58

Nuclear Terrorism *(continued)*

Question	*Yes*	*No*
6. Does nuclear proliferation threaten world peace?	66	32
7. Do these countries' nuclear weapons threaten the U.S.?		
China	57	41
Russia	46	53
Pakistan	43	52
India	26	69
Britain	7	90
8. Would these countries' nuclear weapons threaten the U.S.?		
Iraq	84	14
Iran	80	19
Israel	24	74
Brazil	11	84
9. Do these countries' nuclear weapons threaten world peace?		
Pakistan	66	30
China	61	34
Russia	48	48
India	47	47
Britain	13	83
10. Would these countries' nuclear weapons threaten world peace?		
Iraq	89	9
Iran	83	12
Israel	43	53
Brazil	17	73

SOURCE: From "Many Americans Worry about Nuclear Terrorism." by Keating Holland. **http://www.cnn.com/ALLPOLITICS/1998/06/16/POLL**

In the second scenario, terrorism, low-intensity conflict, and street-level violence replace state-centered, large-scale warfare. As the tools of war slip out of the control of central states and fall into the hands of guerrilla militia forces and terrorist groups, the state will become increasingly powerless.

The third scenario Porter (1994) envisions is entirely different. In this view, independent states are threatened by a return of empire. "In a complex world of economic interdependence and mass communications, large-scale organizations that transcend national borders are growing in importance. . . . By the year 2000 or 2010, it is conceivable that Western Europe will again be a unified empire" (p. 301.)

> Nuclear weapons are inherently dangerous, hugely expensive, militarily inefficient, and morally indefensible.
>
> —G. Lee Butler
> former Commander-in-Chief
> of U.S. Strategic
> Air Command

STRATEGIES FOR ACTION: IN SEARCH OF GLOBAL PEACE

Various strategies and policies are aimed at creating and maintaining global peace. These include the redistribution of economic resources, the creation of a world government, peacekeeping activities of the United Nations, mediation and arbitration, and arms control.

National Data

In 1997, The U.S. federal government spent over $270 billion on national defense. In the same year, U.S. foreign economic aid totaled less than $10 million.

SOURCE: *Statistical Abstract of the United States: 1998,* Tables 569 and 1314

Redistribution of Economic Resources

Inequality in economic resources contributes to conflict and war as the increasing disparity in wealth and resources between rich and poor nations fuels hostilities and resentment. Therefore, any measures that result in a more equal distribution of economic resources are likely to prevent conflict. John J. Shanahan (1995), retired U.S. Navy vice admiral and director of the Center for Defense Information, suggests that wealthy nations can help reduce social and economic roots of conflict by providing economic assistance to poorer countries. Nevertheless, U.S. military expenditures for national defense far outweigh U.S. economic assistance to foreign countries.

As we discussed in Chapter 15, strategies that reduce population growth are likely to result in higher levels of economic well-being. Conversely, unrestrained population growth contributes to economic hardship of impoverished countries. Funke (1994) explains that "rapidly increasing populations in poorer countries will lead to environmental overload and resource depletion in the next century, which will most likely result in political upheaval and violence as well as mass starvation" (p. 326). Although achieving worldwide economic well-being is important for minimizing global conflict, it is important that economic development does not occur at the expense of the environment.

World Government

Some analysts have suggested that world peace might be attained through the establishment of a world government. The idea of a single world government is not new. In 1693, William Penn advocated a political union of all European monarchs, and in 1712, Jacques-Henri Bernardin de Saint-Pierre of France suggested an all-European Union with a "Senate of Peace." Proposals such as these have been made throughout history. President Bush spoke of a new world order that would become possible after the fall of communism in Eastern Europe.

While some commentators are pessimistic about the likelihood of a new world order, Lloyd (1998) identifies three global trends that, he contends, signify the "ghost of a world government yet to come" (p. 28). First, there is the increasing tendency for countries to engage in "ecological good behavior," indicating a concern for a global rather than national well-being. Second, despite several economic crises worldwide, major world players such as the United States and Great Britain have supported rather than abandoned nations in need. And, finally, an International Criminal Court has been created with "powers to pursue, arraign, and condemn those found guilty of war and other crimes against humanity" (p. 28). On the basis of these three trends, Lloyd concludes, "[G]lobal justice, a wraith pursued by peace campaigners for over a century, suddenly seems achievable" (p. 28).

The United Nations

If you want peace, work for justice.

—Pope Paul VI

The United Nations (UN), whose charter begins, "We the people of the United Nations—Determined to save succeeding generations from the scourge of war . . . ," has engaged in over 40 peacekeeping operations since 1948 (United Nations 1998). The UN Security Council can use force, when necessary, to restore international peace and security. Recently, the UN has been involved in overseeing multinational peacekeeping forces in Angola, Lebanon, Bosnia-Herzagovina, and Kosovo. One of its most challenging missions to date is the inspection of weapons facilities in Iraq. When in December of 1998 UN inspectors were refused entry to certain military installations, the United States and Great Britain initiated Operation Desert Fox—4 days of air strikes on Iraqi military targets—at an estimated cost of $450 million (see Table 16.2).

A major problem with the concept of the UN is that its members represent individual nations, not a region or the world. And because nations tend to act in their own best economic and security interests, UN actions performed in the name of world peace may be motivated by nations acting in their own interests. Nevertheless, as the UN celebrates over 50 years of operation, it remains an international agency dedicated to keeping, making, and enforcing peace.

International Data

There are presently 17 United Nations peacekeeping missions underway worldwide, with over 60 countries contributing personnel; the 1997 estimated cost of UN personnel and equipment was $1.2 billion.

SOURCE: UN Chronicle 1997

Mediation and Arbitration

Mediation and arbitration are nonviolent strategies used to resolve conflicts and stop or prevent war. In mediation, a neutral third party intervenes and facilitates negotiation between representatives or leaders of conflicting groups. Mediators do not impose solutions, but rather help disputing parties generate options for resolving the conflict. Ideally, a mediated resolution to a conflict meets at least some of the concerns and interests of each party to the conflict. In other words, mediation attempts to find "win-win" solutions in which each side is satisfied with the solution. Although mediation is used to resolve conflict between individuals, it is also a valuable tool for resolving international conflicts. For example, Irish prime minister Bertie Ahern helped successfully mediate between opposing groups leading to the Good Friday peace agreement in Northern Ireland (Burns 1998).

Nonviolence appeals directly to the mysterious unity among all of us, which is the hidden glory of each of us.

—Michael Nagler
founder of the Peace and Conflict Studies Program at Berkeley

Arbitration also involves a neutral third party who listens to evidence and arguments presented by conflicting groups. Unlike mediation, however, in arbitration the neutral third party arrives at a decision or outcome that the two conflicting parties agree to accept. As Sweet and Brunell (1998) note, **triad dispute resolution,** that is, resolution that involves two disputants and a negotiator—performs "profoundly political functions including the construction, consolidation, and maintenance of political regimes" (p. 64).

Table 16.2 Estimated Cost of Operation Desert Fox

328 tomahawk missles at $750,000 each
90 cruise missles at $1.8 million each
450 laser-guided bombs at $25,000 to $50,000 each
84 gravity bombs at $1,700 each
60 antiradar missles at $320,000 each
2,400 hours flown by aircraft tankers at an average cost per hour of $2,000
Over 300 hours of combat sorties flown at an average cost per hour of $3,500
38 combat and support ships at a cost higher than in peacetime
Estimated Total Cost of Operation Desert Fox to U.S. Taxpayers = More Than $450 Million

SOURCE: Data from "Sticking it to Saddam." 1999. *U.S. News and World Report,* January 11, 37.

In addition to Irish Prime Minister Bertie Ahern, President Clinton and Prime Minister Tony Blair were instrumental in negotiating the Good Friday peace agreement.

Arms Control

In the 1960s, the United States and the Soviet Union led the world in an arms race, each competing to build a more powerful military arsenal than its adversary. If either superpower were to initiate a full-scale war, the retaliatory powers of the other nation would result in the destruction of both nations. Thus, the principle of **mutually assured destruction (MAD)** that developed from nuclear weapons capabilities transformed war from a win-lose proposition to a lose-lose scenario. If both sides would lose in a war, the theory goes, neither side would initiate war.

Due to the end of the Cold War and the growing realization that current levels of weapons literally represented "overkill," governments have moved in the direction of arms control, which involves reducing or limiting defense spending, weapons production, and armed forces. Recent arms-control initiatives include SALT (Strategic Arms Limitation Treaty), START (Strategic Arms Reduction Treaty), NPT (Nuclear Nonproliferation Treaty), and the CTBT (Comprehensive Test Ban Treaty).

National Data

In a recent national survey of U.S. registered voters, 69 percent responded that a goal of the United States should be to reduce or eliminate nuclear weapons, 14 percent favored building "new or better nuclear weapons," 13 percent favored maintaining current levels, and 4 percent were unsure.

SOURCE: Kimball 1998

Strategic Arms Limitation Treaty

Under the 1972 SALT agreement (SALT I), the United States and the Soviet Union agreed to limit both their defensive weapons and their land-based and submarine-based offensive weapons. Also in 1972, Henry Kissinger drafted the Declaration of Principles, known as **detente**, which means "negotiation rather than confrontation." A further arms limitation agreement (SALT II) was reached in 1979, but was never ratified by Congress due to the Soviet invasion of Afghanistan. Subsequently, the arms race continued with the development of new technologies and an increase in the number of nuclear warheads.

> We are closer to nuclear war than we ever were in the last generation of the cold war.
>
> —U.S. Senator Daniel Patrick Moynihan

Strategic Arms Reduction Treaty Strategic arms talks resumed in 1982, but made relatively little progress for several years. During this period, President Reagan proposed the Strategic Defense Initiative (SDI), more commonly known as "Star Wars," which purportedly would be able to block missiles launched by another country against the United States (Brown 1994). Although some research was conducted on the system, Star Wars was never actually built. However, in 1999, President Clinton proposed spending $6.6 billion on further development of a "national missle-defense shield by 2005" (Thompson 1999, 1).

By 1991, the international situation had changed. The communist regime in the Soviet Union had fallen, the Berlin Wall had been dismantled, and many Eastern European and Baltic countries were under self-rule. SALT was renamed START (Strategic Arms Reduction Treaty) and was signed in 1991. A second START agreement, signed in 1993 and ratified by the U.S. Senate, signaled the end of the Cold War. START II calls for the reduction of nuclear warheads to 3,500 by the year 2003, a significant reduction from present levels (Zimmerman 1997). To date, START II awaits ratification by the Russian Parliament, thereby delaying any official negotiations on START III.

> The real danger is that we will be tempted more and more to resort to the use of military forces as we diminish our capacity to respond by other means. As the saying goes, if all you have is a hammer then every problem starts to look like a nail.
>
> –Vice Admiral John J. Shanahan, U.S. Navy (retired) Director of the Center for Defense Information

Nuclear Nonproliferation Treaty The Nuclear Nonproliferation Treaty (NPT) was signed by 156 countries in 1970. The agreement held that countries without nuclear weapons would not to try to get them; in exchange, the nuclear-capable countries (the United States, United Kingdom, France, China, Russia, and India) agreed they would not provide nuclear weapons to countries that did not have them. The 1994 Nuclear Proliferation Act requires that the United States impose economic sanctions on any country that violates the NPT (McGeary 1998).

However, even if military superpowers honor agreements to limit arms, the availability of black market nuclear weapons and materials presents a threat to global security. For example, Kyl and Halperin (1997) note that U.S. security is threatened more by nuclear weapons falling into the hands of a terrorist group than by a nuclear attack from an established government such as Russia. The authors further conclude that if Russia were to launch missiles directed at the United States "the odds are overwhelming that it [would] be a Russian missile fired by accident or without authority" (p. 28).

International Data

The combined total of U.S. and Russian warheads is estimated to be 34,000. Only 500 to 2,000 are needed to induce a nuclear winter and destroy most of life on Earth.

SOURCES: Sagan 1990; McGeary 1998

Comprehensive Test Ban Treaty On September 10, 1996, the UN General Assembly passed the Comprehensive Test Ban Treaty by a vote of 158 to 3. The treaty, a "prime disarmament goal for more than forty years" (UCS 1998, 1), would put an end to underground nuclear testing. President Clinton, the first world leader to sign the ban, submitted it to the U.S. Senate for ratification in September 1997. Britain and France have already ratified the agreement. For the CTBT to be enforced, it must be ratified by the 44 members of the Conference on Disarmament. Both India and Pakistan, which tested nuclear devices in 1998, are members of the Conference (UCS 1998).

> We will never achieve a world free of nuclear weapons unless we first achieve a world free of nuclear explosions.
>
> –John Holum Director U.S. Arms Control and Disarmament Agency

UNDERSTANDING CONFLICT AROUND THE WORLD

As we come to the close of this chapter, how might we have an informed understanding of conflict around the world? Each of the three theoretical positions discussed in this chapter reflects the realities of war. As functionalists argue, war offers societal benefits—social cohesion, economic prosperity, scientific and technological developments, and social change. Further, as conflict theorists contend, wars often occur for economic reasons as corporate elites and political leaders benefit from the spoils of war—land and water resources and raw materials. The symbolic interactionist perspective emphasizes the role that meanings, labels, and definitions play in creating conflict and contributing to acts of war.

Ultimately, we are all members of one community—Earth—and have a vested interest in staying alive and protecting the resources of our environment for our own and future generations. But conflict between groups is a feature of social life and human existence that is not likely to disappear. What is at stake—human lives and the ability of our planet to sustain life—merits serious attention. Traditionally, nations have sought to protect themselves by maintaining large military forces and massive weapons systems. These strategies are associated with serious costs. In diverting resources away from other social concerns, militarism undermines a society's ability to improve the overall security and well-being of its citizens. Conversely, defense-spending cutbacks can potentially free up resources for other social agendas, including lowering taxes, reducing the national debt, addressing environmental concerns, eradicating hunger and poverty, improving health care, upgrading educational services, and improving housing and transportation. Therein lies the promise of a "peace dividend."

Hopefully, future dialogue on the problem of global conflict and war will redefine national and international security to encompass social, economic, and environmental concerns. These other concerns play a vital role in the security of nations and the world. The World Commission on Environment and Development (1990) concluded:

> **The deepening and widening environmental crisis presents a threat to national security—and even survival—that may be greater than well-armed, ill-disposed neighbors and unfriendly alliances. . . . The arms race in all parts of the world—pre-empts resources that might be used more productively to diminish the security threats created by environmental conflict and the resentments that are fueled by widespread poverty. . . . There are no military solutions to environmental insecurity.**

National and global policies aimed at reducing poverty and ensuring the health of our planet and its present and future inhabitants are important aspects of world peace. But as long as we define national and global security in military terms, we will likely ignore the importance of nonmilitary social policies in achieving world peace. According to Funke (1994), changing the definition of national security "is the first step to changing policy" (p. 342).

International Data

Six countries have declared nuclear weapons capabilities (United States, Russia, France, Great Britain, China, and India), two undeclared (Israel and Pakistan), and four are suspected of having nuclear weapons capabilities (Iran, Iraq, North Korea, and Libya).

SOURCE: McGeary 1998, 38–39

> The first victim of war is not truth—it is conscience. Otherwise human beings could not bear the guilt of participating in a savage conflict in which armed force is not a last resort, but a hair trigger.
>
> —William Perdue
> Sociologist

> A human being is part of the whole . . . He experiences himself, his thoughts and feelings, as something separate from the rest, a kind of optical delusion of his consciousness. This delusion is a kind of prison for us, restricting us to our personal desires and to affection for a few persons nearest to us. Our task must be to free ourselves from this prison by widening our circle of compassion to embrace all living creatures.
>
> —Albert Einstein
> Scientist

CRITICAL THINKING

1. Certain actions constitute "war crimes." Such actions include the use of forbidden munitions such as biological weapons, purposeless destruction, killing civilians, poisoning of waterways, and violation of surrender terms. In addition to those listed above, what other actions should constitute "war crimes"?

2. Describe countries that have the highest probability of going to war in terms of their economic, social, and psychological makeup. Now, describe countries that are the least likely to go to war. Does history confirm your hypotheses?
3. Selecting each of the five major institutions in society, what part could each play in attaining global peace?
4. Make a list of famous war movies (e.g., *Apolcolypse Now, Born on the Fourth of July, Platoon*). With specific movies in mind, list media sounds and images of war. Has the portrayal of war in movies changed over time? If so, how and why?

KEY TERMS

Cold War
detente
economic conversion
guerrilla warfare
infowar
military-industrial complex
mutually assured destruction (MAD)
nuclear winter
peace dividend
post-traumatic stress disorder
state
terrorism
triad dispute resolution
war

INTERNET

You can find more information on the military, terrorism, the threat of nuclear war, and nuclear disarmament at the *Understanding Social Problems* web site at **http://sociology.wadsworth.com.**

INFOTRAC COLLEGE EDITION

Either from the Wadsworth sociology resource center at **http://sociology.wadsworth.com** or directly from your web browser, you may access InfoTrac College Edition, an online university library that includes over 700 popular and scholarly journals in which you can find articles related to the topics in this chapter. Suggested articles and questions relating to these articles are listed below.

Laqueur, Walter. 1998. "The New Face of Terrorism." *The Washington Quarterly* 21(4):169–80.

1. What problems are social scientists having with analyzing the mind-set of terrorists today?
2. Why do traditional terrorist groups show no interest in weapons of mass destruction?
3. What three types of terrorist are willing to make use of weapons of mass destruction?

Lloyd, John. 1998. "The Dream of Global Justice." *New Statesman* 127(4404):28–30.

1. What three recent developments seem to be leading to the idea of a world government?

2. What is the ICC, and over what crimes will it have jurisdiction?
3. Why has the United States not signed onto the ICC, and what methods have been proposed to address this?

Maier, Timothy, and Tiffany Danitz. 1997. "PLA Espionage Means Business." *Insight on the News* 13(11):8–15.

1. What are the PLA, the PRC, and the DNC?
2. What do the authors argue is the relationship between the United States and the PLA?
3. What theoretical perspective (structural-functionalism, conflict theory, or symbolic interactionism) best explains the Clinton-Wang incident?

Epilogue

Never doubt that a small group of thoughtful, committed citizens can change the world. Indeed, it's the only thing that ever has.

Margaret Mead,
Anthropologist

Today, there is a crisis—a crisis of faith: faith in the ideals of equality and freedom, faith in political leadership, faith in the American dream, and, ultimately, faith in the inherent goodness of humankind and the power of one individual to make a difference. To some extent, faith is shaken by texts such as this one. Crimes are up; marriages are down; political corruption is everywhere; bigotry's on the rise; the environment is killing us—if we don't kill it first. Social problems are everywhere, and what's worse, many solutions only seem to create more problems.

The transformation of American society in recent years has been dramatic. With the exception of the Industrial Revolution, no other period in human history has seen such rapid social change. The structure of society, forever altered by such macrosociological processes as multinationalization, deindustrialization, and globalization, continues to be characterized by social inequities—in our schools, in our homes, in our cities, and in our salaries.

The culture of society has also undergone rapid change, leading many politicians and lay persons alike to call for a return to traditional values and beliefs and to emphasize the need for moral education. The implication is that somehow things were better in the "good old days," and if we could somehow return to those times, things would be better again. Some things were better—for some people.

A pessimist sees the difficulty in every opportunity: an optimist sees the opportunity in every difficulty.

–Winston Churchill
British statesman

Fifty years ago, there were fewer divorces and less crime. AIDS and crack cocaine were unheard of, and violence in schools was almost nonexistent. At the same time, however, in 1950, the infant mortality rate was over three times what it is today; racial and ethnic discrimination flourished in an atmosphere of bigotry and hate, and millions of Americans were routinely denied the right to vote because of the color of their skin; more than half of all Americans smoked cigarettes; and persons over the age of 25 had completed a median of 6.8 years of school.

The social problems of today are the cumulative result of structural and cultural alterations over time. Today's problems are not necessarily better or worse than those of generations ago—they are different and, perhaps, more diverse as a result of the increased complexity of social life. But, as surely as we brought the infant mortality rate down, prohibited racial discrimination in education, housing, and employment, increased educational levels, and reduced the number of smokers, we can continue to meet the challenges of today's social problems. But how does positive social change occur? How does one alter something as amorphous as society? The answer is really quite simple. All social change takes place because of the acts of individuals. Every law, every regulation and policy, every social movement and media exposé, and every court decision began with one person.

> A human being is part of the whole . . . He experiences himself, his thoughts and feelings, as something separate from the rest, a kind of optical delusion of his consciousness. This delusion is a kind of prison for us, restricting us to our personal desires and to affection for a few persons nearest to us. Our task must be to free ourselves from this prison by widening our circle of compassion to embrace all living creatures.
>
> –Albert Einstein
> Scientist

Sociologist Earl Babbie (1994) recounts how the behavior of one person—Rosa Parks—made a difference. Rosa Parks was a seamstress in Montgomery, Alabama, in the 1950s. Like almost everything else in the South in the 1950s, public transportation was racially segregated. On December 1, 1955, Rosa Parks was on her way home from work when the "white section" of the bus she was riding became full. The bus driver told black passengers in the first row of the black section to relinquish their seats to the standing white passengers. Rosa Parks refused.

She was arrested and put in jail, but her treatment so outraged the black community that a boycott of the bus system was organized by a new minister in town—Martin Luther King Jr. The Montgomery bus boycott was a success. Just 11 months later, in November of 1956, the U.S. Supreme Court ruled that racial segregation of public facilities was unconstitutional. Rosa Parks had begun a process that in time would echo her actions—the civil rights movement, the March on Washington, the 1963 Equal Pay Act, the 1964 Civil Rights Act, the 1965 Voting Rights Act, regulations against discrimination in housing, and affirmative action.

Was social change accomplished? In 1960, the median school years completed by black and white Americans differed by nearly 3 years; by 1995, that difference was less than 1. While many would point out that such changes and thousands like them have created other problems that need to be addressed, who among us would want to return to the "good old days" of the 1950s in Montgomery, Alabama?

> The only thing necessary for the triumph of evil is for good men [and women] to do nothing.
>
> –Edmund Burke
> English political writer/orator

Millions of individuals make a difference daily. Chuck Beattie and Bret Byfield, two social workers in Minneapolis, began Phoenix Group in 1991. Purchasing houses with a few grants and some private donations, they hired "street people" to renovate the houses and then let them move in. Today, Phoenix Group has more than 39 properties, 300 residents, and 11 businesses. Susan Brotchie, deserted by her husband and unable to get child support, began Advocates for Better Child Support, which has helped more than 9,000 people establish and collect child support payments. Pedro José Greer was an intern in Miami when he treated his first homeless patient. Appalled by the fatal incidence of tuberculosis, a curable disease, Dr. Greer opened a medical clinic in a homeless shelter. Today, he heads the largest provider of medical care for the poor in Florida—Camillus Health Concern—annually serving over 4,500 patients (Chinni et al. 1995, 34). After suffering chronic illness from exposure to air pollutants from a sewage plant and home renovation materials, Mary Lamielle founded the National Center for Environmental Health Strategies (NCEHS), a national nonprofit organization dedicated to the development of creative solutions to environmental health problems (Lamielle 1995). Through her work as director of the NCEHS, Lamielle has influenced policy development and research and has provided support and advocacy to sufferers of environmental pollution.

> Some people see things that are and say "Why?" He saw things that never were and said, "Why not?"
>
> –Ted Kennedy of his brother, Bobby Kennedy

College students have also worked to bring about social change. According to Paul Loeb (1995), college students have prompted the adoption of multicultural curriculums, helped the homeless, and made their schools more environmentally accountable. College students also influenced universities to rid themselves of South African investments and played an important role in building the international movement that helped end apartheid.

While only a fraction of the readers of this text will occupy social roles that directly influence social policy, one need not be a politician or member of a social reform group to make a difference. We, the authors of this text, challenge you, the reader, to make individual decisions and take individual actions to make the world a more humane, just, and peaceful place for all. Where should we begin? Where Rosa Parks and others like her began—with a simple individual act of courage, commitment, and faith.

Appendix A
Methods of Data Analysis

There are three levels of data analysis: description, correlation, and causation. Data analysis also involves assessing reliability and validity.

DESCRIPTION

Qualitative research involves verbal descriptions of social phenomena. Having a homeless and single pregnant teenager describe her situation is an example of qualitative research.

Quantitative research often involves numerical descriptions of social phenomena. Quantitative descriptive analysis may involve computing the following: (1) means (averages), (2) frequencies, (3) mode (the most frequently occurring observation in the data), (4) median (the middle point in the data; half of the data points are above, and half are below the median), and (5) range (the highest and lowest values in a set of data).

CORRELATION

Researchers are often interested in the relationship between variables. *Correlation* refers to a relationship among two or more variables. The following are examples of correlational research questions: What is the relationship between poverty and educational achievement? What is the relationship between race and crime victimization? What is the relationship between religious affiliation and divorce?

If there is a correlation or relationship between two variables, then a change in one variable is associated with a change in the other variable. When both variables change in the same direction, the correlation is positive. For example, in general, the more sexual partners a person has, the greater the risk of contracting a sexually transmissible disease. As variable A (number of sexual partners) increases, variable B

Figure A.1 **Positive Correlation**

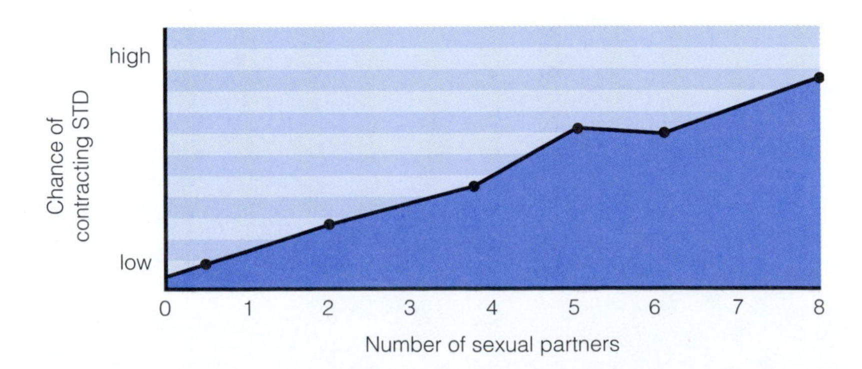

(chance of contracting an STD) also increases. Similarly, as the number of sexual partners decreases, the chance of contracting an STD decreases. Notice that in both cases, the variables change in the same direction, suggesting a positive correlation (see Figure A.1).

When two variables change in opposite directions, the correlation is negative. For example, there is a negative correlation between condom use and contracting STDs. In other words, as condom use increases, the chance of contracting an STD decreases (see Figure A.2).

The relationship between two variables may also be curvilinear, which means that they vary in both the same and opposite directions. For example, suppose a researcher finds that after drinking one alcoholic beverage, research participants are more prone to violent behavior. After two drinks, violent behavior is even more likely, and this trend continues for three and four drinks. So far, the correlation between alcohol consumption and violent behavior is positive. After the research participants have five alcoholic drinks, however, they become less prone to violent behavior. After six and seven drinks, the likelihood of engaging in violent behavior decreases further. Now the correlation betweeen alcohol consumption and violent behavior is negative. Because the correlation changed from positive to negative, we say that the correlation is curvilinear (the correlation may also change from negative to positive) (see Figure A.3).

A fourth type of correlation is called a spurious correlation. Such a correlation exists when two variables appear to be related, but the apparent relationship occurs only because they are both related to a third variable. When the third variable is controlled through a statistical method in which the variable is held constant, the apparent relationship between the variables disappears. For example, blacks have a lower average life expectancy than whites do. Thus, race and life expectancy appear to be related. However, this apparent correlation exists because both race and life expectancy are related to socioeconomic status. Since blacks are more likely than whites to be impoverished, they are less likely to have adequate nutrition and medical care.

CAUSATION

If the data analysis reveals that two variables are correlated, we know only that a change in one variable is associated with a change in another variable. We cannot assume, however, that a change in one variable *causes* a change in the other variable unless our data collection and analysis are specifically designed to assess causation. The research method that best assesses causality is the experimental method (discussed in Chapter 1).

To demonstrate causality, three conditions must be met. First, the data analysis must demonstrate that variable A is correlated with variable B. Second, the data analysis must demonstrate that the observed correlation is not spurious. Third, the

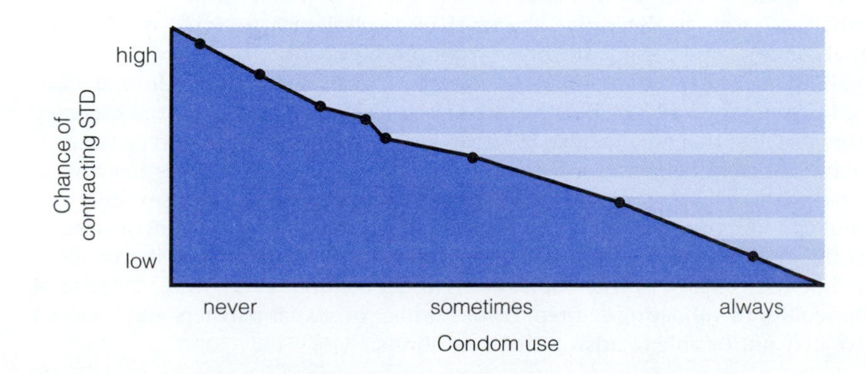

Figure A.2 **Negative Correlation**

Figure A.3 Curvilinear Correlation

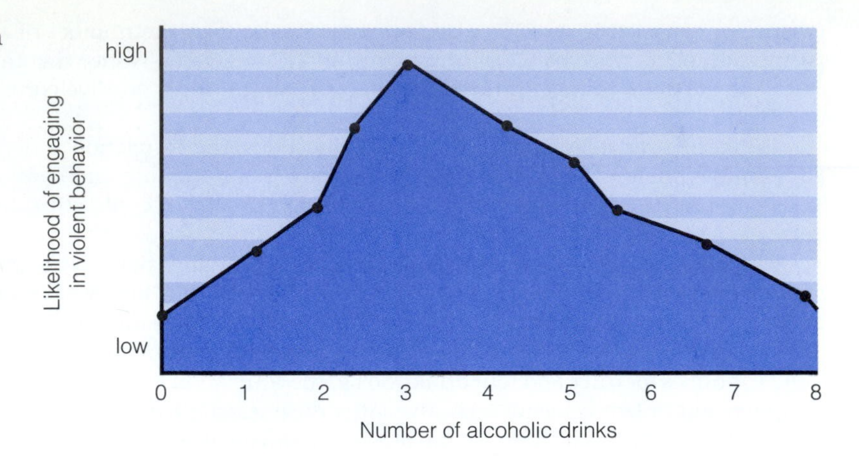

analysis must demonstrate that the presumed cause (variable A) occurs or changes prior to the presumed effect (variable B). In other words, the cause must precede the effect.

It is extremely difficult to establish causality in social science research. Therefore, much social research is descriptive or correlative, rather than causative. Nevertheless, many people make the mistake of interpreting a correlation as a statement of causation. As you read correlative research findings, remember the following adage: "Correlation *does not* equal causation."

RELIABILITY AND VALIDITY

Assessing reliability and validity is an important aspect of data analysis. *Reliability* refers to the consistency of the measuring instrument or technique; that is, the degree to which the way information is obtained produces the same results if repeated. Measures of reliability are made on scales and indexes (such as those in the *Self-Assessments* in this text) and on information-gathering techniques, such as the survey methods described in Chapter 1.

Various statistical methods are used to determine reliability. A frequently used method is called the "test-retest method." The researcher gathers data on the same sample of people twice (usually one or two weeks apart) using a particular instrument or method and then correlates the results. To the degree that the results of the two tests are the same (or highly correlated), the instrument or method is considered reliable.

Measures that are perfectly reliable may be absolutely useless unless they also have a high validity. *Validity* refers to the extent to which an instrument or device measures what it intends to measure. For example, police officers administer "breathalyzer" tests to determine the level of alcohol in a person's system. The breathalyzer is a valid test for alcohol consumption.

Validity measures are important in research that uses scales or indices as measuring instruments. Validity measures are also important in assessing the accuracy of self-report data that are obtained in survey research. For example, survey research on high-risk sexual behaviors associated with the spread of HIV relies heavily on self-report data on such topics as number of sexual partners, types of sexual activities, and condom use. Yet, how valid are these data? Do survey respondents underreport the number of their sexual partners? Do people who say they use a condom every time they engage in intercourse really use a condom every time? Because of the difficulties in validating self-reports of number of sexual partners and condom use, we may not be able to answer these questions.

ETHICAL GUIDELINES IN SOCIAL PROBLEMS RESEARCH

Social scientists are responsible for following ethical standards designed to protect the dignity and welfare of people who participate in research. These ethical guidelines include the following (Nachmias and Nachmias 1987; American Sociological Association 1984; Committee for the Protection of Human Participants in Research 1982):

1. *Freedom from coercion to participate.* Research participants have the right to decline to participate in a research study or to discontinue participation at any time during the study. For example, professors who are conducting research using college students should not require their students to participate in their research.
2. *Informed consent.* Researchers are required to inform participants "of all aspects of the research that might reasonably be expected to influence willingness to participate" (Committee for the Protection of Human Participants in Research 1982, 32). After informing potential participants about the nature of the research, researchers typically ask participants to sign a consent form indicating that the participants are informed about the research and agree to participate in it.
3. *Deception and debriefing.* Sometimes the researcher must disguise the purpose of the research in order to obtain valid data. Researchers may deceive participants as to the purpose or nature of a study only if there is no other way to study the problem. When deceit is used, participants should be informed of this deception (debriefed) as soon as possible. Participants should be given a complete and honest description of the study and why deception was necessary.
4. *Protection from harm.* Researchers must protect participants from any physical and psychological harm that might result from participating in a research study. This is both a moral and a legal obligation. It would not be ethical, for example, for a researcher studying drinking and driving behavior to observe an intoxicated individual leaving a bar, getting into the driver's seat of a car, and driving away.

   Researchers are also obligated to respect the privacy rights of research participants. If anonymity is promised, it should be kept. Anonymity is maintained in mail surveys by identifying questionnaires with a number coding system, rather than with the participants' names. When such anonymity is not possible, as is the case with face-to-face interviews, researchers should tell participants that the information they provide will be treated as confidential. Although interviews may be summarized and excerpts quoted in published material, the identity of the individual participants is not revealed. If a research participant experiences either physical or psychological harm as a result of participation in a research study, the researcher is ethically obligated to provide remediation for the harm.
5. *Reporting of research.* Ethical guidelines also govern the reporting of research results. Researchers must make research reports freely available to the public. In these reports, researchers should fully describe all evidence obtained in the study, regardless of whether the evidence supports the researcher's hypothesis. The raw data collected by the researcher should be made available to other researchers who might request it for purposes of analysis. Finally, published research reports should include a description of the sponsorship of the research study, its purpose, and all sources of financial support.

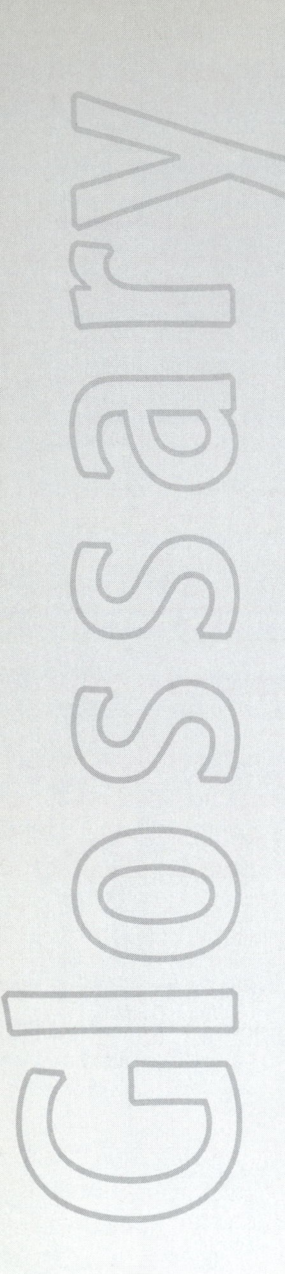

Glossary

abortion The intentional termination of a pregnancy.

absolute poverty The chronic absence of the basic necessities of life, including food, clean water, and housing.

acculturation Learning the culture of a group different from the one in which a person was originally raised.

achieved status A status assigned on the basis of some characteristic or behavior over which the individual has some control.

acid rain The mixture of precipitation with air pollutants, such as sulfur dioxide and nitrogen oxide.

acquaintance rape Rape that is committed by someone known by the victim.

activity theory A theory that emphasizes that the elderly disengage, in part, because they are structurally segregated and isolated with few opportunities to engage in active roles.

acute condition A health condition that can last no more than 3 months.

adaptive discrimination Discrimination that is based on the prejudice of others.

affirmative action A series of programs based on the 1964 Civil Rights Act that provide opportunities or other benefits to persons on the basis of their group membership (e.g., race, gender, ethnicity).

age grading The assignment of social roles to given chronological ages.

age pyramid A graphlike presentation that shows the percentage of a population in various age groups.

ageism The belief that age is associated with certain psychological, behavioral, and/or intellectual traits.

aggregate poverty gap The total amount of income necessary to lift all persons who are below the poverty line up to that level.

Aid to Families with Dependent Children (AFDC) Before 1996, a cash assistance program that provided single parents (primarily women) and their children with a minimum monthly income.

alienation The concept used by Karl Marx to describe the condition when workers feel powerlessness and meaninglessness as a result of performing repetitive, isolated work tasks. Alienation involves becoming estranged from one's work, the products one creates, other human beings, and/or one's self; it also refers to powerlessness and meaninglessness experienced by students in traditional, restrictive educational institutions.

alternative certification programs Programs that permit college graduates, without education degrees, to be certified to teach based on job and/or life experiences.

amalgamation The physical blending of different racial and/or ethnic groups, resulting in a new and distinct genetic and cultural population; results from the intermarriage of racial and ethnic groups over generations.

anomie A state of normlessness in which norms and values are weak or unclear; results from rapid social change and is linked to many social problems, including crime, drug addiction, and violence.

anorexia nervosa An eating disorder characterized by weight loss, excessive exercise, food aversion, distorted body image, and an intense and irrational fear of body fat and weight gain.

Antabuse A prescribed medication, which when combined with alcohol produces severe nausea.

antimiscegenation laws Laws that prohibited interracial marriages.

ascribed status A status that society assigns to an individual on the basis of factors over which the individual has no control.

assimilation The process by which minority groups gradually adopt the cultural patterns of the dominant majority group.

asylees Immigrants who apply from within the United States for admission on the basis of persecution or fear of persecution for their political or religious beliefs.

automation A type of technology in which self-operated machines accomplish tasks formerly done by workers; develops as a society moves toward industrialization and becomes more concerned with the mass production of goods.

beliefs Definitions and explanations about what is assumed to be true.

bilingual education Educational instruction provided in two languages—the student's native language and another language. In the United States, bilingual education involves teaching individuals in both English and their non-English native language.

biodiversity The variability of living organisms on earth.

biphobia Negative attitudes toward bisexuality and people who identify as bisexual.

bisexuality A sexual orientation that involves cognitive, emotional, and sexual attraction to members of both sexes.

blended family A family that consists of remarried spouses with at least one of the spouses having a child from a previous relationship.

bonded labor The repayment of a debt through labor.

bourgeoisie The owners of the means of production.

Brady Bill A law that requires a 5-day waiting period for handgun purchases so that sellers can screen buyers for criminal records or mental instability.

brain drain The phenomenon whereby in developing countries, many individuals with the highest level of skill and education leave the country in search of work abroad.

bulimia nervosa An eating disorder characterized by cycles of binge-eating and purging (self-induced vomiting, use of laxatives, and/or use of diuretics).

burden of disease The number of deaths in a population combined with the impact of premature death and disability on that population.

capitalism An economic system in which private individuals or groups invest capital to produce goods and services, for a profit, in a competitive market.

charter schools Public schools founded by parents, teachers, and communities, and maintained by school tax dollars.

chemical dependency A condition in which drug use is compulsive, and users are unable to stop because of physical and/or psychological dependency.

child abuse The physical or mental injury, sexual abuse, negligent treatment, or maltreatment of a child under the age of 18 by a person who is responsible for the child's welfare.

child labor Children performing work that is hazardous, that interferes with a child's education, or that harms a child's health, or physical, mental, spiritual, or moral development.

chronic condition A long-term health problem, such as a disease or impairment.

classic rape A rape committed by a stranger, with the use of a weapon, resulting in serious bodily injury.

Cold War The state of political tension and military rivalry that existed between the United States and the former Soviet Union from the 1950s through the late 1980s.

colonialism When a racial and/or ethnic group from one society takes over and dominates the racial and/or ethnic group(s) of another society.

coming out Short for "coming out of the closet"; the process in which a person recognizes her or his homosexuality and discloses it to others.

community policing A type of policing in which uniformed police officers patrol and are responsible for certain areas of the city as opposed to simply responding to crimes as they occur.

compressed workweek Workplace option in which employees work full-time, but in 4 rather than 5 days.

computer crime Any violation of the law in which a computer is the target or means of criminal activity.

conflict perspective A sociological perspective that views society as comprising different groups and interests competing for power and resources.

contingent workers (also called "disposable workers") Involuntary part-time workers, temporary employees, and workers who do not perceive themselves as having an explicit or implicit contract for ongoing employment

control theory A theory that argues that a strong social bond between a person and society constrains some individuals from violating norms.

convergence hypothesis The argument that capitalist countries will adopt elements of socialism and socialist countries will adopt elements of capitalism, that is, they will converge.

cooperative learning Learning in which a heterogeneous group of students, of varying abilities, help one another with either individual or group assignments.

corporate downsizing The corporate practice of discharging large numbers of employees. Simply put, the term "downsizing" is a euphemism for mass firing of employees.

corporate multinationalism The practice of corporations to have their home base in one country and branches, or affiliates, in other countries.

corporate violence The production of unsafe products and the failure of corporations to provide a safe working environment for their employees.

corporate welfare Laws and policies that favor corporations, such as low-interest government loans to failing businesses and special subsidies and tax breaks to corporations.

covenant marriage A type of marriage offered in Louisiana that permits divorce only under condition of fault or after a marital separation of more than 2 years.

crime An act or the omission of an act that is a violation of a federal, state, or local law and for which the state can apply sanctions.

cultural imperialism The indoctrination into the dominant culture of a society; when cultural imperialism exists, the norms, values, traditions, and languages of minorities are systematically ignored.

cultural lag A condition in which the material part of the culture changes at a faster rate than the nonmaterial part.

cultural sexism The ways in which the culture of society perpetuates the subordination of individuals based on their sex classification.

culture of poverty The set of norms, values, and beliefs and self-concepts that contribute to the persistence of poverty among the underclass.

cumulative trauma disorders The most common type of workplace illness in the United States; includes muscle, tendon, vascular, and nerve injuries that result from repeated or sustained actions or exertions of different body parts. Jobs that are associated with high rates of upper body cumulative stress disorders include computer programming, manufacturing, meat packing, poultry processing, and clerical/office work.

cybernation The use of machines that control other machines in the production process; characteristic of postindustrial societies that emphasize service and information professions.

de facto segregation Segregation that is not required by law, but exists "in fact," often as a result of housing and socioeconomic patterns.

de jure segregation Segregation that is required by law.

deconcentration The redistribution of the population from cities to suburbs and surrounding areas.

decriminalization The removal of criminal penalties for a behavior, as in the decriminalization of drug use.

Defense of Marriage Act The Defense of Marriage Act states that marriage is a legal union between one man and one woman.

defensible space A neighborhood that is structurally arranged in such a way as to prevent or reduce crime (e.g., adding speed bumps to roads, dividing larger neighborhoods into smaller ones to facilitate community bonding, and closing off selected streets to deter "cruising").

defensive medicine A practice in which health care providers use tests and/or treatments that may be unnecessary in order to guard against malpractice suits.

deforestation The destruction of the earth's rain forests.

deindustrialization The loss and/or relocation of manufacturing industries.

deinstitutionalization The removal of individuals with psychiatric disorders from mental hospitals and large residential institutions and into outpatient community mental health centers.

demographic transition theory A theory that attributes population growth patterns to changes in birthrates and death rates associated with the process of industrialization. In preindustrial societies, the population remains stable because, although the birthrate is high, the death rate is also high. As a society becomes industrialized, the birthrate remains high, but the death rate declines, causing rapid population growth. In societies with advanced industrialization, the birthrate declines, and this decline, in conjunction with the low death rate, slows population growth.

demography The study of the size, distribution, movement, and composition of human populations.

dependent variable The variable that the researcher wants to explain. See also *independent variable*.

deregulation The reduction of government control of, for example, certain drugs.

desertification The expansion of deserts and the loss of usable land due to the overuse of semiarid land on the desert margins for animal grazing and obtaining firewood.

deskilling The tendency for workers in a postindustrial society to make fewer decisions and for labor to require less thought.

detente The philosophy of "negotiation rather than confrontation" in reference to relations between the United States and the former Soviet Union; put forth by Henry Kissinger's Declaration of Principles in 1972.

deterrence The use of harm or the threat of harm to prevent unwanted behaviors.

devaluation hypothesis The hypothesis that argues that women are paid less because the work they perform is socially defined as less valuable than the work performed by men.

dichotomous model of sexual orientation A conceptual model of sexual orientation that views individuals as either homosexual or heterosexual. According to this model of sexual orientation, individuals exhibiting bisexual feelings and behavior are, in actuality, either homosexual or heterosexual and are masking or denying their true sexual orientation.

differential association A theory developed by Edwin Sutherland that holds that through interaction with others, individuals learn the values, attitudes, techniques, and motives for criminal behavior.

disability-adjusted life year (DALY) Years lost to premature death and years lived with illness or disability. More simply, 1 DALY equals 1 year of healthy life.

discrimination Differential treatment of individuals based on their group membership.

discriminatory unemployment High rates of unemployment among particular social groups such as racial and ethnic minorities and women.

disengagement theory A theory claiming that the elderly disengage from productive social roles in order to relinquish these roles to younger members of society. As this process continues, each new group moves up and replaces another, which, according to disengagement theory, benefits society and all of its members.

diversity training Workplace training programs designed to increase employees' awareness of cultural differences in the workplace and how these differences may affect job performance.

divorce law reform Policies and proposals designed to change divorce law. Usually, divorce law reform measures attempt to make divorce more difficult to obtain.

divorce mediation A process in which divorcing couples meet with a neutral third party (mediator) who assists the individuals in resolving such issues as property division, child custody, child support, and spousal support in a way that minimizes conflict and encourages cooperation.

domestic partnership A status that grants legal entitlements such as health insurance benefits and inheritance rights to heterosexual cohabiting couples and homosexual couples.

double jeopardy See *multiple jeopardy.*

doubling time The time it takes for a population to double in size from any base year.

dowry deaths In some countries, the killing of a woman if she brings an insufficient dowry into the marriage.

drug Any substance other than food that alters the structure and functioning of a living organism when it enters the bloodstream.

drug abuse The violation of social standards of acceptable drug use, resulting in adverse physiological, psychological, and/or social consequences.

dumbing down The lowering of educational standards or expectations by students and/or teachers.

Earned Income Tax Credit (EITC) A refundable tax credit based on a working family's income and number of children. In addition to the federal EITC, many states also have a state EITC.

e-commerce The buying and selling of goods and services over the Internet.

ecofeminism A synthesis of feminism, environmentalism, and antimilitarism.

economic conversion The reallocation of resources from military forces and defense industries to other sectors of the economy.

economic institution The structure and means by which a society produces, distributes, and consumes goods and services.

endogamy The social norm that influences people to marry within their social group and discourages interracial and interethnic marriages.

Employment Nondiscrimination Act The Employment Nondiscrimination Act (ENDA) has been promoted by gay rights activists in order to prohibit discrimination, preferential treatment, and quotas in the workplace on the basis of sexual orientation.

empowerment zones/enterprise communities Low-income areas, often in urban areas, that are designated to receive tax incentives for businesses locating there with the stipulation that they must hire a certain proportion of local employees. Enterprise zones/enterprise communities stimulate the local economy, create jobs for the unemployed, and increase city revenues.

environmental injustice (see also *environmental racism*) The tendency for socially and politically marginalized groups to bear the brunt of environmental ills.

environmental racism The tendency for U.S. hazardous waste sites and polluting industries to be located in areas where the surrounding residential population is black, Native American, or Hispanic.

epidemiological transition The shift from a society characterized by low life expectancy and parasitic and infectious diseases to one characterized by high life expectancy and chronic and degenerative diseases.

epidemiologist A scientist who studies the social origins and distribution of health problems in a population and how patterns of illness and disease vary between and within societies.

epidemiology The study of the distribution of disease within a population.

equity principle The belief that those who have more resources in society deserve them because of their greater contributions to society.

ergonomics The designing or redesigning of the workplace to prevent and reduce cumulative stress disorders.

ethnicity A shared cultural heritage and/or national origin.

euthanasia The deliberate taking of an individual's life at her or his request.

experiment A research method that involves manipulating the independent variable to determine how it affects the dependent variable.

expulsion When a dominant group forces a subordinate group to leave the country or to live only in designated areas of the country.

familism A value system that encourages family members to put their family's well-being above their individual and personal needs.

family As defined by the U.S. Census Bureau, a group of two or more persons related by birth, marriage, or adoption who reside together. Some family scholars have redefined the family to include nonrelated persons who reside together and who are economically, emotionally, and sexually interdependent (e.g., cohabiting heterosexual or homosexual couples).

family household As defined by the U.S. Census Bureau, a family household consists of two or more persons related by birth, marriage, or adoption who reside together.

feminization of poverty The disproportionate distribution of poverty among women.

fertility rate The average number of births per woman.

field research A method of research that involves observing and studying social behavior in settings in which it naturally occurs; includes participant observation and nonparticipant observation.

flextime An option in work scheduling that allows employees to begin and end the workday at different times as long as they perform 40 hours of work per week.

folkway The customs and manners of society.

future shock The state of confusion resulting from rapid scientific and technological changes that challenge traditional values and beliefs.

gateway drug A drug (e.g., marijuana) that is believed to lead to the use of other drugs (such as cocaine and heroin).

gender The social definitions and expectations associated with being male or female.

gene monopolies Exclusive control over a particular gene as a result of government patents.

gene therapy The transplantation of a healthy duplicate gene to replace a defective or missing gene.

genetic engineering The manipulation of an organism's genes in such a way that the natural outcome is altered.

genetic screening The use of genetic maps to detect predispositions to human traits or disease(s).

genocide The systematic annihilation of one racial and/or ethnic group by another.

gentrification The process by which private individuals and/or developers purchase and renovate housing in older neighborhoods.

gerontophobia Fear or dread of the elderly.

ghetto A slum section of a city occupied primarily by a minority group.

glass ceiling An invisible, socially created barrier that prevents some women and other minorities from being promoted into top corporate positions.

global economy An interconnected network of economic activity that transcends national borders.

greenhouse effect The collection of increasing amounts of chlorofluorocarbons (CFCs), carbon dioxide, methane, and other gases in the atmosphere, where they act like the glass in a greenhouse, holding heat from the sun close to the earth and preventing the heat from rising back into space.

greenwashing The corporate practice of displaying a sense of corporate responsibility for the environment. For example, many companies publicly emphasize the steps they have taken to help the environment. Another greenwashing strategy is to retool, repackage, or relabel a company's product.

guaranteed annual income A guarantee by the government that a poor person's income will not fall below a specified percentage of the country's median income.

guerrilla warfare Warfare in which organized groups oppose domestic or foreign governments and their military forces; often involves small groups of individuals who use camouflage and underground tunnels to hide until they are ready to execute a surprise attack.

hate crimes Acts of violence motivated by prejudice against different racial, ethnic, religious, gender, or sexual orientation groups.

health A state of complete physical, mental, and social well-being.

health expectancy Number of years an individual can expect to live in good health.

health maintenance organizations (HMOs) Health care organizations that provide complete medical services for a monthly fee.

Healthy Families America A nationwide program modeled after Hawaii's Healthy Start program. The program offers one or two home visits to all new parents, where service providers share information about baby care and development and parenting skills and explain how parents can utilize other organizations and agencies that provide family support.

heterosexism The belief that heterosexuality is the superior sexual orientation; results in prejudice and discrimination against homosexuals and bisexuals.

heterosexuality The predominance of cognitive, emotional, and sexual attraction to persons of the other sex.

home schooling The education of children at home instead of in a public or private school; often part of a fundamentalist movement to protect children from perceived non-Christian values in the public schools.

homophobia Negative attitudes toward homosexuality.

homosexuality The predominance of cognitive, emotional, and sexual attraction to persons of the same sex.

honor murder The practice of wife murder that occurs in some countries when a man suspects his wife of being unfaithful.

household All persons who share occupancy of a housing unit such as a house or an apartment.

human capital The skills, knowledge, and capabilities of the individual.

human capital hypothesis The hypothesis that female-male pay differences are a function of differences in women's and men's levels of education, skills, training, and work experience.

human ecology The study of the relationship between populations and their natural environment.

Human Poverty Index (HPI) A composite measure of poverty based on three measures of deprivation: (1) deprivation of life, which is measured by the percentage of people expected to die before age 40; (2) deprivation of knowledge, which is measured by the percentage of adults who are illiterate; and (3) deprivation in living standards, which is measured as a composite of three variables—the percentage of people without access to health services, the percentage of people without access to safe water, and the percentage of malnourished children under 5.

hypothesis A prediction or educated guess about how one variable is related to another variable.

in-vitro fertilization (IVF) The union of an egg and a sperm in an artificial setting such as a laboratory dish.

incapacitation A criminal justice philosophy that views the primary purpose of the criminal justice system as preventing criminal offenders from committing further crimes against the public by putting them in prison.

incidence The number of new cases of a specific health problem within a given population during a specified time period.

incumbent upgrading Aid programs that help residents of depressed neighborhoods buy or improve their homes and stay in the community.

independent variable The variable that is expected to explain change in the dependent variable.

index offenses Crimes identified by the FBI as the most serious, including personal crimes (homicide, rape, robbery, assault) and property crimes (burglary, larceny, car theft, arson).

individual discrimination Discriminatory acts by individuals that are based on prejudicial attitudes.

individualism A value system that stresses the importance of individual happiness.

Industrial Revolution The period between the mid-eighteenth and the early nineteenth century when machines and factories became the primary means for producing goods. The Industrial Revolution led to profound social and economic changes.

infant mortality rate The number of deaths of infants under 1 year of age per 1,000 live births in a calendar year.

infantilizing elders The portrayal of the elderly in the media as childlike in terms of clothes, facial expression, temperament, and activities.

infotech An abbreviation for "information technology"; any technology that carries information.

infowar The utilization of technology to manipulate or attack an enemy's military and civilian infrastructure and information systems.

institution An established and enduring pattern of social relationships. The five traditional social institutions are family, religion, politics, economics, and education. Institutions are the largest elements of social structure.

institutional discrimination Discrimination in which the normal operations and procedures of social institutions result in unequal treatment of minorities.

integration hypothesis A theory that states that the only way to achieve quality education for all racial and ethnic groups is to desegregate the schools.

intergenerational poverty Poverty that is transmitted from one generation to the next.

internalized homophobia A sense of personal failure and self-hatred among lesbians and gay men due to social rejection and stigmatization.

Internet An international information infrastructure available through many universities, research institutes, government agencies, and businesses; developed in the 1970s as a Defense Department experiment.

Jim Crow laws Laws that separated blacks from whites by prohibiting blacks from using "white" buses, hotels, restaurants, and drinking fountains.

job burnout Prolonged job stress; can cause physical problems, such as high blood pressure, ulcers, and headaches as well as psychological problems.

job exportation The relocation of U.S. jobs to other countries where products can be produced more cheaply.

job sharing A work option in which two people, often husband and wife, share and are paid for one job.

labeling theory A symbolic interactionist theory that is concerned with the effects of labeling on the definition of a social problem (e.g., a social condition or group is viewed as problematic if it is labeled as such) and with the effects of labeling on the self-concept and behavior of individuals (e.g., the label "juvenile delinquent" may contribute to the development of a self-concept and behavior consistent with the label).

latent functions Consequences that are unintended and often hidden, or unrecognized; for example, a latent function of education is to provide schools that function as baby-sitters for employed parents.

law Norms that are formalized and backed by political authority.

legalization Making prohibited behavior legal; for example, legalizing marijuana or prostitution.

life chances A term used by Max Weber to describe the opportunity to obtain all that is valued in society, including happiness, health, income, and education.

life expectancy The average number of years that a person born in a given year can expect to live.

lifestyle A distinct subculture associated with a particular social class.

living wage laws Laws that require state or municipal contractors, recipients of public subsidies or tax breaks, or, in some cases, all businesses to pay employees wages significantly above the federal minimum, enabling families to live above the poverty line.

looking-glass self The idea that individuals develop their self-concept through social interaction.

macro sociology The study of large aspects of society, such as institutions and large social groups.

MADD Mothers Against Drunk Driving. A social action group committed to reducing drunk driving.

Malthusian theory The theory proposed by Thomas Malthus in which he predicted that population would grow faster than the food supply and that masses of people were destined to be poor and hungry. According to Malthus, food shortages would lead to war, disease, and starvation that would eventually slow population growth.

managed care A type of medical insurance plan that controls costs through monitoring and controlling the decisions of health care providers.

manifest functions Consequences that are intended and commonly recognized; for example, a manifest function of education is to transmit knowledge and skills to youth.

marital assimilation Assimilation that occurs when different ethnic or racial groups become married or pair-bonded and produce children.

master status The status that is considered the most significant in a person's social identity.

maternal mortality Deaths that result from complications associated with pregnancy or childbirth.

means-tested programs Public assistance programs that have eligibility requirements based on income.

mechanization The use of tools to accomplish tasks previously done by workers; characteristic of agricultural societies that emphasize the production of raw materials.

Medicaid A jointly funded federal-state-local assistance program designed to provide health care for the poor.

medicalization The tendency to define negatively evaluated behaviors and/or conditions as medical problems in need of medical intervention.

Medicare A national public insurance program created by Title XVIII of the Social Security Act of 1965; originally designed to protect people 65 years of age and older from the rising costs of health care. In 1972, Medicare was extended to permanently disabled workers and their dependents and persons with end-stage renal disease.

Medigap The difference between Medicare benefits and the actual cost of medical care.

Medigap policies Supplementary private insurance policies that cover prescription drugs, long-term nursing home care, and other types of services that Medicare does not cover.

melting pot The product of different groups coming together and contributing equally to a new, common culture.

mental disorder A behavioral or psychological syndrome or pattern that occurs in an individual, and that is associated with present distress or disability, or with a significantly increased risk of suffering, death, pain, disability, or loss of freedom.

metropolis From the Greek meaning "Mother City." See also *metropolitan area.*

metropolitan area A densely populated core area and any adjacent communities that have a high degree of social and economic integration with the core; a large city and its surrounding suburbs; also called a "metropolis."

micropolitan area A small city located beyond congested metropolitan areas.

micro-society school A simulation of the "real" or nonschool world where students design and run their own democratic, free-market society within the school.

micro sociology The study of the social psychological dynamics of individuals interacting in small groups.

military-industrial complex A term used by Dwight D. Eisenhower to connote the close association between the military and defense industries.

minority A category of people who are denied equal access to positions of power, prestige, and wealth because of their group membership.

mixed-use neighborhoods Communities that combine residential and commercial elements along with public and private facilities such as schools, recreation centers, and places of worship. The idea of mixed-use neighborhoods is to provide suburban residents with convenient access to jobs, stores and service providers, schools, and other facilities, thus reducing driving distances.

modern racism A subtle and complex form of racism in which individuals are not explicitly racist, but tend to hold negative views of racial minorities and blame minorities for their social disadvantages.

modernization theory A theory claiming that as society becomes more technologically advanced, the position of the elderly declines.

morbidity The amount of disease, impairment, and accidents in a population.

mores Norms that have moral basis.

mortality Death.

multicultural education Education that includes all racial and ethnic groups in the school curriculum and promotes awareness and appreciation for cultural diversity.

multiculturalism A philosophy that argues that the culture of a society should represent and embrace all racial and ethnic groups in that society.

multidimensional model of sexual orientation A conceptual model of sexual orientation that views sexual orientation as consisting of various independent components, including emotional and social preferences, lifestyle, self-identification, sexual attraction, fantasy, and behavior.

multiple chemical sensitivity (MCS) A controversial health condition in which, after one or more acute or traumatic exposures to a chemical or group of chemicals, people experience adverse effects from low levels of chemical exposure that do not produce symptoms in the general public.

multiple jeopardy The disadvantages associated with being a member of two or more minority groups.

mutually assured destruction (MAD) A perspective that argues that if both sides in a conflict were to lose in a war, neither would initiate war.

naturalized citizen An immigrant who applied and met the requirements for U.S. citizenship.

neglect A form of abuse involving the failure to provide adequate attention, supervision, nutrition, hygiene, health care, and a safe and clean living environment for a minor child or a dependent elderly individual.

neighborhood transit A suburban transportation system that provides transportation on small buses on an on-demand basis.

new ageism The belief that the elderly are a burden on the economy and, specifically, on the youth of America.

New Urbanists A growing group of planners, architects, developers, and traffic engineers who support neighborhood designs that create a strong sense of community by incorporating features of traditional small towns.

no-fault divorce A divorce that is granted based on the claim that there are irreconcilable differences within a marriage (as opposed to one spouse being legally at fault for the marital breakup).

norms Socially defined rules of behavior, including folkways, mores, and laws.

nuclear winter The predicted result of a thermonuclear war whereby dust storms and concentrations of particles would block out vital sunlight, lower temperatures in the Northern Hemisphere, and lead to the death of most living things on earth.

objective element of social problems Awareness of social conditions through one's own life experience and through reports in the media.

occupational sex segregation The concentration of women in certain occupations and of men in other occupations.

one drop of blood rule A rule that specified that even one drop of Negroid blood defined a person as black and, therefore, eligible for slavery.

operational definition In research, a definition of a variable that specifies how that variable is to be measured (or was measured) in the research.

organized crime Criminal activity conducted by members of a hierarchically arranged structure devoted primarily to making money through illegal means.

outing Publicly identifying the sexual orientation of homosexuals without their consent.

overt discrimination Discrimination that occurs because of an individual's own prejudicial attitudes.

patriarchal Literally, rule by father; today, connotes rule by males.

patriarchy A tradition in which families are male-dominated.

peace dividend Resources that are diverted from military spending and channeled into private or public investment or consumption, used to reduce the deficit, and/or used to lower taxes.

Personal Responsibility and Work Opportunity Reconciliation Act (PRWOR) The 1996 legislation that affected numerous public assistance programs, primarily in the form of cutbacks and eligibility restrictions. This law ended Aid to Families with Dependent Children (AFDC) and replaced it with Temporary Aid to Needy Families (TANF).

phased retirement Retirement in which the older worker can withdraw from the workforce gradually.

pink-collar jobs Jobs that offer few benefits, often have low prestige, and are disproportionately held by women.

planned obsolescence The manufacturing of products that are intended to become inoperative or outdated in a fairly short period of time.

pluralism A state in which racial and/or ethnic groups maintain their distinctness, but respect each other and have equal access to social resources.

population transfer See *expulsion.*

postindustrialization The shift from an industrial economy dominated by manufacturing jobs to an economy dominated by service-oriented, information-intensive occupations.

postmodernism A world view that questions the validity of rational thinking and the scientific enterprise.

post-traumatic stress disorder A set of symptoms that may result from any traumatic experience, including crime victimization, war, natural disasters, or abuse.

poverty Lacking resources for an "adequate" standard of living (see also *absolute poverty* and *relative poverty*)

poverty gap The difference between the household income of the poor and the poverty line.

poverty line An annual dollar amount below which individuals or families are considered officially poor by the government.

preferred provider organizations (PPOs) Health care organizations in which employers who purchase group health insurance agree to send their employees to certain health care providers or hospitals in return for cost discounts.

prejudice An attitude or judgment, usually negative, about an entire category of people based on their group membership.

prevalence The total number of cases of a condition within a population that exist at a given time.

primary aging Biological changes associated with aging that are due to physiological variables such as cellular and molecular variation (e.g., gray hair).

primary assimilation The integration of different groups in personal, intimate associations such as friends, family, and spouses.

primary group Small groups characterized by intimate and informal interaction.

primary prevention strategies Family violence prevention strategies that target the general population.

primary work sector The set of work roles in which individuals are involved in the production of raw materials and food goods; develops when a society changes from a hunting and gathering society to an agricultural society.

privatization The use of private services in the public sector, for example, in the educational system.

proletariat Workers, often exploited by the bourgeoisie.

pronatalism A cultural value that promotes having children.

public assistance A general term referring to some form of support by the government to citizens who meet certain established criteria.

public housing An assistance program that provides federal subsidies for low-income housing units built, owned, and operated by local public housing authorities.

race A category of people who are believed to share distinct physical characteristics that are deemed socially significant.

racism The belief that certain groups of people are innately inferior to other groups of people based on their racial classification. Racism serves to justify discrimination against groups that are perceived as inferior.

redlining The practice whereby mortgage companies deny loans for the purchase of houses in minority neighborhoods, arguing that the financial risk is too great.

refugees Immigrants who apply from abroad for admission on the basis of persecution or fear of persecution for their political or religious beliefs.

regionalism A form of collaboration among central cities and suburbs that encourages local governments to share common responsibility for common problems.

regionalization The merging of city and suburban governments to create one government.

regressive taxes Taxes that absorb a much higher proportion of the incomes of lower-income households than of higher-income households.

rehabilitation A criminal justice philosophy that views the primary purpose of the criminal justice system as changing the criminal offender through such programs as education and job training, individual and group therapy, substance abuse counseling, and behavior modification.

relative poverty A deficiency in material and economic resources compared with some other population.

replacement level The average number of births per woman (2.1) in a population below which the population begins to decline.

reverse discrimination The unfair treatment of members of the majority group (i.e., white males) that, according to some, results from affirmative action.

road rage Aggressive and violent driving behavior.

role A set of rights, obligations, and expectations associated with a status.

SADD Students Against Drunk Driving; a student social action group committed to reducing drunk driving.

sample In survey research, the portion of the population selected to be questioned.

sanctions Social consequences for conforming to or violating norms. Types of sanctions include positive, negative, formal, and informal.

sandwich generation The generation that has the responsibility of simultaneously caring for their children and their aging parents.

school vouchers Tax credits that are transferred to the public or private school of a parent's choice.

science The process of discovering, explaining, and predicting natural or social phenomena.

secondary aging Biological changes associated with aging that can be attributed to poor diet, lack of exercise, and increased stress.

secondary assimilation The integration of different groups in public areas and in social institutions, such as neighborhoods, schools, the workplace, and in government.

secondary group A group characterized by impersonal and formal interaction.

secondary prevention strategies Prevention strategies that target groups that are thought to be at high risk for family violence.

secondary work sector The set of work roles in which individuals are involved in the production of manufactured goods from raw materials; emerges when a society becomes industrialized.

Section 8 A rent supplement program targeted at households with incomes below 50 percent of the area median income.

segregation The physical and social separation of categories of individuals, such as racial or ethnic groups.

self-fulfilling prophecy A concept referring to the tendency for people to act in a manner consistent with the expectations of others.

senescence The biology of aging.

sex A person's biological classification as male or female.

sexism The belief that there are innate psychological, behavioral, and/or intellectual differences between females and males and that these differences connote the superiority of one group and the inferiority of another.

sexual harassment When an employer requires sexual favors in exchange for a promotion, salary increase, or any other employee benefit and/or the existence of a hostile environment that unreasonably interferes with job performance, as in the case of sexually explicit remarks or insults being made to an employee.

sexual orientation The identification of individuals as heterosexual, bisexual, or homosexual, based on their emotional and sexual attractions, relationships, self-identity, and lifestyle.

sick building syndrome (SBS) A situation in which occupants of a building experience symptoms that seem to be linked to time spent in a building, but no specific illness or cause can be identified.

single-payer system A single tax-financed public insurance program that replaces private insurance companies.

slavery A condition in which one social group treats another group as property to exploit for financial gain.

slums Concentrated areas of poor housing and squalor in heavily populated urban areas.

smart growth A strategy for managing urban sprawl that serves the economic, environmental, and social needs of communities.

social class Group of people who share a similar position or social status within the stratification system.

social group Two or more people who have a common identity, interact, and form a social relationship. Institutions are made up of social groups.

social problem A social condition that a segment of society views as harmful to members of society and in need of remedy.

social promotion The passing of students from grade to grade even if they are failing.

socialism An economic ideology that emphasizes public rather than private ownership. Theoretically, goods and services are equitably distributed according to the needs of the citizens.

socialized medicine National health insurance systems in other countries such as Canada, Great Britain, Sweden, Germany, and Italy.

sociological imagination A term coined by C. Wright Mills to refer to the ability to see the connections between our personal lives and the social world in which we live.

sociological mindfulness A term used by sociologist Michael Schwalbe to refer to the practice of seeing the many ways in which social conditions shape and are shaped by individuals and groups.

sodomy laws Laws that prohibit what are considered "unnatural acts" such as oral and anal sex.

split-labor market The existence of primary and secondary labor markets. A primary labor market refers to jobs that are stable and economically rewarding and have many benefits; a secondary labor market refers to jobs that offer little pay, no security, few benefits, and little chance for advancement.

state The organization of the central government and government agencies such as the armed forces, police force, and regulatory agencies.

status A position a person occupies within a social group.

stereotypes Oversimplified or exaggerated generalizations about a category of individuals. Stereotypes are either untrue or are gross distortions of reality.

strain theory A theory that argues that when legitimate means of acquiring culturally defined goals are limited by the structure of society, the resulting strain may lead to crime or other deviance.

structural-functionalism A sociological perspective that views society as a system of interconnected parts that work together in harmony to maintain a state of balance and social equilibrium for the whole; focuses on how each part of society influences and is influenced by other parts.

structural sexism The ways in which the organization of society, and specifically its institutions, subordinate individuals and groups based on their sex classification.

structural unemployment Exists when there are not enough jobs available for those who want them; unemployment that results from structural variables such as government and business downsizing, job exportation, automation, a reduction in the number of new and existing businesses, an increase in the number of people looking for jobs, and a recessionary economy where fewer goods are purchased and, therefore, fewer employees are needed.

subcultural theories A set of theories that argue that certain groups or subcultures in society have values and attitudes that are conducive to crime and violence.

subculture The distinctive lifestyles, values, and norms of discrete population segments within a society.

subjective element of social problems The belief that a particular social condition is harmful to society, or to a segment of society, and that it should and can be changed.

subsidized housing The leasing of private housing units by local housing authorities or direct payments of rent supplements to low-income families.

suburbanization The process in which city dwellers move to the suburbs due to concern about the declining quality of life in urban areas.

suburbs The urbanlike areas surrounding central cities.

survey research A method of research that involves eliciting information from respondents through questions; includes interviews (telephone or face-to-face) and written questionnaires.

sustainable development Societal development that meets the needs of current generations without threatening the future of subsequent generations.

sweatshops Work environments that are characterized by less than minimum wage pay, excessively long hours of work often without overtime pay, unsafe or inhumane working conditions, abusive treatment of workers by employers, and/or the lack of worker organizations aimed at negotiating better work conditions.

symbol Something that represents something else.

symbolic interactionism A sociological perspective that emphasizes that human behavior is influenced by definitions and meanings that are created and maintained through symbolic interaction with others.

technological fix The use of scientific principles and technology to solve social problems.

technology Activities that apply the principles of science and mechanics to the solution of specific problems.

technology-induced diseases Diseases that result from the use of technological devices, products, and/or chemicals.

telecommute A work option in which workers complete all or part of their work at home with the use of information technology.

telemedicine Using information and communication technologies to deliver a wide range of health care services, including diagnosis, treatment, prevention, health support and information, and education of health care workers.

telepathology Transmitting images of tissue samples to a pathologist in another location, who can look at the image on a monitor and offer an interpretation.

teleradiology The transmission of radiological images from one location to another for the purpose of interpretation or consultation.

Temporary Aid to Needy Families (TANF) The welfare program that resulted from 1996 welfare reform legislation. Under the TANF program, which replaced Aid to Families with Dependent Children (AFDC), after 2 consecutive years of receiving aid, welfare recipients are required to work at least 20 hours per week or to participate in a state-approved work program (few exceptions are made). A lifetime limit of 5 years is set for families receiving benefits.

terrorism The premeditated use or threatened use of violence by an individual or group to gain a political objective.

tertiary prevention strategies Prevention strategies that target families who have experienced family violence.

therapeutic communities Organizations where approximately 35–100 individuals reside for up to 15 months to abstain from drugs, develop marketable skills, and receive counseling.

tracking An educational practice in which students are grouped together on the basis of similar levels of academic achievement and abilities.

triad dispute resolution Dispute resolution that involves two disputants and a negotiator.

triple jeopardy See *multiple jeopardy.*

underclass A persistently poor and socially disadvantaged group that disproportionately experiences joblessness, welfare dependency, involvement in criminal activity, dysfunctional families, and low educational attainment.

underemployment Employment in a job that is underpaid; is not commensurate with one's skills, experience, and/or education; and/or involves working fewer hours than desired.

under-5 mortality rate The rate of deaths among children under age 5.

unidimensional continuum model of sexual orientation A model of sexual orientation that conceptualizes individuals as having both heterosexual and homosexual elements.

union density The percentage of workers who belong to unions.

upskilling The opposite of deskilling; upskilling reduces employee alienation and increases decision-making powers.

urban population Persons living in cities or towns of 2,500 or more inhabitants.

urban sprawl The ever-increasing outward growth of urban areas.

urban villages Neighborhood communities, within urban settings, that facilitate the formation of intimate and strong social bonds among community members.

urbanism The culture and lifestyle of city dwellers, often characterized by individualistic and cosmopolitan norms, values, and styles of behavior.

urbanization The transformation of a society from a rural to an urban one.

urbanized area One or more places and the adjacent densely populated surrounding territory that together have a minimum population of 50,000.

utilization review In a managed care system, the branch that monitors authorization of a physician's preferred course of action. In many plans, doctors must receive approval from this office before they can hospitalize a patient, perform surgery, or order an expensive diagnostic test.

values Social agreements about what is considered good and bad, right and wrong, desirable and undesirable.

variable Any measurable event, characteristic, or property that varies or is subject to change.

victimless crimes Illegal activities, such as prostitution or drug use, that have no complaining party; also called "vice crimes."

virtual reality Computer-generated three-dimensional worlds that change in response to the movements of the head or hand of the individual; a simulated experience of people, places, sounds, and sights.

war Organized armed violence aimed at a social group in pursuit of an objective.

wealth The total assets of an individual or household, minus liabilities.

wealthfare system Governmental policies and regulations that economically favor the wealthy.

white-collar crime Includes both occupational crime, where individuals commit crimes in the course of their employment, and corporate crime, where corporations violate the law in the interest of maximizing profit.

work-experience unemployment rate The percentage of persons who participated in the labor force in a given year and experienced some unemployment during that year.

working poor Individuals who spend at least 27 weeks a year in the labor force (working or looking for work), but whose income falls below the official poverty line.

work sectors The division of the labor force into distinct categories (primary, secondary, and tertiary) based on the types of goods/services produced.

Y2K (Year 2000) Sometimes called the millennium bug, the concern that computers programmed with double-digit years (e.g., 97, 98, 99) will interpret 00 as 1900, 01 as 1901, and so forth rather than the years 2000, 2001, etc.

References

PREFACE

Safransky, Sy. 1990. *Sunbeams: A Book of Quotations.* Berkeley, CA: North Atlantic Books.

CHAPTER 1

American Council on Education and University of California. 1997. "The American Freshmen: National Norms for Fall 1997." Los Angeles: Los Angeles Higher Education Research Institute.

Blumberg, Paul. 1989. *The Predatory Society: Deception in the American Market Place.* New York: Oxford University Press.

Blumer, Herbert. 1971. "Social Problems as Collective Behavior." *Social Problems* 8(3):298–306.

Catania, Joseph A., David R. Gibson, Dale D. Chitwook, and Thomas J. Coates. 1990. "Methodological Problems in AIDS Behavioral Research: Influences on Measurement Error and Participation Bias in Studies of Sexual Behavior." *Psychological Bulletin* 108:339–62.

Coleman, John R. 1990. "Diary of a Homeless Man." In *Social Problems,* pp. 160–69. Englewood Cliffs, N.J.: Prentice Hall.

Dordick, Gwendolyn A. 1997. *Something Left to Lose: Personal Relations and Survival Among New York's Homeless.* Philadelphia: Temple University Press.

Frank, James G. 1991. "Risk Factors for Rape: Empirical Confirmation and Preventive Implications." Poster session presented at the 99th annual convention of the American Psychological Association, San Francisco, August 16.

Gallup Organization. 1997. **http://198.175.140.8/SpecialReports/racepressrls.htm**

Hewlett, Sylvia Ann. 1992. *When the Bough Breaks: The Cost of Neglecting Our Children.* New York: Harper Perennial.

Hills, Stuart L. 1987. *Corporate Violence: Injury and Death for Profit,* ed. Stuart L. Mills. Lanham, Md.: Rowman & Littlefield.

Human Development Report 1997. United Nations Development Programme. New York: Oxford University Press, Inc.

Jekielek, Susan M. 1998. "Parental Conflict, Marital Disruption and Children's Emotional Well-Being." *Social Forces* 76:905–35.

Kannapell, Andrea. 1997. "Report Says Health of Society Lags Behind That of Economy." *New York Times,* October 12, sec. 1, p. 32, col 1.

Manis, Jerome G. 1974. "Assessing the Seriousness of Social Problems." *Social Problems* 22:1–15.

Merton, Robert K. 1968. *Social Theory and Social Structure.* New York: Free Press.

Mills, C. Wright. 1959. *The Sociological Imagination.* London: Oxford University Press.

Mirowsky, John, Catherine E. Ross, and Marieke Van Willigen. 1996. "Instrumentalism in the Land of Opportunity: Socioeconomic Causes and Emotional Consequences." *Social Psychology Quarterly* 59:322–37.

Nachmias, David, and Chava Nachmias. 1987. *Research Methods in the Social Sciences,* 3d ed. New York: St. Martin's Press.

Peterson, Ruth, and Lauren J. Krivo. 1993. "Racial Segregation and Black Urban Homicide." *Social Forces* 59:131–41.

Roberts, J. Timmons. 1993. "Psychosocial Effects of Workplace Hazardous Exposures: Theoretical Synthesis and Preliminary Findings." *Social Problems* 40:74–89.

Romer, D., R. Hornik, B. Stanton, M. Black, X. Li, I. Ricardo, and S. Feigelman. 1997. "Talking Computers': A Reliable and Private Method to Conduct Interviews on Sensitive Topics with Children." *The Journal of Sex Research* 34:3–9.

Rubington, Earl, and Martin S. Weinberg. 1995. *The Study of Social Problems,* 5th ed. New York: Oxford University Press.

Schwalbe, Michael. 1998. *The Sociologically Examined Life: Pieces of the Conversation.* Mountain View, Calif: Mayfield Publishing Company.

Skeen, Dick. 1991. *Different Sexual Worlds: Contemporary Case Studies of Sexuality.* Lexington, Mass.: Lexington Books.

"This Year's Freshmen: A Statistical Profile." 1999 (Jan. 29). *The Chronical of Higher Education* A48–A49.

Thomas, W. I. [1931] 1966. "The Relation of Research to the Social Process." In *W. I. Thomas on Social Organization and Social Personality,* ed. Morris Janowitz, pp. 289–305. Chicago: University of Chicago Press.

Troyer, Ronald J., and Gerald E. Markle. 1984. "Coffee Drinking: An Emerging Social Problem." *Social Problems* 31:403–16.

Ukers, William H. 1935. *All about Tea,* vol. 1. The Tea and Coffee Trade Journal Co.

Wilson, John. 1983. *Social Theory.* Englewood Cliffs, N.J.: Prentice-Hall.

CHAPTER 2

Abeysinghe, Devinka. 1998. "First Lady Stresses Family Planning on World Health Day." *Popline* (March-April): 3–4.

Academy for Eating Disorders. 1997. **http://www.acadeatdis.org/** (9/13/98).

Alaimo, Katerine, Ronette R. Briefel, Edward A. Frongillo, and Christine M. Olson. 1998. "Food Insufficiency Exists in the United States: Results from the Third National Health and Nutrition Examination Survey (NHANES III)." *American Journal of Public Health* 88:419–26.

American Psychiatric Association. 1994. *Diagnostic and Statistical Manual of Mental Disorders,* 4th ed. (DSM-IV). Washington, D.C.: American Psychiatric Association.

Antezana, Fernando S., Claire M. Chollat-Traquet, and Derik Yach. 1998. "Health for All in the 21st Century." *World Health Statistics Quarterly* 51:3–6.

Centers for Disease Control and Prevention. 1997. "Youth Risk Behavior Surveillance: National College Health Risk Behavior Survey—United States, 1995." *Surveillance Summaries* 46(SS-6):1–54.

Cherner, Linda L. 1995. *The Universal Healthcare Almanac.* Phoenix: Silver & Cherner.

Cockerham, William C. 1998. *Medical Sociology,* 7th ed. Upper Saddle River, N.J.: Prentice Hall.

Common Cause. 1998 (September 3). "Wag the Dollar: Insurance, Securities, & Investments." **http://www.commoncause.org/publications/wagthedollar.htm** (10/22/98).

Conlon, R. T. 1997. "Introducing Technology into the Public STD Clinic." *Health Education & Behavior* 24(1):12–19.

Conrad, Peter, and Phil Brown. 1999. "Rationing Medical Care: A Sociological Reflection." In *Health, Illness, and Healing: Society, Social Context, and Self,* eds. Kathy Charmaz and Debora A. Paterniti, pp. 582–90. Los Angeles: Roxbury Publishing Co.

Creese, Andrew L., John D. Martin, and Jan H.M. Visschedijk. 1998. "Health Systems for the 21st Century." *World Health Statistics Quarterly* 51:21–27.

"Economic Growth Unnecessary for Reducing Fertility." 1998. *Popline* 20:4.

Emmons, Karen M., Henry Wechsler, George Dowdall, and Melissa Abraham. 1998. "Predictors of Smoking among U.S. College Students." *American Journal of Public Health* 88:104–7.

Everett, Sherry A., Rae L. Schnuth, and Joanne L. Tribble. 1998. "Tobacco and Alcohol Use in Top-Grossing American Films." *Journal of Community Health* 23:317–24.

Feldman, Debra S., Dennis H. Novack, and Edward Gracely. 1998. "Effects of Managed Care on Physician Patient Relationships, Quality of Care, and the Ethical Practice of Medicine: A Physician Survey." *Archives of Internal Medicine* 158:1626–32.

Goldstein, Michael S. 1999. "The Origins of the Health Movement." In *Health, Illness, and Healing: Society, Social Context, and Self,* eds. Kathy Charmaz and Debora A. Paterniti, pp. 31–41. Los Angeles: Roxbury Publishing Co.

Health Care Financing Administration. 1997. "Health Insurance Portability and Accountability Act of 1996." **http://www.hcfa.gov/facts/f970sas.htm** (8/1/98).

———. 1998a. "The Children's Health Insurance Program (CHIP)." **http://www.hcfa.gov/facts/f61598.htm** (8/1/98).

———. 1998b. "The Clinton Administration's Comprehensive Strategy to Fight Health Care Fraud, Waste, and Abuse." **http://www.hcfa.gov/facts/f980316.htm** (8/1/98).

———. 1998c. "Highlights: National Health Expenditures, 1996. **http://www.hcfa.gov/stats/nhe-oact/hilites.htm** (9/23/98).

———. 1998d. "Managed Care in Medicare and Medicaid." **http://www.hcfa.gov/facts/f980220.htm** (8/1/98).

Hellinger, F. J. 1998. "Regulating the Financial Incentives Facing Physicians in Managed Care Plans." *American Journal of Managed Care* 4:663–74.

Inciardi, James A., and Lana D. Harrison. 1997. "HIV, AIDS, and Drug Abuse in the International Sector." *Journal of Drug Issues* 27:1–8.

Johnson, Tracy L., and Elizabeth Fee. 1997. "Women's Health Research: An Introduction." In *Women's Health Research: A Medical and Policy Primer,* eds. Florence P. Haseltine and Beverly Greenberg Jacobson, pp. 3–26. Washington, D.C.: Health Press International.

Kessler, Ronald C., Katherine A. McGonagle, Shanyang Zhao, Christopher B. Nelson, Michael Hughes, Suzann Eshleman, Hans-Ulrich Wittchen, and Kenneth S. Kendler. 1994. "Life-time and 12-Month Prevalence of DSM-III-R Psychiatric Disorders in the United States." *Archives of General Psychiatry* 51:8–19.

Lantz, Paula M., James S. House, James M. Lepkowski, David R. Williams, Richard P. Mero, and Jieming Chen. 1998. "Socioeconomic Factors, Health Behaviors, and Mortality: Results from a Nationally Representative Prospective Study of U.S. Adults." *Journal of the American Medical Association* 279:1703–8.

LaPorte, Ronald E. 1997. "Improving Public Health via the Information Superhighway." **http://www.the-scientist.library.upenn.edu/yr1997/august/opin_97018.html** (8/6/98).

Lerer, Leonard B., Alan D. Lopez, Tord Kjellstrom, and Derek Yach. 1998. "Health for All: Analyzing Health Status and Determinants." *World Health Statistics Quarterly* 51:7–20.

Link, Bruce G. and Jo Phelan. 1998. "Social Conditions as Fundamental Causes of Disease." In *Readings in Medical Sociology,* eds. William C. Cockerham, Michael Glasser, and Linda S. Heuser, pp. 23–36. Upper Saddle River, N.J.: Prentice Hall.

Lipton, Rebecca B., Lee M. Losey, Aida Giachello, Joel Mendez, and Mariela H. Girotti. 1998. "Attitudes and Issues in Treating Latino Patients with Type 2 Diabetes: Views of Healthcare Providers." *The Diabetes Educator* 24(1):67–71.

Lorber, Judith. 1997. *Gender and the Social Construction of Illness.* Thousand Oaks, Calif.: Sage Publications.

"Maternal Mortality: A Preventable Tragedy." 1998. *Popline* 20:4.

Miller, Joel E. 1998. "A Reality Check: The Public's Changing Views of Our Health Care System." National Coalition on Health Care. **http://www.americashealth.org/emerge/check0498.html** (9/23/98).

Miller, K., and A. Rosenfield. 1996. "Population and Women's Reproductive Health: An International Perspective." *Annual Review of Public Health* 17:359–82.

Monardi, Fred, and Stanton A. Glantz. 1998. "Are Tobacco Industry Campaign Contributions Influencing State Legislative Behavior?" *American Journal of Public Health* 88:918–23.

Moore, David W. 1997. "AIDS Issue Fades among Americans." *The Gallup Poll.* **http://www.gallup.com/poll/news/971017.html** (4/7/98).

Morse, Minna. 1998. "The Killer Mosquitoes." *Utne Reader* (May–June):14–15.

Murray, C. and A. Lopez, eds. 1996. *The Global Burden of Disease.* Boston: Harvard University Press.

National AIDS Clearinghouse. 1998. "Youth and HIV/AIDS." Centers for Disease Control and Prevention. **http://www.cdcnac.org/youth.htm/** (9/16/98).

National Center for Health Statistics. 1998. *Health, United States, 1998: With Socioeconomic Status and Health Chartbook.* Hyattsville, Md: U.S. Government Printing Office.

National Coalition on Health Care. 1998. "Health Care Coverage and Consumer Impact." **http://www.americashealth.org/know/consumer.html** (9/23/98).

"Older Populations Expanding Rapidly." 1998. *Popline* 20(September–October)1–2.

Ostrof, Paul. 1998. "Readers Write: My Chair." *The Sun* (August): 33–40.

Parsons, Talcott. 1951. *The Social System.* New York: The Free Press.

Pate, Russell R., Michael Pratt, Steven N. Blair, William L. Haskell, Caroline A. Macera, Claude Bouchard, David Buchner, Walter Ettiger, Gregory W. Health, Abby C. King, Andrea Kriska, Arthur L. Leon, Bess H. Marcus, Jeremy Morris, Ralph S. Paffenbarger, Kevin Patrick, Michael L. Pollock, James M. Rippe, James Sallis, and Jack H. Wilmore. 1995. "Physical Activity and Public Health: A Recommendation from the Centers for Disease Control and Prevention and the American College of Sports Medicine." *Journal of the American Medical Association* 273(5):402–6.

PNHP Data Updates. 1997a (July). *Physicians for a National Health Program Newsletter.* **http://www.pnhp.org.Data/dataJL97.html** (9/22/98).

———. 1997b (Dec.). *Physicians for a National Health Program Newsletter.* **http://www.pnhp.org.Data/dataD97.html** (9/22/98).

———. 1998. *Physicians for a National Health Program Newsletter.* **http://www.pnhp.org/dataM98.html** (9/22/98).

PNHP Press Release. 1999. "Less than Half of Americans Have Health Insurance Paid for by a Private Employer." *Physicians for a National Health Program.* **http://www.pnhp.org/press199.html** (2/20/99)

"Poverty Threatens Crisis." 1998. *Popline* 20(May–June):3.

Robins, Douglas N. 1998. "Testimony of Douglas N. Robins, M.D. Physicians for a National Health Program (PNHP)." **http://www.pnhp.org/press/robins.html** (9/22/98).

Safe Motherhood Initiative. 1998. "Fact and Figures." **http://www.safemotherhood.org/init_facts.htm** (9/24/98).

SIECUS Fact Sheet. 1998. "Sexually Transmitted Diseases in the United States." **http://www.nyic.com/siecus/pubs/fact/fact0008.html** (9/15/98).

"Single Payer Fact Sheet." 1999. **http://www.pnhp.org/fctsht.html** (2/20/99).

Statistical Abstract of the United States: 1998, 118th ed. U.S. Bureau of the Census, Washington, D.C.: U.S. Government Printing Office.

Szasz, Thomas. 1970 (orig. 1961). *The Myth of Mental Illness: Foundations of a Theory of Personal Conduct.* New York: Harper & Row.

Thirunarayanapuram, Desikan. 1998. "'Population Explosion' Is Far from Over." *Popline* 20 (January–February):1, 4.

UNICEF. 1998. *The State of the World's Children, 1998.* New York: UNICEF.

"Using Computers to Advance Health Care." 1996. **http://www.ahcpr.gov/research/computer.htm** (7/28/98).

Verbrugge, Lois M. 1999. "Pathways of Health and Death." In *Health, Illness, and Healing: Society, Social Context, and Self,* eds. Kathy Charmaz and Debora A. Paterniti, pp. 377–94. Los Angeles: Roxbury Publishing Co.

Visschedijk, Jan, and Silvere Simeant. 1998. "Targets for Health for All in the 21st Century." *World Health Statistics Quarterly* 51:56–67.

Wasunna, Ambrose E., and D. Y. Wyper. 1998. "Technology for Health in the Future." *World Health Statistics Quarterly* 51:33–43.

Weitz, Rose. 1996. *The Sociology of Health, Illness, and Health Care: A Critical Approach.* Belmont, Calif.: Wadsworth Publishing Co.

Williams, David R., and Chiquita Collins. 1999. "U.S. Socioeconomic and Racial Differences in Health: Patterns and Explanations." In *Health, Illness, and Healing: Society, Social Context, and Self,* eds. Kathy Charmaz and Debora A. Paterniti, pp. 349–76. Los Angeles: Roxbury Publishing Co.

World Health Organization. 1946. *Constitution of the World Health Organization.* New York: World Health Organization Interim Commission.

———. 1997a. Fact Sheet No 170: *Health and Environment in Sustainable Development.* **http://www.cdc.gov/ogh/frames.htm** (7/28/98).

———. 1997b. Fact Sheet No 178: *Reducing Mortality from Major Childhood Killer Diseases.* **http://www.cdc.gov/ogh/frames.htm** (7/28/98).

———. 1997c. *The World Health Report 1997.* **http://www.who.org/whr/1997/exsum97e.htm.**

———. 1998. "Fifty Facts from the World Health Report 1998." **http://www.who.int/whr/1998/factse/htm** (8/8/98).

World Health Organization and United Nations Joint Programme on HIV/AIDS. 1998. "Report on the Global HIV/AIDS Epidemic—June 1998." **http://www.who.int/emchiv/global_report/data/globrep_e/pdf** (8/8/98).

"Youth Suicide Prevention: Warning Signs and How to Respond." 1998. *Every Teen Counts.* Washington State PTA. 2003 65th Ave., West Tacoma, Wash. 98466-6215. **http://www.wastatepta.org**

CHAPTER 3

AAP (Australian Associated Press). 1998. "Genetics of Alcoholism." Institute of Alcohol Studies Update. London. IAS Publications.

American Cancer Society. 1998. "Cancer Facts and Figures: 1998." ACS: National Health Interview Survey. **http://www.cancer.org/statistics/cff98/tobacco.html**

American Heart Association. 1997a. "Cigarette Smoking Statistics." American Heart Association. **http://www.americanheart.org**

———. 1997b. "Tobacco Industry's Economic and Political Influence." American Heart Association. **http://www.americanheart.org**

———. 1997c. "Tobacco Industry's Targeting of Youth, Minorities, and Women." AHA Advocacy Position. American Heart Association. **http://www.americanheart.org**

Becker, H. S. 1966. *Outsiders: Studies in the Sociology of Deviance.* New York: Free Press.

Bennett, William. 1993. "Should Drugs Be Legalized?" In *Social Problems: A Critical Thinking Approach,* 2d ed., eds. Paul J. Baker, Louise E. Anderson, and Dean S. Dorn, pp. 246–48. Belmont, Calif.: Wadsworth.

Brody, Jane E. 1998. "A Fatal Shift in Cancer's Gender Gap." *New York Times,* May 12, 7.

Bureau of Justice Statistics. 1992. "Drugs, Crime, and the Justice System: A National Report from the Bureau of Justice Statistics." U.S. Department of Justice, Office of Justice Programs. Washington, D.C.: U.S. Government Printing Office, Superintendent of Documents.

———. 1999. "Drug Abuse Violations." U.S. Department of Justice. **http://www.ojp.**

usdoj.gov/bjs/glance/drug.html

Califano, Joseph. 1998. Interview by Jim Lehrer. *The NewsHour with Jim Lehrer.* Public Broadcasting System, January 8.

Clinton, William. 1998. "The President's Message to the Congress of the United States." ONDCP. **http://www.whitehouse drugpolicy/gov/policy/98ndcs/message.html**

"Community Drug Prevention Programs Show Progress." 1990. *Public Health Reports* 105: 543.

Crime and Justice International. 1997. "News and Notes: Europe." *Crime and Justice International* 13(6):1–5.

DATOS (Drug Abuse Treatment Outcome Study). 1998. "Representative Drug Treatment Research." ONDCP. **http://www.whitehousedrugpolicy.gov/drugfact/research.html**

DEA (Drug Enforcement Administration). 1997. "Say It Straight: The Medical Myths of Marijuana." Washington, D.C.: U.S. Department of Justice.

Dembo, Richard, Linda Williams, Jeffrey Fagan, and James Schmeidler. 1994. "Development and Assessment of a Classification of High Risk Youths." *Journal of Drug Issues* 24:25–53.

Dennis, Richard J. 1993. "The Economics of Legalizing Drugs." In *Social Problems: A Critical Thinking Approach,* 2d ed., eds. Paul J. Baker, Louise E. Anderson, and Dean S. Dorn, pp. 249–55. Belmont, Calif.: Wadsworth.

Discovery/ABC Poll. 1998. "Medical Marijuana Poll Results: Americans Say 'Let Patients Smoke 'Em.'" ABC News/Discovery Channel. **http://www.dicovery.com/area/discoverynews**

Duke, Steven, and Albert C. Gross. 1994. *America's Longest War: Rethinking Our Tragic Crusade against Drugs.* New York: G. P. Putnam & Sons.

Duster, Troy. 1995. "The New Crisis of Legitimacy in Controls, Prisons, and Legal Structures." *American Sociologist* 26:20–29.

Easley, Margaret, and Norman Epstein. 1991. "Coping with Stress in a Family with an Alcoholic Parent." *Family Relations* 40: 218–24.

ERS (Educational Research Service). 1998. "Drug Use Trends." *ERS Bulletin* 25(60): 1–3.

Feagin, Joe R., and C. B. Feagin. 1994. *Social Problems.* Englewood Cliffs, N.J.: Prentice-Hall.

Foster, Burk, Craig J. Forsyth, and Stasia Herbert. 1994. "The Cycle of Family Violence among Criminal Offenders: A Study of Inmates in One Louisiana Jail." *Free Inquiry in Creative Sociology* 22:133–37.

Gallup Poll. 1997. "Drinking a Cause of Family Problems for 3 of 10 Americans." *Gallup Poll Archives,* September 27, 1–6.

———. 1998. "No Single Problem Dominates Americans' Concerns Today." *Gallup Poll Archives,* May 2, 1–3.

Gentry, Cynthia. 1995. "Crime Control through Drug Control." In *Criminology,* 2d ed., ed. Joseph F. Sheley, pp. 477–93. Belmont, Calif.: Wadsworth.

Gusfield, Joseph. 1963. *Symbolic Crusade: Status Politics and the American Temperance Movement.* Urbana: University of Illinois Press.

HHS (U.S. Department of Health and Human Services). 1997. "National Household Survey on Drug Abuse." Substance Abuse and Mental Health Service Administration. Washington, D.C.: U.S. Government Printing Office.

———. 1998. "Tobacco Use Continues to Rise among High School Students in the U.S." Substance Abuse and Mental Health Service Administration Press Release (April 2). Washington, D.C.

Heroin Drug Conference. 1997. "Administrator's Message." U.S. Department of Justice: Drug Enforcement Administration. **http://udsdoj.gov/dea/pubs/special/heroin.html**

Institute of Alcohol Studies. 1997. "Alcohol and Crime Fact Sheet." **http://www.ias.org.uk/factsheets/crime/html**

Jarvik, M. 1990. "The Drug Dilemma: Manipulating the Demand." *Science* 250:387–92.

Join Together. 1998. "Inhalant Abuse." *Hot Issues,* July 11. **http://www.jointogether.org/sa/issues/hot_issues/inhalants/default.html**

Klonoff-Cohen, Sandra H., S. L. Edelstein, E. S. Lefkowitz, Indu Sriwirasan, David Kaegi, Jae Chun Chang, and Karen Wiley. 1995. "The Effect of Passive Smoking and Tobacco Exposure through Breast Milk on Sudden Infant Death Syndrome." *Journal of the American Medical Association* 273:795–98.

Koch, Wendy. 1998. "Players and Positions." Associated Press, June 11. (appeared in *Daily Reflector,* page D3)

Kort, Marcel de. 1994. "The Dutch Cannabis Debate, 1968–1976." *Journal of Drug Issues* 24: 417–27.

Leonard, K. E., and H. T. Blane. 1992. "Alcohol and Marital Aggression in a National Sample of Young Men." *Journal of Interpersonal Violence* 7:19–30.

Levinson, D. 1989. *Family Violence in Cross-Cultural Perspective.* Newbury Park, Calif.: Sage Publications.

Lindesmith Center, 1998. "Research Brief: Methadone Maintenance Treatment." **http://www.lindesmith.org/cities_sources/brief14.html**

Lipton, Douglas S. 1994. "The Correctional Opportunity: Pathways to Drug Treatment for Offenders." *Journal of Drug Issues* 24:331–48.

MADD (Mothers Against Drunk Driving). 1997. "MADD/Allstate Poll: Americans Fear Drunk Driving, Support .08 Limit." *MADD Alcohol News Online,* December 5. **http://www2.madd.org**

———. 1998a. "Colleges Turning Down Alcohol Ad Proposals." MADD Alcohol News Online, January 9. **http://www2.madd.org**

———. 1998b. "MADD Salutes Congress for Passing Repeat Offender and Open Container Laws." **http//www.madd.org/news/repeatoffenders-opencontainers.html**

Mauro, Tony. 1998. "Fla. Jury Awards Biggest Tobacco Verdict Ever." *Nationline. Daily Reflector wire service.* June 11.

McCaffrey, Barry. 1998. "Remarks by Barry McCaffrey, Director, Office of National Drug Control Policy, to the United Nations General Assembly: Special Session on Drugs." Office of National Drug Control Policy. **http://www.whitehousedrugpolicy.gov/news/speeches**

Moore, Martha T. 1997. "Binge Drinking Stalks Campuses." *USA Today,* October 1, A3.

Morgan, Patricia A. 1978. "The Legislation of Drug Law: Economic Crisis and Social Control." *Journal of Drug Issues* 8: 53–62.

Nylander, Albert, Tuk-Ying Tung, and Xiaohe Xu. 1996. "The Effect of Religion on Adolescent Drug Use in America: An Assessment of Change." American Sociological Association Meetings. San Francisco, Calif., August.

ONDCP (Office of National Drug Control Policy). 1998a. "Focus on the Drug Problem." **http://www.whitehousedrugpolicy.gov/drugfact/drugprob.html**

———. 1998b. "1999 Drug Budget Program Highlights: Overview." **http://www.whitehousedrugpolicy.gov/drugfacts/budget-1.html**

———. 1998c. "Treatment at a Glance." **http://www.whitehousedrugpolicy.gov/treat/glance.html**

———. 1998d. "Trends in Drug Use: Part II: Cocaine." *Pulse Check,* Winter. **http://www.health.org/pulse98/trend2.html**

Perkins, Louise. 1997. "Slaughter of the Innocents." Unpublished manuscript. Lambton Families in Action for Drug Education Inc., January.

Perspective. 1996. "Just Say No." *Investor's Business Daily,* August 30.

Pickens, R. W., and D. S. Svikis. 1988. "Biological Vulnerability in Drug Abuse." National Institute on Drug Abuse (NIDA) Research Monograph No. 88. Washington, D.C.: NIDA.

Reiman, Jeffery. 1998. *The Rich Get Richer and the Poor Get Prison.* Boston: Allyn and Bacon.

Rorabaugh, W. J. 1979. *The Alcoholic Republic: An American Tradition.* New York: Oxford University Press.

Society. 1999. "Drug Education Programs Fail in Houston." *Society.* 36:3–4, Jan/Feb.

Statistical Abstract of the United States: 1997, 117th ed. U.S. Bureau of the Census. Washington, D.C.: U.S. Government Printing Office.

Statistical Abstract of the United States: 1998, 118th ed. U.S. Bureau of the Census. Washington, D.C.: U.S. Government Printing Office.

Straus, Murry, and S. Sweet. 1992. "Verbal/Symbolic Aggression in Couples: Incidence Rates and Relationships to Personal Characteristics." *Journal of Marriage and the Family* 54:346–57.

Sullivan, Thomas, and Kenrick S. Thompson. 1994. *Social Problems.* New York: Macmillan.

Tobacco Wars. 1998. "States Approve Tobacco Deal." *Litigation News,* November 22. **http://www.tobaccowars.com/litigate.html**

Torry, Saundra. 1998. "Tobacco Giants Try to Settle with States." *Washington Post,* July 10, A1.

Tubman, J. 1993. "Family Risk Factors, Parental Alcohol Use, and Problem Behaviors among School-Aged Children." *Family Relations* 42:81–86.

U.S. Department of Justice. 1997. "Ketamine Abuse Increasing." Drug Enforcement Administration. July 4. **http://www.usdoj.gov/programs/diversion/divpub/substance/ketamine.html**

———. 1998. "An Analysis of National Data on the Prevalence of Alcohol Involvement in Crime." Prepared by Laurence Greenfeld for the National Symposium on Alcohol Abuse and Crime. Washington D.C.: U.S. Department of Justice.

U.S. News and Bozell Poll. 1997. "Nature v. Nurture." Poll on American Attitudes toward Human Behavior. Washington, D.C.: KRC Research and Consulting.

Van Dyck, C., and R. Byck. 1982. "Cocaine." *Scientific American* 246:128–41.

Van Kammen, Welmoet B., and Rolf Loeber. 1994. "Are Fluctuations in Delinquent Activities Related to the Onset and Offset in Juvenile Illegal Drug Use and Drug Dealing?" *Journal of Drug Issues* 24:9–24.

Waldorf, D., C. Reinarman, and S. Murphy. 1991. *Cocaine Changes: The Experience of Using and Quitting.* Philadelphia: Temple University Press.

Wechsler, Henry, George Dowdall, Andrea Davenport, and William Dejong. 1998. "Binge Drinking on Campus: Results of a National Study." Higher Education Center. **http://www.edc.org/hes/pubs/binge.html**

White, Helene Raskin, and Erich W. Labouvie. 1994. "Generality versus Specificity of Problem

Behavior: Psychological and Functional Differences." *Journal of Drug Issues* 24:55–74.

Witters, Weldon, Peter Venturelli, and Glen Hanson. 1992. *Drugs and Society,* 3d ed. Boston: Jones & Bartlett.

World Drug Report. 1997. "Report Highlights." United Nations International Drug Control Program. New York: United Nations.

Wysong, Earl, Richard Aniskiewicz, and David Wright. 1994. "Truth and Dare: Tracking Drug Education to Graduation and as Symbolic Politics." *Social Problems* 41:448–68.

Yost, Pete. 1999. "Clinton Reveals New 5-part plan in War on Drugs." *Detnews.* Feb. 8. **http://www.detnews.com/1999/nation/9902/08/02080121.html**

CHAPTER 4

ABA (American Bar Association). 1997. "JJ Hill Watch." ABA Criminal Justice Section: Juvenile Justice Center. **http://www.aba.org/CRIMJUST/JUVJUS/**

ACLU (American Civil Liberties Union). 1996. "Fact Sheet on Juvenile Crime." May 14: **http://www.aclu.org/congress/juvenile.htm**

Albanese, Jay, and Robert D. Pursley. 1993. *Crime in America: Some Existing and Emerging Problems.* Englewood Cliffs, N.J.: Prentice-Hall.

Anderson, Elijah. 1994. "The Code of the Streets: Sociology of Urban Violence." *The Atlantic* 273(5):80–91.

AP (Associated Press). 1996. "Clinton Approves Date-Rape Drug Law." *Greensboro News Record,* October 14, A4.

———. 1998. "Safety Groups Renew Recall against GM Pickups." *USA Today,* May 15, Money section. **http://www.usatoday.com/money/consumer/autos/mauto126.html**

Barkan, Steven. 1997. *Criminology: A Sociological Understanding.* Englewood Cliffs, N.J.: Prentice-Hall.

Barlow, Hugh. 1993. *Introduction to Criminology.* New York: Harper Collins.

Becker, Howard S. 1963. *Outsiders: Studies in the Sociology of Deviance.* New York: Free Press.

Beirne, Piers, and James Messerschmidt. 1995. *Criminology,* 2d ed. Fort Worth: Harcourt Brace.

Califano, Joseph. 1998. Interview by Jim Lehrer. *The NewsHour with Jim Lehrer.* Public Broadcasting System, January 8.

CATW (Coalition Against Trafficking in Women). 1997. "Promoting Sex Work in the Netherlands." *Coalition Report* 4(1): **http://www.uri.edu/artsci/wms/hughes/catw**

Carlson, Peter. 1998. "Corrections Trends for the Twenty-First Century: Our Future behind the Walls and Wire." *The Keeper's Voice:* **http://www.acsp.uic.edu/iaco/kv170105.htm**

Chesney-Lind, Meda. 1996. "Girls, Crime and Women's Place: Toward a a Feminist Model of Female Delinquency." In *Social Deviance: Readings in Theory and Research,* ed. Henry Pontell, pp. 166–80. Englewood Cliffs: N.J.: Prentice-Hall.

Clarke, Ronald, and Patricia Harris. 1992. "Auto Theft and Its Prevention." In *Crime and Justice: A Review of Research,* ed. Michael Tonry, vol. 16, pp. 1–54. Chicago: University of Chicago Press.

Comstock, George. 1993. "The Medium and the Society: The Role of Television in American Life." In *Children and Television: Images in a Changing Sociocultural World,* ed. Gordon L. Berry and Joy Keiko Asamen, pp. 117–31. Newbury Park, Calif.: Sage Publications.

Conklin, John E. 1998. *Criminology,* 6th ed. Boston: Allyn & Bacon.

Cook, Philip, and Jens Ludwig. 1997. "Guns in America: National Survey on Private Ownership and Use of Firearms." National Institute of Justice: Washington, D.C. Research in Brief Series. National Institute of Justice.

COPS. 1998. "About the Office of Community Oriented Policing Services (COPS)." **http://communitypolicing.org/copspage. html**

Decker, Scott, Susan Pennell, and Ami Caldwell. 1997. "Illegal Firearms: Access and Use by Arrestees." NIJ Research in Brief Series. National Institute of Justice: Washington, D.C.

DeJong, William. 1994. "School-Based Violence Prevention: From the Peaceable School to the Peaceable Neighborhood." *Forum,* pp. 8–14. Washington, D.C.: National Institute for Dispute Resolution.

DiIulio, John. 1999. "Federal Crime Policy: Time for a Moratorium." *Brookings Review* 17(1):17. (Winter).

Elliot, D., and S. Ageton. 1980. "Reconciling Race and Class Differences in Self-Reported and Official Estimates of Delinquency." *American Sociological Review* 45:95–110.

Erikson, Kai T. 1966. *Wayward Puritans.* New York: John Wiley & Sons.

Faison, Seth. 1996. "Copyright Pirates Prosper in China despite Promises." *New York Times,* February 20, A1, A6.

Farhi, Paul. 1996. "Researchers Link Psychological Harm to Violence on TV." *Boston Globe,* February 7, D3.

(FBI) Federal Bureau of Investigation. 1995. *Crime in the United States, 1994: Uniform Crime Reports.* Washington, D.C.: U.S. Government Printing Office.

———. 1996. *Crime in the United States, 1995: Uniform Crime*

Reports. Washington, D.C.: U.S. Government Printing Office.
Felson, Marcus. 1998. *Crime and Everyday Life,* 2nd ed. Thousand Oaks, Calif.: Pine Forge Press.
Fox, James A. 1997. "Trends in Juvenile Violence: A Report to the United States Attorney General on Current and Future Rates of Juvenile Offending (1997 Update)." U.S. Department of Justice, Bureau Of Justice Statistics.
Gallup Poll. 1998. "Most Important Problem." April 16–19, 1997. **http://198.175.140.8/Gallup_Poll_Data/mood/problems.htm**
Garey, M. 1985. "The Cost of Taking a Life: Dollars and Sense of the Death Penalty." *U.C. Davis Law Review* 18:1221–73.
Gest, Ted, and Dorian Friedman. 1994. "The New Crime Wave." *U.S. News and World Report,* August 29, 26–28.
Hagan, John, and Ruth Peterson. 1995. "Criminal Inequality in America: Patterns and Consequences." In *Crime and Inequality,* ed. John Hagan and Ruth Peterson, pp 14–36. Stanford, Calif: Stanford University Press.
Harris Poll. 1996. University of North Carolina: Institute for Research in Social Science (IRSS): Chapel Hill, N.C.
———. 1997. University of North Carolina: Institute for Research in Social Science (IRSS).
Henry, Tamara. 1995. "Values Enter Classroom Curriculum." *USA Today,* April 5, D1.
Heubusch, Kevin. 1997. "Teens on the Trigger." *American Demographics,* February. **http://www.demographics.com**
Hills, Stuart. 1987. "Introduction." In *Corporate Violence: Injury and Death for Profit,* ed. Stuart Hills, pp. 1–7. Rowman Publishing Co: Lanham, Md.
Hirschi, Travis. 1969. *Causes of Delinquency.* Berkeley: University of California Press.
Hochstetler, Andrew, and Neal Shover. 1997. "Street Crime, Labor Surplus, and Criminal Punishment, 1980–1990." *Social Problems* 44(3):358–67.
Hughes, Donna M. 1997. "Trafficking in Women on the Internet." CATW: *Coalition Report* 4 (1). **http://www.uri.edu/artsci/ wms/hughes/catw**
ICCA (International Crime Control Act). 1998. "Preface." Washington D.C.: Government Printing Office. p. 1.
INTERPOL. 1998. "INTERPOL Warning: Nigerian Crime Syndicate's Letter Scheme Fraud Takes on New Dimension." Press Releases. **http://www.kenpubs.co.uk/INTERPOL.COM/English/pres/nig.html**
Jacobs, David. 1988. "Corporate Economic Power and the State: A Longitudinal Assessment of Two Explanations." *American Journal of Sociology* 93:852–81.
Lagan, Bernard. 1997. "$3bn Cost of White Collar Criminals." *Sydney Morning Herald,* August 13, 1. **http://www.smh.com.au**
Laub, John, Daniel S. Nagan, and Robert Sampson. 1998. "Trajectories of Change in Criminal Offending: Good Marriages and the Desistance Process." *American Sociological Review* 63 (April):225–38.
Lehrur, Eli. 1999. "Communities and Cops Join Forces." *Insight on the News* 15(3):16. (January 25).
Leidholdt, Dorchen. 1997. "A Fact-Finding Trip to Thailand." *CATW: Coalition Report* 4(1): **http://www.uri.edu/artsci/wms/hughes/catw**
Lewis, Peter H. 1998. "Threat to Corporate Computers Is Often the Enemy Within." *New York Times,* March 2, 1–5.
Lindberg, Kirsten, Joseph Petrenko, Jerry Gladden, and Wayne Johnson. 1997."The Changing Face of Organized Crime in America: Asian Organized Crime." *Crime and Justice International* 13(10): **http://www.acsp.uic.edu/oicj/pvbs/cjintl/310/131005.shtml**
Mannix, Margaret. 1998. "Stolen Identity," *U.S. News and World Report.* June 1:48–51.
Mauer, Marc. 1997. *Americans behind Bars: The International Use of Incarceration.* Washington D.C.: The Sentencing Project. **http://www.sproject.com/press-1.htm**
Merton, Robert. 1957. "Social Structure and Anomie." In *Social Theory and Social Structure.* Glencoe, Ill.: Free Press.
Meyer, Michael, and Anne Underwood. 1994. "Crimes of the 'Net.'" *Newsweek,* November 14, 46–47.
Moore, Elizabeth, and Michael Mills. 1990. "The Neglected Victims and Unexamined Costs of White Collar Crime." *Crime and Delinquency* 36: 408–18.
Morganthau, Tom. 1993. "The New Frontier for Civil Rights." *Newsweek,* November 29, 65–66.
Murray, John P. 1993. "The Developing Child in a Multimedia Society." In *Children and Television: Images in a Changing Sociocultural World,* ed. Gordon L. Berry and Joy Keiko Asamen, pp. 9–22. Newbury Park, Calif.: Sage Publications.
Murray, Mary E., Nancy Guerra, and Kirk Williams. 1997. "Violence Prevention for the Twenty-First Century." In *Enhancing Children's Awareness,* ed. Roger P. Weissberg, Thomas Gullota, Robert L. Hampton, Bruce Ryan, and Gerald Adams, pp. 105–128. Thousand Oaks, Calif: Sage Publications.
Myths and Facts about the Death Penalty. 1998. "Death Penalty: Focus on California." **http://members.aol.com/Dpfocus/facts.htm**
NIJ (National Institute of Justice). 1998. "Research in Brief." October. U.S. Dept. of Justice: Washington D.C.

National Research Council. 1994. *Violence in Urban America: Mobilizing a Response.* Washington, D.C.: National Academy Press.

(NCVS) National Crime Victimization Survey. 1997. "Crime and Victim Statistics." Bureau of Justice Statistics. Washington, D.C.: U.S. Department of Justice.

(OJP) Office of Justice Programs. 1998. "Gallery 37." Innovations in American Government: Award Winners in the Field of Criminal Justice. **http://wwwojp/usdoj/nij/innvprog**

Operation Cease Fire. 1997. Office of Justice Programs: National Institute of Justice. **http://www.ojp.usdoj.gov/nij/innvprog**

Parker, Laura. 1998. "Tests Point to Sheppard's Innocence." *USA Today,* March 5, A3.

Perkins, Craig, and Patsy Klaus. 1996. *Criminal Victimization,* 1994. Washington D.C.: U.S. Department of Justice.

Potok, Mark, and Andrea Stone. 1995. "Values Get a Chance at Comeback." *USA Today,* February 3–5, A1, A2.

Price, Joyce Howard. 1999. "Crime Wave at Ebb Tide." *Insight on the News* 15(5):43 (February 8).

Radelet, M. L., H. A. Bedeau, and C. E. Putnam. 1992. *In Spite of Innocence: Erroneous Convictions in Capital Cases.* Boston: Northeastern University Press.

Randall, Melanie, and Lori Haskell. 1995. "Sexual Violence in Women's Lives." *Violence against Women* 1:6–31.

Reiman, Jeffery. 1998. *The Rich Get Richer and the Poor Get Prison.* Boston: Allyn and Bacon.

Reno, Janet. 1997. "Department of Justice Strategic Plan: 1997–2002." **http://www.usdoj.gov/jmd/mps/plan.htm**

Rideau, William. 1994. "Why Prisons Don't Work." *Time,* March 21, 80.

Rosoff, Stephen, Henry Pontell, and Robert Tillman. 1998. *Profit Without Honor: White Collar Crime and the Looting of America.* Englewood Cliffs, N.J.: Prentice-Hall.

Rubin, Amy Magaro. 1995. "Using Pop Culture to Fight Teen Violence." *Chronicle of Higher Education,* July 21, A5.

Russell, Cheryl. 1995. "Murder Is All American." *American Demographics,* September. **http://www.demographics.com**

Sanday, P. R. 1981. "The Sociocultural Context of Rape: A Cross-Cultural Study." *Journal of Social Issues* 37:5–27.

Shabad, Steven. 1999. "A Himalayan Mafia." *World Press Review* 46(2):20 (February).

Shabalin, Victor, J.J. Albini, and R.E. Rogers. 1995. "The New Stage of the Fight against Organized Crime in Russia." *IASOC: Criminal Organization* 10 (1):19–21.

SSBR (Social Statistics Briefing Room). 1999. "Crime." **http://www.whitehouse.gov/fsbr/crime.html**

Statistical Abstract of the United States: 1997, 117th ed. U.S. Bureau of the Census. Washington, D.C.: U.S. Government Printing Office.

Statistical Abstract of the United States: 1998, 118th ed. U.S. Bureau of the Census. Washington, D.C.: U.S. Government Printing Office.

Steffensmeier, Darrell, and Emilie Allan. 1995. "Criminal Behavior: Gender and Age." In *Criminology: A Contemporary Handbook,* 2d ed., ed. Joseph F. Sheley, pp. 83–113. Belmont, Calif.: Wadsworth.

Sutherland, Edwin H. 1939. *Criminology.* Philadelphia: Lippincott.

The White House. 1998. "International Crime Control Strategy: Goals and Objectives." pp. 12–13. Washington D.C.: Government Printing Office.

United Nations. 1997. "Crime Goes Global." Document No. DPI/1518/SOC/CON/30M. New York: United Nations.

U.S. Department of Justice. 1993. "Highlights of 20 years of Surveying Crime Victims." Bureau of Justice Statistics. Washington, D.C.: U. S. Government Printing Office (NCJ-144525).

———. 1997a. "One in Five U.S. Residents in Contact with Police during the Year." Press Release (November 22). Bureau of Justice Statistics.

———. 1997b. "Prisoners' Executions Rise Significantly." Press Release (May 14). Bureau of Justice Statistics.

———. 1998. "International Crime Statistics." Bureau of Justice Statistics. Washington, D.C. **http://www.ojp.usdoj.gov/bjs/pub**

(VORP) Victim-Offender Reconciliation Program. 1998. "About Victim-Offender Mediation and Reconciliation." **http://www.igc.org/vorp**

Walker, Samuel, Cassia Spohn, and Miriam Delone. 1996. *The Color of Justice: Race, Ethnicity, and Crime in America.* Belmont, Calif.: Wadsworth.

Warner, Barbara, and Pamela Wilcox Rountree. 1997. "Local Social Ties in a Community and Crime Model." *Social Problems* 4(4):520–36.

Williams, Linda. 1984. "The Classic Rape: When Do Victims Report?" *Social Problems* 31: 459–67.

Willing, Richard. 1997. "Tests Exonerate One in Four Suspects." *USA Today,* November 28, A3.

Wolfgang, Marvin, Robert Figlio, and Thorstein Sellin. 1972. *Delinquency in a Birth Cohort.* Chicago: University of Chicago Press.

Wright, James D. 1995. "Guns, Crime, and Violence." In *Criminology: A Contemporary Handbook,* 2d. ed., ed. Joseph F. Sheley, pp. 495–513. Belmont, Calif.: Wadsworth.

Zagaroli, Lisa. 1997. "Consumer Groups Call Side-Mounted GM Truck Fuel Tanks 'lethal.'" *De-*

troit News, October 14. **http://www.detnews.com**

Zimring, F.E., and G. Hawkins. 1997. *Crime Is Not the Problem: Lethal Violence in America.* New York: Oxford University Press.

CHAPTER 5

Amato, Paul R., and Bruce Keith. 1991. "Parental Divorce and Adult Well-Being: A Metaanalysis." *Journal of Marriage and the Family* 53:43–58.

Anderson, Kristin L. 1997. "Gender, Status, and Domestic Violence: An Integration of Feminist and Family Violence Approaches." *Journal of Marriage and the Family* 59:655–69.

Arendell, Terry. 1995. *Fathers and Divorce.* Thousand Oaks, Calif.: Sage Publications.

Berne, L. A., and Huberman, B. K. 1996. "Sexuality Education Works: Here's Proof." *Education Digest,* February, 25–29.

Blankenhorn, David. 1995. *Fatherless America: Confronting Our Most Urgent Social Problem.* New York: Basic Books.

Brienza, Julie. 1996. "At the Fault Line: Divorce Laws Divide Reformers." *Trial* 32(2):12–14.

Browning, Christopher R., and Edward O. Laumann. 1997. "Sexual Contact between Children and Adults: A Life Course Perspective." *American Sociological Review* 62:540–60.

Busby, Dean M. 1991. "Violence in the Family." In *Family Research: A Sixty-Year Review,* vol. 1, ed. S. J. Bahr, pp. 335–85. New York: Lexington Books.

Center on Budget and Policy Priorities. 1998 (Sept. 24). "Poverty Rates Fall, but Remain High for a Period with Such Low Unemployment." **http://www.cbpp.org/9-24-98pov.htm** (September 28, 1998).

Clark, Charles. 1996. "Marriage and Divorce." *CQ Researcher,* 6(18):409–32.

Daro, Deborah. 1998. "Public Opinion and Behaviors Regarding Child Abuse Prevention: 1998 Survey." Chicago, Ill.: National Committee to Prevent Child Abuse. **http://www.childabuse.org/poll98.html**

Demo, David H. 1992. "Parent-Child Relations: Assessing Recent Changes." *Journal of Marriage and the Family* 54:104–17.

———. 1993. "The Relentless Search for Effects of Divorce: Forging New Trails or Tumbling Down the Beaten Path?" *Journal of Marriage and the Family* 55: 42–45.

"Domestic Violence and Homelessness." 1998. NCH Fact Sheet no. 8. National Coalition for the Homeless.

"Domestic Violence Fact Sheet." 1999 (January). Department of Health and Human Services, Administration for Children and Families. **http://www.acf.dhhs.gov/p...pa/facts/domsvio.htm** (February 18, 1999).

Edin, Kathryn, and Laura Lein. 1997. *Making Ends Meet: How Single Mothers Survive Welfare and Low-Wage Work.* New York: Russell Sage Foundation.

Elliott, D. M., and J. Briere. 1992. "The Sexually Abused Boy: Problems in Manhood." *Medical Aspects of Human Sexuality* 26:68–71.

Finkelhor, D., G. Hotaling, I. A. Lewis, and C. Smith. 1990. "Sexual Abuse in a National Survey of Adult Men and Women: Prevalence, Characteristics, and Risk Factors." *Child Abuse and Neglect* 14:19–28.

Ganong, Lawrence H., and Marilyn Coleman. 1994. *Remarried Family Relationships.* Thousand Oaks, Calif.: Sage Publications.

Geile, Janet Z. 1996. "Decline of the Family: Conservative, Liberal, and Feminist Views." In *Promises to Keep: Decline and Renewal of Marriage in America,* eds. D. Popenoe, J. B. Elshtain, and D. Blankenhorn. Lanham, Md: Rowman & Littlefield, pp. 89–115.

Gelles, Richard J. 1993. "Family Violence." In *Family Violence: Prevention and Treatment,* eds. Robert L. Hampton, Thomas P. Gullotta, Gerald R. Adams, Earl H. Potter III, and Roger P. Weissberg, pp. 1–24. Newbury Park, Calif.: Sage Publications.

Gelles, Richard J., and Jon R. Conte. 1990. "Domestic Violence and Sexual Abuse of Children: A Review of Research in the Eighties." *Journal of Marriage and the Family* 52: 1045–58.

———. 1991. "Domestic Violence and Sexual Abuse of Children: A Review of Research in the Eighties." In *Contemporary Families: Looking Forward, Looking Back,* ed. Alan Booth, pp. 327–40. Minneapolis: National Council on Family Relations.

Glenn, Norval D. 1997. *Closed Hearts, Closed Minds: The Textbook Story of Marriage.* New York: Institute for American Values.

Global Study of Family Values. 1998. The Gallup Organization. **http://198.175.140.8/Special_Reports/family.htm** (April 13, 1998).

Harrington, Donna, and Howard Dubowitz. 1993. "What Can Be Done to Prevent Child Maltreatment?" In *Family Violence: Prevention and Treatment,* eds. Robert L. Hampton, Thomas P. Gullotta, Gerald R. Adams, Earl H. Potter III, and Roger P. Weissberg, pp. 258–80. Newbury Park, Calif.: Sage Publications.

Healthy Families America Fact Sheet. 1994 (May). National Committee to Prevent Child Abuse, Healthy Families America. 332 S. Michigan Ave., Suite 1600, Chicago, IL 60604.

Heim, Susan C., and Douglas K. Snyder. 1991. "Predicting Depression from Marital Distress and Attributional Processes."

Journal of Marital and Family Therapy 17: 67–72.
Hewlett, Sylvia Ann and Cornel West. 1998. *The War Against Parents: What We Can Do for Beleaguered Moms and Dads.* Boston: Houghton Mifflin Company.
Hochschild, Arlie Russell. 1997. *The Time Bind: When Work Becomes Home and Home Becomes Work.* New York: Henry Holt and Company.
"In the News." 1998. Family Violence Prevention Fund. **http://www.igc.org/fund/materials/speakup/02_13_98.htm** (May 16, 1998).
Ingrassia, Michelle. 1993. "Daughters of Murphy Brown." *Newsweek,* August 2, 58–59.
Jacobs, C. D., and E. M. Wolf 1995. "School Sexuality Education and Adolescent Risk-Taking Behavior." *Journal of School Health* 65:91–95.
Jekielek, Susan M. 1998. "Parental Conflict, Marital Disruption and Children's Emotional Well-Being." *Social Forces* 76:905–35.
Kaufman, Joan, and Edward Zigler. 1992. "The Prevention of Child Maltreatment: Programming, Research, and Policy." In *Prevention of Child Maltreatment: Developmental and Ecological Perspectives,* ed. Diane J. Willis, E. Wayne Holden, and Mindy Rosenberg, pp. 269–95. New York: John Wiley & Sons.
Kitson, Gay C., and Leslie A. Morgan. 1991. "The Multiple Consequences of Divorce: A Decade Review." In *Contemporary Families: Looking Forward, Looking Back,* ed. Alan Booth, pp. 150–91. Minneapolis: National Council on Family Relations.
Knox, David (with Kermit Leggett). 1998. *The Divorced Dad's Survival Book: How to Stay Connected with Your Kids.* New York: Insight Books.
Knutson, John F., and Mary Beth Selner. 1994. "Punitive Childhood Experiences Reported by Young Adults over a 10-Year Period." *Child Abuse and Neglect* 18: 155–66.
Krug, Ronald S. 1989. "Adult Male Report of Childhood Sexual Abuse by Mothers: Case Description, Motivations, and Long-Term Consequences." *Child Abuse and Neglect* 13: 111–19.
Lachs, Mark S., Christianna Williams, Shelley O'Brien, Leslie Hurst, and Ralph Horwitz. 1997. "Risk Factors for Reported Elder Abuse and Neglect: A Nine-Year Observational Cohort Study." *Gerontologist* 37:469–74.
Levitan, Sar A., Garth L. Mangum, and Stephen L. Mangum. 1998. *Programs in Aid of the Poor,* 7th ed. Baltimore: Johns Hopkins University Press.
Lloyd, S. A., and B. C. Emery. 1993. "Abuse in the Family: An Ecological, Life-Cycle Perspective." In *Family Relations: Challenges for the Future,* ed. T. H. Brubaker, pp. 129–52. Newbury Park, Calif.: Sage Publications.
Louisiana Divorce. 1997. Women's Connection Online, Inc. **http://www.womenconnect.com** (July 21, 1998).
Luker, Kristin. 1996. *Dubious Conceptions: The Politics of Teenage Pregnancy.* Cambridge, Mass.: Harvard University Press.
Marlow, L., and S. R. Sauber. 1990. *The Handbook of Divorce Mediation.* New York: Plenum.
Masheter, C. 1991. "Post-Divorce Relationships between Ex-Spouses: The Roles of Attachment and Interpersonal Conflict." *Journal of Marriage and the Family* 53: 103–10.
Mindel, Charles H., Robert W. Habenstein, and Roosevelt Wright, Jr. 1998. *Ethnic Families in America: Patterns and Variations.* Upper Saddle River, N.J.: Prentice Hall.
Monson, C. M., G. R. Byrd, and J. Langhinrichsen-Rohling. 1996. "To Have and to Hold: Perceptions of Marital Rape." *Journal of Interpersonal Violence,* 11: 410–24.
Morrow, R. B., and G. T. Sorrell. 1989. "Factors Affecting Self-Esteem, Depression, and Negative Behaviors in Sexually Abused Female Adolescents." *Journal of Marriage and the Family* 51: 677–86.
National Parenting Association. 1996. *What Will Parents Vote For?: Findings of the First National Survey of Parent Priorities.* New York: Author.
North Carolina Coalition Against Domestic Violence. 1991 (Spring). *Domestic Violence Fact Sheet.* P.O. Box 51875, Durham, NC 27717–1875.
O'Keefe, Maura. 1997. "Predictors of Dating Violence among High School Students." *Journal of Interpersonal Violence* 12:546–68.
Peterson, Karen S. 1997. "States Flirt with Ways to Reduce Divorce Rate." *USA Today,* April 10, D1–2.
Popenoe, David. 1993. "Point of View: Scholars Should Worry about the Disintegration of the American Family." *Chronicle of Higher Education,* April 14, A48.
Popenoe, David. 1996. *Life without Father.* New York: Free Press.
Prevention Programs. 1998 (May 4). National Committee to Prevent Child Abuse. **www.childabuse.org/8ar97.html** (May 12 1998).
Resnick, Michael, Peter S. Bearman, Robert W. Blum, Karl E. Bauman, Kathleen M. Harris, Jo Jones, Joyce Tabor, Trish Beubring, Renee E. Sieving, Marcia Shew, Marjore Ireland, Linda H. Berringer, and J. Richard Udry. 1997 (September 10). "Protecting Adolescents from Harm." *Journal of the American Medical Association* 278(10):823–32.
Russell, D. E. 1990. *Rape in Marriage.* Bloomington, Ind.: Indiana University Press.

Sapiro, Virginia. 1990. *Women in American Society,* 2d ed. Mountain View, Calif.: Mayfield.

Shapiro, Joseph P., and Joannie M. Schrof. 1995. "Honor Thy Children." *U.S. News and World Report,* February 27, 39–49.

Song, Young I. 1991. "Single Asian American Women as a Result of Divorce: Depressive Affect and Changes in Social Support." *Journal of Divorce and Remarriage* 14:219–30.

Stanley, Scott M., Howard J. Markman, Michelle St. Peters, and B. Douglas Leber. 1995. "Strengthening Marriage and Preventing Divorce: New Directions in Prevention Reseach." *Family Relations* 44:392–401.

Statistical Abstract of the United States: 1998, 118th ed. U.S. Bureau of the Census. Washington, D.C.: U.S. Government Printing Office.

Stein, Theodore J. 1993. "Legal Perspectives on Family Violence against Children." In *Family Violence: Prevention and Treatment,* ed. by Robert L. Hampton, Thomas P. Gullotta, Gerald R. Adams, Earl H. Potter III, and Roger P. Weissberg, pp. 179–97. Newbury Park, Calif.: Sage Publications.

Stock, J. L., M. A. Bell, D. K. Boyer, and F. A. Connell. 1997. "Adolescent Pregnancy and Sexual Risk-Taking among Sexually Abused Girls." *Family Planning Perspectives* 29:200–203.

Stout, James W., and Fredrick P. Rivara. 1989. "Schools and Sex Education: Does It Work?" *Pediatrics* 83:375–79.

Straus, Murray A., David B. Sugarman, and Jean Giles-Sims. 1997. "Spanking by Parents and Subsequent Antisocial Behavior of Children." *Archives of Pediatric Adolescent Medicine* 151:761–67.

"This Year's College Freshmen: A Statistical Profile." 1999 (January 20). *The Chronicle of Higher Education:* A48–A49.

United Nations. 1992. *1991 Demographic Yearbook,* 43d ed. New York: United Nations.

U.S. Census Bureau. 1998. *Poverty in the United States: 1997.* Washington, DC: Government Printing Office.

U.S. Department of Justice. 1998 (March 16). "Murder by Intimates Declined 36 Percent since 1976, Decrease Greater for Male than for Female Victims." Washington, D.C. **http://www.ojp.usdoj.gov/bjs/pub/press/vi.pr**

U.S. Department of Justice. Office of Justice Programs. 1994. "Domestic Violence: Violence between Intimates." Washington, D.C.: Bureau of Justice Statistics.

Ventura, Stephanie J., T. J. Mathews, and Sally C. Curtin. 1998. "Declines in Teenage Birth Rates, 1991–97: National and State Patterns." *National Vital Statistics Report* 47(12).

Viano, C. Emilio. 1992. "Violence among Intimates: Major Issues and Approaches." In *Intimate Violence: Interdisciplinary Perspectives,* ed. C. E. Viano, pp. 3–12. Washington, D.C.: Hemisphere.

Waite, L. J. 1995. "Does Marriage Matter?" *Demography* 32: 483–507.

Wang, Ching-Tung, and Deborah Daro. 1998. "Current Trends in Child Abuse Reporting and Fatalities: The Results of the 1997 Annual 50 State Survey." Chicago, Ill.: National Committee to Prevent Child Abuse. **http://www.childabuse.org/50data97.html** (May 12 1998).

Willis, Diane J., E. Wayne Holden, and Mindy Rosenberg. 1992. "Child Maltreatment Prevention: Introduction and Historical Overview." In *Prevention of Child Maltreatment: Developmental and Ecological Perspectives,* ed. Diane J. Willis, E. Wayne Holden, and Mindy Rosenberg, pp. 1–14. New York: John Wiley & Sons.

CHAPTER 6

AA (Alzheimer's Association) Fact Sheet. 1998. Chicago, Ill: National Alzheimer's Association.

AARP (American Association of Retired Persons) Bulletin. 1998. "Retirement Age: No Easy Answer." September, pp. 1, 11.

Adler, Jerry. 1994. "Kids Growing up Scared." *Newsweek,* January 10, 43–50.

Anetzberger, Georgia J., Jill E. Korbin, and Craig Austin. 1994. "Alcoholism and Elder Abuse." *Journal of Interpersonal Violence* 9:184–93.

AOA (Administration on Aging). 1999. "Profile of Older Americans: 1998." Washington, D.C.: Department of Health and Human Services.

AOA (Administration on Aging). 1998a. "Elder Abuse Prevention." Washington, D.C.: Department of Health and Human Services.

———. 1998b. "Older Women: A Diverse and Growing Population." Washington, D.C.: Department of Health and Human Services.

———. 1998c. "Profile of Older Americans: 1997." Washington, D.C.: Department of Health and Human Services.

AP (Associated Press). 1997. "U.S. Leads in Youth's Deaths by Guns." In *News and Observor* Raleigh, N.C. February 7, A2.

Arluke, Arnold, and Jack Levin. 1990. "'Second Childhood': Old Age in Popular Culture." In *Readings on Social Problems,* ed. W. Feigelman, pp. 261–65. Fort Worth: Holt, Rinehart and Winston.

Bergmann, Barbara R. 1999. "A 'Help for Working Parents' Program Can Reduce Child Poverty." *Brown University Child and Adolescent Behavior Letter* 15(2):1.

Boudreau, Francois A. 1993. "Elder Abuse." In *Family Violence: Prevention and Treatment,* eds. R. L.

Hampton, T. P. Gullota, G. R. Adams, E. H. Potter III, and R. P. Weissberg, pp. 142–58. Newbury Park, Calif.: Sage Publications.

Bradshaw, York, Rita Noonan, Laura Gash, and Claudia Buchmann Sershen. 1992. "Borrowing against the Future: Children and Third World Indebtedness." *Social Forces* 71(3): 629–56.

Brazzini, D. G., W. D. McIntosh, S. M. Smith, S. Cook, and C. Harris. 1997. "The Aging Woman in Popular Film: Underrepresented, Unattractive, Unfriendly, and Unintelligent." *Sex Roles* 36:531–43.

Brody, Elaine M. 1990. *Women in the Middle: Their Parent-Care Years.* New York: Springer.

Brooks-Gunn, Jeanne, and Greg Duncan. 1997. "The Effects of Poverty on Children." *Future of Children* 7 (2):55–70.

Cahill, Spenser. 1993. "Childhood and Public Life: Reaffirming Biographical Divisions." *Social Problems* 37(3): 390–400.

Child and Family Statistics. 1999. "America's Children: Key National Indications of Well Being." Office of Management and Budget. Washington, D.C.: Government Printing Office. **http://childstats.gov**

Children Now. 1998. "Content Analysis on Welfare Reform Reporting: A Report by Children Now." **http://www.childrennow.org/economics/WelRef98/WelfareReformShortDoc.html**

Children's Defense Fund. 1997. "Big Economic Gains Lift Very Few Children Out of Poverty." Washington D.C.: Children's Defense Fund.

———. 1998a. "Children in the United States: A 1998 Profile." Washington D.C.: Children's Defense Fund.

———. 1998b. "Facts on Youth, Violence and Crime." Washington D.C.: Children's Defense Fund.

———. 1998c. "Key Facts about U.S. Children." Washington D.C.: Children's Defense Fund.

"Children's Legal Rights." 1993. *Congressional Quarterly,* April 23, 339–54.

Clements, Mark. 1993. "What We Say about Aging." *Parade Magazine,* December 12: 4–5.

Cowgill, Donald, and Lowell Holmes. 1972. *Aging and Modernization.* New York: Appleton-Century-Crofts.

Crystal, Stephen, and Dennis Shea. 1990. "Cumulative Advantage, Cumulative Disadvantage, and Inequality among Elderly People." *The Gerontologist* 30: 437–43.

Cummings, Elaine, and William Henry. 1961. *Growing Old: The Process of Disengagement.* New York: Basic Books.

DeAngelis, Tori. 1997. "Elderly May Be Less Depressed Than the Young." *APA Monitor,* October. **http://www.apa.org/monitor/oct97/elderly.html**

DHHS (Department of Health and Human Services). 1998a. "Profile of America's Youth." U.S. Department of Health and Human Services. **http:/youth.os.dhhs.gov/youthinf.htm#profile**

———. 1998b. "Statement of Jeanette Takamura." U.S. Department of Health and Human Services, Administration on Aging. June 8. **http://www.aoa.dhhs.gov/pr/graying.html**

Duncan, Greg, W. Jean Yeung, Jeanne Brooks-Gunn, and Judith Smith. 1998. "How Much Does Childhood Poverty Affect the Life Chance of Children?" *American Sociological Review* 63:402–23.

Dychtwald, Ken. 1990. *Age Wave.* New York: Bantam Books.

Elder Fraud. 1999. "Fighting Fraud against Elder Consumers." National Consumers League. **http://www.fraud.org**

Fields, Jason, and Kristin Smith. 1998. "Poverty, Family Structure, and Child Well-Being." Population Division. Washington, D.C.: U.S. Bureau of Census.

Friedan, Betty. 1993. *The Fountain of Age.* New York: Simon & Schuster.

Garrard, Judith, Joan Buchanan, Edward Ratner, Lukas Makris, Hung-Ching Chan, Carol Skay, and Robert Kane. 1993. "Differences between Nursing Home Admissions and Residents." *Journal of Gerontology* 48(6): S310–S309.

George, Christopher. 1992. "Old Money." *Washington Monthly,* pp. 16–21.

Harrigan, Anthony. 1992. "A Lost Civilization." *Modern Age: A Quarterly Review,* Fall, 3–12.

Harris, Diana K. 1990. *Sociology of Aging.* New York: Harper & Row.

Harris, Kathleen, and Jeremy Marmer. 1996. "Poverty, Paternal Involvement and Adolescent Well-Being." *Journal of Family Issues* 17(5):614–40.

Harris Poll. 1998a. "Do You Feel Good about Your Children's Future?" *Institute for Research in Social Science* Study No. S818441 (May), Q200.

———. 1998b. "Many Government Programs Cost a Lot of Money . . ." *Institute for Research in Social Science* Study No. S818418 (April), Q410.

Hewlett, Sylvia. 1992. *When the Bough Breaks: The Cost of Neglecting Our Children.* New York: Basic Books.

Hurd, Michael D. 1990. "The Economic Status of the Elderly." *Science* 244:659–64.

Huuhtanen, P., and M. Piispa. 1996. "Attitudes toward Work and Retirement among Elderly Workers." Helsinki, Finland: Finnish Institute of Occupational Health.

Ingrassia, Michelle. 1993. "Growing up Fast and Frightened." *Newsweek,* November 22, 52–53.

International Labour Organization. 1997. "IPEC Fact Sheet: Action against Most Intolerable Forms of Child Labour." **http://www.ilo.org/public/english/90ipec/factsh/fs96_28.html**

Knoke, David, and Arne L. Kalleberg. 1994. "Job Training in U.S. Organizations." *American Sociological Review* 59: 537–46.

Kolland, Franz. 1994. "Contrasting Cultural Profiles between Generations: Interests and Common Activities in Three Intrafamilial Generations." *Ageing and Society* 14:319–40.

Matras, Judah. 1990. *Dependency, Obligations, and Entitlements: A New Sociology of Aging, the Life Course, and the Elderly.* Englewood Cliffs, N.J.: Prentice-Hall.

Miner, Sonia, John Logan, and Glenna Spitze. 1993. "Predicting Frequency of Senior Center Attendance." *The Gerontologist* 33:650–57.

Minkler, Meredith. 1989. "Gold Is Gray: Reflections on Business' Discovery of the Elderly Market." *The Gerontologist* 29(1):17–23.

Miringoff, Marc. 1989. *The Index of Social Health, 1989: Measuring the Social Well-Being of a Nation.* Tarrytown, N.Y.: Fordham Institute for Innovation in Social Policy.

Moone, Marilyn, and Judith Hushbeck. 1989. "Employment Policy and Public Policy: Options for Extending Work Life." *Generations* 13(3):27–30.

National Council of Senior Citizens. 1998. "Elderly in Dire Need of Housing." **http://www.ncscinc.org/press/housepr.html**

NCCP. 1998. "Columbia University Study Finds More Than 5 Million Young Children in Poverty Despite Economic Growth." **http://cpmcnet.columbia.edu/dept/nccp/98Upress.html**

Peterson, Peter. 1997. "Will America Grow up before It Grows Old?" In *Social Problems: Annual Editions,* ed. Harold A. Widdison, pp. 74–90. Guilford, Conn: Dushkin.

Public Health Reports. 1999. "Report on Older Adults and Substance Abuse." *Public Health Reports* 114(1):11.

Regoli, Robert, and John Hewitt. 1997. *Delinquency in Society.* New York: McGraw-Hill.

Riley, Matilda White. 1987. "On the Significance of Age in Sociology." *American Sociological Review* 52 (February): 1–14.

Riley, Matilda W., and John W. Riley. 1992. "The Lives of Older People and Changing Social Roles." In *Issues in Society,* eds. Hugh Lena, William Helmreich, and William McCord, pp. 220–31. New York: McGraw-Hill.

Russell, Charles. 1989. *Good News about Aging.* New York: Wiley.

Seeman, Teresa E., and Nancy Adler. 1998. "Older Americans: Who Will They Be?" *National Forum,* Spring, 22–25.

Shalala, Donna. 1998. "HHS Approves Nevada Plan to Insure More Children." Press Release. HCFA Press Office: Department of Health and Human Services (August 13).

Sherman, Edmund, and Theodore A. Webb. 1994. "The Self as Process in Later-Life Reminiscence: Spiritual Attributes." *Ageing and Society* 14: 255–67.

Sherman, George. 1998. "Social Security: The Real Story." *National Forum,* Spring, 26–29.

Simon-Rusinowitz, Lori, Constance Krach, Lori Marks, Diane Piktialis, and Laura Wilson. 1996. "Grandparents in the Workplace: the Effects of Economic and Labor Trends." *Generations* 20(1):41–44.

Statistical Abstract of the United States: 1997, 117th ed. U.S. Bureau of the Census. Washington, D.C.: U.S. Government Printing Office.

Statistical Abstract of the United States: 1998, 118th ed. U.S. Bureau of the Census. Washington, D.C.: U.S. Government Printing Office.

Streitfeld, David. 1993. "Abuse of the Elderly—Often It's the Spouse." In *Society in Crisis,* ed. the Washington Post Writers Group, pp. 212–13. Needham, Mass.: Allyn & Bacon.

Taylor, Paul. 1993. "Revamping Welfare with 'Universalism.'" In *Society in Crisis,* ed. the Washington Post Writers Group, pp. 99–104. Needham, Mass.: Allyn & Bacon.

Thurow, Lester C. 1996. "The Birth of a Revolutionary Class." *New York Times Magazine,* May 19, 46–47.

Torres-Gil, Fernando. 1990. "Seniors React to Medicare Catastrophic Bill: Equity or Selfishness?" *Journal of Aging and Social Policy* 2(1):1–8.

UNICEF (United Nations Children's Fund). 1994. "The Progress of Nations." United Nations.

UNICEF (United Nations Children's Fund). 1998. "The First Nearly Universally Ratified Human Rights Treaty in History." *Status.* Washington, D.C.: UNICEF. **http://www.unicef.org/crc/status.html**

Weissberg, Roger P., and Carol Kuster. 1997. "Introduction and Overview: Let's Make *Healthy Children 2010* a National Priority." In *Enhancing Children's Well-Being,* eds. Roger Weissberg, Thomas Gullotta, Robert Hampton, Bruce Ryan, and Gerald Adams, pp. 1–16. Thousand Oaks, Calif.: Sage Publications.

World Congress Against Commercial Sexual Exploitation of Children. 1998. "Regional Profiles." **http://wwwusis.usemb.se/children/csec/226e.html**

Yearbook: The State of America's Children. 1998. Children's Defense Fund. Washington, D.C.: Children's Defense Fund.

CHAPTER 7

American Association of University Women. 1991. *Shortchanging Girls, Shortchanging America.* Washington, D.C.: Greenberg-Lake Analysis Group.

———. 1992. *How Schools Shortchange Girls.* Washington, D.C.: American Association of University Women.

Anderson, John, and Molly Moore. 1998. "The Burden of Womanhood." pp. 170–175 in *Global Issues 98/99.* ed. Robert Jackson Guilford, Conn. Dushkin/McGraw-Hill.

Andersen, Margaret L. 1997. *Thinking about Women,* 4th ed. New York: Macmillan.

Attwood, Lynne. 1996. "Young People's Attitudes towards Sex Roles and Sexuality." In *Gender, Generation and Identity in Contemporary Russia,* ed. Hilary Pilkington, pp. 132–51. London: Routledge.

Bae, Yupin, and Thomas Smith. 1998. "Women and Mathematics and Science." *Issues in Focus.* National Center for Educational Statistics. Washington, D.C: U.S. Department of Education.

Baker, Robin, Gary Kriger, and Pamela Riley. 1996. "Time, Dirt and Money: The Effects of Gender, Gender Ideology, and Type of Earner Marriage on Time, Household Task, and Economic Satisfaction among Couples with Children." *Journal of Social Behavior and Personality* 11:161–77.

Basow, Susan A. 1992. *Gender: Stereotypes and Roles,* 3d ed. Pacific Grove, Calif.: Brooks/Cole.

Beutel, Ann M., and Margaret Mooney Marini. 1995. "Gender and Values." *American Sociological Review* 60:436–48.

Bianchi, Susanne M., and Daphne Spain. 1996. "Women, Work and Family in America." *Population Bulletin* 51(3). Washington D.C.: Population Reference Bureau.

Burger, Jerry M., and Cecilia H. Solano. 1994. "Changes in Desire for Control over Time: Gender Differences in a Ten-Year Longitudinal Study." *Sex Roles* 31:465–72.

Calasanti, Toni, and Carol A. Bailey. 1991. "Gender Inequality and the Division of Labor in the United States and Sweden: A Socialist-Feminist Approach." *Social Problems* 38(1):34–53.

CAWP (Center for the American Women and Politics). 1998. "Women of Color in Elective Office 1998." National Information Bank on Women in Public Office, Eagleton Institute of Politics. New Brunswick, N.J.: Rutgers University.

CEDAW (Convention to Eliminate All Forms of Discrimination against Women). 1998. **http://www.feminist.org/research/cedawhist.html**

Cianni, Mary, and Beverly Romberger. 1997. "Life in the Corporation: A Multi-Method Study of the Experiences of Male and Female Asian, Black, Hispanic and White Employees." *Gender, Work and Organization* 4(2):116–29.

DeStefano, Linda, and Diane Colasanto. 1990. "Unlike 1975, Today Most Americans Think Men Have It Better." *Gallup Poll Monthly* 293:25–36.

Dortch, Shannon. 1997. "Hey Guys, Hit the Books." Tomorrow's Markets. *American Demographics,* September, 2. **http://www.demographics.com**

Educational Indicators. 1998. "Indicator 18: Gender Differences in Earnings." National Center for Educational Statistics. Washington, D.C.: U.S. Department of Education.

FairTest. 1997. "SAT-Math Gender Bias: Causes and Consequences." *FairTest Examiner* (Winter). Cambridge, Mass.: National Center for Fair and Open Testing. **http://www.fairtest.org/examarts/winter97/gender.html**

Faludi, Susan. 1991. *Backlash: The Undeclared War against American Women.* New York: Crown Publishers.

Fiebert, Martin S., and Mark W. Meyer. 1997. "Gender Stereotypes: A Bias against Men." *The Journal of Psychology* 131(4): 407–10.

Fitzgerald, Louise F., and Sandra L. Shullman. 1993. "Sexual Harassment: A Research Analysis and Agenda for the '90s." *Journal of Vocational Behavior* 40:5–27.

GenderGap, 1998. "GenderGap in Government: The Federal Government." **http://www.gendergap.com/governme.htm**

Goldberg, Carey. 1999. "After Girls get Attention, Focus is on Boys' Woes," p. 6 In *Themes of the Times: N.Y. Times* (sociology). Upper Saddle River, N.J.: Prentice Hall.

Goldberg, Stephanie. 1997. "Making Room for Daddy." *American Bar Association Journal* 83:48–52.

Hochschild, Arlie. 1989. *The Second Shift: Working Patterns and the Revolution at Home.* New York: Viking Penguin.

Human Development Report. 1997. United Nations Development Programme. New York: Oxford University Press.

Kate, Nancy. 1998. "Two Careers, One Marriage." *American Demographics,* April, 11.

Kilbourne, Barbara S., Georg Farkas, Kurt Beron, Dorothea Weir, and Paula England. 1994. "Returns to Skill, Compensating Differentials, and Gender Bias: Effects of Occupational Characteristics on the Wages of White Women and Men." *American Journal of Sociology* 100:689–719.

Klein, Matthew. 1998a. "Women Hit the Hoops." *American Demographics,* February, 12.

———. 1998b. "Women's Trip to the Top." *American Demographics,* February, 22.

———. 1998c. "Lifespan Gender Gap Narrows." *American Demographics,* March, 12.

Kopelman, Lotetta M. 1994. "Female Circumcision/Genital Mutilation and Ethical Relativism." *Second Opinion* 20:55–71.

LCCR (Leadership Conference on Civil Rights). 1998a. "A Majority of Americans Support Affirmative Action Programs for Women and People of Color." Leadership Conference on Civil Rights on Line. **http://civilrights.org/aa/polling.html**

———. 1998b. "Talking Points on the McConnell-Canady 'Civil Rights Act of 1997'." Leadership Conference on Civil Rights on Line. **http://civilrights.org/aa/talk.html**

Leo, John. 1997. "Fairness? Promises, Promises." *U.S. News and World Report* 123(4):18.

Long, J. Scott, Paul D. Allison, and Robert McGinnis. 1993. "Rank Advancement in Academic Careers: Sex Differences and the Effects of Productivity." *American Sociological Review* 58: 703–22.

Loo, Robert, and Karran Thorpe. 1999. "Attitudes toward Women's Roles in Society." *Sex Roles* 39(11/12):903–913.

Lorber, Judith. 1998. "Night to his Day." In *Reading Between the Lines.* eds. Amanda Konradi and Martha Schmidt, pp. 213–20. Mountain View, Calif.: Mayfield Publishing.

Luo, Tsun Yin. 1996. "Sexual Harassment in the Chinese Workplace." *Violence against Women* 2(3):284–301.

Marini, Margaret Mooney, and Pi-Ling Fan. 1997. "The Gender Gap in Earnings at Career Entry. *American Sociological Review* 62:588–604.

Martin, Patricia Yancey. 1992. "Gender, Interaction, and Inequality in Organizations." In *Gender, Interaction, and Inequality,* ed. Cecilia Ridgeway, pp. 208–31. New York: Springer-Verlag.

McCammon, Susan, David Knox, and Caroline Schacht. 1998. *Making Choices in Sexuality.* Pacific Grove, Calif.: Brooks/Cole Publishing Co.

Mensh, Barbara, and Cynthia Lloyd. 1997. "Gender Differences in the Schooling Experiences of Adolescents in Low-Income Countries: The Case of Kenya." *Policy Research Working Paper* no. 95. New York: Population Council.

Merida, Kevin, and Barbara Vobejda. 1998. "Role Shift Causes Conflict on the Home Front." *The Washington Post,* April 27, D1.

Mirowsky, John, and Catherine E. Ross. 1995. "Sex Differences in Distress: Real or Artifact?" *American Sociological Review* 60:449–68.

Mogelonsky, Marcia. 1997. "After the Baby." Kaleidiscope. *American Demographics,* December,15. **http://www.demographics.com**

———. 1998. "Where Women Do Better." Kaleidiscope. *American Demographics,* March, 21. **http://www.demographics.com**

National Organization for Women. 1995. "General Information about NOW." **http://www.igc.apc.org./homensnet**

NCFM (National Coalition of Free Men). 1992. "Declaration of the Father's Fundamental Pre-Natal Rights." Manhasset, N.Y.: NCFM. **http://ncfm.org**

———. 1998. "Historical." Manhasset, NY: NCFM. **http://ncfm.org**

Newport, Frank. 1993. "Americans Now More Likely to Say: Women Have It Harder Than Men." *Gallup Poll Monthly* no. 337, October, 11–18.

Nichols-Casebolt, Ann, and Judy Krysik. 1997. "The Economic Well-Being of Never and Ever-Married Mother Families." *Journal of Social Service Research* 23(1):19–40.

O'Kelly, Charlotte G., and Larry S. Carney. 1992. "Women in Socialist Societies." In *Issues in Society,* ed. Hugh F. Lena, William B. Helmreich, and William McCord, pp. 195–204. New York: McGraw-Hill.

Olson, Josephine E., Irene H. Frieze, and Ellen G. Detlefsen. 1990. "Having It All? Combining Work and Family in a Male and a Female Profession." *Sex Roles* 23:515–34.

Peterson, Sharyl B., and Tracie Kroner. 1992. "Gender Biases in Textbooks for Introductory Psychology and Human Development." *Psychology of Women Quarterly* 16:17–36.

Purcell, Piper, and Lara Stewart. 1990. "Dick and Jane in 1989." *Sex Roles* 22:177–85.

Reid, Pamela T., and Lillian Comas-Diaz. 1990. "Gender and Ethnicity: Perspectives on Dual Status." *Sex Roles* 22:397–408.

Rheingold, Harriet L., and Kaye V. Cook. 1975. "The Content of Boys' and Girls' Rooms as an Index of Parent's Behavior." *Child Development* 46:459–63.

Robinson, John P., and Suzanne Bianchi. 1997. "The Children's Hours." *American Demographics,* December, 1–6.

Rosenberg, Janet, Harry Perlstadt, and William Phillips. 1997. "Now That We Are Here: Discrimination, Disparagement, and Harassment at Work and the Experience of Women Lawyers." In *Workplace/Women's Place,* ed. Dana Dunn, pp. 247–59. Los Angeles: Roxbury.

Roth, Rachel. 1998. "Fetal Protection Law Is Threat to Women." *New York Times,* May 27, 20.

Rubenstein, Carin. 1990. "A Brave New World." *New Woman* 20(10):158–64.

Sadker, Myra, and David Sadker. 1990. "Confronting Sexism in the College Classroom." In *Gender in the Classroom: Power and Pedagogy,* eds. S. L. Gabriel and I. Smithson, pp. 176–87. Chicago: University of Illinois Press.

Sanchez, Laura. 1997. "Similarities in Women's and Men's Perceptions of the Societal Gender Division of Labor." *Sociological Spectrum* 17:389–416.

Sapiro, Virginia. 1994. *Women in American Society*. Mountain View, Calif.: Mayfield.

Schneider, Margaret, and Susan Phillips. 1997. "A Qualitative Study of Sexual Harassment of Female Doctors by Patients." *Social Science and Medicine* 45:669–76.

Schroeder, K. A., L. L. Blood, and D. Maluso. 1993. "Gender Differences and Similarities between Male and Female Undergraduate Students regarding Expectations for Career and Family Roles." *College Student Journal* 27:237–49.

Schur, Edwin. 1984. *Labeling Women Deviant: Gender, Stigma, and Social Control.* New York: Random House.

Schwalbe, Michael. 1996. *Unlocking the Iron Cage: The Men's Movement, Gender Politics, and American Culture.* New York: Oxford University Press.

Signorielli, Nancy. 1998. "Reflections of Girls in the Media: a Content Analysis across Six Media." *Overview.* **http://childrennow.org/media/mc97/ReflectSummary.html**

Statistical Abstract of the United States: 1998, 118th ed. U.S. Bureau of the Census. Washington, D.C.: U.S. Government Printing Office.

Steiger, Thomas L., and Mark Wardell. 1995. "Gender and Employment in the Service Sector." *Social Problems* 42(1):91–123.

Stoneman, Z., G. H. Brody, and C. E. MacKinnon. 1986. "Same-Sex and Cross-Sex Siblings: Activity Choices, Behavior and Gender Stereotypes." *Sex Roles* 15:495–511.

Tam, Tony. 1997. "Sex Segregation and Occupational Gender Inequality in the United States: Devaluation or Specialized Training?" *American Journal of Sociology* 102(6):1652–92.

Tannen, Deborah. 1990. *You Just Don't Understand: Women and Men in Conversation.* New York: Ballantine Books.

Tomaskovic-Devey, Donald. 1993. "The Gender and Race Composition of Jobs and the Male/Female, White/Black Pay Gap." *Social Forces* 72(1):45–76.

Tserkonivnitska, Marina. 1997. "Where Have All the Women Gone?" *Gender and Global Change Newsletter,* Spring, 1–4.

United Nations. 1997. "Report on the World Social Situation: 1997." Department for Economic and Social Information and Public Analysis. New York: United Nations.

UP (University of Philippines). 1997. "Teachers, Textbooks Reinforce Gender Bias—UP Study." *University of Phillipines Newsletter* 21 (4) **http://www.icpd.edu.ph/newsletter**

Van Willigen, Marieke, and Patricia Drentea. 1997. "Benefits of Equitable Relationships: The Impact of Sense of Failure, Household Division of Labor, and Decision-Making Power on Social Support." Presented at the American Sociological Association, Toronto, Canada, August.

Williams, Christine L. 1995. *Still a Man's World: Men Who Do Women's Work.* Berkeley: University of California Press.

Williams, John E., and Deborah L. Best. 1990a. *Measuring Sex Stereotypes: A Multination Study.* London: Sage Publications.

———. 1990b. *Sex and Psyche: Gender and Self Viewed Cross-Culturally.* London: Sage Publications.

Wilmot, Alyssa. 1999. "First National Love Your Body Day a Big Success." Press Release. National Organization for Women Newsletter (Winter). **http://www.now.org**

WOC Alert. 1998. "Tell the Senate to Join the World; Good News." Women Leaders Online. **http://wlo.org/alert/021398.html**

Wolfe, Jessica, Ertica Sharkansky, Jennifer Read, and Ree Dawson. 1998. "Sexual Harrassment and Assault as Predictors of PTSD Symptomology among U.S. Female Persian Gulf War Military Personnel." *Journal of Interpersonal Violence* 13(1):40–57.

Wootton, Barbara. 1997. "Gender Differences in Occupational Employment." *Monthly Labor Review,* April, 15–24.

Wright, Erik Olin, Janeen Baxter, with Gunn Elisabeth Birkelund. 1995. "The Gender-Gap in Workplace Authority: A Cross-National Study." *American Sociological Review* 60:407–35.

Yoder, Janice D., and Patricia Aniakudo. 1997 "Outsiders within the Firehouse: Subordination and Difference in the Social Interactions of African American Women Firefighters." *Gender and Society* 11(3):324–41.

Zimmerman, Marc A., Laurel Copeland, Jean Shope, and T.E. Dielman. 1997. "A Longitudinal Study of Self-Esteem: Implications for Adolescent Development." *Journal of Youth and Adolescence* 26 (2):117–41.

CHAPTER 8

Bobo, Lawrence, and James R. Kluegel. 1993. "Opposition to Race-Targeting: Self Interest, Stratification Ideology, or Racial Attitudes?" *American Sociological Review* 58(4):443–64.

Bogardus, Emory. 1968. "Comparing Racial Distance in Ethiopia, South Africa, and the United

States." *Sociology and Social Research* 52(January):149–56.
Cohen, Mark Nathan. 1998. "Culture, Not Race, Explains Human Diversity." *Chronicle of Higher Education* 44(32): B4–B5.
Current Population Survey, U.S. Bureau of the Census. 1997 (March). "Country of Origin and Year of Entry into the U.S. of the Foreign Born, by Citizenship Status: March 1997." **http://www.bls.census.gov/cps/pub/1997/for_born.htm** (April 6, 1998).
Dovidio, John F., and Samuel L. Gaertner. 1991. "Changes in the Expression and Assessment of Racial Prejudice." In *Opening Doors: Perspectives on Race Relations in Contemporary America*, eds. Harry J. Knopke, Robert J. Norrell, and Ronald W. Rogers, pp. 119–48. Tuscaloosa: University of Alabama Press.
Etzioni, Amitai. 1997. "New Issues: Rethinking Race." *The Public Perspective* (June–July):39–40. **http://www.ropercenter.unconn.edu/pubper/pdf/!84b.htm** (May 11, 1998).
Feagin, Joe R., and Clairece Booher Feagin. 1993. *Racial and Ethnic Relations*. Englewood Cliffs, N.J.: Prentice-Hall.
Gallup Organization. 1997 (June 10). "Black/White Relations in the U.S." **http://198.175.140.8/Special-Reports/racepressrls.htm** pp. 1–7. (April 6, 1998).
Galper, Josh. 1998 (April). "Population Update for April." *American Demographics* 20(4):26–28. **http://www.demographics.com/publications/AD/98_ad/ad98049.htm** (May 4, 1998).
Glynn, Patrick. 1998. "Racial Reconciliation: Can Religion Work Where Politics Has Failed?" *American Behavioral Scientist* 41:834–41.
Goldstein, Joseph. 1999 (January). "Sunbeams." *The Sun* 277:48.
Guinier, Lani. 1998. Interview with Paula Zahn. *CBS Evening News*, July 18.
Halton, Beau. 1998 (March 26). "City's Housing Bias Called 'Abysmal.'" **http://www.jacksonville.c...98/met_2blhousi.html** (February 8, 1999).
Healey, Joseph F. 1997. *Race, Ethnicity, and Gender in the United States: Inequality, Group Conflict, and Power*. Thousand Oaks, Calif.: Pine Forge Press.
Hodgkinson, Harold L. 1995. "What Should We Call People?: Race, Class, and the Census for 2000." *Phi Delta Kappan*, October, 173–79.
Hull, Jon D. 1994. "Do Teachers Punish according to Race?" *Time*, April 4, 30–31.
Immigration and Naturalization Service. 1997 (November 19). "Characteristics of Legal Immigrants." **http://www.ins.usdoj.gov/stats/annual/fy96/979.html** (April 6, 1998).
Immigration and Naturalization Service. 1998 (February 17). "General Naturalization Requirements." **http://www.ins.usdoj.gov/natz/general.html** (April 6, 1998).
Immigration and Naturalization Service Statistics Division. 1997 (May). "Immigration Overview." **http://www.fairus.org/04121604.htm** (April 6, 1998).
"In Our Own Words." 1996 (Fall). *Teaching Tolerance*. Montgomery, Ala.: Southern Poverty Law Center.
Intelligence Report. 1998 (Winter, issue no. 89). Montgomery, Ala.: Southern Poverty Law Center.
Jones, Charisse. 1998. "Race Killing in Texas Fuels Fear and Anger." *USA Today*, June 11, 1A–2A.
Keita, S. O. Y., and Rick A. Kittles. 1997. "The Persistence of Racial Thinking and the Myth of Racial Divergence." *American Anthropologist* 99(3):534–44.
Kleg, Milton. 1993. *Hate, Prejudice and Racism*. Albany: State University of New York Press.
Kozol, Jonathan. 1991. *Savage Inequalities: Children in America's Schools*. New York: Crown.
Landau, Elaine. 1993. *The White Power Movement: America's Racist Hate Groups*. Brookfield, Conn.: Millbrook Press.
Langone, John. 1993. *Spreading Poison*. Boston: Little Brown.
Lawrence, Sandra M. 1997. "Beyond Race Awareness: White Racial Identity and Multicultural Teaching." *Journal of Teacher Education* 48(2):108–17.
Levin, Jack, and Jack McDevitt. 1995. "Landmark Study Reveals Hate Crimes Vary Significantly by Offender Motivation." *Klanwatch Intelligence Report*, August, 7–9.
Lichter, Robert S., Linda S. Lichter, Stanley Rothman, and Daniel Amundson. 1987. "Prime-Time Prejudice: TV's Images of Blacks and Hispanics." *Public Opinion*, March/April, 12–13.
Lieberman, Leonard. 1997. "Gender and the Deconstruction of the Race Concept." *American Anthropologist* 99(3):545–58.
Lofthus, Kai R. 1998. "Swedish Biz Decries Racist Music." *Billboard*, January 24, 71, 73.
Massey, Douglas, and Nancy Denton. 1993. *American Apartheid: Segregation and the Making of an American Underclass*. Cambridge, Mass.: Harvard University Press.
Mishel, Lawrence, Jared Bernstein, and John Schmitt. 1999. *The State of Working America 1998–99*. Ithaca NY: Cornell University Press.
Molnar, Stephen. 1983. *Human Variation: Races, Types, and Ethnic Groups*, 2d ed. Englewood Cliffs, N.J.: Prentice-Hall.
Nash, Manning. 1962. "Race and the Ideology of Race." *Current Anthropology* 3:258–88.
Neckerman, Kathryn, and Joleen Kirschenman. 1991. "Hiring

Strategies, Racial Bias, and Inner-City Workers." *Social Problems* 38(4):433–47.

Pascarella, Ernest T., Marcia Edison, Amaury Nora, Linda Serra Hagedorn, and Patrick T. Terenzini. 1996. "Influences on Students' Openness to Diversity and Challenge in the First Year of College." *Journal of Higher Education* 67(2):174–93.

Pescosolido, Bernice A., Elizabeth Grauerholz, and Melissa A. Milkie. 1997. "Culture and Conflict: The Portrayal of Blacks in U.S. Children's Picture Books Through the Mid- and Late-Twentieth Century." *American Sociological Review* 62:443–64.

Reid, Evelyn. 1995. "Waiting to Excel: Biraciality in the Classroom." In *Educating for Diversity: An Anthology of Multicultural Voices,* ed. Carl A. Grant, pp. 263–73. Needham Heights, Mass.: Allyn & Bacon.

Schaefer, Richard T. 1998. *Racial and Ethnic Groups*, 7th ed. New York: HarperCollins.

Schwartz, Herman. 1992. "In Defense of Affirmative Action." In *Taking Sides*, 7th ed., pp. 189–94. Guilford, Conn.: Dushkin Publishing Co.

Schwartz, Joe, and Thomas Exter. 1989. "All Our Children." *American Demographics* 11 (May): 34–37.

Shipler, David K. 1998 (March 15). "Subtle vs. Overt Racism." *Washington Spectator* 24(6):1–3.

Spencer, Jon Michael. 1997. *The Mixed-Race Movement in America*. New York: New York University Press.

Statistical Abstract of the United States: 1995, 115th ed. U.S. Bureau of the Census. Washington, D.C.: U.S. Government Printing Office.

Statistical Abstract of the United States: 1998, 118th ed. U.S. Bureau of the Census. Washington, D.C: U.S. Government Printing Office.

"Study Finds Benefits From Immigration." 1997 (May 18). *Minneapolis Star Tribune*. p. 4A.

"This Year's Freshmen: A Statistical Profile." 1999 (Jan. 29). *The Chronicle of Higher Education,* pp. A48–A49.

U.S. Bureau of the Census. 1996 (March). "Resident Population of the United States: Middle Series Projections, 2001–2005, by Sex, Race, and Hispanic Origin, with Median Age." **http://www.census.gov/population/projections/nation/nsrh/nprh0105.txt** (May 1, 1998).

———. 1998 (April 9). "United States Department of Commerce News." **http://www.census.gov/Press-Release/cb98-57.html** (June 3, 1998).

———. 1998 (May 29). "Resident Population of the United States: Estimates, by Sex, Race, and Hispanic Origin, with Median Age." **http://www.census.gov/population/estimates/nation/intfile3-1.txt** (June 3, 1998).

U.S. Department of Justice, Federal Bureau of Investigation. 1998 (January 8). "Hate Crime Statistics 1996." **http://www.fbi.gov/pressrel/hate96/hate.htm** (April 23, 1998).

Wheeler, Michael L. 1994. *Diversity Training: A Research Report*. New York: The Conference Board.

Williams, Eddie N., and Milton D. Morris. 1993. "Racism and Our Future." In *Race in America: The Struggle for Equality*, eds. Herbert Hill and James E. Jones Jr., pp. 417–24. Madison: University of Wisconsin Press.

Wilson, William J. 1987. *The Truly Disadvantaged: The Inner City, the Underclass and Public Policy.* Chicago: University of Chicago Press.

Zack, Naomi. 1998. *Thinking about Race*. Belmont, Calif.: Wadsworth Publishing Co.

Zinn, Howard. 1993. "Columbus and the Doctrine of Discovery." In *Systemic Crisis: Problems in Society, Politics, and World Order,* ed. William D. Perdue, pp. 351–57. Fort Worth, Tex.: Harcourt Brace Jovanovich.

Zwerling, Craig, and Hilary Silver. 1993. "Race and Job Dismissals in a Federal Bureaucracy." *American Sociological Review* 57(5):651–60.

CHAPTER 9

"1998 in Review." 1999. *Out,* January:6.

ACLU Fact Sheet: Overview on Lesbian and Gay Parenting. 1997 (May). American Civil Liberties Union. **http://www.aclu.org/issues/gay/parenting.html** (May 9 1998).

"Anti-Marriage Bills Fact Sheet." 1998 (April 22). The Human Rights Campaign. 1101 14th St., NW Washington, DC 20005.

Bayer, Ronald. 1987. *Homosexuality and American Psychiatry: The Politics of Diagnosis,* 2d ed. Princeton, N.J.: Princeton University Press.

Beauvais-Godwin, Laura, and Raymond Godwin. 1997. *The Complete Adoption Book*. Holbrook, Mass.: Adams Media Corporation.

Brannock, J. C., and B. E. Chapman. 1990. "Negative Sexual Experiences with Men among Heterosexual Women and Lesbians." *Journal of Homosexuality* 19:105–10.

Button, James W., Barbara A. Rienzo, and Kenneth D. Wald. 1997. *Private Lives, Public Conflicts: Battles over Gay Rights in American Communities*. Washington, D.C.: CQ Press.

Cipriaso, Jon Christian. 1999 (Feb.). "The Wages of Hate." *Out,* February:28.

"Comparative Survey of the Legal Situation for Homosexuals in Europe." 1998. Gay and Lesbian International Lobby. **http://inet.uni2.dk/~steff/ghl.htm**

"Constitutional Protection." 1999. GayLawNet. **http://www.nexus.net.au/~dba/news.html#top** (February 14, 1999).

De Cecco, John P., and D. A. Parker. 1995. "The Biology of Homosexuality: Sexual Orientation or Sexual Preference? *Journal of Homosexuality* 28:1–28.

D'Emilio, John. 1990. "The Campus Environment for Gay and Lesbian Life." *Academe* 76(1):16–19.

"Discharges of Gay Troops Up." 1998. *U.S. News and World Report,* April 20, 35.

Doell, R.G. 1995. "Sexuality in the Brain." *Journal of Homosexuality* 28:345–56.

Durkheim, Emile. 1993. "The Normal and the Pathological." Originally published in *The Rules of Sociological Method,* 1938. In *Social Deviance,* ed. Henry N. Pontell, pp. 33–63. Englewood Cliffs, N.J.: Prentice-Hall.

Esterberg, K. 1997. *Lesbian and Bisexual Identities: Constructing Communities, Constructing Selves.* Philadelphia: Temple University Press.

Faulkner, Anne H., and Kevin Cranston. 1998. "Correlates of Same-Sex Sexual Behavior in a Random Sample of Massachusetts High School Students." *Journal of Public Health* 88 (February):262–66.

"Fighting Anti-Gay Hate Crimes." 1998. Human Rights Campaign. **http://www.hrc.org**

Frank, Barney. 1997. Foreword to *Private Lives, Public Conflicts: Battles over Gay Rights in American Communities,* by J. W. Button, B. A. Rienzo, and K. D. Wald. Washington D.C.: CQ Press.

Franklin, Sarah. 1993. "Essentialism, Which Essentialism? Some Implications of Reproductive and Genetic Techno-Science." *Journal of Homosexuality* 24:27–39.

Freedman, Estelle B., and John D'Emilio. 1990. "Problems Encountered in Writing the History of Sexuality: Sources, Theory, and Interpretation." *Journal of Sex Research* 27:481–95.

Garnets, L., G. M. Herek, and B. Levy. 1990. "Violence and Victimization of Lesbians and Gay Men: Mental Health Consequences." *Journal of Interpersonal Violence* 5:366–83.

Goode, Erica E., and Betsy Wagner. 1993. "Intimate Friendships." *U.S. News and World Report,* July 5, 49–52.

Greenhouse, Linda. 1998. "Gay Rights Case Fails in Bid for Supreme Court Hearing." *New York Times,* January 13, 15, late edition, East Coast.

Herek, Gregory M. 1989. "Hate Crimes against Lesbians and Gay Men." *American Psychologist* 44:948–55.

———. 1990. "The Context of Anti-gay Violence: Notes on Cultural and Psychological Heterosexism." *Journal of Interpersonal Violence* 5:316–33.

Human Rights Campaign News. 1999. "Discharges of Gay and Lesbian Soldiers Sky-Rocket on Eve of 5th Anniversary of Don't Ask, Don't Tell, Don't Pursue Policy." **http://www.hrc.org/hrc/hrcnews/990122.txt** (February 16, 1999).

Idaho Court to Take Up First Case of Adoption by Same-Sex Couple." 1999. GayLawNet. **http://www.labyrinth.net.au/~dba/ch1999.html#supreme_court_rules** (January 16, 1999).

ILGA Annual Report. 1996/1997. **http://www.pangea.org/org/cgl/ilga/repilga97e.html** (February 14, 1999).

Ingrassia, Michele, and Melissa Rossi. 1994. "The Limits of Tolerance?" *Newsweek,* February 14, 47.

Kite, M. E., and B. E. Whitley, Jr. 1996. "Sex Differences in Attitudes toward Homosexual Persons, Behavior and Civil Rights: A Meta-analysis." *Personality and Social Psychology Bulletin* 22:336–52.

Klassen, Albert D., Colin J. Williams, and Eugene E. Levitt. 1989. *Sex and Morality in the United States.* Middletown, Conn.: Wesleyan University Press.

Knox, David. 1998. *The Divorced Dad's Survival Book.* New York: Insight Publishing Co.

Lambda Legal Defense and Education Fund. 1998. "Summary of States Which Prohibit Discrimination Based on Sexual Orientation." **www.lambdalegal.org** (February 10,1998).

Lambda Legal Defense and Education Fund. 1997. **http://www.lambdalegal.org./cgi-bin/pages/documents/record?** (February 17, 1999).

Lawbriefs. 1999 (Winter). Vol. 2(1). Human Rights Campaign. **http://www.hrc.org/pubs/lawb0201.html** (February 16, 1999).

Lever, Janet, David E. Kanouse, William H. Rogers, Sally Carson, and Rosanna Hertz. 1992. "Behavior Patterns and Sexual Identity of Bisexual Males." *Journal of Sex Research* 29:141–167.

Louderback, L. A., and B. E. Whitley. 1997. "Perceived Erotic Value of Homosexuality and Sex-Role Attitudes as Mediators of Sex Differences in Heterosexual College Students' Attitudes toward Lesbians and Gay Men." *Journal of Sex Research* 34:175–82.

Mathison, Carla. 1998. "The Invisible Minority: Preparing Teachers to Meet the Needs of Gay and Lesbian Youth." *Journal of Teacher Education* 49:151–55.

Michael, Robert T., John H. Gagnon, Edward O. Laumann, and Gina Kolata. 1994. *Sex in America: A Definitive Survey.* Boston: Little, Brown.

Mills, Kim I. 1997. "Stop the Hate." *HRC Quarterly* (Publication of the Human Rights Campaign and the Human Rights Campaign Foundation). Washington, D.C., Summer, 8–9.

Mohr, Richard D. 1995. "Anti-Gay Stereotypes." In *Race, Class, and Gender in the United States,* 3d ed., ed. P. S. Rothenberg, pp. 402–8. New York: St. Martin's Press.

Moore, David W. 1993. "Public Polarized on Gay Issue." *Gallup Poll Monthly* 331 (April):30–34.

Moser, Charles. 1992. "Lust, Lack of Desire, and Paraphilias: Some Thoughts and Possible Connections." *Journal of Sex and Marital Therapy* 18:65–69.

The National Coalition of Anti-Violence Programs. 1998 (March 4). *Anti-Lesbian, Gay, Bisexual and Transgendered Violence in 1997.* New York: The New York City Gay & Lesbian Anti-Violence Project.

Nugent, Robert, and Jeannine Gramick. 1989. "Homosexuality: Protestant, Catholic, and Jewish Issues: A Fishbone Tale." *Journal of Homosexuality* 18:7–46.

Patterson, Charlotte. 1997. "Children of Lesbian and Gay Parents: Summary of Research Findings." In *Same-Sex Marriage: Pro and Con,* ed. A. Sullivan, pp. 146–54. New York: Vintage Books.

Paul, J. P. 1996. "Bisexuality: Exploring/Exploding the Boundaries." In R. Savin-Williams and K. M. Cohen (Eds.), *The Lives of Lesbians, Gays, and Bisexuals: Children to Adults,* pp. 436–61. Fort Worth, Tex.: Harcourt Brace.

Pillard, Richard C., and J. Michael Bailey. 1998. "Human Sexuality Has a Heritable Component." *Human Biology* 70 (April): 347–65.

Price, Jammie, and Michael G. Dalecki. 1998. "The Social Basis of Homophobia: An Empirical Illustration." *Sociological Spectrum* 18:143–59.

Rosin, Hanna, and Richard Morin. 1999 (January 11). "In One Area, Americans Still Draw a Line on Acceptability." *The Washington Post National Weekly Edition* 16(11):8.

Saad, Lydia. 1996. "Americans Growing More Tolerant of Gays." *Gallup Poll Archives,* Princeton, N.J.: The Gallup Organization. **http://198.175.140.8/POLL-ARCHIVES/961214.htm**

Salkind, Susanne, and Kevin Layton. 1997. "Leader of the PACs." *HRC Quarterly,* Summer, p. 11. Washington, D.C.: Human Rights Campaign and the Human Rights Campaign Foundation.

Sanday, Peggy. R. 1995. "Pulling Train." In *Race, Class, and Gender in the United States,* 3d ed., ed. P. S. Rothenberg, pp. 396–402. New York: St. Martin's Press.

Sarda, Alejandra. 1998. "Lesbians and the Gay Movement in Argentina." *NACLA Report on the Americas* 31(4):40–41.

"SIECUS Position Statements on Human Sexuality, Sexual Health and Sexuality Education and Information, 1995–1996." 1996. *SIECUS Report* 24:21–23.

Simon, A. 1995. "Some Correlates of Individuals' Attitudes toward Lesbians." *Journal of Homosexuality* 29:89–103.

Stachelberg, Winnie. 1997. "Position of Strength." *HRC Quarterly,* Summer, 12. Washington, D.C.: Human Rights Campaign and the Human Rights Campaign Foundation.

"Status of U.S. Sodomy Laws." 1999. American Civil Liberties Union. **http://www.aclu.org/issues/gay/sodomy.html** (February 16, 1999).

Sullivan, A. 1997. "The Conservative Case." In *Same-Sex Marriage: Pro and Con,* ed. A. Sullivan, pp. 146–54. New York: Vintage Books.

Tasker, Fiona L., and Susan Golombok. 1997. *Growing Up in a Lesbian Family: Effects on Child Development.* New York: Guilford.

"This Year's Freshmen: A Statistical Profile." 1999. *The Chronicle of Higher Education,* January 29, pp. A48–A49.

Thompson, Cooper. 1995. "A New Vision of Masculinity." In *Race, Class, and Gender in the United States,* 3d ed., ed. P.S. Rothenberg, pp. 475–81. New York: St. Martin's Press.

The United Methodist Church and Homosexuality. 1999. **http://religioustolerance.org/hom_umc.htm** (February 16, 1999).

U.S. Department of Health and Human Services. 1989. "Report of the Secretary's Task Force on Youth Suicide, Volume 3: Prevention and Interventions in Youth Suicide." Rockville, Md.

CHAPTER 10

Albelda, Randy, and Chris Tilly. 1997. *Glass Ceilings and Bottomless Pits: Women's Work, Women's Poverty.* Boston, Mass.: South End Press.

Albelda, Randy, Nancy Folbre, and the Center for Popular Economics. 1996. *The War on the Poor.* New York: The Free Press.

Alex-Assensoh, Yvette. 1995. "Myths about Race and the Underclass." *Urban Affairs Review* 31:3–19.

Barlett, Donald L., and James B. Steele. 1998. "The Empire of the Pigs." *Time,* November 30, 52–64.

Bergmann, Barbara R. 1996. *Saving Our Children from Poverty: What the United States Can Learn from France.* New York: Russell Sage Foundation.

"Bill Gates Tops Forbes Billionaires for 5th Year." 1998 (September 28). PointCast Network (Online News Channel).

Briggs, Vernon M. Jr. 1998. "American-Style Capitalism and Income Disparity: The Challenge of Social Anarchy." *Journal of Economic Issues* 32(2):473–81.

Bureau of Labor Statistics. 1997 (December). "A Profile of the Working Poor, 1996." U.S. Department of Labor, Report 918. **http://stat.bls.gov/cpswp96.htm** (December 1, 1998).

Card, David, and Alan Krueger. 1995. *Myth and Measurement: The New Economics of the Minimum Wage*. Princeton, N.J.: Princeton University Press.

Center on Budget and Policy Priorities. 1998a (March 9). "Strengths of the Safety Net: How the EITC, Social Security, and Other Government Programs Affect Poverty." **http://www.cbpp.org/snd98-rep.htm** (September 28, 1998).

______. 1998b (September 24). "Poverty Rates Fall, but Remain High for a Period with Such Low Unemployment." **http://www.cbpp.org/9-24-98pov.htm** (September 28, 1998).

Children's Defense Fund. 1998a. "Facts about Child Care in America." **http://www.childrensdefen...1dcare/cc_facts.html** (December 5, 1998).

Children's Defense Fund. 1998b. *Locked Doors: States Struggling to Meet the Child Care Needs of Low-Income Working Families*. Washington, D.C.: Children's Defense Fund.

Children's Defense Fund. 1998c. *Poverty Matters: The Cost of Child Poverty in America*. Washington, D.C.: Children's Defense Fund.

Children's Defense Fund and the National Coalition for the Homeless. 1998. *Welfare to What: Early Findings on Family Hardship and Well-Being*. Washington, D.C.: Children's Defense Fund.

Chossudovsky, Michel. 1998. "Global Poverty in the Late 20th Century." *Journal of International Affairs* 52(1):293–303.

Collins, Chuck. 1995. "Aid to Dependent Corporations: Exposing Federal Handouts to the Wealthy." *Dollars and Sense*, 199(May–June), 15.

Common Cause. 1997. *Common Cause Report: Return on Investment*. **http://www.commoncause/org/publications/return_1htm** (October 26, 1998).

Common Cause. 1998. *Pocketbook Politics: How Special-Interest Money Hurts the American Consumer.* **http://www.commoncause.org/publications/pocketbook1.htm** (October 26, 1998).

Corcoran, Mary, and Terry Adams. 1997. "Race, Sex, and the Intergenerational Transmission of Poverty." In *Consequences of Growing Up Poor*, eds. Greg J. Duncan and Jeanne Brooks-Gunn, pp. 461–517. New York: Russell Sage Foundation.

Davis, Kingsley, and Wilbert Moore. 1945. "Some Principles of Stratification." *American Sociological Review* 10:242–49.

Deng, Francis M. 1998. "The Cow and the Thing Called 'What': Dinka Cultural Perspectives on Wealth and Poverty." *Journal of International Affairs 52*(1):101–15.

DeParle, Jason. 1999. "In Blooming Economy, Poor Still Struggle to Pay the Rent." *Themes of the Times (Sociology), New York Times* Spring. p. 3.

Duncan, Greg J., and Jeanne Brooks-Gunn. 1997. "Income Effects across the Life Span: Integration and Interpretation." In *Consequences of Growing Up Poor,* eds. Greg J. Duncan and Jeanne Brooks-Gunn, pp. 596–610. New York: Russell Sage Foundation.

Edin, Kathryn, and Laura Lein. 1997. *Making Ends Meet: How Single Mothers Survive Welfare and Low-Wage Work.* New York: Russell Sage Foundation.

Freeman, Richard B. 1997. "When Earnings Divide: Causes, Consequences, and Cures for the New Inequality in the U.S." Washington, D.C.: National Policy Association.

Gans, Herbert J. 1972. "The Positive Functions of Poverty." *American Journal of Sociology* 78 (September): 275–388.

Handler, Joel F., and Yeheskel Hasenfeld. 1997. *We the Poor People: Work, Poverty, and Welfare*. New Haven, Conn.: Yale University Press.

Hernandez, Donald J. 1997. "Poverty Trends." In *Consequences of Growing Up Poor*, eds. Greg J. Duncan and Jeanne Brooks-Gunn, pp. 18–34. New York: Russell Sage Foundation.

Hill, Lewis E. 1998. "The Institutional Economics of Poverty: An Inquiry into the Causes and Effects of Poverty." *Journal of Economic Issues 32*(2):279–86.

Hodgkinson, Virginia A., and Murray S. Weitzman. 1996. *Giving and Volunteering in the United States, 1996*. Washington, D.C.: Independent Sector. **http://www.indepsec.org/p...public_attitude.html** (November 23, 1998).

"Homelessness and Poverty in America." 1998. National Law Center on Homelessness and Poverty in America. **http://www.nlchp.org/h&pusa.htm** (November 18, 1998).

Human Development Report 1997. 1997. United Nations Development Programme. New York: Oxford University Press.

Jencks, Christopher. 1997. Foreword to *Making Ends Meet: How Single Mothers Survive Welfare and Low-Wage Work,* by Kathryn Edin and Laura Lein. New York: Russell Sage Foundation.

Johnson, Nicholas, and Ed Lazere. 1998. "Rising Number of States Offer EITCs." Center on Budget and Policy Priorities. **http://www.cbpp.org/9-14-98sfp.htm** (September 28, 1998).

Kennedy, Bruce P., Ichiro Kawachi, Roberta Glass, and Deborah Prothrow-Stith. 1998. "Income Distribution, Socioeconomic Status, and Self-Rated Health in the U.S.: Multilevel Analysis." *British Medical Journal* 317(7163):917–22.

Kim, Marlene, and Thanos Mergoupis. 1997. "The Working Poor and Welfare Recipiency: Participation, Evidence, and Policy Directions." *Journal of Economic Issues* 31:707–28.

Kingsley, Thomas. 1998. "Federal Housing Assistance and Welfare Reform: Uncharted Territory." The Urban Institute. **http://www.urban.org/uiporder.htm** (November 20, 1998).

Larin, Kathryn. 1998. "Should We Be Worried about the Widening Gap between the Rich and the Poor?" *Insight on the News* 14(5):24–28.

Levitan, Sar A., Garth L. Mangum, and Stephen L. Mangum. 1998. *Programs in Aid of the Poor.* 7th ed. Baltimore: Johns Hopkins University Press.

Lewis, Oscar. 1966. "The Culture of Poverty." *Scientific American* 2(5):19–25.

Lewis, Oscar. 1998. "The Culture of Poverty: Resolving Common Social Problems." *Society* 35(2):7–10.

Lord, Mary. 1998. "Growing Apart." *U.S. News & World Report,* May 11, 51.

Luker, Kristin. 1996. *Dubious Conceptions: The Politics of Teenage Pregnancy*. Cambridge, Mass.: Harvard University Press.

Massey, D. S. 1991. "American Apartheid: Segregation and the Making of the American Underclass." *American Journal of Sociology* 96:329–57.

Mayer, Susan E. 1997a. *What Money Can't Buy: Family Income and Children's Life Chances.* Cambridge, Mass.: Harvard University Press.

Mayer, Susan E. 1997b. "Trends in the Economic Well-Being and Life Chances of America's Children." In *Consequences of Growing Up Poor,* eds. Greg J. Duncan and Jeanne Brooks-Gunn, pp. 49–69. New York: Russell Sage Foundation.

McBride, Ann. 1998. "Big Money Big Benefits." Common Cause. **http://www.commoncause.org/publications/072498.htm** (October 26, 1998).

Mead, L. 1992. *The New Politics of Poverty: The Non-Working Poor in America*. New York: Basic Books.

Michel, Sonya. 1998. "Childcare and Welfare (In)justice." *Feminist Studies* 24:44–54.

Office of Management and Budget. 1998. "Budget of the United States Government Fiscal Year 1999." **http://www.access.gpo.gov/su_docs/budget99/index.htm** (December 1, 1998).

Parenti, Michael. 1998. "The Super Rich Are Out of Sight." *Dollars and Sense* 217(May–June):36–37.

Pitcoff, Winton. 1998 (May/June). "Closing the Wage Gap." National Housing Institute. **http://www.nhi.org/online/issues/99/pitcoff.html** (October 22, 1998).

Piven, Frances Fox. 1996. "Scapegoating the Poor." In *The War on the Poor,* eds. Randy Albelda, Nancy Folbre, and the Center for Popular Economics, pp. 112–14. New York: The New Press.

Pressman, Steven. 1998. "The Gender Poverty Gap in Developed Countries: Causes and Cures." *The Social Science Journal* 35(2):275–87.

Reich, Robert B. 1997. "The Unfinished Agenda." Speech of the U.S. Secretary to the Council on Excellence in Government, Washington, D.C., January 9. Reprinted in the *Daily Labor Report,* January 10, pp. E-13–E-17.

Ryan, William. 1992. "Blaming the Victim." In *Taking Sides,* 7th ed., eds. Kurt Finsterbusch and George McKenna, pp. 155–62. Guilford, Conn.: Dushkin Publishing Group.

Smallwood, David M. 1998. "Food Assistance and Welfare Reform." *Food Review* 21:2–3.

Sorensen, E. 1995. "The Benefits of Increased Child Support Enforcement." In *Welfare Reform: An Analysis of Issues,* ed. I. Sawhill. Washington, D.C.: Urban Institute Press.

Speth, James Gustave. 1998. "Poverty: A Denial of Human Rights." *Journal of International Affairs* 52(1):277–86.

Streeten, Paul. 1998. "Beyond the Six Veils: Conceptualizing and Measuring Poverty." *Journal of International Affairs* 52(1):1–8.

"The Unfinished Agenda." 1997. Speech of the U.S. Secretary to the Council on Excellence in Government, Washington, D.C., January 9. Reprinted in the *Daily Labor Report,* January 10, pp. E-13–E-17.

United Nations. 1997. *Report on the World Social Situation, 1997.* New York: United Nations.

U.S. Census Bureau. 1999 (February 3). "Poverty Thresholds by Size of Family and No. of Children: 1998." **http://www.census.gov/hhes/poverty/threshold/thresh98.html** (February 18, 1999).

______. 1998a. "Annual Demographic Survey, March Supplement. Table 1" **http://ferret.bls.census.gov/macro/031998/pov/new1_001.htm** (November 20, 1998).

______. 1998b. "Annual Demographic Survey, March Supplement. Table 7." **http://ferret.bls.census.gov/macro/031998/pov/new7_000.htm** (November 20, 1998).

______. 1998c. *Poverty in the United States: 1997.* Washington, D.C.: U.S. Government Printing Office.

U.S. Department of Health and Human Services. 1999a (January). "Changes in Welfare Caseloads since Aug.' 96." **http://www.acf.dhhs.gov/news/stats/aug-sep.htm** (February 18, 1999).

U.S. Department of Health and Human Services. 1999b (January). "Welfare Caseloads: Families and Recipients 1960–1998." **http://www.acf.dhhs.gov/news/stats/6097rf.htm** (February 18, 1999).

U.S. Department of Health and Human Services. 1999c (February 1). "Characteristics and Financial Circumstances of TANF Recipients." **http://www.acf.dhhs.gov/programs/opre/particip/revsum.htm** (February 18, 1999).

U.S. Department of Labor. 1998 (February). "About Welfare: Myths, Facts, Challenges and Solutions." **http://www.wtw.dolcta.gov/resources/myths.htm** (October 23, 1998).

Van Kempen, Eva T. 1997. "Poverty Pockets and Life Chances: On the Role of Place in Shaping Social Inequality." *American Behavioural Scientist* 41(3):430–50.

Weinstein, Michael M. 1999 (January 4, late edition). "5 Problems Tarnishing a Robust Economy." *New York Times,* p. 10-2.

Weisbrot, Mark. 1998. "Globalization for Whom?" Preamble Center. **http://www.preamble.org/Globalization.html** (October 22, 1998).

Weisbrot, Mark, and Michelle Sforza-Roderick. 1996. "Baltimore's Living Wage Law: An Analysis of the Fiscal and Economic Costs of Baltimore City Ordinance 442." Preamble Center. **http://www.rtk./net/preamble/baltimore.html** (October 22, 1998).

Wilson, William J. 1987. *The Truly Disadvantaged: The Inner City, the Underclass, and Public Policy.* Chicago: University of Chicago Press.

Wilson, William J. 1996. *When Work Disappears: The World of the New Urban Poor.* New York: Knopf.

CHAPTER 11

AFL-CIO. 1998. "The Union Difference: Fast Facts on Union Membership and Pay." **http://www.aflcio.org/uniondifference/index.htm** (December 22, 1998).

Ambrose, Soren. 1998. "The Case against the IMF." *Campaign for Human Rights Newsletter* no. 12. **http://www.summersault.co...wsletter/news12.html** (December 8, 1998).

Barlett, Donald L., and James B. Steele. 1998. "Corporate Welfare: First in a Series." *Time* 152(19):36–39.

Bell, Daniel. 1973. *The Coming of Post-Industrial Society.* New York: Basic Books.

Bettelheim, Adriel. 1998. "Sleep Deprivation." *CQ Researcher* 8(24):553–76.

Bond, James T., Ellen Galinsky, and Jennifer E. Swanberg. 1997. *The 1997 National Study of the Changing Workforce.* New York: Families and Work Institute.

Bracey, Gerald W. 1995. "The Fifth Bracey Report on the Condition of Public Education." *Phi Delta Kappan,* October, 149–62.

Brecher, Jeremy. 1996. "Countering Corporate Downsizing: A Survey of Proposals to Halt Layoffs and Job Degradation." The Preamble Collaborative Center. **http://www.preamble.org/intro.html** (December 8, 1998).

Bureau of Labor Statistics. 1998a (January 30). "Union Members in 1997." U.S. Department of Labor. **http://stats.bls.gov/newsrels.htm** (December 12, 1998).

______. 1998b (February 27). "State and Regional Unemployment, 1997 Annual Averages." **http://stats.bls.gov/newsrels.htm** (December 8, 1998).

______. 1998c (November 25). "Work Experience of the Population in 1997." **http://stats.bls.gov/newsrels.htm** (December 8, 1998).

______. 1998d (December 4). "Unemployment Rates in 9 Countries: 1997." **http://www.bls.gov/flsdata.htm** (December 15, 1998).

______. 1998e (December). "Workplace Injuries and Illnesses in 1997." **http://www.osha.gov/oshstats/bls/osnr0007.txt** (February 23, 1999).

______. 1998f (May 21). "Employment Characteristics of Families Summary." **http://stat.bls.gov.news.release/famee.nws.htm** (February 23, 1999).

Caston, Richard J. 1998. *Life in a Business-Oriented Society: A Sociological Perspective.* Boston: Allyn and Bacon.

Center for Workplace Health Information. 1998. "Treating and Preventing CTDs." *CTDNewsonline.* **http://ctdnews.com/suffercare.html** (December 12, 1998).

Child Labour. 1997. *UN Chronicle,* no. 4, 52.

Chossudovsky, Michel. 1998. "Global Poverty in the Late 20th Century." *Journal of International Affairs* 52(1):293–303.

Cockburn, Alexander. 1992. "Clinton, Labor, and Free Trade." *Nations,* November 2, 508–9.

Coleman, James. 1994. *The Criminal Elite: The Sociology of White Collar Crime,* 3d ed. New York: St. Martins Press.

Commission on Leave, U.S. Department of Labor. 1996. "A Workable Balance: Report to Congress on Family and Medical Leave Policies." Washington, D.C.: Government Printing Office.

Conrad, Peter. 1999. "Wellness in the Work Place: Potentials and Pitfalls of Work-Site Health Promotion." In *Health, Illness, and Healing: Society, Social Context, and Self,* eds. K. Charmaz and D. A. Paterniti, pp. 263–275, Los Angeles: Roxbury Publishing Company.

Danaher, Kevin. 1998. "Are Workers Waking Up?" Global Exchange: Education for Action. **http://www.globalexchange.org/education/econnomy/laborday.html** (December 17, 1998).

Davis, Letitia. 1997. "Youth Employment versus Exploitative Child Labor." *Public Health Reports* 113(January/February):3–4.

Downs, Donald G. 1997. "Nonspecific Work-Related Upper Extremity Disorders." *American Family Physician* 55(4): 1296–1302.

Durkheim, Emile. [1893] 1966. *On the Division of Labor in Society*, trans. G. Simpson. New York: Free Press.

Eitzen, Stanley, and Maxine Baca Zinn, eds. 1990. *The Reshaping of America: Social Consequences of the Changing Economy*. Englewood Cliffs, N.J.: Prentice-Hall.

Epstein, Gerald, Julie Graham, and Jessica Nembhard, eds. 1993. "Third World Socialism and the Demise of COMECON." In *Creating a New World Economy: Forces of Change and Plans of Action*, pp. 405–20. Philadelphia: Temple University Press.

Feather, Norman T. 1990. *The Psychological Impact of Unemployment*. New York: Springer-Verlag.

Galinsky, Ellen, and James T. Bond. 1998. *The 1998 Business Work-Life Study*. New York: Families and Work Institute.

Galinsky, Ellen, James E. Riesbeck, Fran S. Rodgers, and Faith A. Wohl. 1993. "Business Economics and the Work-Family Response." In *Work-Family Needs: Leading Corporations Respond*, pp. 51–54. New York: The Conference Board.

Global March against Child Labor. "Global March against Child Labor." 1998. **children@globalmarch-us.org** (December 12, 1998).

Gordon, David M. 1996. *Fat and Mean: The Corporate Squeeze of Working Americans and the Myth of Managerial "Downsizing."* New York: The Free Press.

Hewlett, Sylvia Ann. 1992. *When the Bough Breaks: The Cost of Neglecting Our Children*. New York: HarperCollins.

Hewlett, Sylvia Ann, and Cornell West. 1998. *The War against Parents: What We Can Do for America's Beleaguered Moms and Dads*. Boston: Houghton Mifflin Company.

Hochschild, Arlie Russell. 1997. *The Time Bind: When Work Becomes Home and Home Becomes Work*. New York: Henry Holt and Company.

Hodson, Randy, and Teresa A. Sullivan. 1990. *The Social Organization of Work*. Belmont, Calif.: Wadsworth.

International Labor Rights Fund. 1997. "Congress Acts to Ban Imports Made with Child Labor." **www.laborrights.org** (December 7, 1998).

Kennedy, Joseph II. 1996. "Keynote Address." In *Forced Labor: The Prostitution of Children*, pp. 1–6, Washington, D.C.: U.S. Department of Labor, Bureau of International Labor Affairs.

Kenworthy, Lane. 1995. *In Search of National Economic Success*. Thousand Oaks, Calif.: Sage Publications.

Koch, Kathy. 1998a. "High-Tech Labor Shortage." *CQ Researcher* 8(16):361–84.

______. 1998b. "Flexible Work Arrangements." *CQ Researcher* 8(30):697–720.

"Labor's 'Female Friendly' Agenda." 1998. *Labor Relations Bulletin no.* 690, 2.

Lenski, Gerard, and J. Lenski. 1987. *Human Societies: An Introduction to Macrosociology,* 5th ed. New York: McGraw-Hill.

Leonard, Bill. 1996 (July). "From School to Work: Partnerships Smooth the Transition." *HR Magazine* (Society for Human Resource Management). **http://www.shrm.org/hrmag...articles/0796cov.htm** (December 8, 1998).

Levitan, Sar A., Garth L. Mangum, and Stephen L. Mangum. 1998. *Programs in Aid of the Poor*, 7th ed. Baltimore: Johns Hopkins University Press.

Liem, Joan H., and G. Ramsey Liem. 1990. "Understanding the Individual and Family Effects of Unemployment." In *Stress between Work and Family*, eds. J. Eckenrode and S. Gore, pp. 175–204. New York: Plenum Press.

Loomis, Dana, and David Richardson. 1998. "Race and the Risk of Fatal Injury at Work." *American Journal of Public Health* 88(1):40–44.

Mishel, Lawrence, Jared Bernstein, and John Schmitt. 1999. *The State of Working America, 1998–99*. Economic Policy Institute Series. Ithaca, N.Y.: Cornell University Press.

Mokhiber, Russell, and Robert Weissman. 1998. *Focus on the Corporation*. Multinational Monitor. **http://www.essential.org/...ocus/focus.9806.html** (January 14, 1999).

National Center for Policy Analysis. 1997. "Falling Union Membership Prompting Mergers." **http://www.ncpa.org/pd/unions/pduni/pduni1.html** (December 8, 1998).

National Safety Council. 1997. *Accident Facts 1997 Edition*. Itasca Ill.: National Safety Council.

"New OSHA Policy Relieves Employees." 1998. *Labor Relations Bulletin no.* 687, 8.

Occupational Safety and Health Administration. 1997. "National Census of Fatal Occupational Injuries, 1996." U.S. Department of Labor. **http://www.osha.gov/oshstats/cfoi.nws.html** (December 12, 1998).

Parker, David L. (with Lee Engfer and Robert Conrow). 1998. *Stolen Dreams: Portraits of Working Children.* Minneapolis: Lerner Publications Company.

Reed, Stanley. 1999. "Now the Dollar Has a Co-Star on the World Stage." *Business Week,* January 18, 36–37.

Report on the World Social Situation. 1997. New York: United Nations.

Robie, Chet, Ann Marie Ryan, Robert A. Schmieder, Luis Fernando Parra, and Patricia C. Smith. 1998. "The Relation between Job Level and Job Satisfaction." *Group and Organizational Management* 23(4):470–86.

Ross, Catherine E., and Marylyn P. Wright. 1998. "Women's Work, Men's Work, and the Sense of Control." *Work and Occupations 25*(3):333–55.

Silvers, Jonathan. 1996. "Child Labor in Pakistan." *Atlantic Monthly* 277(2):79–92.

Straus, Murray A. 1980. "A Sociological Perspective on the Prevention of Wife-Beating." *The Social Causes of Husband-Wife Violence,* eds. M. A. Straus and G. T. Hotaling, pp. 211–32. Minneapolis: University of Minnesota Press.

Thurow, Lester. 1996. *The Future of Capitalism: How Today's Economic Forces Shape Tomorrow's World.* New York: Morrow.

UNICEF. *1997 State of the World's Children.* 1997. New York: United Nations.

U.S. Department of Labor. 1995. *By the Sweat and Toil of Children.* Vol. 2, *The Use of Child Labor in U.S. Agricultural Imports and Forced and Bonded Child Labor.* Washington, D.C.: U.S. Department of Labor, Bureau of International Labor Affairs.

Western, Bruce. 1995. "A Comparative Study of Working Class Disorganization: Union Decline in Eighteen Advanced Capitalist Countries." *American Sociological Review* 60:179–201.

"Work Stoppages Drop to Record Low in 1997." 1998. *Labor Relations Bulletin* no. 690, 8.

CHAPTER 12

AAUW (American Association of University Women). 1998. "Report Finds Separating by Sex Not the Solution to Gender Inequity in School." Press Release (2300). **http://aauw.org**

Apple, Michael W. 1995. "The Politics of a National Curriculum." In *Transforming Schools,* eds. Peter W. Cookson Jr. and Barbara Schneider, pp. 345–70. New York: Garland Publishing Co.

Arenson, Laren W. 1998. "More Colleges Plunging into the Unchartered Waters of On-Line Courses." *New York Times,* November 2, A14.

Ascher, Carol, Norm Fruchter, and Robert Berne. 1997. *Hard Lessons: Public Schools and Privatization.* New York: Twentieth Century Fund.

Associated Press. 1998. "Education Becomes Major Political Issue as Candidates Listen to Voters." *New York Times,* September 20, A5.

Bailey, Susan M. 1993. "The Current Status of Gender Equity Research in American Schools." *Educational Psychologist* 28:321–40.

Baker, David P., and Deborah P. Jones. 1993. "Creating Gender Equality: Cross-National Gender Stratification and Mathematical Performance." *Sociology of Education* 66:91–103.

Banks, James A., and Cherry A. Banks, eds. 1993. *Multicultural Education,* 2d ed. Boston: Allyn & Bacon.

Bankston, Carl, and Stephen Caldas. 1997. "The American School Dilemma: Race and Scholastic Performance." *Sociological Quarterly* 38(3):423–29.

Barnett, Steven. 1995. "Long Term Effects of Early Childhood Programs on Cognitive and School Outcomes." *The Future of Children* 5:25–50.

Barton, Paul. 1992. *America's Smallest School: The Family.* Princeton, N.J.: Educational Testing Service.

Belluck, Pam. 1998. "School Ruling Shakes Milwaukee." *New York Times,* June 15, 12.

Bolick, Clint. 1998. "School Choice Can Broaden the GOP's Base." *Wall Street Journal* June 3, A18.

Bulkeley, William M. 1998. "Education: Kaplan Plans a Law School via the Web." *Wall Street Journal,* September 16, B1.

Burd, Stephen. 1998. "Congress Passes Bill to Extend the Higher Education Act." *Chronicle of Higher Education,* October 9, A40.

Bushweller, Kevin. 1995. "Turning Our Backs on Boys." *Education Digest,* January, 9–12.

Call, Kathleen, Lorie Grabowski, Jeylan Mortimer, Katherine Nash, and Chaimun Lee. 1997. "Impoverished Youth and the Attainment Process." Presented at the annual meeting of the American Sociological Association, Toronto, Canada,. August.

Carvin, Andy. 1997. EdWeb: *Exploring Technology and School Reform.* **http://edweb.gsn.org**

Celis, William. 1993. "10 Years after a Scathing Report, Schools Show Uneven Progress." *New York Times,* April 28.

Clinton, William. 1998a. "Statement by the President." Press Release, Office of the Press Secretary (October 22).

______. 1998b. "Statement by the President: School Modernization Tax Credits." Press Release, Office of the Press Secretary (October 22).

Cohen, Warren. 1998. "Vouchers for Good and Ill." *U.S. News and World Report,* April 27, 46.

Coleman, James S., J. E. Campbell, L. Hobson, J. McPartland,

A. Mood, F. Weinfield, and R. York. 1966. *Equality of Educational Opportunity.* Washington, D.C.: U.S. Government Printing Office.

College Board. 1997. "1997 Profile of College Bound Seniors." In *Fair Test Examiner:SAT/ACT Scores 1996* **http://fairtest.org**

Condition of Education. 1997. Department of Education. National Center on Education Statistics, Indicator 23. Washington, D.C.: U.S. Government Printing Office.

Condition of Education. 1998. Department of Education. National Center on Education Statistics, Indicators 2, 6, 16, 20, 21, 32. Washington, D.C.: U.S. Government Printing Office.

Digest of Education Statistics. 1997. "Problems Facing Local Public Schools: 1970–1996." **http://nces.ed.gov/pubs/digest97/d97t023.html**

ED Initiatives. 1998. "Helping All Children Reach High Standards." September 25. Department of Education. **http://www.ed.gov/pubs/EDInitiatives/98/98-09-25.html**

ED Initiatives. 1999. "2000 Budget." February 5. Department of Education. **http://www.ed.gov/pubs/EDInitiatives/99/99-02-05.html**

Elam, Stanley M., Lowell C. Rose, and Alec M. Gallup. 1994. "The 26th Annual Phi Delta Kappa/Gallup Poll of the Public's Attitudes toward the Public Schools." *Phi Delta Kappan,* September, 41–56.

Embry, Dennis, and Daniel Flannery, T. Alexander-Vazsonvi, Kenneth Powell, and Henry Atha. 1996. "PeaceBuilders: A Theoretically Driven School Based Model for Early Violence Prevention." *American Journal of Preventive Medicine* 12(5):91–100.

Executive Summary. 1998. "Financing Schools." *Center for the Future of Children* 7(3):1–8.

Fact File. 1999. "This Year's Freshmen: A Statistical Profile." *The Chronicle of Higher Education,* January 29, A48.

FairTest. 1997. "Leading Civil Rights and Education Reform Groups Oppose President Clinton's National Testing Plan." Press Release (October 20).

______. 1998. "FairTest Fact Sheet on National Testing." **http://www.fairtest.org/facts/ntfacts.html**

Finsterbusch, Kurt. 1999. *Taking Sides: Clashing Views on Controversial Social Issues.* Guildford, Conn.: Dushkin/McGraw Hill.

Fletcher, Robert S. 1943. *History of Oberlin College to the Civil War.* Oberlin, Ohio: Oberlin College Press.

Flexner, Eleanor. 1972. *Century of Struggle: The Women's Rights Movement in the United States.* New York: Atheneum.

Freitas, Frances Anne, Scott Meyers, and Theodore Avtgis. 1998. "Student Perceptions of Instructor Immediacy in Conventional and Distributed Learning Classrooms." *Communication Education* 47(4):366–72.

Gallup Organization. 1997. "Briefing Paper: National School Testing." February. **http://198.175.140.8/Briefing_Papers/9802bp.html**

Goldberg, Carey. 1999. "After Girls Get Attention, Focus is on Boys' Woes." In *Themes of the Times: New York Times.* Upper Saddle River, N.J.: Prentice-Hall, p. 6.

Goldring, Ellen B., and Anna V. Shaw Sullivan. 1995. "Privatization: Integrating Private Services in Public Schools." In *Transforming Schools,* eds. Peter W. Cookson Jr. and Barbara Schneider, pp. 537–59. New York: Garland Publishing Co.

Guernsey, Lisa. 1998. "An Unusual Graduate Program Requires Students to Find and Pay Their Professors." *Chronicle of Higher Education* 44:A14–A16.

Harris Poll. 1998a. "A 21st Century Juxtaposition: Grandma, Grandpa and High Technology." *The Harris Poll* 12 (February 26). **http://www.louisharris.com**

______. 1998b. "Question 415." Study No. 5818418: April. Chapel Hill, N.C.: Institute for Research in Social Sciences.

Hewlett, Sylvia Ann. 1992. *When the Bough Breaks: The Cost of Neglecting Our Children.* New York: HarperPerennial.

Hindustan Times. 1999. "40% Men and 60% Women Illiterate in India." September 19. **http://www.hindutimes.com**

IHAD (I Have a Dream). 1999. "FAQs." **http://www.ihad.org/faq.html**

IRSS (Institute for Research in Social Science). 1998. "Southern Focus Poll: Question 27, Spring." Chapel Hill: University of North Carolina.

Jencks, Christopher, and Meredith Phillips. 1998. "America's Next Achievement Test: Closing the Black-White Test Score Gap." *The American Prospect,* September/October, 44–53.

Kanter, Rosabeth Moss. 1972. "The Organization Child: Experience Management in a Nursery School." *Sociology of Education* 45:186–211.

Koch, James V. 1998. "How Women Actually Perform in Distance Education." *Chronicle of Higher Education* 45:A60.

Koeppel, David. 1998. "Easy Degrees Proliferate on the Web." *New York Times,* August 2, 17.

Kozol, Jonathan. 1991. *Savage Inequalities: Children in America's Schools.* New York: Crown Publishers.

Kuh, George, Ernest Pascarella, and Henry Wechsler. 1996. "The Questionable Value of Fraternities." *Chronicle of Higher Education,* April 19, A68.

Lam, Julia. 1997. "The Employment Activity of Chinese-American High School

Students and Its Relationship to Academic Achievement." Presented at the annual meeting of the American Sociological Association, Toronto, Canada, August.

Lareau, Annette. 1989. *Home Advantage: Social Class and Parental Intervention in Elementary Education.* Philadelphia: Falmer Press.

Legislative Action. 1998. "Public Law 105-277." **http://www.house.gov/eeo/testindex.htm**

Leo, John. 1998. "Dumbing Down Teachers." *U.S. News and World Report,* August 3, 15.

Marlein, Mary Beth. 1997. "SAT Scores Up, But So Is Grade Inflation." *USA Today,* August 27.

Mensh, Barbara, and Cynthia Lloyd. 1997. "Gender Differences in the School Experiences of Adolescents in Low-income Countries." *Policy Research Working Paper* no. 95. New York: Population Council.

Merton, Robert K. 1968. *Social Theory and Social Structure.* New York: Free Press.

Murnane, Richard J. 1994. "Education and the Well-Being of the Next Generation." In *Confronting Poverty: Prescriptions for Change,* eds. Sheldon H. Danziger, Gary D. Sandefur, and Daniel H. Weinberg, pp. 289–307. New York: Russell Sage Foundation.

Natriello, Gary. 1995. "Dropouts: Definitions, Causes, Consequences, and Remedies." In *Transforming Schools,* eds. Peter W. Cookson Jr. and Barbara Schneider, pp. 107–28. New York: Garland Publishing Co.

NCES (National Center for Education Statistics). 1997a. "Social Statistics: Briefing Room." **http://www.whitehouse.gov/fsbr/eduation.html**

______. 1997b. "Survey and Analysis of Salary Trends." *Digest of Education Statistics 1997.* Washington, D.C.: U.S. Department of Education.

______. 1998. "Projections of Education Statistics to 2008." **http://nces.ed.gov/pubs98/pj2008/foresumm.html**

______. 1999. "Teachers Report Need for More Preparation." **http://nces.ed.gov/pressrelease/tq1.html**

Noddings, Nel. 1995. "A Morally Defensible Mission for Schools in the 21st Century." *Phi Delta Kappan,* January, 365–68.

OJJDP (Office of Juvenile Justice and Delinquency Prevention). 1997. "Juvenile Offenders and Victims: 1997 Update on Violence." U.S. Department of Justice: Office of Justice Programs. Pittsburgh, Pa: National Center for Juvenile Justice.

Orfield, G., and S.E. Eaton. 1997. *Dismantling Desegregation: The Quiet Reversal of Brown v. Bd. Of Education.* New York: New Press.

Pinkerton, Jim. 1998. "Here's How to Pass While Schools Fail." *USA Today,* April 23, A11.

Ratnesar, Romesh. 1998. "Lost in the Middle." *Time,* September 14, 60–64.

Ray, Carol A., and Roslyn A. Mickelson. 1993. "Restructuring Students for Restructured Work: The Economy, School Reform, and Non-College-Bound Youths." *Sociology of Education* 66:1–20.

Richmond, George. 1989. "The Future School: Is Lowell Pointing Us toward a Revolution in Education?" *Phi Delta Kappan,* November, 232–36.

Rodriquez, Richard. 1990. "Searching for Roots in a Changing World." *In Social Problems Today,* ed. James M. Henslin, pp. 202–13. Englewood Cliffs, N.J.: Prentice-Hall.

Roscigno, Vincent. 1998. "Race and the Reproduction of Educational Disadvantage." *Social Forces* 76(3):1033–60.

Rosenthal, Robert, and Lenore Jacobson. 1968. *Pygmalion in the Classroom: Teacher Expectations and Pupils' Intellectual Development.* New York: Holt, Rinehart & Winston.

Rudich, Joe. 1998. "Internet Learning." *Link-Up* 15:23–25.

Rumberger, Russell W. 1987. "High School Dropouts: A Review of Issues and Evidence." *Review of Educational Research* 57:101–21.

Sautter, R. Craig. 1995. "Standing Up to Violence: Kappan Special Report." *Phi Delta Kappan,* January, K1–K12.

Shanker, Albert. 1996. "Mythical Choice and Real Standards." In *Reducing Poverty in America,* ed. Michael Darby, pp. 154–72. Thousand Oaks, Calif.: Sage.

Silber, John. 1998. "The Correct Answer: Too Many Can't Teach." *The News & Observer* July 8, A2.

Simpson, Janice. 1993. "Adding Up the Under Skilled." *Time,* September 20, 75.

Sklar, Debbie. 1998. "Wisconsin Beefs Up Distance Learning Initiative." *America's Network* 102:19.

Sommerfeld, Meg. 1998. "Micro-Society Schools Tackle Real World Woes." *Teacher Magazine.* vol. 12, December 2. **http://www.teachermag.org**

"State of the Union: Key Items." 1999. *Time,* vol. 153, no. 4. **http://cgi.pathfinder.com/time/magazine/articles/0,3266,19227,00**

Statistical Abstract of the United States: 1998, 118th ed. U.S. Bureau of the Census. Washington, D.C.: U.S. Government Printing Office.

Summary Report. 1998. "Doctorate Recipients from United States Universities." Office of Scientific and Engineering Personnel. National Research Council. Washington D.C.: National Academy Press.

Svestka, Sherlie. 1996. "Head Start and Early Head Start Programs." *International Journal of Early Childhood* 28:59–62.

Tauscher, Ellen. 1997. "Tauscher Introduces School Construction

Bill." **http://house.gov/tauscher/6-5-97.html**

Teachman, Jay D., Kathleen Paasch, and Klaren Carver. 1997. "Social Capital and the Generation of Human Capital." *Social Forces* 75(4):1343–59.

Toch, Thomas. 1992. "Schools for Scandal." *U.S. News and World Report,* April 27, 66–70.

______. 1998. "Education Bazaar." *U.S. News and World Report,* April 27, 35–46.

Trout, Paul A. 1998. "Incivility in the Classroom Breeds 'Education Lite.'" *Chronicle of Higher Education* July 24, A40.

United Nations Population Fund. 1999. "Campaign Issues: Facing the Facts." *Face to Face* **http://www.facecampaign.org**

USA Today. 1997. "Schools Going Down the Drain." USA Snapshot, January 16, A1.

______. 1998. "Better Schools for Teachers Mean Better Teachers for Kids." Editorial, April 29, A12.

U.S. Bureau of the Census. 1998a. "Back to School." *Grassroots*. The Pointcast Network. September 25.

______. 1998b. "Back to School Teachers." *Grassroots*. The Pointcast Network. September 25.

U.S. Department of Education. 1995. Press release. **http://www.ed.gov/**

______. 1998a. "Help Meet the Technology Challenge." **http://www.ed.gov/Technology/challenge**

______. 1998b. "School District Fiscal Data." National Center for Education Statistics. Washington, D.C.: U.S. Government Printing Office.

Webb, Julie. 1989. "The Outcomes of Home-Based Education: Employment and Other Issues." *Educational Review* 41:121–33.

Weinstein, Michael M. 1999. "5 Problems Tarnishing a Robust Economy." *New York Times,* January 4, A10.

Wells, Amy, and Robert Crain. 1997. *Stepping over the Color Line: African American Students in White Suburban Schools*. New Haven, Conn.: Yale.

WGU (Western Governors University). 1998. **http://www.heatcenter.org/colleges/wgu/index.html**

White House Conference on School Violence. 1998. "Press Release." October 15. **http://www.ed.gov/PressReleases/10-1998/wh1015.html**

Youth Indicators. 1996. U.S. Department of Education. National Center for Education Statistics: International Assessment of Educational Progress. Indicator 38. **http://nces.ed.gov/pubs/yi/y9638a.html**

Zigler, Edward, and Sally Styfco. 1996. "Reshaping Early Childhood Intervention to Be a More Effective Weapon against Poverty." In *Reducing Poverty in America: Views and Approaches,* ed. Michael R. Darby, pp. 310–33. Thousand Oaks, Calif.: Sage.

CHAPTER 13

Accordino, John. 1998. "The Consequences of Welfare Reform for Central City Economies." *Journal of the American Planning Association* 64(1):11–15.

Archer, Dennis W. 1998. "The Lesson of Detroit: Never Underestimate a City." *Vital Speeches* 64(11):340–43.

Bickford, Adam, and Douglas S. Massey. 1991. "Segregation in the Second Ghetto: Racial and Ethnic Segregation in American Public Housing, 1977." Social *Forces* 69(4):1011–36.

"The Bridge to the 21st Century Leads to Gridlock in and around Decaying Cities." 1997. *The Washington Spectator* 23(12), The Public Concern Foundation, Inc.

Brockerhoff, Martin, and Ellen Brennan. 1997. "The Poverty of Cities in the Developing World." Policy Research Division Working Paper no. 96. New York: Population Council.

Clark, David. 1998. "Interdependent Urbanization in an Urban World: An Historical Overview." *The Geographical Journal* 164(1): 85–96.

"Crisis in Low-Income Rental Housing." 1998. *America* 179(1):3.

Davis, Robert. 1994. "We Live in an Age of Exotic Defenses." *USA Today*, November 22, A1.

Durning, Alan. 1996. *The City and the Car*. Northwest Environment Watch. Seattle: Sasquatch Books.

"Empowerment Zone/Enterprise Community Initiative." 1998 (August 28). Department of Housing and Urban Development. **http://www.hud.gov/ezec/ezecinit.html** (January 12, 1999).

Fisher, Christy. 1997 (October). "What We Love and Hate about Cities." *American Demographics*. **http://www.demographics.com/publications/AD/97_ad/9710_ad/ad971029.htm** (May 4, 1998).

Fisher, Claude. 1982. *To Dwell among Friends: Personal Networks in Town and City*. Chicago: University of Chicago Press.

Friedman, Dorian. 1998. "The Draw of Downtown: Big Growth Predicted for Many U.S. Cities." *U.S. News & World Report* 125(13):63.

Froehlich, Maryann. 1998. "Smart Growth: Why Local Governments Are Taking a New Approach to Managing Growth in Their Communities." *Public Management* 80(5):5–9.

Gans, Herbert. [1962] 1984. *The Urban Villagers*, 2d ed. New York: Free Press (first edition published in 1962).

Geddes, Robert. 1997. "Metropolis Unbound: The Sprawling American City and the Search for Alternatives." *The American Prospect* 35 (November–December):40–46.

Goozner, Merrill. 1998. "The Porter Prescription." *The American Prospect* 38 (May-June): 56–64.

Gordon, Peter, and Harry W. Richardson. 1998. "Prove it: The Costs and Benefits of Sprawl." *Brookings Review* 16(4):23–26.

Heubusch, Kevin. 1998 (January). "Small Is Beautiful." *American Demographics*. **http://www.demographics.com/publications/AD/98_ad/9801_ad/ad980130.htm** (May 4, 1998).

Ingram, Gregory K. 1998. "Patterns of Metropolitan Development: What Have We Learned?" *Urban Studies* 35(7):1019–36.

Jargowsky, Paul A., and William Julius Wilson. 1997. *Poverty and Place: Ghettos, Barrios, and the American City*. New York: Russell Sage Foundation.

Johnson, William C. 1997. *Urban Planning and Politics*. Chicago: American Planning Association, Planners Press.

Kelbaugh, Douglas. 1997. *Common Place: Toward Neighborhood and Regional Design*. Seattle: University of Washington Press.

Kessler, Ronald C., Katherine A. McGonale, Shanyang Zhao, Christopher B. Nelson, Michael Hughes, Suzann Eschleman, Hans-Ulrich Wittchen, and Kenneth S. Kendler. 1994. "Lifetime and 12-Month Prevalence of DSM-III-R Psychiatric Disorders in the United States." *Archives of General Psychiatry* 51:18–19.

Kozol, John. 1991. *Savage Inequalities: Children in American Schools*. New York: Crown Publishing.

Le Roux, Johann, and Cheryl Sylvia Smith. 1998. "Causes and Characteristics of the Street Child Phenomenon: A Global Perspective." *Adolescence* 33 (131):683-89.

Lord, George F., and Albert C. Price. 1992. "Growth Ideology in a Period of Decline: Deindustrialization and Restructuring, Flint Style." *Social Problems* 39(2):155–69.

McDonough, Kate. 1998. "Constructive Financing." *American City & County* 113(6):18–27.

McNulty, Robert. 1993. "Quality of Life and Amenities as Urban Investment." In *Interwoven Destinies: Cities and the Nation*, ed. Henry Cisneros, pp. 231–49. New York: W. W. Norton.

National League of Cities. 1998 (January 21). "Cities Are Increasing Service Levels and Range of Services as Problem Solvers for Citizens; Most Local Leaders Optimistic about Conditions and Outlook; Concerned about Mandates; Annual NLC Survey Finds Many Cities Successfully Involved with Information Technology." **http://www.nlc.org/pres-opn.htm** (January 18, 1999).

Nelessen, Anton C. 1997. "The Computer Commuter: Neighborhood Transit for the 21st Century." Urban Design, Telecommunication and Travel Forecasting Conference. **http://www.bts.gov/tmip/p...ip/udes/nelessen.htm** (January 12, 1999).

"News Briefs." 1999. *Environmental Science & Technology* 33(3):63A.

Orfield, Myron. 1997. *Metropolitics: A Regional Agenda for Community and Stability*. Washington, D.C.: Brookings Institution Press and Cambridge, Mass.: The Lincoln Institute of Land Policy.

Pack, Janet Rothenberg. 1998. "Poverty and Urban Expenditures." *Urban Studies* 35(11): 1995–2015.

Pelley, Janet. 1999 (January 1). "Building Smart-Growth Communities." *Environmental Science & Technology News* 33(1): 28A–32A.

Schmoke, Kurt L. 1998. "Ingredients for a Successful City: Variety Is the Spice of Urban Life." *Vital Speeches of the Day* 65(4):110–17.

Shoop, Julie Gannon. 1993. "Criminal Lawyers Develop 'Urban Psychosis' Defense." *Trial*, August, 12–13.

Smart Growth Network. 1999 (January 14). "About the Smart Growth Network: Mission Statement and Principles." **http://www.smartgrowth.org/** (January 23, 1999).

Statistical Abstract of the United States: 1993, 113th ed., U.S. Bureau of the Census. Washington, D.C.: U.S. Government Printing Office.

Statistical Abstract of the United States: 1994, 114th ed., U.S. Bureau of the Census. Washington, D.C.: U.S. Government Printing Office.

Statistical Abstract of the United States: 1998, 118th edition, U.S. Bureau of the Census. Washington, D.C.: U.S. Government Printing Office.

"Technology Smooths the Ride for Santa Ana Commuters." 1998. *American City & County* 113(1): 17–18.

Tittle, Charles. 1989. "Influences on Urbanism: A Test of Predictions from Three Perspectives." *Social Problems* 36(3):270–88.

Union of Concerned Scientists. 1999. "The Hidden Costs of Transportation." **http://www.ucsusa.org/transportation/hidden.html** (January 19, 1999).

United Nations. 1994. "Programme of Action." The United Nations International Conference on Population and Development (ICPD). Cairo, Egypt. September 5–13, 1994.

United Nations Population Fund. 1996. *1996 State of the World Population Report*. New York: United Nations.

Warren, Roxanne. 1998. *The Urban Oasis: Guideways and Greenways in the Human Environment*. New York: McGraw-Hill.

Wilson, Thomas C. 1993. "Urbanism and Kinship Bonds." *Social Forces* 71(3):703–12.

Wirth, Louis. 1938. "Urbanism as a Way of Life." *American Journal of Sociology* 44:8–20.

Wolch, Jennifer R. 1998. "America's New Urban Policy: Welfare Reform and the Fate of American Cities." *Journal of the American Planning Association* 64(1):8–11.

Wolff, Kurt H. 1978. *The Sociology of George Simmel*. Toronto: Free Press.

World Health Organization and United Nations Joint Programme on HIV/AIDS. 1998. "Report on the Global HIV/AIDS Epidemic—June 1998." **http://www.who.int/emc-hiv/global_report/data/globrep-e.pdf** (August 8, 1998).

CHAPTER 14

American Council on Education and University of California. 1997. "The American Freshmen: National Norms for Fall, 1997." Los Angeles: Los Angeles Higher Education Research Institute.

Associated Press. 1998. "Japanese Pushing Cloning Efforts." *Greensboro News Record*, November 9, A5.

Bajak, Frank. 1998. "Brace Yourself for Year 2000 Glitches: Much of the World Hasn't." Associated Press, July 5.

Begley, Sharon. 1997. "The Science Wars." *Newsweek*, April 21, 54–57.

______. 1998. "Designer Babies." *Newsweek*, November 9, 61–62.

Bell, Daniel. 1973. *The Coming of Post-Industrial Society: A Venture in Social Forecasting*. New York: Basic Books.

Beniger, James R. 1993. "The Control Revolution." In *Technology and the Future*, ed. Albert H. Teich, pp. 40–65. New York: St. Martin's Press.

Boles, Margaret, and Brenda Sunoo. 1998. "Do Your Employees Suffer from Technophobia?" *Workforce* 77(1):21.

Brin, David. 1998. *The Transparent Society: Will Technology Force Us to Choose between Privacy and Freedom?* Reading, Mass.: Addison Wesley.

Burstein, Rachel. 1998. "Regulating the Researchers." *Mother Jones*, May/June, 57.

Bush, Corlann G. 1993. "Women and the Assessment of Technology." In *Technology and the Future*, ed. Albert H. Teich, pp. 192–214. New York: St. Martin's Press.

Carey, Patricia M. 1998. "Sticking It Out in the Sticks." *Home Office Computing* 16:64–69.

Ceruzzi, Paul. 1993. "An Unforeseen Revolution: Computers and Expectations, 1935–1985." In *Technology and the Future*, ed. Albert H. Teich, pp. 160–74. New York: St. Martin's Press.

Clarke, Adele E. 1990. "Controversy and the Development of Reproductive Sciences." *Social Problems* 37(1):18–37.

Clayton, Gary. 1998. "Manager's Journal: Eurocrats Try to Stop Data at the Border." *Wall Street Journal*, November 2, A34.

Cohen, Adam. 1999. "Cyberspeech on Trial." *Time Daily*, 153(6): 1–2. **http://www.time.com**

"College Technology." 1998. *USA Today*, October 19, D1.

Conrad, Peter. 1997. "Public Eyes and Private Genes: Historical Frames, New Constructions, and Social Problems." *Social Problems* 44:139–54.

Durkheim, Emile. [1925] 1973. *Moral Education*. New York: Free Press.

Ehrenfeld, David. 1998. "A Techno-Pox upon the Land." *Harper's*, October, 13–17.

Eibert, Mark D. 1998. "Clone Wars." *Reason* 30(2):52–54.

Elmer-Dewitt, Philip. 1994. "The Genetic Revolution." *Time*, January 17, 46–53.

Executive Summary. 1998. "Public Perceptions of Genetic Engineering." International Social Science Survey. Sydney, Australia: Australian National University.

Facts in Brief: Induced Abortion. 1996. New York: Alan Guttmacher Institute.

Fang, Bay. 1998. "Chinese 'Hacktivists' Spin a Web of Trouble: The Regime is Unable to Control the Net." *U.S. News and World Report*, September 28, 47.

FAS (Federation of American Scientists). 1998a. "Cyberstrategy Project: U.S. Households." **http://www.fas.org/cp/house.gif?69,34**

______. 1998b. "FAS Cyberstrategy Project." **http://www.fas.org/cp/nrtestats.html**

Find/SVP. 1997. "The Market for Gene Therapy." (April). **http://www.findsvp.com/tocs/ML0497.htm**

Fix, Janet L. 1994. "Automation Makes Bank Branches a Liability." *USA Today*, November 28, B1.

Fox, Maggie. 1998. "Spooky Teleportation Study Brings Future Closer." *Reuters: The PointCast Network*, October 23.

General Social Survey. 1993. Storrs, Conn.: Roper Organization, Inc.

GIP (Global Internet Project). 1998a. "Introduction: The Internet Today and Tomorrow." **http://www.gip.org/gip2a.html**

______. 1998b. "Press Alert." **http://www.gip.org**

______. 1998c. "The Workplace." **http://www.gip.org/gip2g.html**

Glendinning, Chellis. 1990. *When Technology Wounds: The Human Consequences of Progress*. New York: William Morrow.

Global Statistics. 1998. "Global Internet Statistics by Language." **http://www.euromktg.com/globalstats/**

Goodman, Paul. 1993. "Can Technology Be Humane?" In *Technology and the Future,* ed. Albert H. Teich, pp. 239–55. New York: St. Martin's Press.

"Group Seeks Standards for Infertility Services." 1998. *American Medical News* 21(40):12.

Hafner, Katie. 1999. "Horse and Blender, Car and Crockpot." In *Themes of the Times: N.Y. Times,* p. 1. Upper Saddle River, N.J.: Prentice Hall.

Hancock, LynNell. 1995. "The Haves and the Have-Nots." *Newsweek,* February 27, 50–53.

Harris Poll. 1998a. "The Remorseless Rise of the Internet." *The Harris Poll* 9 (February 18). **http://www.louisharris.com/poll/1998pols**

______. 1998b. "A 21st Century Juxtaposition: Grandma, Grandpa and High Technology." *The Harris Poll* 12 (February 26). **http://www.louisharris.com/poll/1998pols**

Hayes, Frank. 1998. "Age Bias an IT Reality." *Computerworld* 32(46):12.

Headcount. 1998. "Profile of U.S. Users." **http://www.headcount.com/globalsource/profile/index.htm?choice=the_us&id-144**

Hosmer, Ellen. 1986. "High Tech Hazards: Chipping Away at Workers' Health." *Multinational Monitor* 7 (January 31):1–5.

IFR (International Federation of Robotics). 1997. "1997 Key Data for the World Robot Market." **http://www.ifr.org**

"Influencing Net Use." 1998. *USA Today,* September 14, A1.

IRSS. 1998. "Southern Focus Poll." Spring. Institute for Social Science Research. Chapel Hill: University of North Carolina.

Johnson, Jim. 1988. "Mixing Humans and Nonhumans Together: The Sociology of a Door-Closer." *Social Problems* 35:298–310.

Kahin, Brian. 1993. "Information Technology and Information Infrastructure." In *Empowering Technology: Implementing a U.S. Strategy,* ed. Lewis M. Branscomb, pp. 135–66. Cambridge, Mass.: MIT Press.

Kahn, A. 1997. "Clone Mammals . . . Clone Man." *Nature,* March 13, 119.

Kate, Nancy Ten. 1998. "PCs at Home." *American Demographics,* February, 1.

Kelly, Jason. 1997. "Technophobia." *Atlanta Business Chronicle,* August 18. **http://www.amcity.com/atlanta/stories/1997/08/18/focus1.html**

Kennedy, Paul. 1998. "Preparing for the 21st Century: Winners and Losers." In *Global Issues,* ed. Robert Jackson, pp. 10–26. Guilford, Conn.: Dushkin/McGraw Hill.

Klein, Matthew. 1998. "From Luxury to Necessity." *American Demographics* 20(8):8–12.

Kluger, J. 1997. "Will We Follow the Sheep?" *Time,* March 10, 67, 70–72.

Kuhn, Thomas. 1973. *The Structure of Scientific Revolutions.* Chicago: Chicago University Press.

Lash, Alex. 1998. "Clinton's New Crypto Advisor." **http://cnet.com** (January 15).

Lemonick, Michael, and Dick Thompson. 1999. "Racing to Map Our DNA." *Time Daily,* 153:1–6. **http://www.time.com**

Leslie, Jacques. 1998. "Computer Visions." *Modern Maturity,* November-December, 36–39.

Levy, Pierre. 1997. "Cyberculture in Question: A Critique of the Critique." *Revue-du-Mauss* 9:111–26.

Levy, Steven. 1995. "TechnoMania." *Newsweek,* February 27, 25–29.

Lewin, Tamar. 1998. "Serious Gender Gap Remains in Technology." N.Y. Times News Service. October 18.

Lohr, Steve. 1999. "Bold Vision Propels Phone-TV Mergers Media Convergence." In *Themes of the Times: N.Y. Times,* p. 8. Upper Saddle River, N.J.: Prentice Hall.

Macklin, Ruth. 1991. "Artificial Means of Reproduction and Our Understanding of the Family." *Hastings Center Report,* January/February, 5–11.

Makris, Greg. 1996. "The Myth of a Technological Solution to Television Violence." *Journal of Communication Inquiry* 20: 72–91.

Marshall, Eliot. 1998. "Iceland's Blond Ambition." *Mother Jones,* May/June, 53–56.

McCormick, S. J. Richard A. 1994. "Blastomere Separation." *Hastings Center Report,* March/April, 14–16.

McDermott, John. 1993. "Technology: The Opiate of the Intellectuals." In *Technology and the Future,* ed. Albert H. Teich, pp. 89–107. New York: St. Martin's Press.

McFarling, Usha L. 1998. "Bioethicists Warn Human Cloning Will Be Difficult to Stop." *Raleigh News and Observer,* November 18, A5.

Mehlman, Maxwell H., and Jeffery R. Botkin. 1998. *Access to the Genome: The Challenge to Equality.* Washington, D.C.: Georgetown University Press.

Merton, Robert K. 1973. "The Normative Structure of Science." In *The Sociology of Science,* ed. Robert K. Merton. Chicago: University of Chicago Press.

Mesthene, Emmanuel G. 1993. "The Role of Technology in Society." In *Technology and the Future,* ed. Albert H. Teich, pp. 73–88. New York: St. Martin's Press.

Negroponte, Nicholas. 1995. "Nicholas Negroponte: The Multimedia Today Interview." *Multimedia Today,* July–September, 86–88.

NSTC (National Science and Technology Council). Executive Secretariat Office. 1994. "PCAST Fact Sheet." August 3, 1994.

Washington, D.C.: U.S. Government Printing Office.
Ogburn, William F. 1957. "Cultural Lag as Theory." *Sociology and Social Research* 41:167–74.
Pascal, Zachary G. 1996. "The Outlook: High Tech Explains Widening Wage Gap." *Wall Street Journal,* April 22, A1.
Perrolle, Judith A. 1990. "Computers and Capitalism." In *Social Problems Today,* ed. James M. Henslin, pp. 336–42. Englewood Cliffs, N.J.: Prentice-Hall.
Postman, Neil. 1992. *Technopoly: The Surrender of Culture to Technology.* New York: Alfred A. Knopf.
Powers, Richard. 1998. "Too Many Breakthroughs." Op-Ed. *New York Times,* November 19, 35.
Quick, Rebecca. 1998. "Technology: Pieces of the Puzzle—Not So Private Lives: Will We Have Any Secrets in the Future?" *Wall Street Journal,* November 13, R27.
Rabino, Isaac. 1998. "The Biotech Future." *American Scientist* 86(2):110–12.
Reuters. 1998. "U.S. Senate Takes up 'Partial Birth' Again." *PointCast Network,* September 17.
Rifkin, Jeremy, 1996. *The End of Work: The Decline of the Global Labor Force and the Dawn of Post-Market Era.* Berkeley, Calif.: Putnam.
Rosenberg, Jim. 1998. "Troubles and Technologies." *Editor and Publisher* 131(6):4.
Scout Report. 1998. "PC Sales Heat up in First Quarter of 1998." *RetailVision,* Fall, 12, 102.
Shand, Hope. 1998. "An Owner's Guide." *Mother Jones,* May/June, 46.
Shellenbarger, Sue. 1998. "Flood of Pagers, E-mail Gives Families Means to Communicate More." *Wall Street Journal,* Nov. 18, B1.
Statistical Abstract of the United States: 1998, 118th ed. U.S. Bureau of the Census. Washington, D.C.: U.S. Government Printing Office.
Toffler, Alvin. 1970. *Future Shock.* New York: Random House.
United Nations Population Fund. 1991. *Population Policies and Programmes: Lessons Learned from Two Decades of Experience,* ed. Nafis Sadik. New York: New York University Press.
Walshok, Mary Lindenstein. 1993. "Blue Collar Women." In *Technology and the Future,* ed. Albert H. Teich, pp. 256–64. New York: St. Martin's Press.
Weil, Nancy. 1998. "Congress Approves Tech Bills." *InfoWorld,* November 2:59–60.
Weinberg, Alvin. 1966. "Can Technology Replace Social Engineering?" *University of Chicago Magazine* 59 (October):6–10.
Weinberg, Robert A. 1993. "The Dark Side of Genome." In *Technology and the Future,* ed. Albert H. Teich, pp. 318–28. New York: St. Martin's Press.
Welter, Cole H. 1997. "Technological Segregation: A Peek through the Looking Glass at the Rich and Poor in an Information Age." *Arts Education Policy Review* 99(2):1–6.
Whine, Michael. 1997. "The Far Right on the Internet." In *The Governance of Cyberspace,* ed. Brian D. Loader, pp. 209–27. London: Routledge.
Winner, Langdon. 1993. "Artifact/Ideas as Political Culture." In *Technology and the Future,* ed. Albert H. Teich, pp. 283–94. New York: St. Martin's Press.
Wiseman, Paul, and Dottie Enrico. 1994. "Techno Terror Slows Info Highway Traffic." *USA Today,* November 14, B1.
"World Abortion Policies 1994." 1997. United Nations Department for Economic and Social Information and Policy Analysis. **gopher//gopher.undp.org:70/00/~ungophers/popin/wdtrends/charts**
WRC (World Risk Consulting). 1998. "The Growing Epidemic." **http://www.wrcservice.com/**

CHAPTER 15

Agbese, Pita Ogaba. 1995. "Nigeria's Environment: Crises, Consequences, and Responses." In *Environmental Policies in the Third World: A Comparative Analysis,* ed. O. P. Dwivedi and Dhirendra K. Vajpeyi, pp. 125–44. Westport, Conn.: Greenwood Press.
Bergman, Lester V. 1998 (December). "Cataract Development: It's Cumulative." *Environmental Health Perspectives* 106(12). **http://ehpnet1.niehs.nih.gov/docs/1998/06-12/forum.html** (February 1, 1999).
Black, Harvey Karl. 1999 (February). "Complex Cleanup." *Environmental Health Perspectives* 107(2). **http://ehpnet1.niehs.nih.gov/docs/1999/07-2/focus-abs.html** (February 1, 1999).
Boland, Reed, Sudhakar Rao, and George Zeidenstein. 1994. "Honoring Human Rights in Population Policies: From Declaration to Action." In *Population Policies Reconsidered: Health, Empowerment, and Rights,* eds. Gita Sen, Adrienne Germain, and Lincoln C. Chen, pp. 89–105. Boston: Harvard School of Public Health.
Bongaarts, John, and Susan Cotts Watkins. 1996. "Social Interactions and Contemporary Fertility Transitions." *Population and Development Review* 22(4):639–82.
Brown, Lester R. 1995. "The State of the World's Natural Environment." In *Seeing Ourselves: Classic, Contemporary, and Cross-Cultural Readings in Sociology,* 3d ed., eds. John J. Macionis and Nijole V. Benokraitis, pp. 411–16. Englewood Cliffs, N.J.: Prentice-Hall.
———. 1998a. "The Future of Growth." In *State of the World 1998,* eds. Lester R. Brown, Christopher Flavin, and Hilary French, pp. 3–20. New York: W.W. Norton & Co.

———. 1998b. "Overview: New Records, New Stresses." In *Vital Signs 1998,* ed. Lester R. Brown, Michael Renner, and Christopher Flavin, pp. 15–24. New York: W. W. Norton & Co.

Brown, Lester R., and Jennifer Mitchell. 1998. "Building a New Economy." In *State of the World 1998,* ed. Lester R. Brown, Christopher Flavin, and Hilary French pp. 168–187. New York: W. W. Norton & Co.

Brown, Lester R., Christopher Flavin, and Hilary French. 1998. Foreword. In *State of the World 1998,* ed. Lester R. Brown, Christopher Flavin, and Hilary French, pp. xvii–xix. New York: W.W. Norton & Co.

Brown, Lester R., Gary Gardner, and Brian Halweil. 1998. *Beyond Malthus: Sixteen Dimensions of the Population Problem.* World Watch Paper 143. Washington, D.C.: World Watch Institute.

Bullard, Robert D. 1996. *Unequal Protection.* Sierra Club Books.

Bullard, Robert D., and Glenn S. Johnson. 1997. "Just Transportation." In *Just Transportation: Dismantling Race and Class Barriers to Mobility,* eds. Robert D. Bullard and Glenn S. Johnson, pp. 1–21. Stony Creek, Conn.: New Society Publishers.

Catley-Carlson, Margaret, and Judith A. M. Outlaw. 1998. "Poverty and Population Issues: Clarifying the Connections." *Journal of International Affairs* 52(1):233–43.

Cooper, Mary H. 1998. "Population and the Environment." *The CQ Researcher* 8(26):601–24.

"The Delicate Balance." 1994. The National Center for Environmental Health Strategies, vol. 5, nos. 3–4. 1100 Rural Avenue, Voorhees, N.J. 08043.

DesJardins, Andrea. 1997. "Sweet Poison: What Your Nose Can't Tell You about the Dangers of Perfume." **http://members.aol.com/enviroknow/perfume/sweet_poison.htm** (January 20, 1999).

Dionis, Joanna. 1999. "Handle with Care." *Mother Jones,* January–February, 25.

Edwards, Bob, and Anthony Ladd. 1998. "Where the Hogs Are '97: Environmental Justice and Farm Loss in North Carolina, 1980–1997." Paper presented at the 2nd National Black Land Loss Summit in Tillery, N.C., February 1998.

"Fertility Declines Reported." 1997. *Popline,* May–June, 3.

Fisher, Brandy E. 1998 (December). "Scents and Sensitivity." *Environmental Health Perspectives* 106(12). **http://ehpnet1.niehs.nih.gov/docs/1998/106-12/focus-abs.html** (February 1, 1999).

Fisher, Brandy E. 1999 (January). "Focus: Most Unwanted." *Environmental Health Perspectives* 107(1). **http://ehpnet1.niehs.nih.gov/docs/1999/107-1/focus-abs.html** (February 1, 1999).

Gallup, Alec, and Lydia Saad. 1997. "Public Concerned, Not Alarmed about Global Warming." *Gallup Poll Archives.* The Gallup Organization. **http://198.175.140.8/poll%5Farchives/1997/971202.htm** (January 8, 1999).

Gardner, Gary. 1998. "Sanitation Access Lacking." In *Vital Signs 1998,* eds. Lester R. Brown, Michael Renner, and Christopher Flavin, pp. 70–1. New York: W.W. Norton & Co.

Greenhalgh, Susan, Zhu Chujuzhu, and Li Nan. 1994. "Restraining Population Growth in Three Chinese Villages, 1988–93." *Population and Development Review* 20:365–95.

Holcombe, Randall G. 1995. *Public Policy and the Quality of Life: Market Incentives versus Government Planning.* Westport, Conn.: Greenwood Press.

Jan, George P. 1995. "Environmental Protection in China." *Environmental Policies in the Third World: A Comparative Analysis,* eds. O. P. Dwivedi and Dhirendra K. Vajpeyi, pp. 71–84. Westport, Conn.: Greenwood Press.

Jensen, Derrick. 1999. "The War on Truth: The Secret Battle for the American Mind: An Interview with John Stauber." *The Sun* 279(March):6–15.

Karliner, Joshua. 1997. *The Corporate Planet: Ecology and Politics in the Age of Globalization.* Sierra Club Books.

Karliner, Joshua. 1998. "Corporate Greenwashing." *Green Guide* 58 (August):1–3.

Kemps, Dominic. 1998. "Deaths, Diseases Traced to Environment." *Popline* 20 (May–June): 3.

Koenig, Dieter. 1995. "Sustainable Development: Linking Global Environmental Change to Technology Cooperation." *Environmental Policies in the Third World: A Comparative Analysis,* eds. O. P. Dwivedi and Dhirendra K. Vajpeyi, pp. 1–21. Westport, Conn.: Greenwood Press.

Lindauer, Wendy. 1999 (September). "Fact Sheet: Sick Building Syndrome." Environmental Health Center. **http://www.nsc.org/ehc/indoor/sbs.htm** (February 1, 1999).

Livernash, Robert, and Eric Rodenburg. 1998. "Population Change, Resources, and the Environment." *Population Bulletin* 53(1):1–36.

Mason, Karen Oppenheim. 1997. "Explaining Fertility Transition." *Demography* 34(4): 443–54.

"Matters of Scale." 1999 (January/February). "Spending Priorities." *World Watch.* **http://www.worldwatch.org/mag/1999/99-1b.html** (January 29, 1999).

Mattoon, Ashley T. 1998. "Paper Recycling Climbs Higher." In *Vital Signs 1998,* eds. Lester R.

Brown, Michael Renner, and Christopher Flavin, pp. 144–45. New York: W. W. Norton & Co.

McIntosh, C. Alison, and Finkle, Jason L. 1995. "The Cairo Conference on Population and Development: A New Paradigm?" *Population and Development Review* 21:223–60.

Mead, Leila. 1998. "Radioactive Wastelands." *The Green Guide* 53(April 14):1–3.

Mitchell, Jennifer D. 1998. "Before the Next Doubling." *World Watch* 11(1):21–29.

Murray, C. J. L., and A. D. Lopez. 1996. *The Global Burden of Disease.* Geneva: World Health Organization.

National Oceanic and Atmospheric Administration (NOAA). 1999 (January 11). "1998 Warmest Year on Record, NOAA Announces." **http://www.publicaffairs.noaa.gov/stories/sir45.html** (January 29, 1999).

National Solid Wastes Management Association. 1991. "New Landfills Can Solve the Garbage Crisis." In *The Environmental Crisis,* eds. David L. Bender and Bruno Leone, pp. 122–27. San Diego, Calif.: Greenhaven Press.

"New Rules for Feedlots." 1998. *Environmental Health Perspectives* 106(12). **http://ehpnet1.niehs.nih.gov/docs/1998/106-12/forum.html** (February 1, 1999).

"A New Way to Make a Six-Pack Disappear." 1998. *Utne Reader,* May–June, p. 18.

"Ninety-Eight Percent of Growth in Developing Countries." 1997. *Popline* 19(May–June):6.

O'Meara, Molly. 1998a. "CFC Production Continues to Plummet." In *Vital Signs 1998,* eds. Lester R. Brown, Michael Renner, and Christopher Flavin, pp. 70–71. New York: W. W. Norton & Co.

———. 1998b. "Sales of Compact Fluorescents Surge." In *Vital Signs 1998,* ed. Lester R. Brown, Michael Renner, and Christopher Flavin, pp. 62–63. New York: W. W. Norton & Co.

Pimentel, David, and Anthony Greiner. 1997. "Environmental and Socio-Economic Costs of Pesticide Use." In *Techniques for Reducing Pesticide Use,* ed. D. Pimentel, pp. 50–78. New York: John Wiley & Sons.

Pimentel, David, Maria Tort, Linda D'Anna, Anne Krawic, Joshua Berger, Jessica Rossman, Fridah Mugo, Nancy Doon, Michael Shriberg, Erica Howard, Susan Lee, and Jonathan Talbot. 1998. "Ecology of Increasing Disease: Population Growth and Environmental Degradation." *BioScience* 48(October):817–27.

Population Institute. 1998. "1998 World Population Overview and Outlook 1999." **http://www.populationinstitute.org/overview98.html** (January 29, 1999).

Population Reference Bureau. 1998. "World and Regional Population." 1998 World Population Data Sheet. **http://www.prb.org/prb/** (January 28, 1999).

Reid, T. R. 1998. "Feeding the Planet." *National Geographic,* October, 56–74.

Renner, Michael. 1996. *Fighting for Survival: Environmental Decline, Social Conflict, and the New Age of Insecurity.* New York: W. W. Norton & Co.

Roodman, David Malin. 1998. "Taxation Shifting in Europe." In *Vital Signs 1998,* eds. Lester R. Brown, Michael Renner, and Christopher Flavin, pp. 140–41. New York: W. W. Norton & Co.

"San Francisco Bans Pesticides." 1997. *Green Guide* 35(February 7):1.

Statistical Abstract of the United States: 1995, 115th ed. U.S. Bureau of the Census. Washington, D.C.: U.S. Government Printing Office.

Statistical Abstract of the United States: 1998, 118th ed. U.S. Bureau of the Census. Washington, D.C.: U.S. Government Printing Office.

Stephens, Sharon. 1998. "Reflections on Environmental Justice: Children as Victims and Actors." In *Environmental Victims,* eds. Christopher Williams, pp. 48–71. London: Earthscan Publications.

Stiefel, Chana. 1998. "Population Puzzle: Is the World Big Enough?" *Science World* 54(13): 17–20.

Stover, Dawn. 1995. "The Nuclear Legacy." *Popular Science,* August, 52–83.

Switzer, Jacqueline Vaughn. 1997. *Green Backlash: The History and Politics of Environmental Opposition in the U.S.* Boulder, Colo.: Lynne Rienner Publishers.

Tuxill, John. 1998. *Losing Strands in the Web of Life: Vertebrate Declines and the Conservation of Biological Diversity.* Worldwatch Paper 141. Washington, D.C.: Worldwatch Institute.

United Nations. 1999. "Below-Replacement Fertility." **http://www.popin.org/pop1998/7.htm** (March 1, 1999).

United Nations Population Fund. 1996. *1996 State of the World Population.* New York: United Nations.

United Nations Population Fund. 1997. *1997 State of the World Population.* New York: United Nations.

Vajpeyi, Dhirendra K. 1995. "External Factors Influencing Environmental Policymaking: Role of Multilateral Development Aid Agencies." *Environmental Policies in the Third World: A Comparative Analysis,* eds. O. P. Dwivedi and Dhirendra K. Vajpeyi, pp. 24–45. Westport, Conn.: Greenwood Press.

"Water Wars Forecast if Solutions Not Found." 1999 (January 1). Environment News Service. **http://ens.lycos.com/ens/jan99/1999L-01-01-02.html** (January 8, 1999).

World Health Organization. 1997 (June). "Health and Environment in Sustainable Development." **http://www.cdc.gov/ogh/frames.htm** (July 28, 1998).

World Resources Institute. 1998a. *Building a Safe Climate, Sound Business Future.* Baltimore, Md.: World Resource.

———. 1998b. *Climate, Biodiversity, and Forests: Issues and Opportunities Emerging from the Kyoto Protocol.* Baltimore, Md.: World Resource.

Zabin, L. S., and K. Kiragu. 1998. "The Health Consequences of Adolescent Sexual and Fertility Behavior in Sub-Saharan Africa." *Studies in Family Planning* 2(June 29):210–32.

Zhang, Junsen, and Roland Strum. 1994. "When Do Couples Sign the One-Child Certificate in Urban China?" *Population Research and Policy Review* 13:69–81.

Zwingle, Erla. 1998. "Women and Population." *National Geographic,* October, 35–55.

CHAPTER 16

Amnesty International. 1995. *Human Rights Are Women's Right.* New York: Amnesty International USA.

Barke, Richard P., Hank Jenkins-Smith, and Paul Slovic. 1997. "Risk of Perceptions of Men and Women Scientists." *Social Science Quarterly* 78:167–76.

BICC (Bonn International Center for Conversion). 1998. "Chapter Six." *Conversion Survey, 1998.* Bonn, Germany: BICC.

Broder, John. 1999. "President Steps up War on New Terrorism." *New York Times,* January 23, 14.

Brown, Seyom. 1994. *The Causes and Prevention of War.* New York: St. Martin's Press.

Burns, Mike. 1998. "The Irish Peacemaker: Prime Minister Bertie Ahern Negotiates Historic Agreement." *Europe* 379:10–15.

Calhoun, Martin L. 1996. "Cleaning up the Military's Toxic Legacy." *USA Today Magazine* 124:60–64.

Card, Claudia. 1997. "Addendum to 'Rape as a Weapon of War.'" *Hypatia* 12:216–18.

Carneiro, Robert L. 1994. "War and Peace: Alternating Realities in Human History." In *Studying War: Anthropological Perspectives,* eds. S. P. Reyna and R. E. Downs, pp. 3–27. Langhorne, Pa.: Gordon & Breach Science Publishers.

Cauffman, Elizabeth, Shirley Feldman, Jaime Waterman, and Hans Steiner. 1998. "Posttraumatic Stress Disorder among Female Juvenile Offenders." *Journal of the American Academy of Child and Adolescent Psychiatry* 37:1209–17.

CNN. 1997. "Unabomber Suspect Is Caught Ending 18 Year Manhunt." *CNN Year in Review: 1996.* **http://www.cnn.com/EVENTS/1996/year.in.review**

Cohen, Ronald. 1986. "War and Peace Proness in Pre- and Postindustrial States." In *Peace and War: Cross-Cultural Perspectives,* eds. M. L. Foster and R. A. Rubinstein, pp. 253–67. New Brunswick, N.J.: Transaction Books.

Cohen, William. 1998. "We Are Ready to Act Again." *The Washington Post,* August 23, C1.

Coker, W. J., B. Bhalt, N. F. Blatchley, and J. T. Graham. 1999. "Clinical Findings for the First 1,000 Gulf War Veterans in the Ministry of Defence's Medical Assessment Programme." *British Medical Journal* 318(7179):290–91.

Cooney, Mark. 1997. "From Warre to Tyranny: Lethal Conflict and the State." *American Sociological Review* 62:316–38.

Danitz, Tiffany. 1997. "Drowning the Demons of War" *Insight on the News* 13:14–16.

DefenseLink. 1999. "Military Strength Figures." Press Release, Office of Assistant Secretary of Defense (February 8). **http://www.defenselink.mil.news**

Dixon, William J. 1994. "Democracy and the Peaceful Settlement of International Conflict." *American Political Science Review* 88(1):14–32.

Doyle, Michael. 1986. "Liberalism and World Politics." *American Political Science Review* 80(December):1151–69.

Eddy, Melisa. 1999. "Violence Explodes in Kosovo." *The Daily Reflector* January 18, A1, A9.

Enders, Walter, and Todd Sandler. 1993. "The Effectiveness of Anti-Terrorism Policies: A Vector-Autoregression-Intervention Analysis." *American Political Science Review* 87(4):829–44.

"Fast New Gizmos Can Unearth Them Safely: How to Reduce Land Mine Causalties." 1998. *U.S. News and World Report,* December 27, 60.

Forrow, Lachlan, Bruce Blair, Ira Hefland, George Lewis, Theodore Postol, Victor Sidel, Barry Levy, Herbert Abrams, and Christine Cassel. 1998. "Accidental Nuclear War—A Post–Cold War Assessment." *The New England Journal of Medicine* 338:1326–31.

Funke, Odelia. 1994. "National Security and the Environment." In *Environmental Policy in the 1990s: Toward a New Agenda,* 2d ed., eds. Norman J. Vig and Michael E. Kraft, pp. 323–45. Washington, D.C.: Congressional Quarterly, Inc.

Gallup Poll. 1998. "Nuclear Tests by India and Pakistan Have Little Effect on Public Opinion." *Gallup Poll Archives,* June 19. **http:198.175.140.8/poll%5farchives/980619.htm**

Gardner-Outlaw, Tom, and Robert Engelman, 1997. "Sustaining Water, Easing Scarcity: A Second Update." *Population Action International Report.* **http://www.populationaction.org/why_pop/water/water.html**

Gentry, John. 1998. "Military Force in an Age of National Cowardice." *The Washington Quarterly* 21:179–92.

Gibbs, Nancy. 1994. "Cry the Forsaken Country." *Time,* August 1, 27–37.

Gioseffi, Daniela. 1993. Introduction to *On Prejudice: A Global Perspective,* ed. Daniela Gioseffi, pp. xi–1. New York: Anchor Books, Doubleday.

Hagan, Frank. 1997. *Political Crime: Ideology and Criminality.* Boston: Allyn and Bacon.

Harris, Paul. 1996. "Military Advising is Growing Industry." *Insight on the News* 12:12–14.

Harris Poll. 1998. "Dramatic Increase in Confidence in Leadership of Nation's Major Institutions." *The Harris Poll* 8:1–3.

Hayman, Peter, and Douglas Scaturo. 1993. "Psychological Debriefing of Returning Military Personnel: A Protocol for Post-Combat Intervention." *Journal of Social Behavior and Personality* 8(5):117–30.

Hooks, Gregory, and Leonard E. Bloomquist. 1992. "The Legacy of World War II for Regional Growth and Decline: The Effects of Wartime Investments on U.S. Manufacturing, 1947–72." *Social Forces* 71(2): 303–37.

ICCA (International Crime Control Act). 1998. "Goals and Objectives." The White House. Washington, D.C.: U.S. Government Printing Office.

INTERPOL. 1998. "Frequently Asked Questions about Terrorism." **http://www.kenpubs.co.uk/INTERPOL.COM/English/faq**

Kimball, David. 1998. "Majority of Americans Support Nuclear Weapons Reductions/Elimination." News Release, (August 27). Washington D.C.: Coalition to Reduce Nuclear Dangers.

Klingman, Avigdor, and Zehava Goldstein. 1994. "Adolescents' Response to Unconventional War Threat Prior to the Gulf War." *Death Studies* 18:75–82.

Koerner, Brendan. 1998. "Creepy Crawly Spies: Tiny Robot Insects May Soon Serve as Military Scouts." *U.S. News and World Report,* September 14, 48–50.

Kyl, Jon, and Morton Halperin. 1997. "Q: Is the White House's Nuclear-Arms Policy on the Wrong Track?" *Insight on the News* 42:24–28.

Laqueur, Walter. 1998. "The New Face of Terrorism." *The Washington Quarterly* 21: 169–79.

Lewis, Bernard. 1990. "The Roots of Islamic Rage." *The Atlantic,* September, 47–60.

"Light of Day in Bloody Eire." 1998. *U.S. News and World Report,* December 28, 78–79.

Lloyd, John. 1998. "The Dream of Global Justice." *New Statesman* 127:28–30.

McGeary, Johanna. 1998. "Nukes . . . They're Back." *Time,* May 25, 34–39.

Militia Task Force. 1995. "Emergency Update of Militia Terrorists." The Southern Poverty Law Center, 400 Washington Avenue, Montgomery, AL 36104.

Miller, Susan. 1993. "A Human Horror Story." *Newsweek,* December 27, 17.

Moaddel, Mansoor. 1994. "Political Conflict in the World Economy: A Cross-National Analysis of Modernization and World-System Theories." *American Sociological Review* 59 (April): 276–303.

Mylvaganam, Senthil. 1998. "The LTTE: A Regional Problem or a Global Threat?" *Crime and Justice International* 14:1–2.

National Gulf War Resource Center. 1997. "Who Has Gulf War Syndrome?" Fall, 7–8. **http://www.gulfwar.self2/page7.html**

Nelson, Murry R. 1999. "An Alternative Medium of Social Education—the 'Horrors of War' Picture Cards." *The Social Studies* 88:100–108.

Ngowi, Rodrique. 1999. "Tanzania Bombing Details Uncovered." *Yahoo News.* Associated Press, September 25. **http://dailynews.yahoo.com**

Novac, Andrei. 1998. "Traumatic Disorders—Contemporary Directions." *The Western Journal of Medicine* 169:40–42.

Office of Management and Budget. 1998. **http://www.access.gpo.gov**

Olick, Jeffery, and Daniel Levy. 1997. "Collective Memory and Cultural Constraints: Holocaust Myth and Rationality in German Politics." *American Sociological Review* 62:921–36.

O'Prey, Kevin P. 1995. *The Arms Export Challenge: Cooperative Approaches to Export Management and Defense Conversion.* Washington, D.C.: The Brookings Institute.

"The Patriot Movement." 1998. *Intelligence Report,* Spring, 6–7. Montgomery, Ala.: The Southern Poverty Law Center.

Paul, Annie Murphy. 1998. "Psychology's Own Peace Corps." *Psychology Today* 31:56–60.

Perdue, William Dan. 1993. *Systemic Crisis: Problems in Society, Politics and World Order.* Fort Worth, Tex.: Harcourt Brace Jovanovich.

Pfefferbaum, Betty. 1997. "Posttraumatic Stress Disorder in Children: A Review of the Past 10 Years." *Journal of the American Academy of Child and Adolescent Psychiatry* 36:1503–12.

Porter, Bruce D. 1994. *War and the Rise of the State: The Military Foundations of Modern Politics.* New York: Free Press.

Renner, Michael. 1993a. "Environmental Dimensions of Disarmament and Conversion." In *Real Security: Converting the Defense Economy and Building Peace,* eds. Kevin J. Cassidy and Gregory A. Bischak, pp. 88–132. Albany: State University of New York Press.

______. 1993b. "National Insecurity." In *Systematic Crisis: Problems in Society, Politics and World Order,* ed. William D. Perdue, pp. 136–41. Fort Worth, Tex.: Harcourt Brace Jovanovich.

Sagan, Carl. 1990. "Nuclear War and Climatic Catastrophe: Some Policy Implications." In *Readings on Social Problems,* ed. William Feigelman, pp. 374–88. Fort Worth, Tex.: Holt, Rinehart & Winston.

Scheff, Thomas. 1994. *Bloody Revenge.* Boulder, Colo.: Westview Press.

Shanahan, John J. 1995. "Director's Letter." *The Defense Monitor* 24(6)8. Washington, D.C.: Center for Defense Information.

SIPRI (Stockholm International Peace Research Institute). 1998. *SIPRI Yearbook 1998: Armaments, Disarmament and International Security.* Oxford: Oxford University Press.

Solomon, Jay. 1999. "Religious Violence Reignites in Indonesia." *Wall Street Journal,* March 3, A16.

Starr, J. R., and D. C. Stoll. 1989. "U.S. Foreign Policy on Water Resources in the Middle East." Washington, D.C.: The Center for Strategic and International Studies.

Statistical Abstract of the United States: 1998, 118th ed. U.S. Bureau of the Census. Washington, D.C.: U.S. Government Printing Office.

Sutker, Patricia B., Madeline Uddo, Kevin Brailey, and Albert N. Allain, Jr. 1993. "War-Zone Trauma and Stress-Related Symptoms in Operation Desert Shield/Storm (ODS) Returnees." *Journal of Social Issues* 49(4):33–50.

Sweet, Alec Stone, and Thomas L. Brunell. 1998. "Constructing a Supranational Constitution: Dispute Resolution and Governance in the European Community." *American Political Science Review* 92:63–82.

Tharp, Mike, and William Holstein. 1997. "Mainstreaming the Militia." *U.S. News and World Report,* April, 21:24–37.

Thompson, Mark. 1999. "Star Wars: The Sequel." *Time Daily,* February 22, 153(7):1–3. **http://www.time.com**

UCS (Union of Concerned Scientists). 1998. "Comprehensive Test Ban Treaty." **http://www.ucsusa.org/arms/ctbt.top.html**

United Nations. 1998. "UN Peacekeeping Operations: 50 years, 1948–1998." **http://www.uno.org/Depts/dpko/mail.html**

"UN Peacekeeping: Some Questions and Answers." 1997. *UN Chronicle* 64:64–66.

USIS (United States Information Source). 1998. "Patterns of Global Terrorism: 1997." **http://www.usis.usemb.se/terror/rpt1997/review.html**

Waller, Douglas. 1995. "Onward Cyber Soldiers." *Time,* August 21, 38–44.

World Commission on Environment and Development (Brundtland Commission). 1990. *Our Common Future.* New York: Oxford University Press.

Zimmerman, Tim. 1997. "Just When You Thought You Were Safe . . . Could a False Alarm Still Start a Nuclear War?" *U.S. News and World Report,* February 10, 38–40.

EPILOGUE

Babbie, Earl. 1994. *The Sociological Spirit: Critical Essays in a Critical Science.* Belmont, Calif.: Wadsworth.

Chinni, Dante, Marc Peyser, John Leland, Annetta Miller, Tom Morganthau, Peter Annin, and Pat Wingert. 1995. "Everyday Heroes." *Newsweek,* May 29, 26–39.

Lamielle, Mary. 1995. Personal communication. National Center for Environmental Health Strategies. 1100 Rural Avenue, Voorhees, NJ 08043.

Loeb, Paul Rogat. 1995. "The Choice to Care." *Teaching Tolerance,* Spring, 38–43.

APPENDIX A

American Sociological Association. 1984. "Code of Ethics" Washington, D.C.: Author.

Committee for the Protection of Human Participants in Research. 1982. "Ethical Principles in the Conduct of Research with Human Participants." Washington, D.C.: American Psychological Association.

Nachmias. David and Chava Nachmias. 1987. *Research Methods in the Social Sciences.* 3d ed., New York: St. Martin's Press.

Name Index

Subject Index

Photo Credits

This page constitutes an extension of the copyright page. We have made every effort to trace the ownership of all copyrighted material and to secure permission from copyright holders. In the event of any question arising as to the use of any material, we will be pleased to make the necessary corrections in future printings. Thanks are due to the following authors, publishers, and agents for permission to use the material indicated.

Chapter 1: 3: © David Rae Morris/ Impact Visuals. **4:** © Michael Okoniewski/ Liaison Agency; **23:** AP/ Wide World Photos. **Chapter 2: 32:** Courtesy of World Health Magazine, World Health Organization; **36:** AP/ Wide World Photos; **44:** © J. Berndt/ Stock Boston. **59:** © Georges Merillon/ Gamma Liaison. **Chapter 3: 71:** © Alan S. Weiner/ Gamma Liaison. **75:** AP/ Wide World Photos. **80:** AP/ Wide World Photos. **Chapter 4: 86:** © Patrick Chauvel/ Sygma; **97:** © William Campbell/ Sygma; **100:** © A. Ramey/ PhotoEdit; **109:** © Billy E. Barnes/ PhotoEdit. **Chapter 5: 122:** Courtesy of American Amusement Machine Association. **Chapter 6: 141:** © David Falconer/ David R. Frazier Photolibrary; **144:** © Dana Fineman/ Sygma. **149:** © Wally McNamee/ Woodfin Camp & Associates; **159:** ©Tom Miner/ The Image Works. **Chapter 7: 170:** © Leroy Cath/ SIPA Press; **171:** Brown Brothers; **179:** © David Young-Wolff/ PhotoEdit; **183:** © M. Abramson/ Woodfin Camp & Associates; **192:** AP/ Wide World Photos. **Chapter 8: 196:** © Pam Francis/ Liaison Agency; **198:** © AP/ Wide World Photos. **201:** © Brad Markel/ Liaison Agency; **213:** AP/ Wide World Photos; **220:** © Wernher Krutein/ Liaison Agency. **Chapter 9: 224:** © Evan Agostini/ Liaison Agency; **235:** © From the authors' files: Used with permission; **237:** © Elena Olivo/ Brooklyn Image Group; **238:** AP/ Wide World Photos. **Chapter 10: 255:** Corbis. **260:** © Elena Rooraid/ PhotoEdit; **276:** © Katalin Arkell/ Liaison Agency. **Chapter 11: 283:** © Bonnie Kamin/ PhotoEdit. **292:** © Jim Bouorg/ Liaison Agency. **298:** © PhotoEdit. **300:** © Erik S. Lesser/ Liaison Agency. **Chapter 12: 311:** © Karen Kasmauski/ Woodfin Camp & Associates. **314:** © Allen Tannenbaum/Sygma; **324:** © AP/ Wide World Photos; **334:** © Eastcott/ Momatiuk/ Woodfin Camp & Associates. **Chapter 13: 340:** Corbis; **346:** © Yvonne Hemsey/ Liaison Agency; **350:** © Mark Richards/ PhotoEdit; **355:** © Mark Richards/ PhotoEdit. **Chapter 14: 368(left):** © Corbis; **368(right):** © Douglas Mason/ Woodfin Camp & Associates; **371:** © Robert Burke/ Liaison Agency; **378:** © DeKeerle/ Liaison Agency; **381:** © Retro de L,Annee/ Liaison Agency. **Chapter 15: 403:** © David Austen/ Stock Boston; **405:** © Manoocher/ Sygma. **406:** © Owen Franken/ Sygma; **410:** © From the authors' files: Used with permission. **Chapter 16: 426:** © T. Hartwell/ Sygma; **429:** Corbis. **440:** © S. Compoint/ Sygma; **442:** Liaison Agency; **448:** © AP/ Wide World Photos.

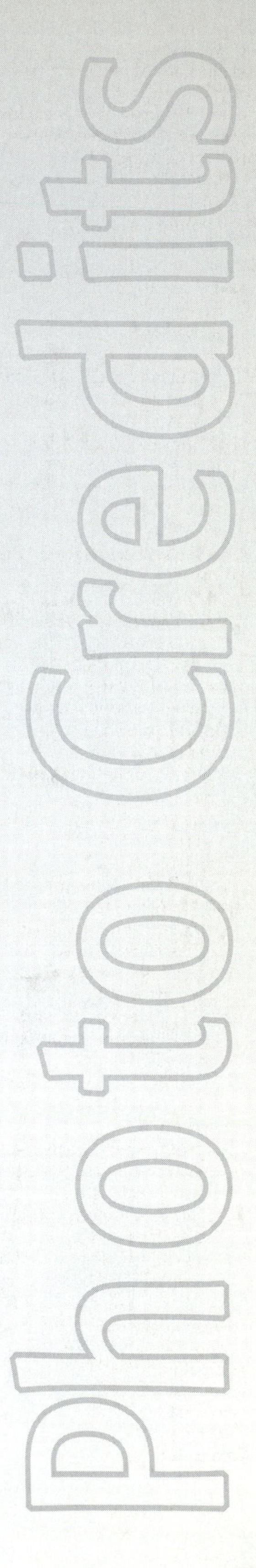